Krister T. Smith, Stephan F. K. Schaal, Jörg Habersetzer (Eds)

MESSEL – An Ancient Greenhouse Ecosystem

Krister T. Smith, Stephan F. K. Schaal, Jörg Habersetzer (Eds)

MESSEL – An Ancient Greenhouse Ecosystem

Senckenberg Gesellschaft für Naturforschung

Imprint Senckenberg-Buch 80

Publisher
Prof. Dr. Dr. h.c. Volker Mosbrugger, Senckenberg Gesellschaft für Naturforschung, Senckenberganlage 25, 60325 Frankfurt am Main, Germany

Editors of this Volume
Dr. Krister T. Smith, Senckenberg Research Institute and Natural History Museum Frankfurt, Messel Research and Mammalogy, Head of Section Palaeoherpetology, Senckenberganlage 25, 60325 Frankfurt am Main, Germany

Dr. Stephan F. K. Schaal, Senckenberg Research Institute and Natural History Museum Frankfurt, Departmental Leadership and Messel Research and Mammalogy, Senckenberganlage 25, 60325 Frankfurt am Main, Germany

Dr. Jörg Habersetzer, Senckenberg Research Institute and Natural History Museum Frankfurt, Messel Research and Mammalogy, Head of Section Vertebrate Radiography and Paleobiology (until 2017), Senckenberganlage 25, 60325 Frankfurt am Main, Germany

Cover figures
Background image: modern-day maar lake Ranu Lading (Java) / Photo: Georg Büchel
SMF-Mel 8226 / Photo: Uta Kiel, SMF-ME 11534a / Photo: Anika Vogel, SM.B.Me 19838 / Photo: Volker Wilde

Copy Editor
Susanne Warmuth, Lektorat und Redaktion, Darmstadt, Germany

Translation
Actitis Translations, Hendrik Herlyn, Corvallis, OR, USA

Production
E. Schweizerbart'sche Verlagsbuchhandlung (Nägele u. Obermiller), Stuttgart, Germany

Layout
DTP + TEXT Eva Burri, Stuttgart, Germany

Print
Gulde Druck, Tübingen, Germany

Distribution
E. Schweizerbart'sche Verlagsbuchhandlung (Nägele u. Obermiller), Johannesstraße 3A, 70176 Stuttgart, Germany
www.schweizerbart.de, E-Mail: mail@schweizerbart.de

Information on this title: www.schweizerbart.de/9783510614110

This title is also available in German:
Stephan F. K. Schaal, Krister T. Smith, Jörg Habersetzer: MESSEL – ein fossiles Tropenökosystem
Informationen zur deutschen Ausgabe: www.schweizerbart.de/9783510614103

© 2018 E. Schweizerbart'sche Verlagsbuchhandlung (Nägele u. Obermiller) and
Senckenberg Gesellschaft für Naturforschung

Senckenberg-Buch 80

ISBN 978-3-510-61411-0
ISSN 0341-4108

The authors and editors are solely responsible for the content of this volume.
All rights reserved. No part of this publication may be reproduced, stored in a retrieval system, or transmitted, in any form or by any means, electronic, mechanical photocopying, recording, or otherwise, without the prior written permission.

www.senckenberg.de

Printed in Germany

Leibniz Association

This book is dedicated to the memory of

Dr. Gerhard Storch
May 21, 1939 – August 11, 2017

The paleobiologist Gerhard Storch dedicated a significant portion of his working life at Senckenberg to different mammals from Messel, such as pangolins, bats and rodents. These interests are reflected in numerous publications. For many years he was a valued colleague and for many of us a good friend as well. He always respected his colleagues on a personal level, and it was always easy to engage with him.

Dr. Gregg F. Gunnell
July 19, 1954 – September 20, 2017

The paleontologist Gregg Gunnell was internationally known for his worldwide studies of Eocene mammals, particularly primates, carnivores and bats. These studies attracted considerable scientific attention to these groups, even outside paleontology. Visits to the Department of Messel Research and Mammalogy were a customary part of his work over the last 20 years, where he never failed to inspire and motivate us.

The years of pleasant collaboration on common interests and their calm and pleasant demeanor bound these exceptional researchers to many of us as friends, nor were they neglectful of younger colleagues. Their friendship, personal enthusiasm and professional expertise will be sorely missed even as they remain in respectful memory.

Foreword

Since 1995, the oil shale deposits of the Messel Pit have been included in the UNESCO World Heritage List – and for a good reason. With the Messel Pit, the State of Hesse is in possession of a unique window into geological history, a site that today offers a fascinating look into the world 48 million years ago. Nowhere else in the world can there be found a quantity and quality of fossils from that epoch that will reveal such a wealth of detail to the experienced observer.

Therefore, the now petrified habitat at Messel is a significant treasure for the State of Hesse. Today, our state owes its high quality of life in no small part to the widely protected natural landscapes, which make up one third of the state's land area. Does it thus not stand to reason to also protect the valuable fossil habitat of "Messel" – in particular, since it is of such global scientific significance?

After a tumultuous history that, for several decades, was marked by the clashing interests of various user groups, the spectacular paleontological research findings from the Messel Pit played a significant role in favor of the pit's preservation for science, and thereby the recultivation of the former oil shale surface mining. The State of Hesse supported the preservation by purchasing the mining operation in 1991, along with the permit rights for the extraction of oil shale.

Now it is up to us to preserve this magnificent piece of natural history. Together with the Senckenberg Gesellschaft für Naturforschung, the Hessian State Museum in Darmstadt and the Visitor Information Center Messel Pit GmbH, the State of Hesse pursues this challenging task. Today, exhibitions in Frankfurt, Darmstadt and Messel as well as in major natural history museums abroad display fossils from our Hessian world heritage. In this context, we are particularly pleased with the intensive civic involvement. I extend my heartfelt thanks to all of you for your commitment to the preservation and exploration of the Messel Pit.

For several decades now, scientists of the Senckenberg Gesellschaft für Naturforschung have been conducting research at the Messel Pit, frequently in conjunction with national and international experts. They perform extensive paleontological detective work, thereby helping to open a large window for the public into the history of our earth and its life. With impressive images of rare fossils and with varied and exciting texts, the present book draws a vivid picture of the world in Messel.

My thanks go to all who contributed to this book. I hope the readers will find enjoyment and, above all, many new insights while perusing this volume.

Volker Bouffier
Minister President of the State of Hesse

Foreword

When the discovery of a fossil crocodile was reported near Messel in the year 1876, nobody could foresee that this location would one day become an important World Heritage Site. More than 100 years had to pass before this status was achieved. Oil shale surface mining for the production of fuel gas was established and later abandoned, followed by almost 20 years of discussions regarding the future use of the mine. At the same time, efforts began to extract the real treasure of Messel: unique fossils, 48 million years old and of high scientific value. These fossils and their history form the ontent of the present book.

On 9 December 1995, the Messel fossil site was added to the UNESCO World Heritage List, which currently includes 1052 sites in 165 counties, among them 41 World Heritage Sites in Germany. This is impressive proof of the great universal significance of the Messel site as a heritage for humanity and its worldwide recognition. Since 1975, Senckenberg has been conducting research in the oil shale deposits, taking over the surface mining operation at the Messel Pit on 1 July 1992. That same year, a special department was established for this purpose. Senckenberg decided to make the Messel Pit a research focus and, together with the Hessian State Museum in Darmstadt, another Hessian institution, to open a window for the general public into the history of our earth and its life.

But why is Messel so valuable for science? With its uniquely preserved, 48-million-year-old fossil remains, the site offers an unusually complete insight into a long vanished ecosystem, which existed under conditions very different from those encountered today. Back then, the area was dominated by a greenhouse climate with extremely high atmospheric carbon levels far in excess of 1000 ppm, with high precipitation and mean annual temperatures of $18 \pm 2.5°C$ – i.e., conditions we are approaching once again today, due to the anthropogenic carbon dioxide emissions. And of equal importance: Messel offers us a unique look at the early evolution of mammals, including our own ancestors, which was only made possible by the extinction of the dinosaurs 66 million years ago.

The year 2017 marks a special anniversary of the Senckenberg Gesellschaft für Naturforschung: We look back at 200 years of exciting nature research and are very pleased to discover that there were many high points, to which the Messel research contributed greatly in the past decades. This is impressively documented by the large number of scientific publications by German and international Messel researchers, and we have been successful in our attempt to make these results available to the public by all suitable means. In the context of our museum construction project in Frankfurt, a new, expanded Messel exhibition is in the works for the near future.

Following the 25th anniversary of the Department of Messel Research and Mammalogy, this second, popular-scientific book about Messel is being published by Senckenberg. With impressive images of rare fossils from public and private collections and with diverse texts, it demonstrates the impressive advances in Messel research, while at the same time pointing out the potential yet to be tapped.

With this book, published in English and German, we attempt to live up to our goal of imparting this knowledge in an understandable and adept manner. At the same time, we would like to increase the interest in the world heritage and offer fascinating insights into our prehistoric world.

I wish our readers much enjoyment when immersing in the Eocene epoch, a heartfelt "Glück Auf!" to the excavation and preparation team of the Messel surface mining operation, and exciting findings and lively discussions for all Messel researchers. And if reading this book whets your appetite for more, simply stop by the Visitor Information Center at the Messel Pit, which is always worth a visit.

Prof. Dr. Dr. h.c. Volker Mosbrugger
General Director of the Senckenberg Gesellschaft für Naturforschung

Foreword

The Messel Pit, one of the world's most important cultural and natural monuments, was added to the World Heritage List on 9 December 1995 as Germany's first UNESCO Natural Monument. Since the fossil site's discovery in the late 19th century, outstanding fossils have been recovered in southern Hesse.

If you look at the outstandingly well-preserved fossils, it is hard to imagine that they owe their preservation to the destructive force of a volcanic eruption. But it was this very force that created the deep crater in which the Messel Lake formed. Thus, a basically horrible event created the perfect environment for allowing this diverse primeval forest habitat to endure throughout the ages, to be unearthed again 48 million years later, intact with its entire original inventory. And the discovery of this unique fossil habitat was again connected to an act of destruction – in this case by the hand of man. For it was the mining for lignite and, later, oil shale that enabled the uncovering of this treasury in historic times.

Following Rudolf Ludwig's notice in 1876 regarding the first discovery of crocodile bones, it became increasingly clear that the Messel Pit carried a significant scientific potential. The Hessisches Landesmuseum Darmstadt started the first scientific excavations in 1966, but they were only conducted on a regular basis since 1971, after cessation of the mining operation. In the early years, a number of other museums and university institutions were involved in the digs, e.g., Karlsruhe, Tübingen, and Brussels. Currently, the excavations are primarily run by the Hessian State Museum in Darmstadt and the Senckenberg Gesellschaft für Naturforschung in Frankfurt. The rich, globally unique fauna and flora unearthed in the process has provided such a wealth of data that the time has come to summarize the latest findings.

Thankfully, with the present book, the colleagues at the Senckenberg Gesellschaft für Naturforschung direct the focus back to the unique archive of geological history. Together with numerous other researchers, they present the new scientific results regarding the fossils, their habitat and the living conditions in and around the Messel Lake as well as the climate and evolution, based on a number of outstanding examples.

Dr. Gabriele Gruber
Head of the Department of Natural History at the Hessisches Landesmuseum Darmstadt

Preface

This second Messel book by Senckenberg, published simultaneously in German and English, will introduce readers with an interest in nature to the results of the research conducted since the 1970s by biologists and geo-scientists on the unique fossils from the Messel oil shale deposits. Alternatively, you can simply peruse the illustrations and take pleasure from the beauty of the 48-million-year-old fossils.

Most of the illustrated fossils are part of the Senckenberg collection and thus primarily originate from our own research excavations. Additional fossils illustrated here come from the Hessian State Museum in Darmstadt as well as other museums and private collectors. We are extremely grateful to them for making their fossils and photographs available. A complete list of the institutions and collections can be found in the image captions and sources.

We would like to pay tribute to all who were involved in the campaign to permanently preserve the Messel Pit, from the cessation of the mining activities until its purchase by the State of Hesse, and who helped to unearth the traces of a buried world and make them widely available for future generations. Besides volunteers and honorary fossil hunters, this also includes hundreds of interns who supported the institutes during their annual digs. Our technical assistants and taxidermists always assured the required care and quality during the excavations and preparation. We express our heartfelt thanks to all of the authors who, with their respective chapters, contributed to the book's success. The large number of contributions adds a personal touch to each individual chapter and impressively demonstrates the international co-operation. The up-to-date, first-hand information makes the topics accessible to the discerning, interested layperson without abbreviating the scientific content. In the two editions of this book, the change of primary editor represents the competency of the respective native speaker.

Twenty-five years ago, the Department of Messel Research was established at the Senckenberg Research Institute in Frankfurt am Main. For Senckenberg, which had already been conducting scientific excavations in the pit since 1975, this solidification represented an important step and an acknowledgement of the Messel research. The research, carried out in conjunction with many German and international scientists, produces exciting results but also illustrates that the species diversity at the site has only been incompletely recorded to date. Based on extrapolations, it is likely that there are still many species of invertebrates waiting to be discovered, and that the known diversity of seed-bearing plants and several groups of vertebrates may increase significantly. Therefore, it stands to reason that several future generations will continue to be involved in exploring the fossil site.

New methods established in paleontology during the past decades, e.g., 3-D computer tomography for non-destructive analysis of bone structures, have significantly improved the examination of Messel fossils in regard to both qualitative and quantitative data. Accordingly, more exciting research findings can be expected in the future, which will further increase our understanding of the Messel ecosystem.

The inclusion of the Messel Pit Fossil Site in the UNESCO World Heritage List in 1995 represents a global acknowledgement and appreciation of the work done by many involved persons. In the year 2020, the Messel Pit will celebrate its 25-year anniversary as a World Heritage Site. This success story gives all of us a reason to be proud.

Krister T. Smith Stephan F. K. Schaal Jörg Habersetzer

Contents

Dedication	V
Forewords	VII
Preface	X

Chapter 1 Messel – Eventful Past, Exciting Future ... 1

Chapter 2 The Formation of the Messel Maar ... 7
   The volcano and the maar at Messel ... 8
      The Middle Messel Formation with oil shale ... 9
      Sand and ash: the Lower Messel Formation ... 11
   What did the Messel Maar look like? ... 12
   The crater's history ... 13

Chapter 3 Paleoclimate – Learning from the Past for the Future ... 17
   Pollen and spores – A means for documenting climate fluctuations ... 18
   Varves – "Annual rings" in the lake sediment ... 20
   The oil shale – A unique Eocene climate archive ... 21

Chapter 4 Joined in Death – the Burial Community of Messel ... 25
   Distortion in the course of time ... 26
   The mystery of the bats ... 28
   Fossil color preservation ... 30
   Cause of death: Unknown ... 32

Chapter 5 Messel Research – Methods and Concepts ... 35
   Excavation, conservation, preparation ... 35
   Examination by means of X-ray techniques and electron microscopy ... 37
   Taxonomy and Phylogeny ... 38
   Species diversity, viewed mathematically ... 40

Chapter 6 The Fossil Flora of Messel ... 43
   History of study ... 43
      The state of preservation of plant remnants ... 46
   Systematics of the flora ... 48
      Algae, mosses, ferns ... 48
      Gymnosperms ... 50
      Primitive flowering plants or basal angiosperms ... 51
      Monocotyledonous flowering plants or monocots ... 52
      Higher flowering plants or eudicotyledons ... 54
   The vegetation surrounding the maar lake ... 59

Chapter 7 Jewels in the Oil Shale – Insects and Other Invertebrates 63
- Sponges (Porifera) 64
  - Paleobiogeography and paleoenvironment 65
- Mollusks (Mollusca) 65
  - Mystery snails (Viviparidae) 66
  - Ramshorn snails (Planorbidae) 66
- Arthropods (Arthropoda) 66
  - Spiders (Araneae) 67
  - Harvestmen (Opiliones) 69
- Crustaceans (Crustacea) 69
  - Water fleas (Cladocera) 69
  - Seed shrimp (Ostracoda) 69
  - Decapods (Decapoda) 69
- Insects (Insecta, Hexapoda) 70
  - Abundance of the different insect groups in Messel 71
  - Mayflies (Ephemeroptera) 72
  - Dragonflies and damselflies (Odonata) 72
  - Stoneflies (Plecoptera) 72
  - Earwigs (Dermaptera) 73
  - Grasshoppers, crickets and katydids (Orthoptera) 74
  - Stick insects (Phasmatodea) 74
  - Cockroaches and termites (Blattodea) 75
  - Thrips (Thysanoptera) 76
  - Cicadas and "hoppers" (Auchenorrhyncha) 76
  - Plant lice, scale insects and whiteflies (Sternorrhyncha) 76
  - True bugs (Heteroptera) 77
  - Hymenopterans (Hymenoptera): Sawflies and parasites 79
  - Hymenopterans (Hymenoptera): Bees and wasps 82
  - Hymenopterans (Hymenoptera): Ants 84
  - Net-winged insects (Neuroptera) 88
  - Twisted-wing parasites (Strepsiptera) 89
  - Beetles (Coleoptera): Primitive groups 90
  - Beetles (Coleoptera): Rove beetles, water dwellers and other handsome beetles 91
  - Beetles (Coleoptera): Various plant eaters 95
  - Caddisflies (Trichoptera) 97
  - Butterflies and moths (Lepidoptera) 99
  - Flies (Diptera) 100
  - Scorpionflies (Mecoptera) 101
- Paleobiogeography of the insects in Messel 101

Chapter 8 Actinopterygians – the Fishes of the Messel Lake 105
- Range of species 105
- Paleobiology 109
- Paleogeography 110

Chapter 9 Amphibians in Messel – in the Water and on Land 113
    Frog fauna 113
        Terrestrial: *Eopelobates wagneri* *113*
        Aquatic: *Palaeobatrachus tobieni* *114*
        Lutetiobatrachus gracilis, an almost blank canvas *117*
    Salamanders 117

Chapter 10 Amniotes – Mammals, Birds and Reptiles 121

Chapter 10.1 Lizards and Snakes – Warmth-loving Sunbathers 123
    The Messel gecko 123
    Ornatocephalus 124
    Lacertiformes: the early success 125
    Iguanidae: Immigrants from the New World 132
    Creepers in the underbrush 134
    Eurheloderma: an early Gila Monster 136
    The semi-aquatic shinisaurs 138
    Necrosaurs: the "death lizards" 139
    Small and large boas 140
    Palaeopython 144
    The squamate community 145

Chapter 10.2 Turtles – Armored Survivalists 149
    Palaeoemys messeliana 151
    Neochelys franzeni 153
    Allaeochelys crassesculpta 154
    Palaeoamyda messeliana 154

Chapter 10.3 Crocodyliforms – Large-bodied Carnivores 159
    Diplocynodon darwini 159
    Diplocynodon deponiae 160
    Hassiacosuchus haupti 160
    Asiatosuchus germanicus 164
    Tomistominae – Gharials in Europe 164
    Boverisuchus – the "hoofed" crocodyliform 165
    Bergisuchus – a southern immigrant 166
    The crocodyliform community 167

Chapter 11 Birds – the Most Species-rich Vertebrate Group in Messel 169
    Large ratites and other terrestrial species 170
        The palaeognathous birds in the Messel forest 171
        Gastornithidae 174
        The gallinaceous bird *Paraortygoides* 174
        Seriemas 174
        Strigogyps 176
        The Messel rail 177
    Bird life at water's edge 181
    The aerial insect hunters 182
    Nightjars and allies 182
    Swifts and early relatives of the hummingbirds 185

Scaniacypselus 186
Parargornis 187
The arboreal birds of the Messel forest 188
Mousebird diversity 190
Parrots and passerines 194
Surprising relationships 195
Trogons and Coraciiformes 199
Trogons 199
The Messel hoopoes 200
Rollers 200
A kingfisher relative 203
Several mystery birds 204
Biogeographic connections 206
Messel birds and tropical avifaunas 209
What remains to be discovered 211

Chapter 12 Mammalia – Another Success Story 215

Chapter 12.1 Marsupials – a Surprise in Messel 217
Anatomy and morphology 217
Paleoecology 219
Evolution and biogeography of the marsupials from Messel 221

Chapter 12.2 Four Archaic Yet Highly Specialized Mammals 223
The remarkable adaptations of *Leptictidium* 224
The piscivore *Buxolestes* 227
The tree-climbing *Kopidodon macrognathus* 229
The long-fingered *Heterohyus nanus* 231
Paleobiogeography 232

Chapter 12.3 With and Without Spines – the Hedgehog Kindred from Messel 235
A fish-loving hedgehog 236
Macrocranion tenerum: the smallest lipotyphlan from Messel 237
A spiny, strong-headed, and scaly-tailed hedgehog 238
Paleobiogeography and Paleoenvironment 239

Chapter 12.4 Primates – Rarities in Messel 241
The first discoveries 242
Ida, the little diva of Messel 244
Further discoveries 246

Chapter 12.5 Bats – Highly Specialized Nocturnal Hunters with Echolocation 249
The bats at the Messel Lake 249
Wing shapes and hunting modes 250
Stomach contents 251
What the cochlea reveals 254
The evolution of echolocation 257
Summary of Eocene bats worldwide 261

Chapter 12.6 Rodents – Gnawing Their Way to Success ... 263
Systematics ... 263
The large leaf-eater *Ailuravus* ... 265
The short-legged climber *Masillamys* ... 266
Hartenbergeromys: a still enigmatic rodent ... 267
Eogliravus: The oldest dormouse ... 267
Paleobiogeography and paleoenvironment ... 268

Chapter 12.7 Ferae – Animals that Eat Animals ... 271
Systematics of Carnivoraformes and Pholidotamorpha ... 271
Lesmesodon: the Messel hyaenodontan ... 272
Paroodectes feisti: an agile climber ... 274
Messelogale kessleri: a small predator ... 276
Eomanis waldi: the oldest pangolin ... 277
Euromanis krebsi: the headless anteater ... 279
Eurotamandua joresi: a doubtful South American ... 281
Paleogeography ... 283

Chapter 12.8 The Advent of Even-toed Hoofed Mammals ... 285
Messelobunodon: a primitive even-toed ungulate ... 285
Aumelasia: a cousin from France ... 287
Eurodexis: the smallest artiodactyl from Messel ... 288
Masillabune: a robust browser ... 289
Paleobiogeography and Paleoenvironment ... 290

Chapter 12.9 Odd-toed Ungulates – Early Horses and Tapiromorphs ... 293
The early horses (Equoidea) ... 293
   The life of the early horses ... 295
   From leaf browser to grass eater ... 298
The tapir-like animals (Tapiromorpha) ... 299

Chapter 13 The Messel Ecosystem ... 303
Topography and lake chemistry ... 303
The aquatic ecosystem ... 305
The shore and possible tributaries ... 305
The terrestrial ecosystem ... 309
Reasons for the great species diversity in Messel ... 309
The role of niches ... 311
Future prospects ... 313

References ... 315

List of Authors ... 339

Index ... 343

Acknowledgments and Image Credits ... 349

Chapter 1
Messel – Eventful Past, Exciting Future

Stephan F. K. Schaal

About 48,270,000 years ago Volcanism in Europe causes the formation of maars – differently shaped craters of variable depth – thereby creating space for maar lakes and the deposit of lake sediments. According to the plate tectonic model by Scotese (2013), at the time of its formation, the Messel Maar was located at the current geographic latitude of the Alps. The crater quickly filled with water and sediments began to accumulate as Messel drifted northward with the European continental plate (Chapter 2).

About 47,300,000 years ago Over a period of about 1 million years, the Messel oil shale was formed. Upon filling up with sediments, the maar is silted up and from now on holds the fossilized remains of animals and plants from paratropical habitats (Chapter 3).

About 30,000,000 years ago The lifting and folding of the Alps begins. Central Europe – an island archipelago at the time – becomes dry land as the Antarctic binds significant amounts of water as ice and sea level drops. The Alpine region as well as the area including the Messel Maar located farther north are affected by tectonic uplift; this is followed by erosion (continuing to the present day). Several hundred vertical meters of material are eroded in the Messel region.

Start of modern chronology The Messel Maar has reached its current position, now located farther to the north. From the time of its origin until today it moved approximately 500 km northward, due to the shift of the European continental plate (Fig. 2.2; Chapter 2).

18th century Oil shale is discovered south of the village of Messel, located between Frankfurt am Main and Darmstadt in the state of Hesse.

19th century The first fossil crocodylian is recovered from the oil shale layer in 1875, and the discovery is described shortly afterwards (Ludwig 1877). The first paper offering an overview of the fossil site's geology and paleontology is published (Wittich 1898). Following the awarding of the Messel claim, the Messel Union is founded in 1884 for the exploitation of ironstone and lignite. In addition, a pyrolization plant is operated for the production of mineral oil and paraffin from the oil shale.

1912 A contractual agreement with the mine owner transfers the sole rights of the Messel fossils to the Grand Ducal Museum at Darmstadt (the future Hessian State Museum Darmstadt). Henceforth, until 1973, all fossils recovered in Messel find their way into the Darmstadt fossil collection.

1945–1959 The production facilities are destroyed during World War II, but are immediately rebuilt after the end of the war. The facility is seized by the American military government and, until 1953, is under the auspices of the I.G. Farbenindustrie AG, Frankfurt am Main. The facility is converted into a paraffin and mineral oil factory. A mining and extraction agreement is entered into with the state of Hesse.

1959 The YTONG concern, an aerated concrete manufacturer from Sweden, takes over the paraffin and mineral oil factory Messel GmbH (Beeger in Schaal & Schneider 1995).

Fig. 1.1: View into the Messel Pit, summer of 2017.

Fig. 1.2: Model of the carbonization plant with the circular kilns at Messel.

1961 For the first time it is possible to successfully transfer the highly delicate Messel fossils to synthetic resins. A scientist from Berlin uses various materials in order to fixate the original bones in a way that allows subsequent examination at any time and permanent storage (Kühne 1961). The transfer method described by him is later regularly applied and individually developed by private fossil collectors and the preparators of research museums active in Messel.

1966–1967 Initiated by the Hessian State Museum in Darmstadt, the first two systematic fossil digs are conducted in Messel (Kuster-Wendenburg 1969).

Late 1960s to late 1980s Still tolerated to a certain degree, private collectors continue to successfully search for fossils in the oil shale (Behnke et al. 1986), unearthing significant specimens in the process. Thus, the work of private collectors also led to an important contribution toward the exploration of the now world-famous fossil site.

1971 Final cessation of the industrial oil shale extraction. The fossils' remarkable state of preservation and the synthetic resin transfer method developed in the 1960s cause a renewed increase in digging activity by private collectors. These activities take place within a "gray zone" without proper permits by the mining authority, yet they are tolerated by the responsible parties.

Upon the announcement of initial plans to fill the Messel Pit with garbage from the region of Darmstadt, Dieburg, Offenbach and Frankfurt, citizens of Messel organize a citizens' initiative against the planned landfill. At the same time, scientists caution against the loss of the fossil site and the consequences for paleontological research.

1975 Upon receiving the official excavation permit, the two large Hessian institutions – the Hessian State Museum (Hessisches Landesmuseum) Darmstadt (HLMD) and the Senckenberg Natural History Society (Senckenbergische Naturforschende Gesellschaft) in Frankfurt am Main – begin conducting regular scientific digs at Messel. In addition, there are short-term activities by various other institutions, in particular during the 1980s.

The "Citizens' Initiative for the Prevention of a Landfill at the Messel Pit" founds an association and, according to its bylaws, "...makes it their mission to preserve the natural and social environment of the citizens of Messel, the citizens of Darmstadt and other municipalities, therefore working decisively against the establishment of a major landfill in the Messel Pit."

1977 The Senckenberg Museum in Frankfurt puts on a special exhibition entitled "Prehistoric Horses and Crocodiles" in close co-operation with participating private collectors.

1980 Two years after its foundation, the Museumsverein Messel e.V. (Messel Museum Association) opens a fossil and local history museum in the old city hall. An impressive model of the pyrolization plant with circular kilns, built especially for the exhibition, brings the local industry of the past back to life (Fig. 1.2).

1981 Upon receiving the planning approval decision for the conversion work and the filling of the pit with garbage, the Higher Mining Authority orders the immediate implementation. Since legal remedies do not have a suspensory effect, the conversion of the surface mine into a landfill may commence despite the pending appeal to the Administrative Court in Darmstadt. The "Citizens' Initiative for the Prevention of a Landfill at the Messel Pit" and the municipality of Messel file a suit against the decision. The ensuing years are characterized by legal disputes and continuing construction work in the abandoned mining pit.

1983 Based on its successful digs, the Senckenbergische Naturforschende Gesellschaft establishes a research station in Messel.

1984 Following an order of immediate implementation by the Administrative Court of Kassel, the construction work continues. In the course of the ongoing conversion from surface mining to a landfill, from 1984 until the end of the 1980s an access road, a drainage system with pump shaft and water treatment plant, a garbage transfer station and administrative buildings are constructed.

1988 Compiled by formerly active Messel researchers, the first book is published that presents all important research results known by then in a popular scientific fashion (Schaal & Ziegler 1988; English version 1992).

1990–1991 Following the initiation of a planning permission procedure for a pile landfill next to the pit in 1988, which was later abandoned, the final out for the landfill in Messel does not arrive until 1990. Subsequently, the Environmental Ministry in Wiesbaden orders the Hessian Higher Mining Authority to cancel the planning permission decision. The appeals pending at the Federal Administrative Court are halted as a consequence. After a decade, the legal battle surrounding a landfill at the Messel Pit is finally over.

1992 Following the purchase of the Messel Pit, in 1991 the State of Hesse concludes a contract with the Senckenbergische Naturforschende Gesellschaft regarding the operation of the Messel Pit. According to the Federal Mining Act, effective immediately, Senckenberg becomes responsible for the surface mining operations. Its takeover as operator in 1992 heralds a new era for Messel research at Senckenberg (Schaal & Schneider 1995). And after many years of oil shale mining for the production of crude phenols, for the first time the extraction of fossils takes priority. On the occasion of the 175th anniversary of the Senckenbergische Naturforschende Gesellschaft, a new permanent Messel exhibition is opened in Frankfurt am Main.

1993 Establishment of a scientific advisory board for the Messel Pit and founding of a Messel Administrative Association. A specimen from the Senckenberg dig, a bat, accompanies the German Spacelab Mission D-2 as the mascot of German astronaut Ulrich Walter (Fig. 1.3).

1994 The high number of significant fossil discoveries and the popular scientific presentation of the Messel research results in radio and television have created a public awareness for paleontology – in particular since the first special exhibition and the publication of commemorative postage stamps. The scientific results and the fossils' uniqueness lead to an application for acceptance of the Messel oil shale deposits into the UNESCO's World Heritage List.

1995 The mining rights to the "Messel Pit" are transferred into the ownership of the State of Hesse, which may immediately begin with the extraction of the natural resources. On 9 December, the UNESCO decides to accept the Messel oil shale deposits onto the World Heritage List as "Messel Pit Fossil Site" (Fig. 1.4).

1997 It becomes possible for anyone to take a look inside the Messel Pit: the Messel Administrative Association builds a viewing platform for visitors (Fig. 1.5).

1998 By decision of the general assembly of the "Citizens' Initiative for the Prevention of a Landfill at the Messel Pit e.V.," after more than 25 years the association self-dissolves in accordance with its by-laws (Mößle & Pagnia 2000), once the association's purpose has been achieved.

Fig. 1.3: A reminder of the D 2 Spacelab mission with bat SMF-ME 10 (NASA Space Shuttle flight STS-55).

Fig. 1.4: UNESCO certificate dated 9 Dec 1995: the Messel Pit fossil site becomes a World Heritage Site.

2001 An exploration drilling to a depth of 433 m is conducted. It aims at answering questions about the origin of the oil shale deposit. By means of corresponding rock samples, it can be shown that the oil shale deposits formed in a maar lake, and that the volcanism in Messel dates back to an age of 47.8 ± 0.2 million years (Mertz & Renne 2005). The age is currently given as 48.27 ± 0.22 (Lenz et al. 2015). In addition to the Senckenberg Research Institute, the Institute for Geoscientific Joint Tasks and the Hessian State Authority for the Environment and Geology participate in the joint project.

2002 Designs and studies for a visitor and information center at the south edge of the Messel Pit mark the kickoff for the planning of the future center. It is commissioned by the Hessian Ministry for Science and the Arts.

2010 Opening of the visitor and information center at the Messel Pit. Fig. 1.6 shows a view toward the north, including the visitor center, molded into the never-used garbage transfer station in a step-like design (right), with the Messel Pit in the background. To the left is the access road to the pit, with the observation platform at the pit edge.

2011 World-famous Messel fossils travel to China as part of a Chinese-German Senckenberg project, where they will be on display as a permanent exhibition in the Paleontological Museum of Liaoning, China's largest paleontological museum.

2014 The Hessian State Museum in Darmstadt is reopened after several years of extensive renovation. This marks the end of the special exhibition "Messel on Tour" (Hessian State Museum Darmstadt 2007), which becomes an impressive new permanent exhibition in the museum.

2017 The Department of Messel Research and Mammalogy of the Senckenberg Institute celebrates its 25-year anniversary.

Future perspectives The overall success of research on Messel becomes apparent based on the large number of scientific publications. To date, there are almost 2,000 publications, plus intensive public relations work by all involved institutions.

By now, original Messel fossils are on display in national and international special exhibitions as "world heritage ex situ."

Worldwide, about one hundred biologists, paleontologists and geologists are involved in Messel research. Using the latest methods, they explore old, already known as well as brand-new fossil discoveries. Thus, to date there is no end in sight for the exciting history of the Messel fossil site.

Fig. 1.5: Visitor platform at the southern edge of the Messel Pit.

Fig. 1.6: Visitor and information center at the Messel Pit.

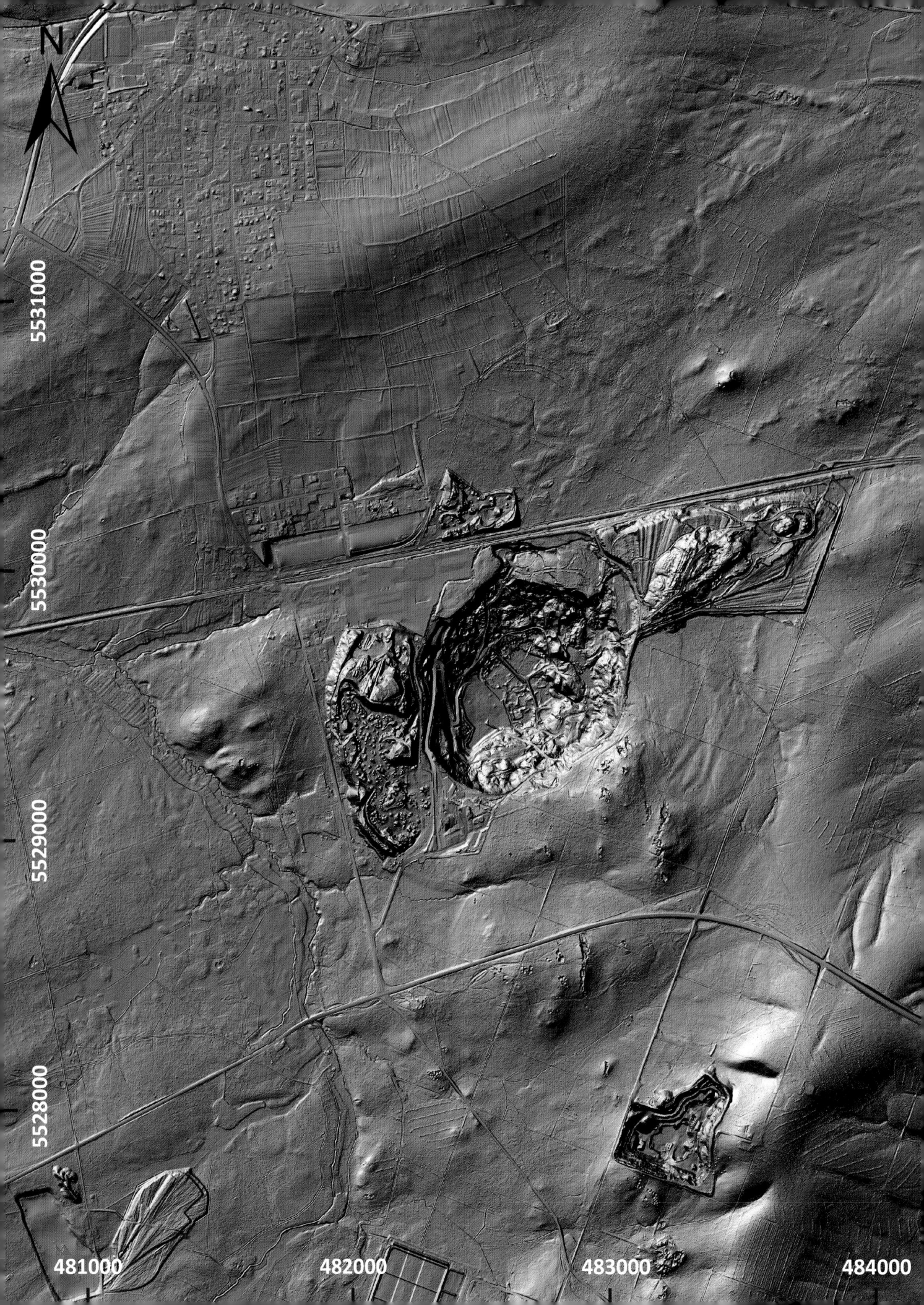

Chapter 2
The Formation of the Messel Maar

Georg N. Büchel & Stephan F. K. Schaal

The Messel Pit is embedded in a topography with a shallow relief, which rises 80–120 m above the Upper Rhine lowlands. The gentle gradients are indicative of the strong Cenozoic geochemical weathering and erosion (denudation) that led to a leveling. Volcanic deposits only remain in the vents, where they were protected from erosion and survived until today.

The Messel Pit at the center of the map section in Fig. 2.1 originated as a maar volcano, which was significantly larger than the present-day pit. A maar volcano consists of a large crater that has formed into the past land surface. The crater is surrounded by a shallow embankment consisting of loose volcanic materials. Shortly after its formation, the crater fills with water above a deep-reaching tuff vent (diatreme). At the time of its origin, the maar lake was situated at approximately 46° latitude N, corresponding to a modern location in the French part of the western Alps. Since then – according to the reconstruction by Scotese (2013) – the Messel Pit has been carried by continental drift some 500 km northward (Fig. 2.2).

At the time of the maar formation in Messel, Europe constituted an island archipelago (Fig. 2.3), similar to modern-day Indonesia. Large parts were covered by shallow water zones. The Atlantic already separated North America from Europe, and the ocean to the south, named after Tethys, the Greek goddess of the sea, was slowly closing due to the collision with the African plate. In addition, this collision led to the formation of the Alps, which during that time were likely only visible as a chain of islands. There is a strong indication that the Tethys Sea was connected to the Polar Sea via the Turgai Strait (Scotese 2013). The extent to which this separation between Europe and Asia served as a barrier for faunal interchange is subject to discussion (e.g., Solé et al. 2016). The primeval North Sea extended up to the modern-day low mountains.

A faunal and floral interchange between the African, Asian and North American continents was largely precluded during the formation of the Messel fossil site. The European island archipelago was clearly demarcated from other land masses, allowing for the development of a fauna with an endemic character, i.e., without external influences. The De Geer route from Canada across Ellesmere Island and Svalbard was only temporarily available as a land bridge for the spread of organisms during the early Eocene (McKenna 1983). Until the early Eocene, another land bridge, the so-called Thule Bridge, formed a connection between Canada and Europe. For several millions of years, this served as a path that enabled a faunal interchange via Greenland, Iceland and the Faroe Islands, which were otherwise separated by oceans (Brikiatis 2014).

Due to the movements of the earth's crust and the associated continental drift, Europe moved noticeably from southwest to northeast, eventually assuming its current position. The oceans to the south, east and north were displaced, and a new land bridge once again enabled the immigration of animals and plants.

Fig. 2.1: Modern land surface in the area around the Messel Pit following extensive erosion (denudation). The digital elevation model is based on LIDAR data.

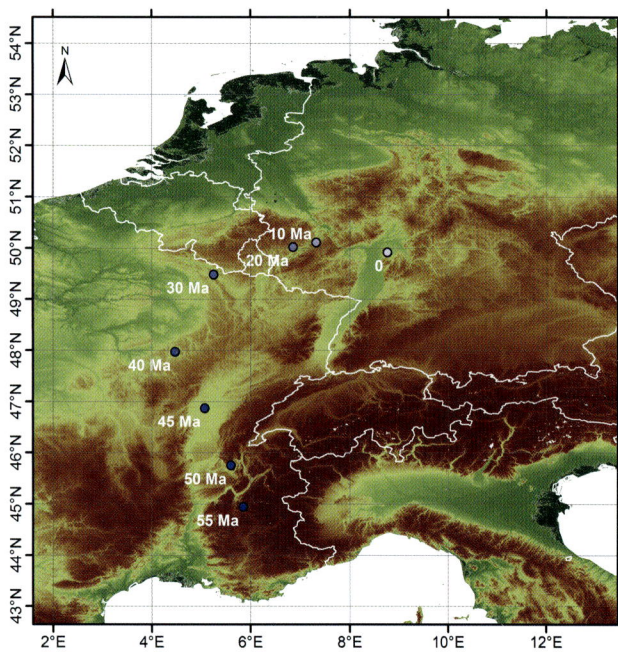

Fig. 2.2: Shift of the geographic position of the Messel fossil site during the Cenozoic.

The volcano and the maar at Messel

The Messel Maar is located in the southeastern part of the Sprendlinger Horst, which is covered by a loose association of additional Eocene volcanoes extending toward Frankfurt (Fig. 2.4). These likely involve 20 former cinder cones, approximately four additional maar volcanoes and three trachyte deposits, whose volcanic rocks buckled upward to form dome-like protrusions (Negendank 1975, Lippolt et al. 1975). All of these volcanic phenomena were formed 50–45 million years ago (Mertz & Renne 2005).

The age of eruption of the Messel volcano was determined to be 48.2 million years ago by means of $^{40}Ar/^{39}Ar$ dating (Mertz & Renne 2005, Fig. 2.5). Based on pollen analyses and the correlation with the astronomical time scale (periodic change in the earth's orientation relative to the sun), the length of the sedimentation period of the Lower Messel Formation and approximately two-thirds of the Middle Messel Formation is estimated at roughly 0.74–0.94 million years (Lenz et al. 2015). Adding the sedimentation period for the upper part of the Middle Messel with an average sedimentation rate of 0.14 mm per year assumed for the lower part results in a time

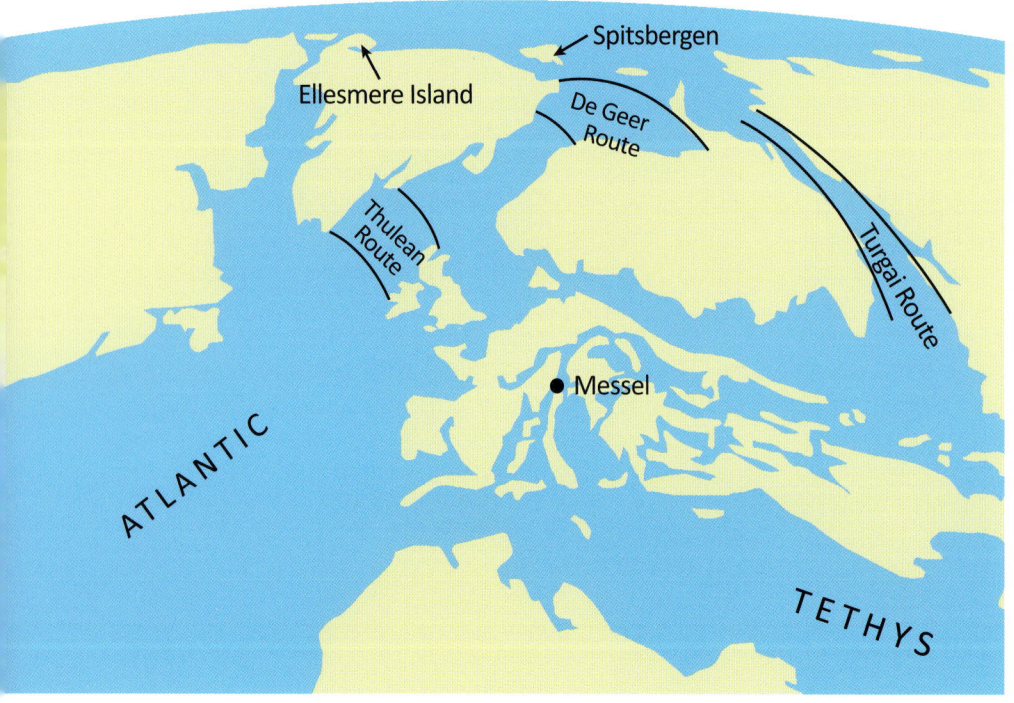

Fig. 2.3: European island archipelago and the land-sea distribution 48 million years ago. Simplified geological reconstruction including the occasionally present land bridges.

period of clearly more than one million years during which the Messel Maar is assumed to have existed.

Prior to the formation of the Messel Maar, there existed a massive sequence of breccia, conglomerates and sandstone from the Upper Rotliegend (Permian). Today, all that remains is a sediment layer of 66 m thickness near the pit's eastern edge (Felder & Harms 2004, Marell 1989). The crystalline Odenwald to the south served as the delivery area. Below the Rotliegend layers, the Sprendlinger Horst consists of igneous rock such as granites of the Hercynian bedrock. Chronologically, they can be correlated with the Upper Devonian/Lower Carboniferous (Stein 2001, Mezger et al. 2013, Ogg et al. 2016, Fig. 2.5). It is in these Rotliegend and granite rocks that the crater and vent of the Messel Maar were formed. The three-part sediment layers that formed in the crater during the early to middle Eocene are referred to as Lower, Middle and Upper Messel Formation.

Almost gone: the Upper Messel Formation

At the end of the 19th century the "oil shale" was discovered, which was mined for crude oil production until 1971 by means of surface mining down to a depth of 60 m (Matthess 1966). In the process, approximately 55 m of laminated black pelites were removed from the Middle Messel Formation as well as the sediments of the Upper Messel Formation, except for a few remnants. The basis of the Upper Messel Formation, with a maximum thickness of 40 m, is located at a depth of approximately 120 m above sea level. The depressions occur in the eastern and southeastern area around the pit's outcrop. The Upper Messel Formation consists of black clay with lignite seams and colorful claystones, overlain by clayey sands. Using the existing dating methods, they are indistinguishable from the underlying laminated black pelites of the Middle Messel Formation.

At this point, the description of the Middle and Lower Messel Formation is augmented by the research drilling conducted in 2001 (position: R 3482757.75, H 5531296.42), which at 105.9 m above sea level was located centrally at the lowest point of today's pit. The reconstructed former land surface above the drilling showed an elevation of approximately 162 m above sea level. Accordingly, this re-

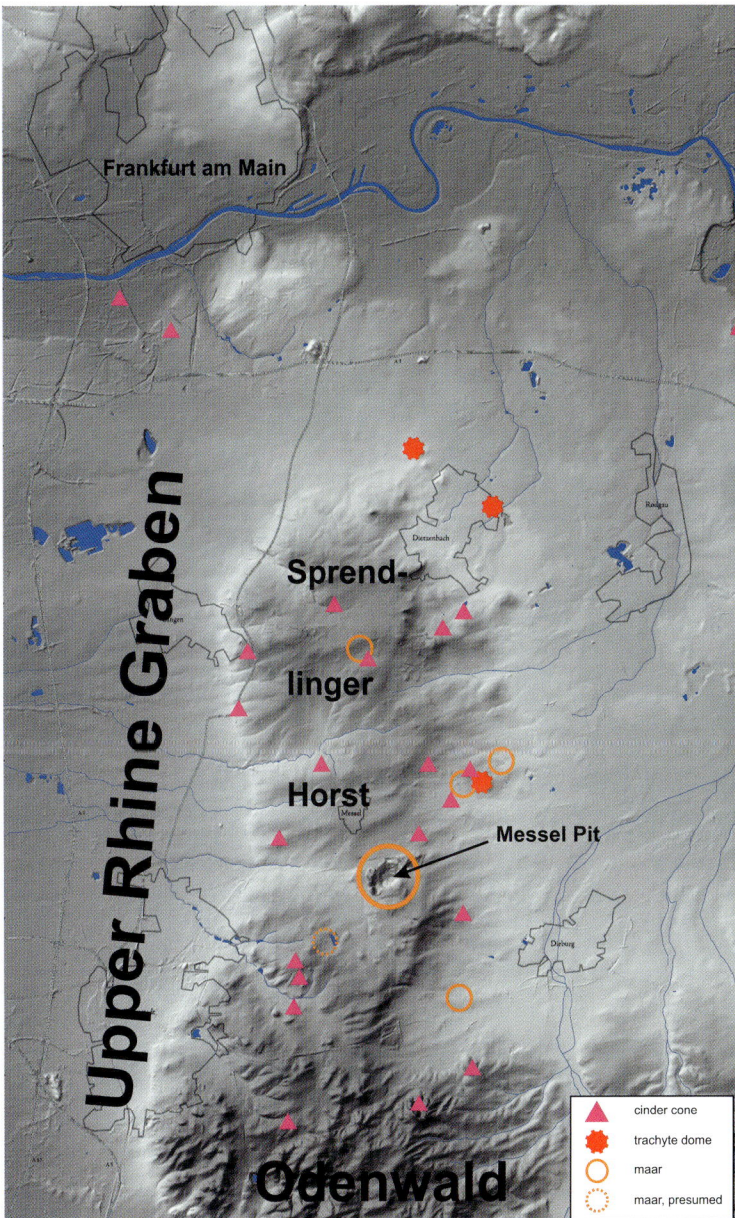

Fig. 2.4: Location of the Messel Pit within Germany.

sults in the presence of a 54 m thick oil shale layer, after subtracting the thin layer of Quaternary wind-blown sand overlying the Rotliegend layers.

Chapter 2 The Formation of the Messel Maar

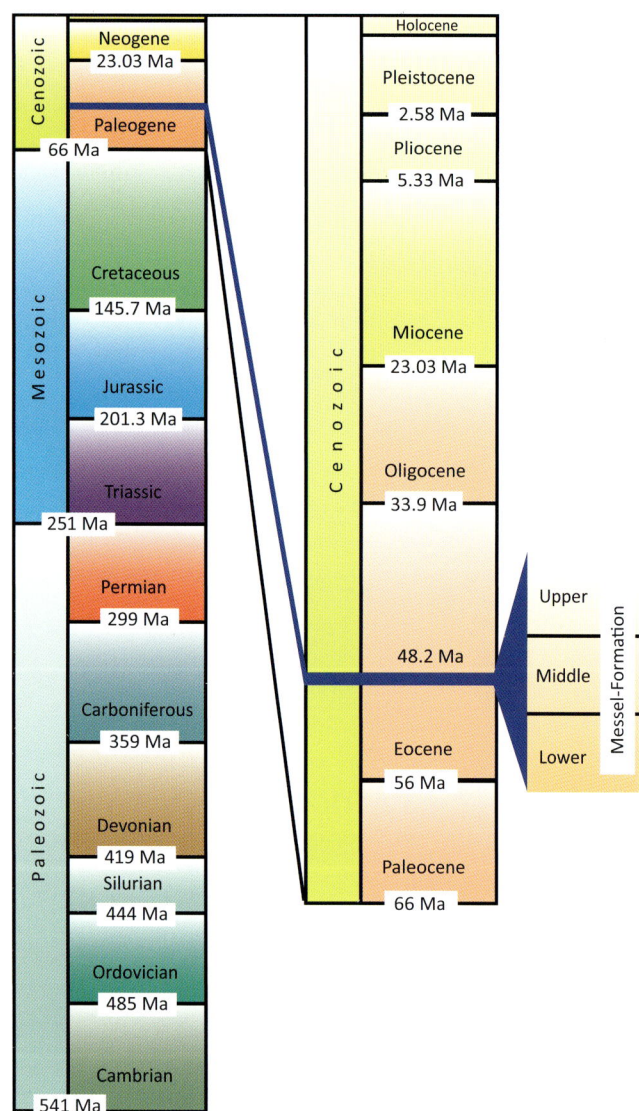

Fig. 2.5: Position of the Messel Formation in the geological time table.

The Middle Messel Formation with oil shale

The layer between 0–94 m (106–12 m above sea level) shows the presence of dark, bituminous, often finely laminated dark olive-gray to dark brownish-gray clay- and siltstones with a high water content (Fig. 2.6 top). These clay- and siltstones are referred to as "oil shale" by the miners, since they can be split like roofing slate, and crude oil can be extracted by means of pyrolysis. The sequence of black pelite consists of finely laminated areas alternating with coarsely laminated layers up to 20 cm thick. Closer to the edges, the grain size increases, while the organic components decrease at the same time (Pirrung 1998). These sandy-gravelly inclusions represent volcanic (pyroclastic) sediments that slid into the lake from the maar's tuff embankment.

Due to its high content of organic substances (crude phenols up to 25 % by weight), for many years, oil shale was considered worth mining. The majority of these so-called kerogens originates from green algae of the species *Tetraedron minimum* und from other plants. Animals played no role in the formation of kerogens. The green algae are the product of algal blooms, and their sedimentation was interrupted by clay particles introduced during seasonal rainy periods. The resulting double layers are interpreted as an annual event, based on which the lake's lifespan can be determined. Due to its large water depth, the Messel maar lake was a meromictic lake, i.e., subject to partial mixing. Its body of water was layered, and only the lake's upper 10 to maybe 30 m were subject to temperature and oxygen fluctuations (Chapter 13). Below this and at the lake's bottom there was no circulation and mixing with the surface water. The oxygen contained in the water was used up during the decomposition of dead organisms, creating ideal conditions at the lake bottom for the formation of crude phenols and the preservation a fossilized fauna and flora (Chapter 4).

The finely laminated sections of the Middle Messel Formation frequently show eyelet-shaped drainage structures, rip marks and faults (Fig. 2.6 top). The edge areas show an increasing extent of slippage structures (Pirrung 1998) that indicate settling phenomena of the underlying diatreme and the hardening (diagenesis) of the black pelites themselves. According to Rullkötter et al. (1988), the thickness of the overlaying sediments of the Upper Messel Formation must have been less than a few hundred meters.

The oil shale primarily consists of smectite (up to 35 % by weight), an expandable clay mineral formed during the weathering of volcanic rocks, and of water (up to 40 % by weight). Components such as iron, calcium and phosphate are only present in minute amounts. They are occasionally concentrated in layers and can be easily recognized when cutting the rock. Fig. 2.6 shows a drill core half from the drilling in 2001 with striking yellow and brown siderite layers (iron carbonate) in the oil shale. A high iron content in the sludge may have led to an increased propagation of microbes, which can explain the occasional formation of extremely thin to several millimeter-thick siderite layers (Felder & Harms 2004). Phosphate minerals such as messelite and montgomeryite can also be found irregularly distributed, e.g., in the marker horizon m (Schaal et al. 1987). In addition to other marker horizons (Alpha, Beta and Gamma), this horizon makes it possible to relate the discovery locations within the fossil site to each other.

Sand and ash: the Lower Messel Formation

The noticeably sandy Lower Messel Formation was explored by drilling between 94–228 m (12–122 m above sea level) (Schulz et al. 2002). The base of this formation shows the first layered, subaquatic sediments, which contain a significantly lower amount of erupted magma components (juvenile components) than the underlying lapilli tuffs (Fig. 2.6, center top). The sequence of the Lower Messel Formation can be divided into three parts. The lower part consists of sand layers, breccia and xenoliths (secondary rock clasts) up to block size from crystalline and Rotliegend rocks. The middle part consists of finely stratified sand, silt and clay layers interspersed with two layers containing lapilli tuffs. The upper lapilli tuff layer (162.6–159 m) contains black pelites and coaled wood remnants (Felder & Harms 2014).

Between 373–228 m, a total of 145 m of ash and lapilli tuff layers can be found, which only contain a small number of crystalline and Rotliegend fragments. The lapilli contain few bubbles and are frequently of a spherical and zoned design, with a diameter between 0.1–2 cm (Fig. 2.6, center bottom). Moreover, this layer contains slightly larger, drossy lapilli. Wood remnants in the lapilli's core are not uncommon. The high proportion of juvenile, rounded lapilli with single pieces of dross indicate that the eruptions near the

Fig. 2.6: Drill core halves of the Messel drilling in 2001. Top: oil shale, center top: sandy-gravelly sediments, center bottom: lapilli tuff, bottom: volcanic clastic rock; length of frame edges 5 cm. The three bottom units originate from the Lower Messel Formation, the top unit from the Middle Messel Formation.

end of the eruption activity in the Messel Maar took place closer to the earth's surface. Similar phenomena can also be identified in the Ranu Agung Maar in the eastern part of the island of Java (Pirrung et al. 2004) and in the Laach Maar (Büchel 1988) in the Eifel.

The lowest layers explored by drilling (433–373 m) consist of diatreme breccia with crystalline and bleached Rotliegend sediment blocks (Fig. 2.6, bottom). There is no reliable information regarding the extension of the diatreme toward the Rotliegend past the final depth at 433 m. The geomagnetic and gravimetric investigations by diploma students in W. Jacoby's working group at the University of Mainz in the 1990s were unable to unravel this deep area (oral communication W. Jacoby). The reflection-seismic investigations by Buness et al. (2004) lead to the expectation of a stratified diatreme structure for an additional 150 m below the final exploration drilling depth after approximately 50 m of breccia, at least down to these depths. Based on the examination of deeply cut diatremes, e.g., from the diamond industry, or from cuttings of deeply eroded diatremes in Cenozoic volcano fields, it may well be expected that the unlayered diatreme tephra extends to a depth of an additional kilometer (Lorenz 2000).

What did the Messel Maar look like?

In order to give an idea of the Messel Maar's original shape, Fig. 2.7 illustrates the Dalongwan Maar (China) with its maar lake and the cut tuff embankment. The outwardly inclined ash- and block-bearing lapilli tuffs, which are abruptly cut off toward the maar lake, are clearly visible. The edge of the maar thus intrudes rather steeply into the lake and surrounds the body of water on all sides. This steepness and seclusion likely existed for some time at the Messel Maar, as well. The reconstruction of the Messel Maar is based on an assumed thickness of the tuff wall of 80 m. The Laach Maar in the Western Eifel, with a diameter of 1,000 m and a tuff wall with a total thickness of approx. 40 m, may serve as an indicator for this estimate. In all likelihood, the wall at Messel was at least twice as thick.

This leaves the question about the diameter of the original crater lake, and to what extent the relief was lowered by the Cenozoic denudation. Based on the pool of existing data regarding the Messel volcano, some of which we have introduced here, and the comparison of old and young maar volcanoes around the world, we have attempted to visualize the former Messel volcano. In the process, the more strongly structured volcano landscape was simulated by means of a twofold super-elevation of the topography. The extensive amount of erosion was assumed to be approximately 300 m (Fig. 2.8). To arrive at this

Fig. 2.7: The 80 m-deep Chinese maar lake Dalongwan (approximately 1 km diameter) with ring wall in cross-section; the enlarged section shows the typical layers.

Fig. 2.8: Reconstruction of the Messel ring wall at the time of the maar's formation 48 million years ago (top), in comparison to the present land surface at Messel (bottom).

amount, and thus the elevation of the immediate surroundings of the original crater, a post-sedimentary reduction of the diatreme tephra's pore volume and the crater lake sediments from 30 % to 20 % was assumed. The diatreme's depth was calculated at 1 km below the base of the lapilli tuffs. The total thickness of the Lower, Middle and Upper Messel Formation was extrapolated at approximately 360 m (80 m Upper, 156 m Middle und 127 m Lower Messel Formation). In addition, it was assumed that the sediments of the Upper Messel Formation completely filled the crater, thus extending to the level of the surrounding land surface. This results in an elevation of the previous land surface during the time of eruption of the Messel Maar of 460 m, as opposed to 162 m above sea level today.

The preservation of the potentially almost complete crater sediments is therefore due to an elevator-like shift of the depth from the crater to the upper part of the vent (diatreme), based on the compaction of the formerly loose diatreme tuffs and breccia, as demonstrated by Suhr et al. (2006) at the maars of Baruth and Kleinsaubernitz. This depth shift of a central part of the formerly far-reaching crater sediments into a much smaller diatreme narrowing downward in a funnel shape was associated with a bowl-shaped alignment of the crater sediments. Thus, at the modern excavation level there is a centrally directed inclination of black pelites of 2°–13° at the pit's center, with an inclination of 20°–30° near the edges. Based on drilling results at the north edge, the modern-day rim of the diatreme shows an inclination of approximately 60o (Pirrung 1998). According to the seismic examinations, the rim of the diatreme appears to become steeper towards the bottom (Buness et al. 2004). On the other hand, for the crater walls of the original maar crater we have considered an inclination of 30°–40° in the crater's reconstruction.

The crater's history

Magma rises through the crystalline rocks of the Sprendlinger Horst, repeatedly encountering groundwater-filled fractures on its way to the top. Due to the pore water's high pressure, groundwater comes into contact with the rising magma, where it spreads out and creates water vapor films due to heating by the magma (Fig. 2.9, step 1). From a certain depth on, perhaps a few hundred meters, the vapor films from the water intrusions collapse. This leads to immediate contact between the magma and the water

14 Chapter 2 The Formation of the Messel Maar

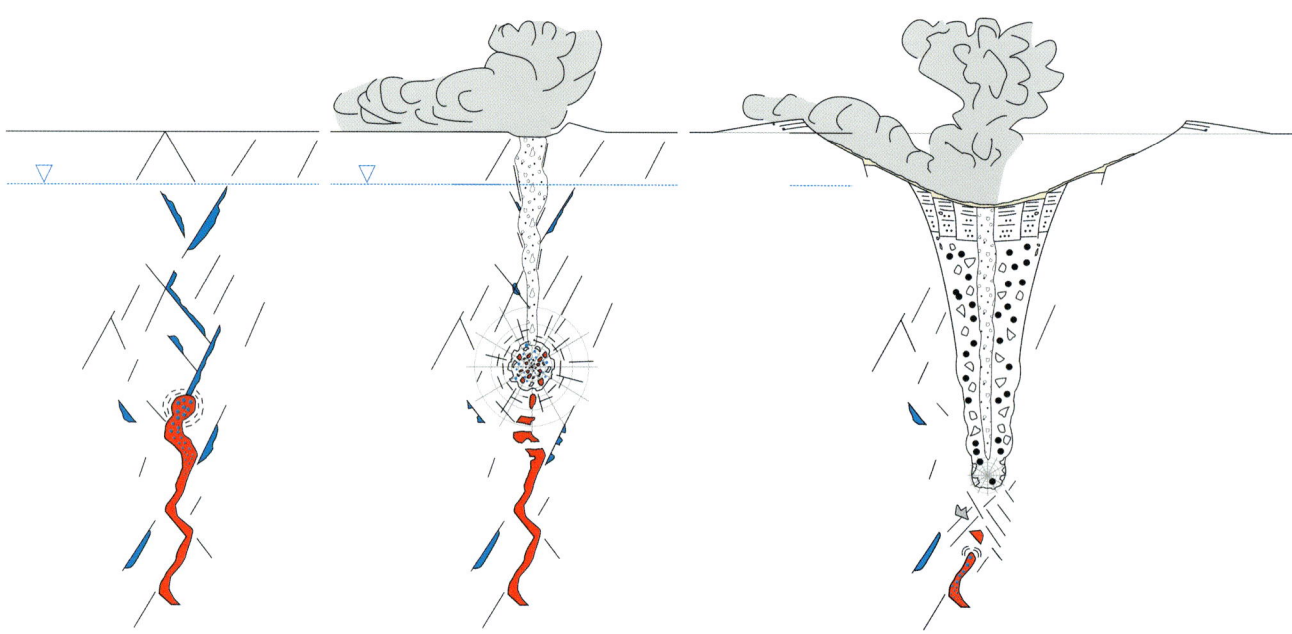

Fig. 2.9: The main steps in the formation of the Messel Maar (from left to right).

Fig. 2.10: Hopi Buttes volcano field (Arizona) with feeder dike exposed by erosion in the foreground and tuff vents at middle distance.

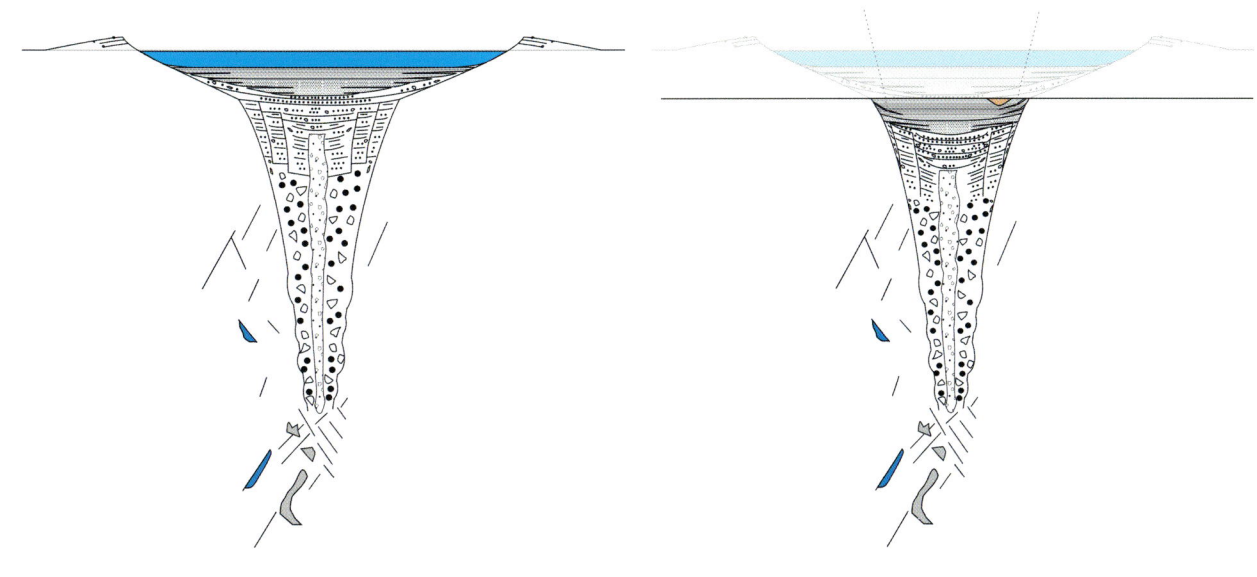

introduced from outside, an immediate crushing of the bedrock stones due to shock waves, and a subsequent expansion of the superheated water vapor. In volcanology, these recurring events are called phreatomagmatic eruptions (Fig. 2.9, step 2). Through a narrow vent, the bedrock that falls into the explosion chamber is fragmented, and the turbulent eruption system unloads in a large, strongly expanding eruption cloud. This so-called "base surge" races across the densely tree-covered land surface at supersonic speed, tearing down the giants of the primeval forest in the vicinity of the crater's rim. At the same time, large expelled bedrock blocks crash down onto the freshly deposited pyroclastics. Higher air currents move ash particles over many kilometers. Additional eruptions ensue, and in the process, the events keep shifting farther downward (Fig. 2.9, step 3). An increasing number of bedrock slabs falls into the hollow spaces caused by explosions and eruptions, widening the vent. The expelled pyroclastics fill up the resulting hollow spaces from above. This leads to a deepening and widening of the vent and a widening of the crater. Several weeks later, the eruption activity finally ceases. Over time, the large crater's lower part fills with intruding groundwater (Fig. 2.9, step 4). At the same time, sedimentation sets in. The lower part of the crater is filled with clastic sediments, while bituminous clays are deposited higher up under still water conditions. Meanwhile, the tuff wall is eroded. The upper part of the crater contains sediments from the time when the crater was almost completely filled up. Over time, the surrounding areas of the crater also become subject to intense geochemical weathering. Several hundred meters of the Rotliegend sediments and the crystalline rocks are eroded. Simultaneously, the middle part of the crater sediments collapses into the diatreme, since the loose volcanic rocks located in the vent become compacted (Fig. 2.9, step 5).

In the case of Messel, no information is available to us regarding the lower part of the diatreme. The diatreme in the volcano region of the Hopi Butte, shown in the background in Fig. 2.10, is a known example of compacted diatreme rocks. Via the chute-like intrusion visible in the foreground, the magma reaches the upper region of the crust, where it made contact with the groundwater.

Chapter 3
Paleoclimate – Learning from the Past for the Future

Olaf K. Lenz, Volker Wilde, Walter Riegel

Throughout its history, the earth has been subject to steadily alternating glacial periods and greenhouse phases. In order to make reliable predictions regarding the future global climate change and its effects on our species and our environment, it is necessary to develop an understanding of the processes that took place over periods of time far beyond the instrumental record of climate data during the past decades.

With the aid of numerous deep sea drillings, scientists in recent years developed an oxygen isotope curve that reflects the development of deep sea temperatures in the past 65 million years (Zachos et al. 2001, 2008; Fig. 3.2). In the meantime, it has also been accepted as a global temperature development curve. It shows a warming trend since the middle Paleocene, which reached its high point with the "Early Eocene Climatic Optimum" (EECO) approx. 50–52 million years ago. This was followed by a gradual, initially very slow cooling, until the occurrence of a drastic drop in temperature around the turn from the Eocene to the Oligocene. This started the transition from the Paleogene greenhouse climate to the glacial climate of the Oligocene and Neogene (Zachos et al. 2001).

The long-term climate development was only interrupted by a few short-term fluctuations. The most prominent of these warming events ("hyperthermals"), the "Paleocene-Eocene Thermal Maximum" (PETM; McInerney & Wing 2011), occurred at the Paleocene-Eocene boundary, approx. 56 million years ago. The total duration of this event is estimated to have lasted about 170,000 years (Röhl et al. 2007). It was caused by a short-term massive input of greenhouse gases in the atmosphere and was accompanied by an average temperature increase of 5–6 °C. This increase was felt most noticeably in the high latitudes, where, with a rise of 7–10 °C, it far exceeded these average values.

The Eocene – the time when the Messel oil shale was deposited – thus falls into a period when the earth was dominated for the last time by a greenhouse climate not caused by man. For example, the report of the Intergovernmental Panel on Climate Change (IPCC 2014) predicts a CO_2 level for the year 2100 that is comparable to the levels in the Eocene (Parrish & Soreghan 2013). The Eocene is thus particularly well-suited as a reference period.

During the Eocene, the area around Messel was covered by a forest characterized by a subtropical to

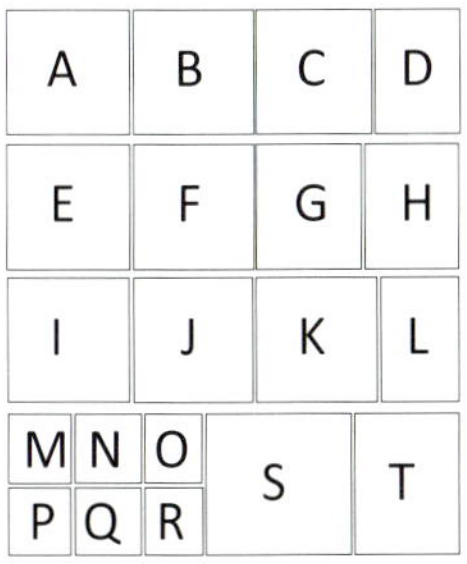

Fig. 3.1 Examples of pollen and spores from the scientific drilling at Messel in 2001.
A: spore of a climbing fern (Schizaeaceae), B and C: different types of fern spores, D and F: spores of polypod ferns (Polypodiaceae), E: spore of a club moss (Selaginellaceae), G: pollen grains of a water lily (Nymphaeaceae), H: pollen grain of a sapotaceous plant (Sapotaceae), I: pollen grain of a restio plant (Restionaceae), J: pollen grain of a plant from the kapok family (Bombacaceae), K: pollen grain of a witch hazel plant (Hamamelidaceae), L: pollen grain of a vine plant (Vitaceae), M, P and Q: pollen grains of extinct walnut plants (Juglandaceae), N: pollen grain of an unknown extinct plant, O: pollen grain of an oleaceous plant (Oleaceae), R: pollen grain of an heather plant (Ericacea), S: colony of the "oil alga" *Botryococcus*, T: shell of the dinoflagellate *Messelodinium* (Dinophyceae). Scales: 10 µm.

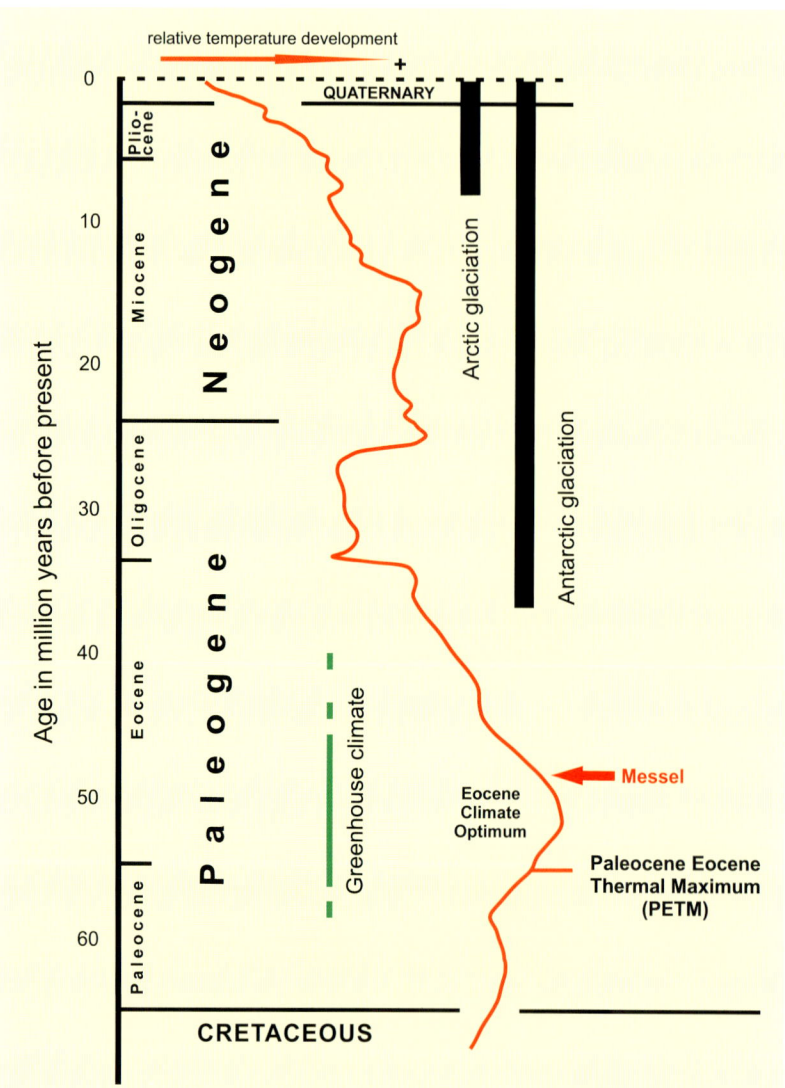

Fig. 3.2: Temperature development in the deep sea during the Cenozoic. The Messel oil shale originates from the time of the Paleogene greenhouse climate.

be assigned to modern families. Since a particularly species-rich fauna and a paratropical flora from the Eocene have been preserved in the Messel Pit, this fossil site plays a key role in the exploration of the Eocene paleoclimate in Europe. With the aid of floristic macro- and microfossils, and taking into account the plant families' modern climatic requirements, it was possible to calculate a mean annual temperature of 16.8–23.9 °C (Grein et al. 2011). Measurements of oxygen isotopes in vertebrate bones led to almost identical values, at 18 ± 2.5 °C (Tütken 2014). Even during the coldest month, the temperature did not drop below 10 °C. This is also confirmed by the presence of warmth-loving crocodylians (Grein et al. 2011; Markwick 1998) that inhabited the surface water of the Messel lake, which reached temperatures of 25 ± 3 °C (Tütken 2014). In addition, with an annual precipitation between 803 and 2,540 mm, the climate was very wet (Grein et al. 2011). The characteristic, finely laminated deposits of the oil shale even reflect seasonal fluctuations. Thus, in regard to their temporal resolution, these deposits show close similarities to the so-called varves (sediments deposited in thin annual layers) from glacial meltwater lakes or the growth rings of trees. Both count among the climate archives that can be analyzed by measuring their thickness and counting their frequency. Besides these varves, pollen and spores of plants serve as particularly important climate proxies, since plants are highly sensitive indicators for the continental paleoenvironment and the paleoclimate (Traverse 2007).

Pollen and spores – A means for documenting climate fluctuations

The flora surrounding the maar lake at Messel during the Eocene can be reconstructed by means of embedded remnants, such as leaves, fruits, seeds, flowers, and occasionally, wood (Wilde 2004; Collinson

tropical vegetation. However, it contained a diversity of deciduous trees and shrubs primarily found in the temperate climate zones today (Wolfe 1979). At annual mean temperatures between 20–25°C, this vegetation grew in the Northern Hemisphere in the transitional zone between the globally dominating tropical rainforests and the temperate mixed deciduous forests widespread in the polar region. With the predominance of mammals and flowering plants, the flora and fauna of that time already resembled that of today, and many of the plants existing at that time can easily

et al. 2012; Chapter 6). Contrary to macrofossils, floristic microfossils can be found in each rock sample from the Messel oil shale. These are microscopic pollen grains and spores or algae, which, after chemical treatment, can be extracted in large numbers from the sediment and subsequently be analyzed under the microscope (Thiele-Pfeiffer 1988).

By means of statistical analysis, it is possible to recognize and analyze changes in the frequency of individual taxa over the time that elapsed during the deposition. While the initial deposition rate was still very high and no oil shale was formed in the beginning (Fig. 3.3), seasonally layered oil shale was later deposited very slowly and continuously, thereby showing an unusually high temporal resolution. At an average sedimentation rate of 0.14 mm per year, determined on the basis of various methods (Irion 1977; Goth 1990; Lenz et al. 2010), it was possible by means of the scientific drilling at Messel in 2001 (Chapter 2) to continually analyze a time period of about 640,000 years (Lenz et al. 2011). With the deposition of oil shale, which also housed the macrofossils, the maar had stabilized, and a climax vegetation had replaced the pioneer vegetation from the recolonization phase in the crater area (Chapter 13). This led to the central question in the investigation regarding whether and to what extent the climate cycles that are predominant during the current glacial period of the Quaternary are also reflected by the vegetation under the greenhouse conditions of the Eocene (Lenz et al. 2011, 2015).

Following a successful preliminary study on the basis of old data from a drilling in 1980 (Thiele-Pfeiffer 1988; Lenz et al. 2005), the section of the 2001 Messel drill core that lay above 94.60 m was therefore sampled at regular intervals (every 10 cm). This is the section that contains the continuous oil shale that was deposited without major disturbances.

Subsequently, the samples, which were taken at every 20 cm – corresponding to a time interval of about 1,400 years between samples – were examined for the presence of plant microfossils. As a result, it could be clearly demonstrated that regular vegetation changes occurred in the early Middle Eocene at Messel, which apparently correspond to Milankovitch cycles (Box 3.1). However, the climate changes were always so small that they did not lead to a complete displacement or immigration of new plants within the climax community, but only to shifts in the spatial distribution and spread of individual plant species or communities (Lenz et al. 2011).

Next, the sample density was increased to 10 cm in a 10 m-long segment of the drill core, thereby reducing the time interval between two samples to 700 years. In addition, an oil shale block of 1 m thickness was extracted from the pit (Fig. 3.4). In the laboratory, samples were taken from this block at 1 cm intervals, so that the time period between samples was reduced to 70 years. The statistical analyses of these time series led to the conclusion that cyclic vegetation changes also occurred in the range of centuries to a few millennia. Such sub-Milankovitch cycles precisely correspond to the short-term periodicities known from the glacial period of the Quaternary.

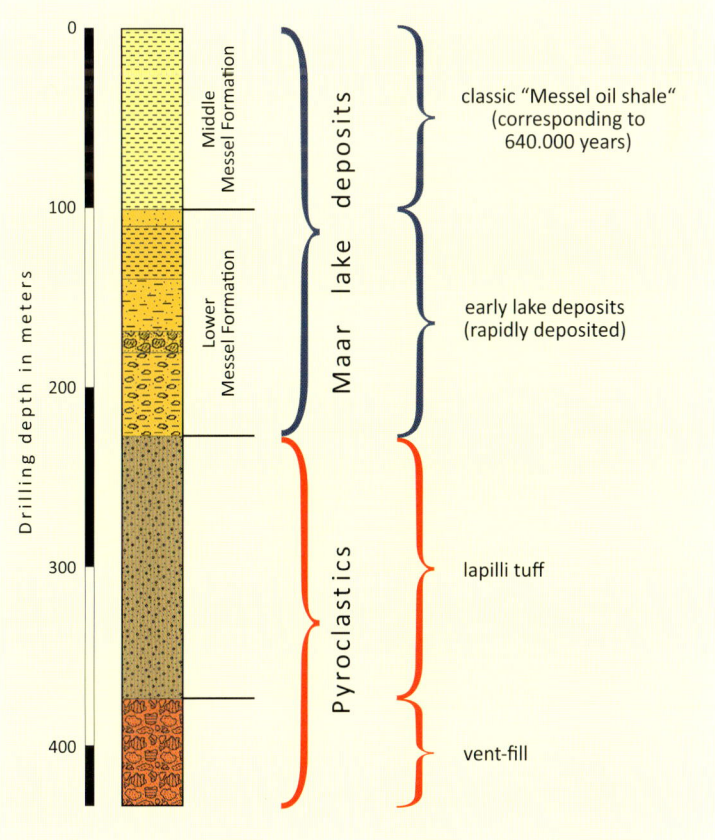

Fig. 3.3: Schematic profile of the Messel scientific drilling in 2001; the examined deposits of the maar lake are shown in the upper section.

Box 3.1: Milankovitch and sub-Milankovitch cycles

Even without human impact, the earth's climate undergoes changes and follows regular cycles. Long-term periodic climate fluctuations are particularly characteristic for the glacial period of the Quaternary and cause alternating glacials ("ice ages") and interglacial periods. At the end of the 1940s, the Serbian mathematician Milutin Milankovitch correlated these cycles to cyclic changes in astronomical ("orbital") parameters; in honor of their discoverer, these are referred to as "Milankovitch cycles." Thus, the earth's orbit around the sun changes within approx. 100,000 and 400,000 years from almost circular to slightly elliptical and back to circular ("eccentricity"). The tilt of the earth axis in relation to the level of the earth's orbit ("obliquity") fluctuates over a cycle of approx. 40,000 years between 22.1° and 24.5°, and the rotation of the earth's axis, which rotates like the swaying axis of a spinning top on the level of the earth's orbit ("precession"), also fluctuates in a cycle lasting about 21,000–26,000 years. All of these cyclic processes lead to changes in solar radiation and thus ultimately to climate changes. Through the overlapping of individual cycles, these effects can be additionally strengthened or weakened.

Even below the Milankovitch cycles, there are obvious cyclical climate changes, which fluctuate in a range between a few decades to several thousand years. These so-called sub-Milankovitch cycles are primarily due to changes in sun spot activity. An immediate consequence is changes in the ozone content, the temperature, cloud formation or circulation within the stratosphere. This may result in noticeable effects on the vegetation, e.g., during the so-called "Little Ice Age," which lasted from the 14th to the early 19th century.

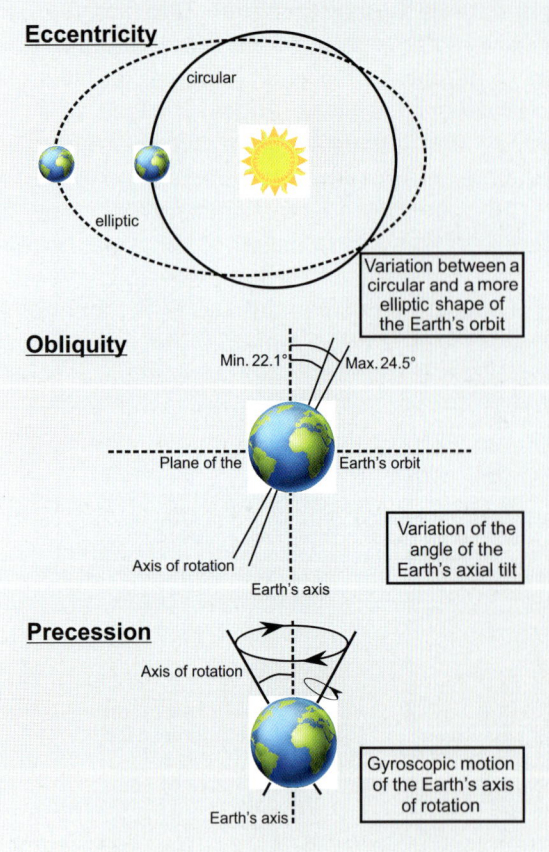

Thus, it could be documented for the first time that even the greenhouse regime of the Eocene was subject to fluctuations, e.g., regarding solar activity, that are comparable to the recent geological past, and that these fluctuations had an effect on the vegetation even under the stable greenhouse climate (Lenz et al. 2017).

Varves – "Annual rings" in the lake sediment

The quantitative analysis of pollen and spores has a lower limit in regard to the temporal resolution. That is because there is a minimum amount of material in individual samples that is required for the statistical analysis. It is not possible to take a sufficiently large sample size below a layer thickness of one centimeter, since this results in an insufficient number of pollen grains and spores for an analysis.

However, the oil shale is almost continually characterized by a fine light-dark stratification; it has long been known that this fine lamination is due to annual blooms of the green alga *Tetraedron minimum* (Goth 1990). The light-colored layers in the oil shale are almost exclusively composed of the preserved remains of this single-cell green alga, whereas the dark layers are primarily composed of clay minerals (Fig. 3.5). The light-colored layers thus constitute the deposit of algal blooms that likely formed during spring and summer, whereas the dark layers were deposited in the fall and winter (Chapter 2). The thickness of the individual layers that make up the oil shale is not constant. In particular, the algal layers show strong fluctuations in thickness, since the algae's growth directly depends on the nutrient supply and the temperature and is thus controlled by climate. The layers of clay minerals also show changes in thickness, albeit to a lesser extent, which may be due to variations in the sediment input into the lake as a result of fluctuations in precipitation. The individual layers (laminae) could therefore be measured, and in turn, variation statistically analyzed.

It became apparent that various periodic fluctuations occur within a range of 2–82 years. Those in a range between two to six years can be clearly correlated to the temporal scales that today govern the fluctuations for El Niño/La Niña events (Lenz et al. 2010). The corresponding long-distance effects during the time of deposition of the Messel oil shale in the area of modern-day Europe were probably much more strongly pronounced than today. The Atlantic had not yet reached its current width, and the climate effects resulting from the Pacific therefore had a more pronounced impact in the Messel area (Box 3.2).

The oil shale – A unique Eocene climate archive

With the aid of our statistical analyses regarding the frequency of pollen and spores as well as certain algal remains in the seasonally laminated oil shale of Messel, we were able to document that noticeable changes in the vegetation followed cyclic and periodic climate fluctuations. In this, all cycles known from the current glacial period of the Quaternary with a frequency below 100,000 yr can be clearly documented (Box 3.3). These include, in particular, the

Fig. 3.4: To allow the examination of the oil shale in the laboratory down to the centimeter scale, a block of 1 m thickness was cut out (top) and wrapped in household plastic foil (bottom).

Box 3.2: El Niño Southern Oscillation (ENSO)

Another phenomenon with a significant impact on the environment and humans are the El Niño/La Niña events that occur more or less regularly every few years, which lead to a short-term reversal of the oceanic current in the Pacific Ocean. El Niño is triggered by changes in the water circulation in the Pacific. In the process, for reasons yet unknown, the warm equatorial current that leads to cloud formation and thus to high precipitation, and which usually follows a westerly direction at the surface, is temporarily reversed. This repeatedly leads to catastrophic effects on the climate in the Pacific region: cyclones, flooding on the Eastern Pacific side in South America and extended drought periods in the Western Pacific, from the Philippines to Indonesia and Australia. The opposite occurs during a La Niña phase. The interchange between these events during a period of 2-7 years, which may also be felt on a global scale, is referred to as ENSO. Longer, ENSO-like temperature fluctuations over a period of several decades are described as PDO (Pacific Decadal Oscillation).

One of the most important and controversial questions in climate research is whether such ENSO events also occurred in the geological past and in a warmer world. The fact that we can document this climate phenomenon for the Eocene therefore offers an important contribution toward this discussion.

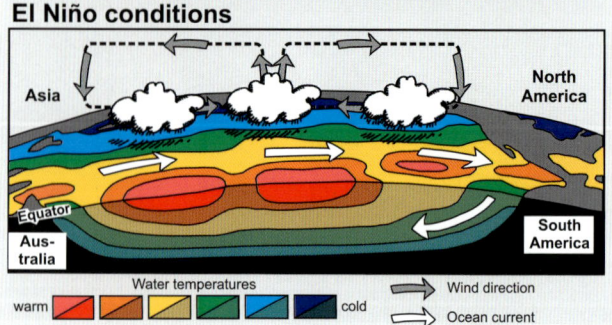

Box 3.3: Time series analyses

Cyclic or periodic events that occur in the course of geological history can be recognized by means of statistical methods. In the process, the changes in a wide variety of parameters are analyzed over time regarding the occurrence of regular repetitions. This is referred to as a "time series analysis." On the one hand, a prerequisite for this is an even and well-known deposition rate in the examined sediments. On the other hand, the respective layers must represent sufficiently long time periods, in order to allow for statistically sound periodicity. These conditions exist primarily in the sediments of stable lakes. Such a system can be found in Messel, and it was examined for the first time for the Eocene using the methods developed for Quaternary lake deposits. In our study, we considered changes in the vegetation that can be revealed by the deposited pollen, spores and certain algal remains in the sediment. No pollen or spore types disappeared in the examined profile section, and no new ones were added; however, there were obvious fluctuations in the frequency of individual species or species groups. This indicates the complex and multifaceted climate changes that occurred during the deposition time of the lake sediments.

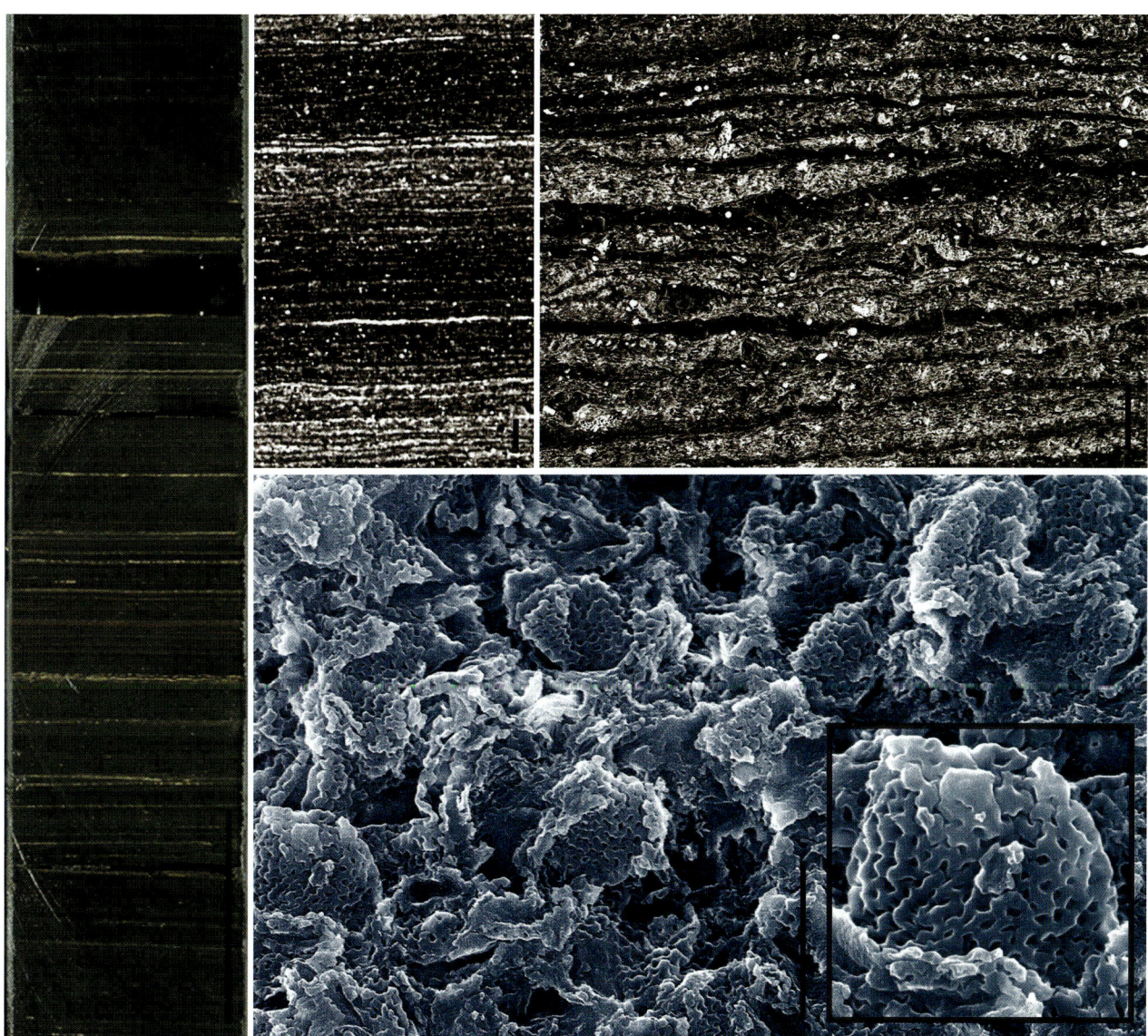

Fig. 3.5: The seasonal stratification in the oil shale: left: drilling core (scale: 10 cm). Center top: Alternating light (green alga *Tetraedron*) and dark (clay minerals) layers (scale: 1 mm). Right top: SEM image of the stratification (scale: 0.1 mm). Images below: SEM images of a *Tetraedron* layer (scale: 5 µm). Detail enlargement: single *Tetraedron* vesicle with a characteristic, reticulated surface.

Milankovitch cycles caused by variation in orbital parameters. However, there are several apparent cycles below the Milankovitch periodicity that were also known already from the Quaternary. In addition, the precise measurement of the oil shale layers revealed periodicities following the regular changes in solar activity and oceanic circulation in the Pacific. Thus, we were able for the first time to prove that the same mechanisms that are at work in the ongoing glacial period of the Quaternary were also active in the Eocene greenhouse system, which was not influenced by humans. These results make the Messel oil shale one of the best known pre-Quaternary sedimentary climate archives with the highest resolution. Future research must now clarify which individual climate factors (temperature, precipitation or humidity) led to the documented effects on the vegetation and sedimentation.

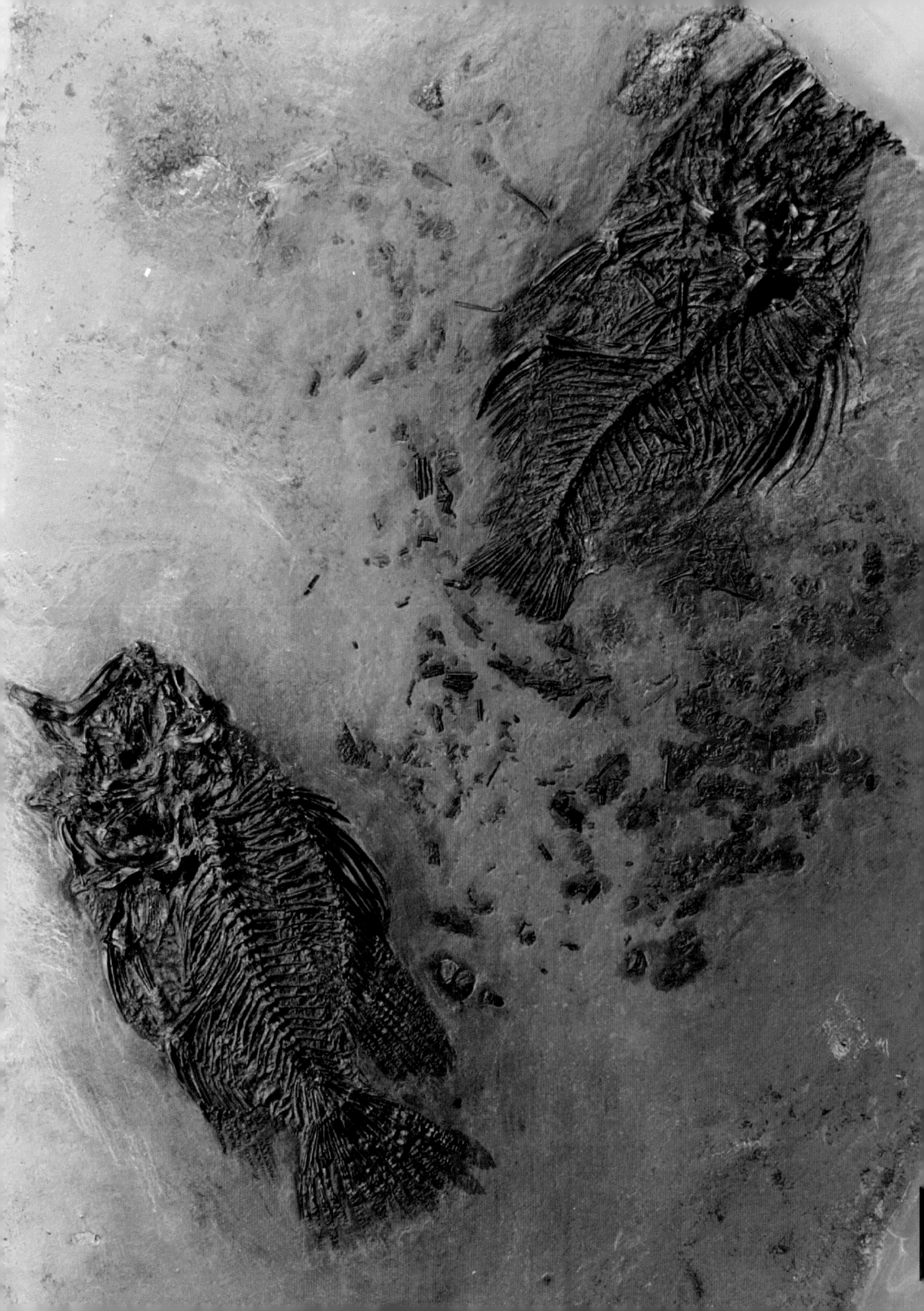

Chapter 4
Joined in Death: the Burial Community of Messel

Michael Wuttke, Renate Rabenstein, Jakob Vinther

Fossilized vertebrate skeletons that are still anatomically articulated (Fig. 4.1), the preservation of ephemeral organic structures down to the finest hairs, or the possibility to reconstruct the former coloration of plumage and fur – these are some of the reasons why Messel counts among the world's most important fossil sites. Sites of this kind allow deep insights into the evolutionary history of life and the environmental conditions during fossilization. As easily gathered when reading the following chapters, the Messel oil shale contains a veritable burial community (taphocenosis; from the Greek "taphos": tomb, burial) of plants and animals. They inhabited both the maar lake itself as well as its banks, or they found their way into the lake from the hinterland. Some of the fossils appear to have been struck down in the middle of life, such as the pregnant mare of an early horse or badly chewed-up frogs and birds. A specimen of the snake *Palaeopython fischeri* constitutes a particular highlight, as it was preserved complete with its prey in the gut tract: a lizard of the species *Geiseltaliellus maarius*, which in turn had just consumed an insect. This discovery thus serves as direct evidence of a food chain (Chapter 10.1; Smith & Scanferla 2016).

These fascinating and even touching "dramas of life" raise the following question for paleontologists: How was such a fantastic degree of preservation even possible? To what circumstances do we owe this knowledge of individual details regarding the fossil?

The natural course of microbial tissue breakdown in a carcass on land only leaves behind carbon dioxide and a few minerals; the bones can weather away or be transported and separated by water and wind. Therefore, in the case of Lake Messel, there must have been specific environmental conditions present in the water that partially or completely prevented such processes from occurring. These types of questions are subject of a special branch of geosciences, the field of taphonomy (from Greek "nomos": law). It encompasses all circumstances from the death of an organism until its final embedding and the associated mechanical and biological-chemical processes. Based on the realization that the present is the key to the past, there exist various working methods. Taphonomists work in a similar fashion as forensic pathologists who reconstruct a crime scene and the criminal act, attempting to determine a fossil's individual taphonomic history (Wuttke & Reisdorf 2012). Since the site of death and embedding of a former organism are not always identical, it is important to record the geological and paleontological context, such as the embedding environment under water or the absence of organisms in the sediment. Actualistic experiments regarding the decomposition of vertebrates (Fig. 4.7) can serve as an additional key, offering insights into the normal temporal process of decomposition. This decomposition depends on the individual tissue types, which show varying resis-

Fig. 4.1: Decomposed perch *Amphiperca multiformes* and current-related distribution of its scales and skeletal parts next to a complete specimen deposited at a later date. Scale: 2 cm.

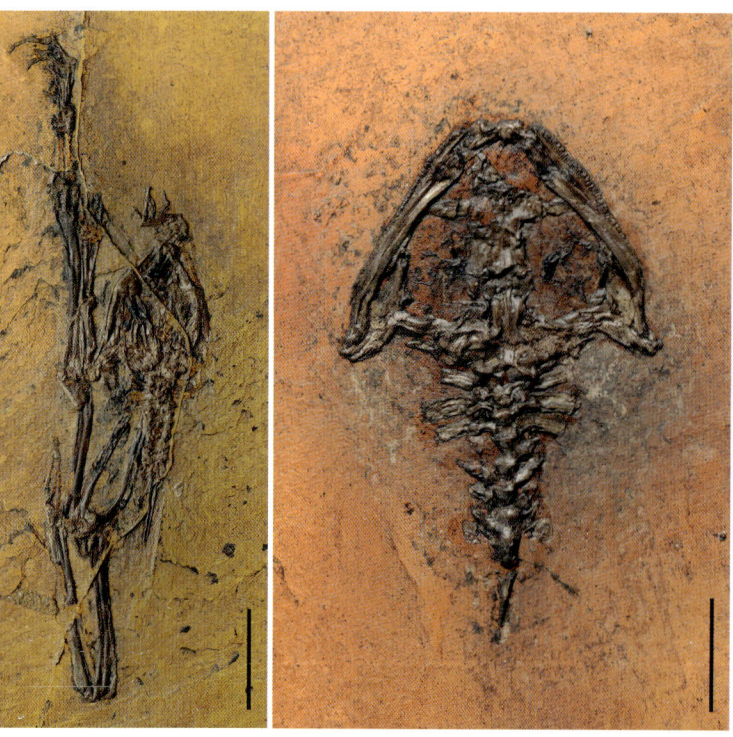

Fig. 4.2: Fossil discoveries unusual for Messel: Regurgitated pellet (left, scale: 2 cm) and disarticulated frog (right, scale: 1 cm).

These are numerous in Messel, mainly in the form of the remains of frogs (Fig. 4.2) and birds (Mayr 2016). Characteristic in this regard are isolated or missing skeleton parts that always show splintered bone ends. To date, few records of such mammals are known. On the other hand, the partial skeleton of the primate *Europolemur koenigswaldi* contains a crocodile tooth that is stuck in a bone (Franzen & Frey 1993). In the skeleton of a pregnant mare of a Messel horse, a large crocodile tooth cut through both tibias and fibulas (Franzen et al. 2017), most likely when the crocodile dragged its prey into the water to drown it. For some reason, it then abandoned the prey, so that the mare was able to sink to the lake bottom without further injuries. In relation to all vertebrate discoveries, however, such feeding traces constitute an exception. Based on the level of preservation of the majority of Messel fossils, it can be inferred that they reached the lake bottom intact or – much more rarely – as the remains of carcasses drifting on the water's surface for a long time and decaying in the process.

Distortion in the course of time

An essential indicator for delimiting the time period during which the vertebrates at Messel sank to the lake bottom following their death is found in the relative position of the bones to each other in the fossil. This can be clearly demonstrated by the example of the Messel early horse (Fig. 4.3). In vertebrates, ligaments serve to hold movable joints together, which they enclose in the form of joint capsules. When a joint remains in the so-called rest position, all involved ligaments of the joint are equally relaxed or tensed, and the bones joined by the ligaments are situated at a specific angle toward each other. This rest position may be actively assumed; upon death, it automatically occurs after the slackening of the musculature if the animal ends up in a body of water, since the buoyancy counteracts gravity (Reisdorf & Wuttke 2012). This rest position can easily be recognized in the Messel early horse by the slightly backward curve of the neck and the relative position of the leg bones toward each other (Fig. 4.3). This means that the animal reached the lake bottom before putrefaction had proceeded so far that the ligaments' tension could no longer keep the joints in the rest position. As

tance against microbial breakdown. The embedding environment also influences the composition of the biocenosis of the decomposing microorganisms.

Additional influencing factors are processes inside the carcass itself, such as the formation of putrefaction gases (e.g., Smith & Wuttke 2012), that cause the displacement of skeleton parts from their anatomical articulation. For the longest time, taphonomic research failed to consider processes that are based on the release of forces from preloaded ligaments due to putrefaction of less resilient tissue, e.g., muscles. Once the resistance of the remaining tissue weakens, these tensile forces can lead to the displacement of entire skeleton segments (Reisdorf & Wuttke 2012). Therefore, it is possible, based on the skeleton's completeness, the degree of decomposition and the preserved relative position of the skeleton components in relation to each other to infer whether the presence of putrefaction gases, preloaded ligaments or currents was responsible for any shifts.

Inferences as to the underlying cause of death are most easily drawn in case of feeding traces.

visible in the image, some of the skeleton's elements are no longer within their natural association. This is particularly true for those elements that formerly pointed toward the water's surface (the large Messel vertebrates are prepared in such a fashion that the eventual display side is the side that used to rest on the lake bottom). The illustration clearly shows the impact of putrefaction gases – the body was bloated so severely that individual skeleton parts were shifted from their original position and subsequently, after additional decomposition, sank back down and came to rest at a different angle.

The effect of the preloaded ligaments even after an animal's death can be particularly well demonstrated by the example of long-necked birds, such as this Messel Rail (*Messelornis cristata*) (Fig. 4.4). In fossils of this species, the neck is usually bent so far backward that the skull almost comes to rest on the vertebral column. The reason for this is a preloaded ligament extending along the upper side of the spinal column (Ligamentum elasticum interlaminare) and connecting the individual vertebrae with each other. Here, "pretension," a term from the field of mechanics, means a mechanical tension already in the ligaments (without external stress) necessary to keep the long neck upright against gravity in the rest position. This helps save muscle effort, and thereby energy. Under water, gravity is greatly reduced and the neck starts to bend backward, until the force of mechanical resistance of the lower neck musculature equals the ligament's pretension. Only in the course of the carcass's continuing decomposition, in the process of which the muscle proteins can be broken down more easily by microorganisms than the ligament's collagen, can the resulting released forces of the ligament continue to have an effect (Fig. 4.4). Slowly, i.e., over the course of days or weeks, the neck thus keeps bending farther and farther backward (Reisdorf & Wuttke 2012).

Fig. 4.3 Displacement of skeletal parts in the torso area of an early horse. Scale: 5 cm.

Fig. 4.4: Hyperextension of the neck of a Messel rail. Scale: 2 cm.

The mystery of the bats

Along the biggest mysteries in the composition of the burial community of the Messel sediments is the unusually high number of bat specimens (more than 700). At the same time, they represent a paleobiologically as well as taphonomically extremely well-studied group. Bat fossils are generally rare; most come from sediments in their former daytime quarters – natural rock caves or crevasses (Habersetzer et al. 1992). They are extremely rarely found in lake sediments: Among 1,900 vertebrate specimens from the Eocene crater lake of Mahenge (Tanzania) there was only one single bat, the holotype of *Tanzanycteris mannardi* (Fig. 12.5.13; Kaiser et al. 2006). Only the Eocene lignite mine at Vastan (India) has yielded the same diversity as Messel, with seven recorded species in four families, albeit only in the form of jaw fragments and isolated teeth (Smith et al. 2007). In Messel, on the other hand, the skeletons are often completely preserved, many with additional soft parts (Chapter 12.5). In *Palaeochiropteryx tupaiodon* and *Hassianycteris messelensis* this even led to the first evidence of fur color in fossil mammals (Colleary et al. 2015). We now know that these bats had a reddish-brown fur; however, the reason and manner of their death remains unknown.

If we compare this to the discovery situation of other fossil maar sediments with a similarly long history of continual scientific excavations (25 years or more), the differences to Messel become even more apparent: In the Eckfelder Maar (Eocene; Eifel, Germany), there were three discoveries of isolated bat skeleton remains during this time period; with none in the Enspel Maar (Upper Oligocene; Westerwald, Germany (Lutz & Kaulfuss 2006; Schindler & Wuttke 2015).

The Messel bats make particularly interesting subjects for taphonomic investigations for three reasons. First, it is surprising that flying mammals are preserved in the sediments of this freshwater lake, of all places. Second, with several hundred individual specimens, they occupy the top position among mammal groups. And third, the analysis of Senckenberg's bat collection revealed that almost 90 % are available as more or less complete skeletons (Habersetzer & Rabenstein 2011). Examples of such complete skeletons, often with articulated bones, are illustrated in Chapter 12.5. Sometimes, the head lies separately next to the otherwise intact skeleton (Fig. 4.5, *H. messelensis*), and even in disarticulated (broken-down) skeletons, all bones are usually present (Fig. 4.6, top: *P. spiegeli*). Only about 10 % of all discoveries consist of isolated body parts, mainly wings and heads of *H. messelensis* (Fig. 4.6, center, bottom left), and very rarely of individual bones (Fig. 4.6, bottom right: right lower mandible of *P. tupaiodon*).

Habersetzer & Rabenstein (2011) published information regarding the chronological sequence of skeletal decomposition in bats, based on the examination of more than 50 captive specimens of Seba's Short-tailed Bat (*Carollia perspicillata*) immediately after their death. Under controlled laboratory conditions, they allowed the animals' bodies to decompose in shallow, water-filled experimental containers (water height 8 cm). The water temperature was based on that of modern-day tropical lakes, according to their own measurements and data published by other sci-

The mystery of the bats 29

of the decomposition experiments the experimental containers were located directly in the X-ray laboratory, thus requiring only minimal transportation.

The results showed that dead bats initially floated on the water's surface. Although, like all mammals, they have a higher specific weight than water, the amount of air trapped in the hair, along with the air-filled lungs, is sufficient to keep the animals afloat at the surface. Based on this, it is possible that the lungs acted as an additional buoyancy body when no water was aspirated, i.e., the animals did not drown. In addition, putrefaction gases escaped from the mouth and anus, so that half of the bodies sank to the bottom

Fig. 4.5: Bat with a disarticulated head. Scale: 2 cm.

Fig. 4.6: Rare state of preservation at Messel: partially disarticulated bat, individual skeleton parts. Scale: top and center: 2 cm, bottom left: 1 cm, bottom right: 2 mm.

entists (Rabenstein et al. 2004, Ruttner 1931). The decomposition at low, medium and high water temperatures (20, 25 und 30 °C) was documented by means of digital radiographs. This method and standardized processes to avoid artifacts due to undesired movements of the liquid had been designed in advance. This ensured that besides the cadavers' natural decomposition there were no additional changes in position during the X-ray process, since for the duration

> **Box 4.1: Ancient Maar Lakes Compared**
> In the Eckfelder Maar, a lake with a smaller diameter than the Messel Maar, about 6,000 years of the lake's history could be explored through excavations to date. Many of the vertebrate species recorded there originated from a mud stream of sediments formerly located near the bank, which were deposited at the lake's deepest point. A possible trigger for this type of slippage could have been earthquakes, which may lead to reduced static friction between the deposited sediment grains due to water-filled pore spaces. Since the vertebrate discoveries in the mud stream primarily constitute scattered skeleton parts, it appears likely that the animals initially decomposed along the banks of the lake and were embedded in the sediments there. Such concentrations may be best explained by the fact that land animals that drowned in the lake were transported by currents or wind drift toward the bank areas, where they began to break down. As can be seen from the explanations below, even dead bats will drift on the surface for up to three days before they sink to the bottom – sufficient time to reach the nearshore areas.
>
> In the case of the Enspel Maar, which had a diameter similar to that of the Messel Maar, finds of land-dwelling vertebrates are generally very rare. To date, there is no truly plausible explanation for this. There is no documentation of a mass die-off of aquatic vertebrates, whether caused by circulation of the water body that brought toxic water from the depth to the surface, or by the toxins from bacteria or algae. The maar lake's sediments that were explored by the digs were deposited about 600,000 years after the eruption of the maar. It is impossible to determine to what extent volcanic gases continued to exude from the diatreme after such a long period of time, since – similar to the Messel Maar – the former bank areas have not been preserved and there are no traces of gas bubbles that passed through the finely laminated sediment. Thus, unlike the hypotheses discussed in the case of Messel regarding potential cause of death for land vertebrates, no such documentation is available for Enspel.

between day 7 and 14; the remaining bodies reached the bottom no later than by day 17. In the latter, the inner joint articulations already showed advanced decomposition (Fig. 4.7 right: day 14). Thus, it can be inferred from this experiment that the large number of fully articulated Messel bats indicates that their bodies sank within only a few days to an oxygen-free depth in the lake. In this context, the high water pressure also played a role, as it partially compressed or dissolved putrefaction gases, generally preventing the vertebrate bodies from floating back to the surface.

Fossil color preservation

The unique preservation of fossilized soft tissue in Messel allowed hitherto unknown insights into geological history: the decoding of fossil color patterns. Scientific discoveries in the past ten years have shown that the unusual preservation of feathers, hairs, skin and internal organs such as eyes or liver can be traced back to the fossilization of the pigment melanin, which survived in these soft parts (Colleary et al. 2015). In vertebrates, melanin is found in cell organelles called melanosomes. These are small, ovoid to elongated objects of 400 nm to 2 μm in length, which can be seen with the aid of a high-resolution scanning electron microscope (Fig. 4.8). In the early 1980s, such tiny objects had already erroneously been described at Messel as autolithified bacteria due to their remarkable similarity (Wuttke 1983).

A more recent study revealed that these small, bacteria-like bodies actually constitute melanosomes (Vinther et al. 2008). This study not only showed that these micro-bodies are similar to melanosomes, but also that they are arranged in such a way as to produce color patterns, in a similar fashion as modern-day melanosomes in feathers. This scientific breakthrough was based on the examination of ink from fossil cephalopods (Glass et al. 2012), which was

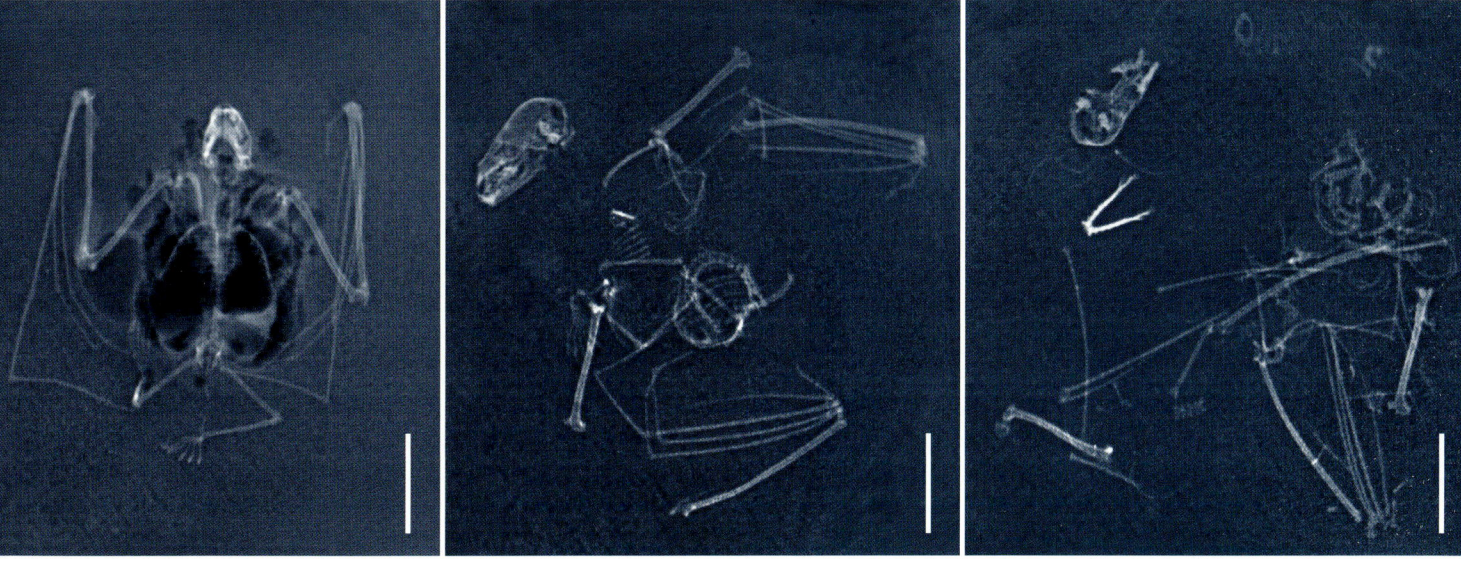

Fig. 4.7: Radiographic documentation of the decomposition of a bat that was placed in a water basin with the temperature of a tropical lake immediately following its death. Scale: 2 cm.

known to be preserved in fossilized form and to also consist of melanin. That fact that melanosomes can fossilize gave rise to an entirely new scientific discipline: the study of fossil colors. Even the original coloration of dinosaurs – including the iridescence – could be reconstructed based on preserved melanosomes.

Melanin occurs in two chemical variants, eumelanin and phaeomelanin. Eumelanin creates black colors and is widespread in the animal kingdom, from the ink of a squid to the skin of hair and skin in humans. Phaeomelanins create rusty-red colors, or yellowish-brown in lower concentrations, like the colors found in the Labrador retriever dog breed. Interestingly, the melanosomes not only differ in their chemistry but also in their shape. The elongated eumelanosomes are reminiscent of small sausages, while phaeomelanosomes show globular to ovoid shapes. Over millions of years, this regularity remained intact in both mammals and birds. Therefore, it is possible to analyze fossil animals (including dinosaurs) that originate from a common ancestor in regard to their coloration.

Put in simple terms, the shape of the melanosomes allows inferences about the original colors of a fossil. Melanosomes also play a role in the production of iridescent colors, e.g., the metallic gloss found in the vivid colors of many birds such as peacocks, hummingbirds or male mallards. The regular arrangement of melanosomes in the finest feather barbs allows the coherent scattering of the light's different wavelengths at different angles, which is what we perceive as the metallic gloss. The invention of iridescent feathers is of great importance to modern-day birds. It enables an enormous diversity of coloration and the development of unique courtship behaviors. The outstanding preservation of fossils at Messel means that the melanosomes are still present in their original arrangement; i.e., the same way as in the original feather – even after the feather's main component, the protein keratin, in which the melanosomes were embedded, had been decomposed by microbes. Based on a well-preserved feather from Messel, it was even possible to show that the original arrangement of the melanosomes that created the iridescence was still preserved (Vitek et al. 2013). This feather is particularly noteworthy, as it shows a silvery shimmer in the outer areas of the feather vane. It was only present in the finest branches, the so-called feather barbs (Box 11.6).

Under the scanning electron microscope, it became apparent that the melanosomes are arranged in shallow layers near the feather barbs' former surface, lending the surface a sheen referred to as layer iridescence. This type of iridescence is common among birds and occurs when two different materials are arranged in homogeneous layers, like when gasoline is poured on water. The visible, vivid play of colors is caused by a thin-film interference; in the case of gasoline due to the most minute changes in thickness on the water's surface. Unfortunately, it is not possible to infer the original color of the iridescent feather from Messel, since the keratin layer above the melanosomes, which ultimately determines the exact color, has not been preserved. Nevertheless, the results indicate that the extinct birds had a colorful plumage and possessed an expressive display behavior.

Some of the Messel birds showed color patterns reminiscent of modern-day birds. Two animals belongs to the Strisores, a group which contains, among others, the modern nightjars (*Caprimulgus*) (Chapter 11). Similar to modern owls, they have a grayish-brown plumage, affording them camouflage during the daytime. *Hassiavis* (Fig. 11.20) had a barred tail, while the potoo (*Nyctibius*) shows a striped pattern on the wings' flight feathers. *Messelirissor* is a remote relative of modern-day hoopoes (*Upupa*) and wood hoopoes (*Phoeniculus*); it definitely has a banded tail (Fig. 11.40). As mentioned above, the two most common bat species in Messel, *Palaeochiropteryx* and *Hassianycteris*, had small, ovoid melanosomes that indicate a brownish coloration, as typical for modern bats and many mammals in general, which often show a rather monotone and camouflaging coloration.

The preservation of melanin offers a valuable new window into the fossil record, since it makes it possible to reconstruct the coloration of extinct animals, but the extent of preservation of even subcellular structures at Messel is only beginning to be explored. Cadena (2016), for instance, discovered that bone cells (osteocytes) and collagen fibers may be preserved in bones from Messel.

Fig. 4.8: Melanosomes (SEM images) of a mammal (top) and a bird (bottom). Scale: 10 µm.

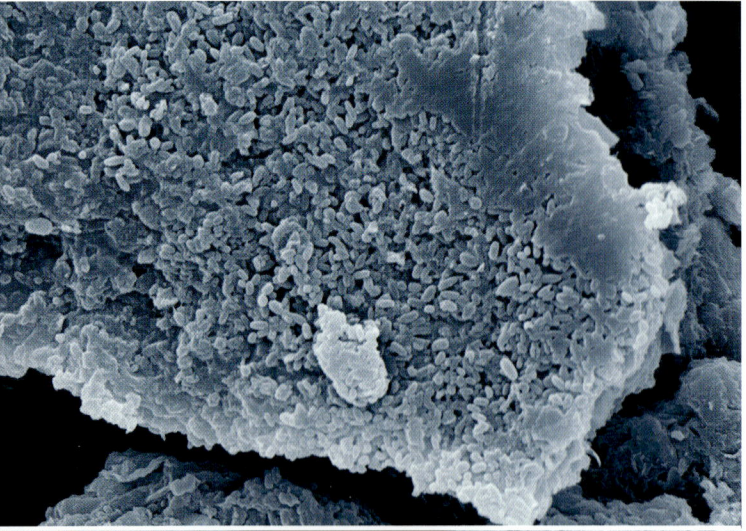

Cause of death: Unknown

All described phenomena and methods applied in an attempt to understand the origins of fossilization only rarely solve the underlying mystery: How and why did the non-water-dwelling vertebrates end up in the lake, and were they dead or alive? Was there a mass die-off of lake dwellers, or does the extremely low sedimentation rate (compacted 0.14 mm per year) lead to the false impression that fossils are abundant, e.g., in case of the bats?

The discussion of potential causes of death for terrestrial vertebrates arises in particular due to the conspicuously frequent discovery of bats. To date, two possible causes of death have been discussed: Drowning after being rendered unconscious by toxic gases (Habersetzer et al. 1994), or immediate death after consuming a film of cyanobacteria (scum) floating on the water's surface (Koenigswald et al. 2004)

– both methods consistent with the bats' lifestyle. However, the possibility of accidental drowning has also been raised (Smith & Wuttke 2012). Many bats show a preference for bodies of water. Certain insectivorous species prefer to hunt at low elevations above calm water surfaces, where they encounter a rich source of food at night. They are also able to grab larger insects drifting on the water surface with their hind feet (skimming). The fossil species of *Palaeochiropteryx* were low-flying and agile insect hunters, as well, which is clearly evidenced by their wing morphology and stomach contents (Habersetzer et al. 1994). Apparently, no old or sick bats died in Lake Messel, since the fossils show neither any injuries nor heavily worn teeth. The intestinal tracts, frequently filled with insect remnants (Fig. 12.5.11), indicate a sudden accident after a successful hunt. In this regard, the hunting behavior of modern bats also offers an important indication. They only drink in flight from the water surface prior to returning to their daytime quarters. We can assume the same for the bats in Messel, both the low-flying *Palaeochiropteryx* species and the species of *Hassianycteris* and *Tachypteron* that hunted higher up in the open air (Chapter 12.5). Observations in the wild (Habersetzer, oral communication) show that bats can fall in the water during drinking or skimming, but are able to take flight again. Therefore, this does not offer any indication as to a specific taphonomic filter that could explain the high rate of discovery of the bat species hunting close to the lake's surface.

Various hypotheses are being discussed regarding the cause of death for land vertebrates in Lake Messel; for some, there are even examples from nature. Thus, it may be possible that gases formed deep in the lake due to bacterial decomposition, which subsequently reached the surface due to mixing of the water column. In principle, volcanic gases are also conceivable (Pfanz 2008), which, in case of Lake Messel, may have continued to exude from the ground even hundreds of thousands of years after the eruption of the maar volcano, coming to the surface near the edge of the lake (Franzen et al. 2017). The literature also contains an example of a mass die-off of bats that ingested too many toxic cyanobacteria while drinking (Pybus et al. 1986); it is known that even large mammals can die only minutes after ingesting such toxins. However, in the meantime, some of the groups of animals that formed the basis for this hypothesis must be evaluated differently (Chapter 10.2; Joyce et al. 2012). In those cases, as well, accidental drowning must be considered as the null hypothesis (Smith & Wuttke 2012).

To date, there is no possible way to test the various hypotheses on the basis of the discoveries, since neither gases nor bacterial toxins leave traces on the skeletons. Moreover, the Messel oil shale includes no indication that volcanic gases from the ground percolated through it, as known from younger maars with an age of only a few hundred to a few thousand years. It was also not possible to document the presence of toxins from cyanobacteria in the older sediments. This means that additional research will be required to achieve better supported insights regarding the circumstances of the death of vertebrates at Messel.

Chapter 5
Messel Research – Methods and Concepts

Sonja Wedmann, Jörg Habersetzer, Thomas Lehmann, Irina Ruf, Stephan F. K. Schaal, Krister T. Smith

Before fossils can be studied scientifically, they must first be discovered, extracted and prepared. Therefore, we will begin by introducing the various steps and methods used to prepare a fossil for scientific examination and study. Theoretical aspects such as systematic placement, the evolutionary relationships and origin or a species' function in its habitat play a central role when studying fossils. Therefore, we will also offer a brief explanation of basic terms from the fields of taxonomy, phylogeny, systematics and biodiversity research.

Excavation, conservation, preparation

The excavations in the oil shale layers in the Messel Pit are carried out with permission from the Hessian State Office for the Preservation of Monuments. All fossils that are found and prepared are inventoried in the respective institutions' research collections. In this regard, professional work in the technical and scientific fields, both during the excavation and the preparation of the fossils, is an essential prerequisite (Keller et al. 1991).

The excavations are conducted by small teams of six to ten persons. They consist of experienced excavation leaders as well as student interns, aided by additional volunteer workers. First, oil shale blocks are extracted with the aid of sledge hammers, wedges, crowbars and even chain saws (Fig. 5.1, 5.2). Large specimens like early horses or crocodiles are usually discovered at this early stage, since massive bones become apparent through unusual bulging in the rock.

Since the moist oil shale is subject to rapid drying and subsequent disintegration, only a few blocks are extracted for additional processing at one time. The blocks are finely split with knives and examined for fossils, first with the naked eye and later with a magnifying glass. Often fossils become divided between the two oil shale surfaces that surrounded them, so that one fossil is preserved on two plates, named A and B. Any discovered fossils are recorded in an excavation log. In order to protect the finds from destruction through desiccation, they must be kept moist at all times. For this purpose, vertebrate specimens and large plant parts are covered with a thin plastic foil, wrapped in wet paper and packed in plastic bags. Insects and smaller plant fragments are stored in water-filled buckets.

In the winter, when excavations are not possible, the specimens are prepared. Insects, invertebrates

Fig. 5.1: An excavation team at work.

Fig. 5.2: Oil shale extraction in the Messel Pit.

Fig. 5.3: Preparation of a giant ant.

and plants are painstakingly exposed under the binocular microscope, using a special preparation kit (Fig. 5.3). For permanent storage, the specimens are then kept in glycerin to avoid desiccation.

Vertebrate fossils are preserved by means of the synthetic resin transfer method (Kühne 1961), since earlier preparation methods, e.g., the lacquer film method or transfer to liming wax, did not prove successful in the long run. First, the overlying oil shale and occasional siderite (iron carbonate) is carefully removed from the bones and any potential soft parts (Fig. 5.4). The surrounding oil shale is leveled and a clay frame is modelled around the resulting plate (Fig. 5.5, preparation scheme). During careful drying with a hair dryer, the plate is covered with a thin layer of synthetic resin, which partially impregnates the exposed bones. In several steps, further layers of synthetic resin are poured in to harden, limited by the surrounding clay frame. The fossil is now embedded on one side. Once the resin has set, the plate is turned over and the side of the fossil still enclosed in the oil shale is completely exposed. This method can also be used to transfer and conserve soft parts, which are frequently preserved in Messel fossils. For long-term storage of fossils in a water bath, it is advisable to create a so-called embedded specimen, in which the surface of the fossil exposed during excavation is fully prepared and sealed with synthetic resin, while the other side remains entirely embedded in oil shale (Schaal & Habersetzer 1991).

For particularly compact vertebrates, e.g., turtles, a preparation method is advisable in which the fossil is completely extracted and hardened with superglue. After the preparation, both sides of the fossil are thus accessible. The preservation of soft tissue traces is not possible with this method.

After successful preparation, the details of the external and internal physical structures are recorded by means of microphotography, fluorescence microscopy, scanning electron and transmission electron microscopy as well as X-ray techniques. The destruction-free methods of micro-X-ray and micro-computed tomography have been used and further developed by Senckenberg for 40 years.

Fig. 5.4: Preparation of an early horse.

Examination by means of X-ray techniques and electron microscopy

Micro-X-ray and micro-computed tomography (µCT) are non-invasive methods and thus of invaluable advantage for the study of the often unique fossils from Messel. Compared to medical X-ray images, the resolution of microradiographs is 100 times higher. The use of a special X-ray tube, the so-called micro-focus tube, took on great importance in this regard. Due to the virtually point-shaped radiation source it is possible to enter microscopic dimensions (Habersetzer 1995, 2004).

For over two decades, a technical variant of medical computed tomography (CT) has been available for scientific applications: micro-computed tomography (µCT) (e.g., Rowe et al. 1995). In medical computed tomography, the patient is at rest, while the X-ray tube and the opposing detector rotate around the person. In contrast, in micro-computed tomography the investigated object rotates inside the X-ray beam on a turntable, while the X-ray tube and the radiation detector are virtually stationary. With special mechanical improvements, it is further possible to set different magnification factors. Here, as well, extreme magnification was only made possible through the use of the micro-focus tube (Habersetzer et al. 2012, Ruf et al. 2016).

Applications of these X-ray techniques can be found in various contributions in this volume. To exemplify µCT, we take the famous and controversial Messel "anteater" *Eurotamandua joresi*. In the Senckenberg Research Institute, small objects can be examined with a µCT device (maximum resolution 5 µm), while a nanoCT device (maximum resolution 0.9 µm) is used for very small objects. For fossil plates larger than approximately 20 cm, an industrial µCT scanner must be employed. In the case of *Eurotamandua*, this is a high-tech device with a special focusing procedure that allows an additional increase in resolution. Fig. 5.6 shows the RayScan µCT scanner (AUDI laboratory, Neckarsulm), where the fossil was fixed with a specially designed mount. Fig. 5.7 shows a 2D image of the forelimbs in the standard µCT procedure (left). The center image shows the same region scanned using the special procedure, in which the effective resolution is almost doubled by means of a horizontal expansion of the measuring

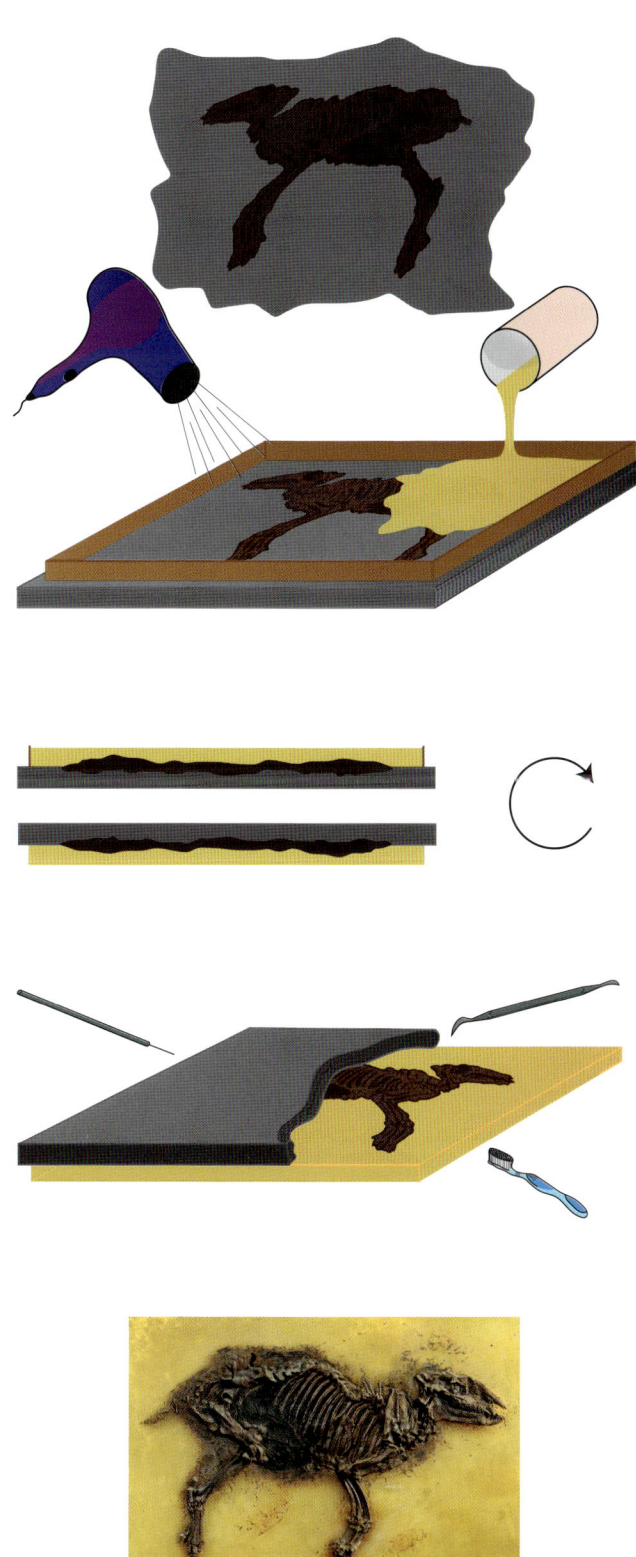

Fig. 5.5: Preparation scheme for the transfer method (details see text).

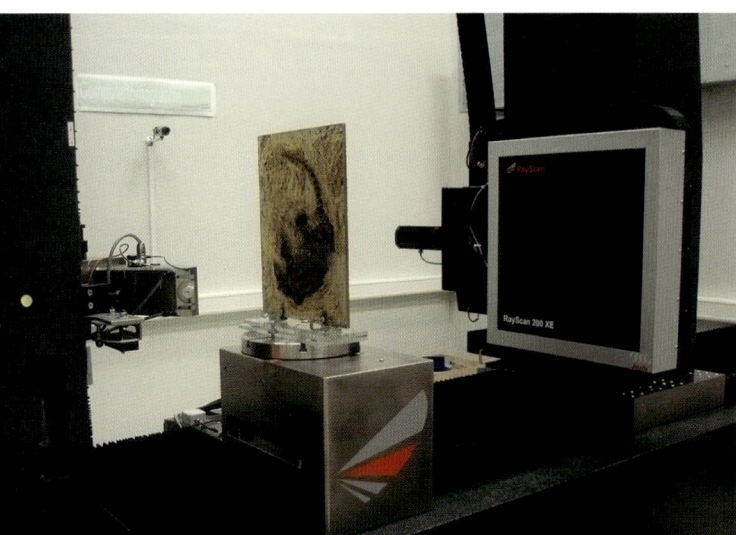

Fig. 5.6: View inside the µCT apparatus with mounted *Eurotamandua joresi* (holotype) in the AUDI laboratory (Neckarsulm).

The scanning electron microscope (SEM) is an additional important aid in examining fossils. Scanning with an electron beam creates images of the sample's surface. This allows high magnification of tiny, square-millimeter-sized pieces of biological and geological samples. In Messel research, the SEM has been used since the 1980s for the examination of soft parts and sediments (Richter & Storch 1980, Wuttke 1983, Goth 1990, Chapter 4). It has also been applied to study the gut contents of vertebrates. As an example, we refer to the images of the gut contents of bats (Chapter 12.5).

Taxonomy and Phylogeny

A formal system for classifying and naming species (i.e., for taxonomy) was first developed by Carl von Linné (Linnaeus 1758). The version of his system currently in use follows several foundational principles: 1) each species is assigned a two-part (binominal), usually Latinized name, consisting of the genus, followed by the species name (e.g., *Homo sapiens*); 2) a species' first published name has priority over any subsequent names; 3) the name is tied to a particular specimen called the "holotype."

Genera are assigned to a family, which in turn is assigned to a higher systematic category (or rank). In case of doubt regarding the identification, question marks are used. Irrespective of its rank, each unit

field that significantly reduces the veil artifacts (inside and outside of the fossil). On the right, the 3D model based on this special procedure is shown. A comparison reveals that the 2D presentation illustrates the joint surfaces and bone trabeculae much more precisely, which is also reflected in a highly detailed, life-like surface representation in the 3D model. The difference in quality becomes particularly apparent in the detail enlargements, which show the metacarpal bones in twofold magnification.

Fig. 5.7: Quality comparison of the different µCT procedures (left, center) and virtual 3D reconstruction (right) shown for the hand skeleton of *Eurotamandua joresi*. Scale: 2 cm.

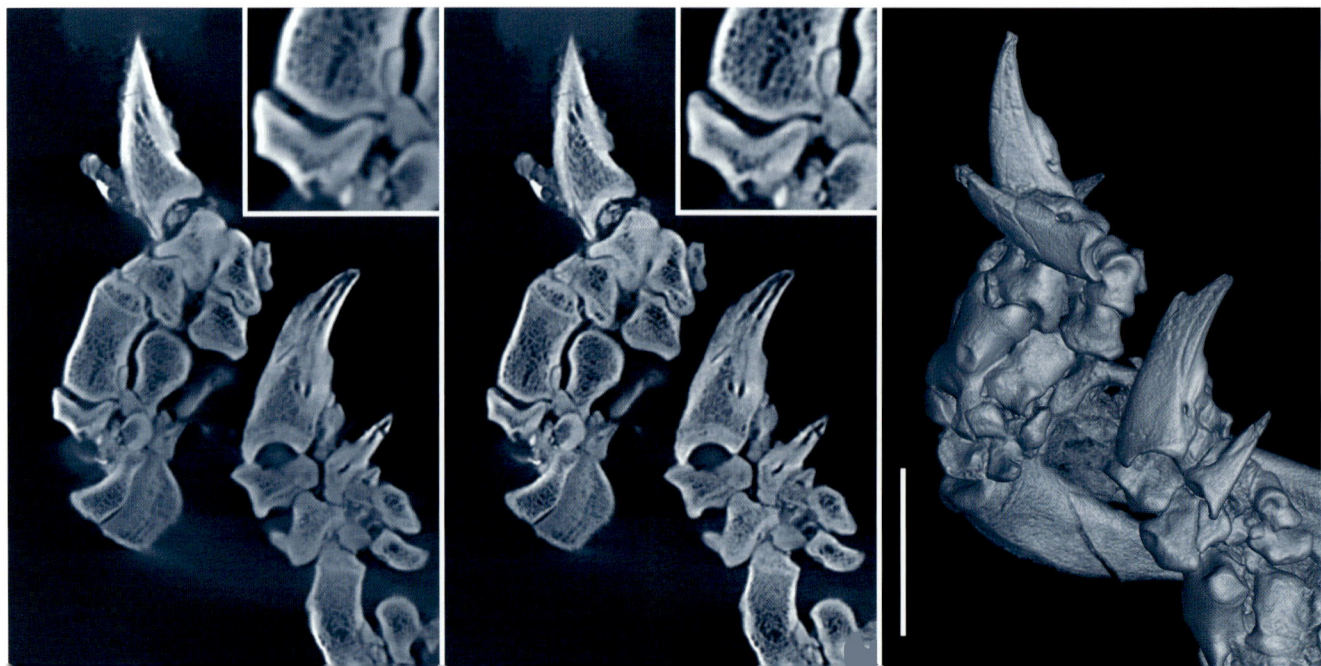

is referred to as a taxon (plural: taxa). The allocation of a rank to a respective taxon cannot be objectively justified (e.g., subfamily instead of family). Therefore, units with the same rank are not always directly comparable, and Linnean categories (ranks), if they are used at all, are rather considered an aid.

Today, most biologists accept that classification should be based on common ancestry. Irrespective of the observer, there is a single real tree of life, which is the result of evolutionary events and links all living and extinct organisms. This tree is the most objective basis for the classification of organisms. A phylogeny is a kind of family tree that represents the relationships among organisms. The branching points are the result of splitting events, and the branches represent real lineages (i.e., a common ancestor and all of its descendants). Systematic biology (systematics for short) is the field of biology concerned with the classification of organisms according to their phylogeny.

Systematists reconstruct the phylogeny by examining various characteristics in living and fossil organisms in search of shared derived features. However, shared traits are not always homologous (i.e., inherited from a common ancestor). They can also be the result of convergent evolution, where similar selective conditions may lead to similar-looking characteristics that developed independently of each other (analogous) (Fig. 5.8). The methods of phylogenetic

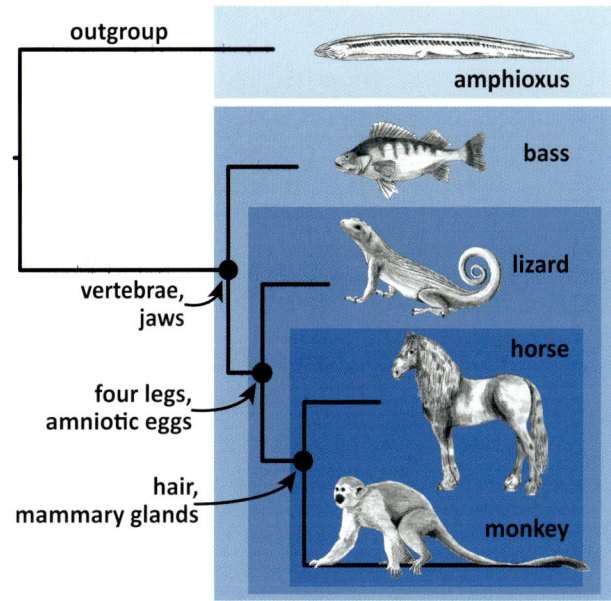

Fig. 5.9: Simplified phylogeny of vertebrates. Clades are highlighted by colored boxes, and some synapomorphies are noted at the basis of the clades.

systematics (also known as "cladistics"), introduced by Hennig (1966), are the most widely applied today to distinguish between homologous and analogous traits.

If evolution is defined as "descent with modification," then we are interested in the modifications, i.e., the evolutionary innovations. Technically, these derived characters are called apomorphies, whereas the primitive (ancestral) state of the character is called plesiomorphic. We generally look to species outside the group under study (the "outgroup") to determine whether features are apomorphic or plesiomorphic. Using different phylogenetic methods, we then search for the underlying tree that best fits the character data. This tree is also called a phylogeny or sometimes a cladogram (Fig. 5.9). A frequently used method for finding the "best" tree is the principle of parsimony, which prefers the branching pattern that requires the fewest number of evolutionary changes. More complicated methods based on statistical models are also widely used. A derived feature shared by two or more species is called a synapomorphy.

The identification and naming of taxa is then based on the best tree. A group of species that includes a common ancestor and all of its descendants is called

Fig. 5.8: Homologous versus analogous/convergent. The arms of these different vertebrates are derived from a common ancestor, and are therefore homologous. However, the adaptations for flight in birds and bats arose independently by convergence.

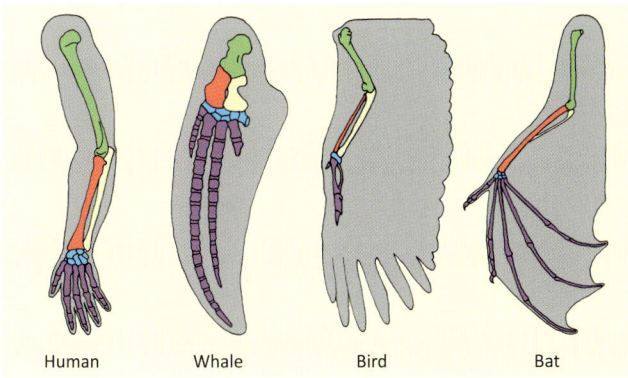

a monophyletic group or clade (Fig. 5.10). In general, only clades are considered valid taxa. A group that includes some, but not all of the descendants of the most recent common ancestor is called paraphyletic. Presumably paraphyletic taxa are written in quotation marks. For example, the great apes ("Pongidae") represent a paraphyletic taxon if humans are excluded. The descendant branches from a node on the tree are called sister taxa. In Fig. 5.10 for instance, *Homo* is the sister taxon of *Pan*. The clade X comprising the most recent common ancestor of two or more living species and its descendants is called "crown group." Its "stem" comprises all extinct species down to the last common ancestor of clade X and its sister-group (Fig. 5.11).

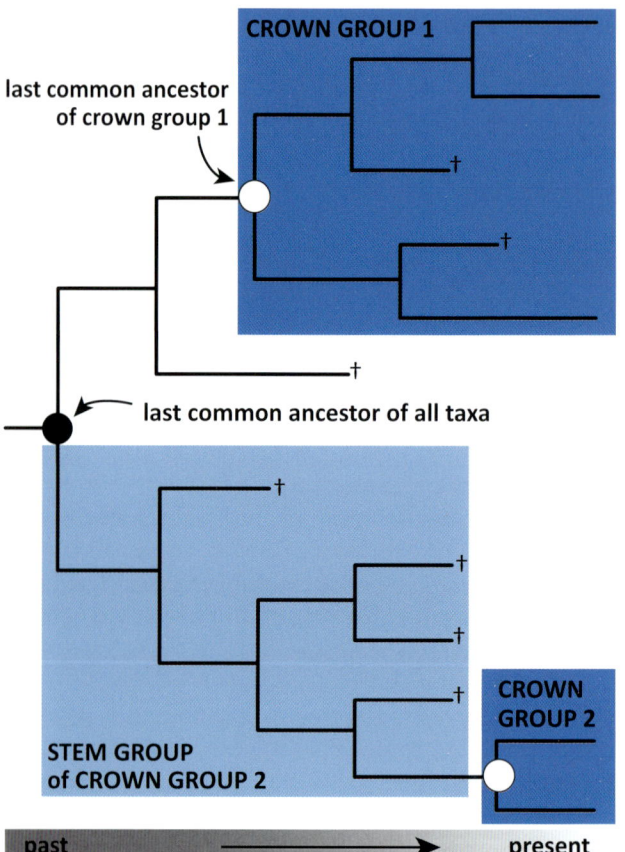

Fig. 5.11: Stem and crown group concept. The two dark blue clades, including their basal node, are crown groups. The stem of crown group 2 (light blue) includes all extinct taxa more closely related to it than to its living sister taxon, crown group 1.

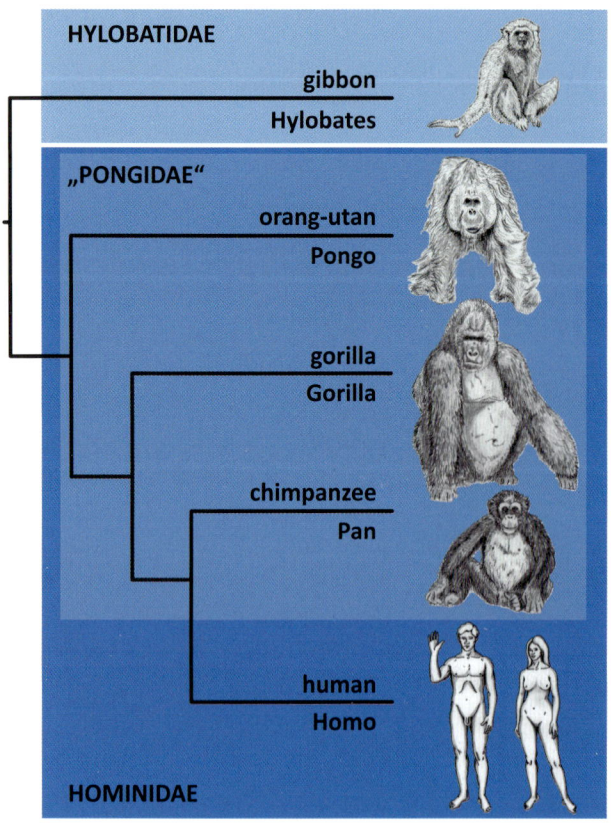

Fig. 5.10: Simplified phylogeny of Hominoidea. Hominidae is a valid monophyletic group, consisting of an ancestor (at the node) and all of its descendants. "Pongidae" is an invalid paraphyletic group, since only some, but not all of the descendants of the last common ancestor were included.

Species diversity, viewed mathematically

Since the 1980s, "biodiversity" has become a buzzword, and for a good reason. It encompasses the entire diversity of life: genetic and morphological diversity within and among species, the number of species in different groups, and the diversity of the higher communities they constitute. A high biodiversity ensures the stability of ecosystems. Irrespective of the question whether biodiversity is a good thing in itself and worthy of preservation, it has a significant financial impact on human society.

Paleontology distinguishes two basic concepts of biodiversity: species richness and disparity. The for-

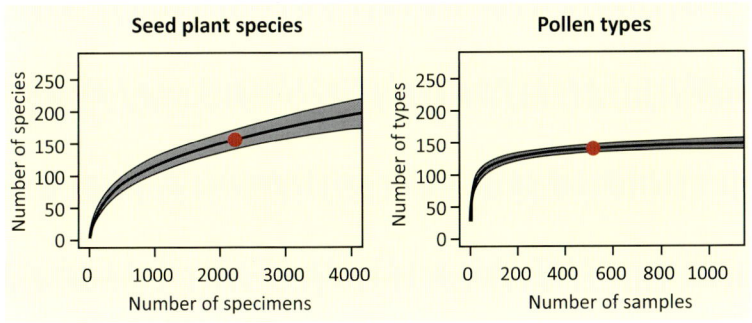

Fig. 5.12: Computation of species diversity in seed-bearing plants and pollen from the Messel Pit.

mer concept simply refers to the number of species. This concept may appear very simple, but this does not hold true upon closer scrutiny (Gotelli & Colwell 2001).

The first problem is the "equity" or balance of the various species. Imagine a small forest with 100 trees of 10 species. Most people would agree that forest A, with 91 trees belonging to one species and the remaining nine species represented by one tree each, is less diverse than forest B, where each species is represented by ten trees. Various mathematical methods have been developed that represent this situation.

The second problem is the sample size, and this is not independent of the first problem. Ecologists only realized rather late that it is not possible to compare the species diversity of two studied ecosystems without taking sample size into consideration. Imagine this: You just discovered a specimen of an insect. Since this specimen necessarily belongs to a particular species, it follows that you discovered one species. But when you collect a second specimen, there are two possibilities. Either, it is a second specimen of the first species, or the first specimen of a second species. The expected value of species diversity – how many species would you find on average if the experiments were repeated many times – for two specimens varies between 1 and 2. It turns out that the expected value heavily depends on how common the different species are in an ecosystem. In the case of forest A (see previous paragraph), there is a high probability that the second specimen will belong to the same species as the first; in the case of forest B, the probability is much higher that the second specimen will represent another species. The larger the sample size, the more species can be discovered; however, when there are many rare species, an even larger sample size is required to discover all species.

Mathematically inclined biologists have developed statistical methods that allow the comparison of species diversity with different sample sizes and even make it possible to predict how many additional species will be discovered, if another X number of specimens is collected (Colwell et al. 2012). These methods can also be applied in Messel in order to find out how many species are yet to be discovered.

Let's take plants as an example (Fig. 5.12). To date, 2,222 specimens of seeds and fruits have been discovered in Messel, representing 157 species (Collinson et al. 2012). If we double the number of specimens, we can expect a species diversity of 201.8 species, i.e., about 29 % more. On the other hand, if we consider the types of pollen, which currently amount to about 141, we see that the expected value will not increase significantly, regardless of how many additional samples we examine. This difference can be explained by the fact that many plant species produce similar pollen types, which cannot be differentiated under the microscope. The extrapolation of expected species richness in almost all of the groups from Messel (e.g., Fig 5.12, left: seed types) has not yet leveled off and suggests that additional excavations will continue to unearth many other new species.

"Disparity" – the second concept of biodiversity – refers to the differences in shape (Foote 1997). It is obvious that two species of rats (*Rattus*) look more similar to each other and therefore show a lower disparity than a rat and a capybara (*Hydrochoerus*). In principle, disparity can vary in entirely different ways than species diversity. A community of rodents, e.g., could show high species diversity and a low disparity, while another community shows a low species diversity and a high disparity. To date, disparity does not play a major role in Messel; for this reason, it is not further considered here.

Chapter 6
The Fossil Flora of Messel

Volker Wilde

The so-called green plants (Viridiplantae), which engage in photosynthesis by means of chlorophyll and are thus autotrophic, first evolved in the water. Therefore, single-celled, aquatic green algae are their most primitive representatives. They share an evolutionary innovation (synapomorphy) with the terrestrial plants: the chloroplasts, where photosynthesis takes place.

The higher terrestrial plants or embryophytes originated from a group of freshwater algae related to the modern-day charophytes (Leliaert et al. 2012) and conquered the land more than 475 million years ago. Primitive terrestrial plants such as mosses and ferns are still highly dependent on a moist environment. Club mosses, horsetails and ferns, along with the seed-bearing plants, are referred to as vascular plants or tracheophytes. They are characterized by the presence of special conducting tissues (phloem and xylem), which enable the efficient transport of water and nutrients through the plant's body. Compared to club mosses and horsetails, ferns are more "highly developed", since they possess more highly differentiated organs, including, in particular, leaves or fronds with an extensive surface area (Kenrick & Crane 1997).

The development of seeds allowed the terrestrial plants to leave behind their dependency on water for sexual reproduction. In the seed-bearing plants (spermatophytes), fertilization requires that the pollen is transported to the ovules (pollination); in this, both pollen and ovule are protected from desiccation.

The seed-bearing plants are divided into the gymnosperms ("naked seed") and the angiosperms (flowering plants). Today, angiosperms encompass 80 % of all species of green plants (Christenhusz & Byng 2016). They are primarily characterized by their flowers. The ovule, which later develops into a seed, is completely enclosed by at least one additional integument; together they later become the fruit. The perianth can be colorful, thereby attracting animals that act as pollinators. The oldest seed-bearing plants known to date are about 130 million years old and originate from the Lower Cretaceous. They underwent a rapid subsequent development, already dominating the vegetation in many parts of the world as early as the Upper Cretaceous. Main groups within the flowering plants (APG IV 2016) are the magnoliids, the monocots, the rosids and the asterids. Fig. 6.2 only lists the scientific names; the common names are shown below.

History of study

Due to their sessile lifestyle and their physical build, the fossil record of plants differs significantly from that of most animals. Plants show a distinct modular composition with clearly differentiated organs or organ systems. At various points in time, the individual modules such as leaves, fruits, seeds (Fig. 6.3), flowers and pollen grains become separated from the respective

Fig. 6.1: Leaf of a water lily-like plant (Nymphaeales); length: 14.3 cm.

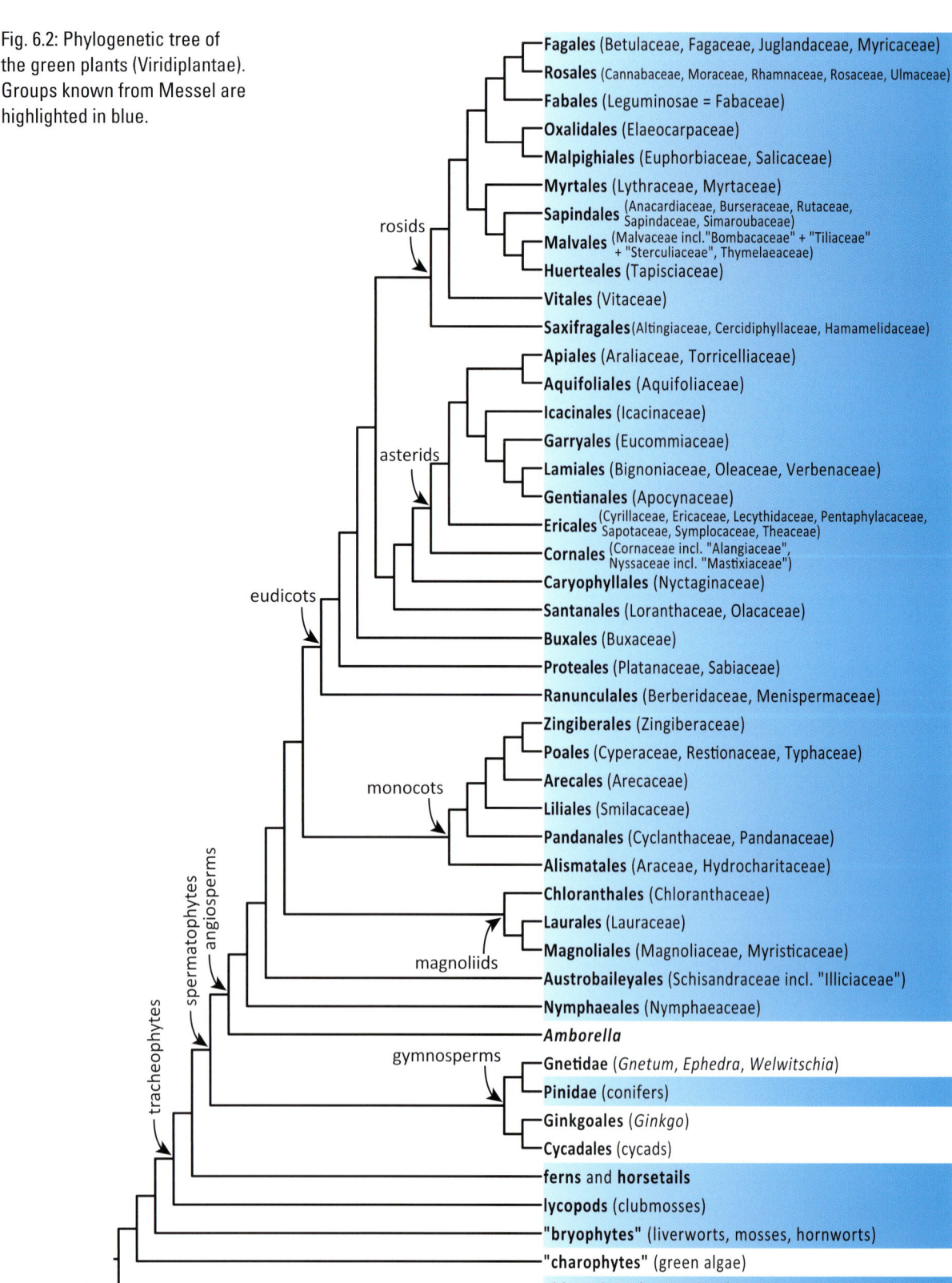

Fig. 6.2: Phylogenetic tree of the green plants (Viridiplantae). Groups known from Messel are highlighted in blue.

was undertaken by Engelhardt; however, it was only published posthumously by Menzel (1922). It primarily included leaves, with only a few seeds and fruits mentioned. As was usual at the time, the identification of the leaves (Fig. 6.4) was based on general morphological criteria such as shape and primary venation, compared to recent leaves. Schweitzer (in Matthess 1966) and Sturm (in Tobien 1969) undertook a reexamination of the old materials, culminating in the selection of leaves that were obviously part of

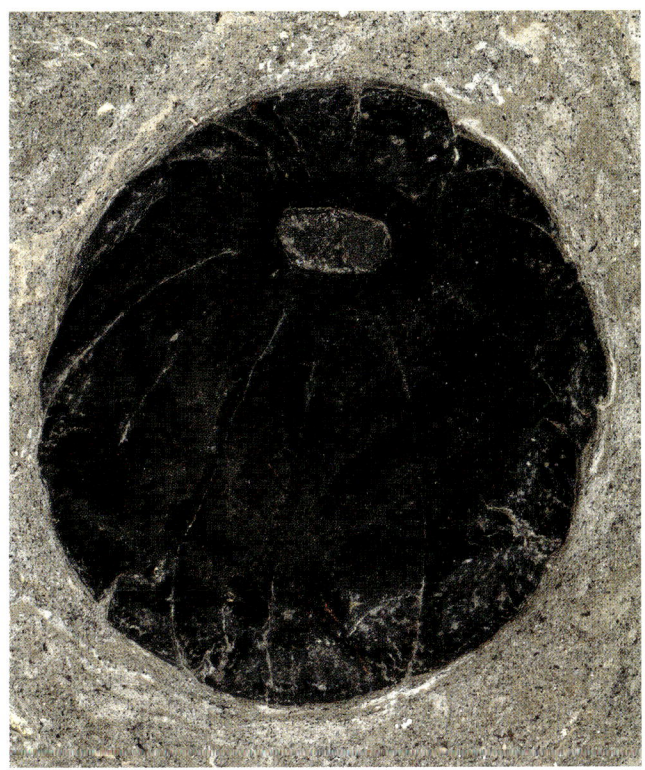

Fig. 6.3: Fruit of a flowering plant of unknown affinity; diameter: 1.5 cm.

Fig. 6.4: Serrated leaf of a flowering plant of unknown affinity on a twig; length of leaf including petiole: 5.5 cm.

mother plant. As a consequence, they are frequently preserved apart from each other in the fossil record. Therefore, the individual modules are not only discovered separately, but are frequently described separately and given different names. This problem is exacerbated by the fact that certain leaves or pollen grains, in particular, may strongly resemble each other, even though they belong to very different plant species or even families. Thus, their biological affiliation remains unknown in many cases. For this reason, scientists dealing with fossilized plants, and especially flowering plants, early on specialized on particular organs, e.g., leaves, fruits or seeds, pollen and spores, or wood. Unfortunately, many of the plant remnants from Messel are lacking certain characteristics that would enable their unambiguous systematic placement. This is also reflected in their history of study.

Remnants of plants from the Messel oil shale deposit were already mentioned early on by Chelius (1886). The first monographic treatment of the flora

the Lauraceae for a monographic treatment by Sturm (1971). In this process, the so-called cuticular analysis was systematically applied for the first time at Messel. This process involves the chemical isolation of the cuticle (or cuticula), which covers the leaf's outer layer (epidermis) and is frequently well-preserved. Subsequently, the cell pattern of the epidermis is examined under the microscope, which further aids in the identification. Additional leaf remnants became available during the processing of the gut contents of the Messel horses (Chapter 10.9; Franzen 1977; Sturm 1978). Aided by reflective light fluorescence microscopy as employed by Schaarschmidt (1982) for this purpose, the examination of cuticular structures later enabled an extensive review of the Messel leaf flora (Wilde 1989). This was followed by detailed work on individual taxa (palms: Schaarschmidt & Wilde 1986; Lauraceae: Kvaček 1988; Myricaceae: Wilde & Frankenhäuser 2000; Myrtaceae: Glinka & Walther 2003; Ulmaceae: Wilde & Manchester 2003; Araceae: Wilde et al. 2005; Malvales: Kvaček & Wilde 2010).

The classic work by Thomson & Pflug (1953) includes the first description of pollen types from the oil shale deposit at Messel. Later, Thiele-Pfeiffer (1988) used the core material from a research drilling for their systematic monographic treatment of the middle Eocene pollen flora. An additional research drilling provided material for extensive statistical analyses of the microflora in regard to climate cycles (Chapter 3) and the development of faunal biodiversity throughout time. Additional pollen grains, which adhered to the remains of bees from Messel, were described by Grimsson et al. (2017).

The fruits and seeds from Messel (Fig. 6.5) were only considered rather late (Collinson 1982, 1986, 1988). Following initial monographs of the individual taxa (Rutaceae: Collinson & Gregor 1988; Palms: Schaarschmidt & Wilde 1986; Juglandaceae: Manchester et al. 1994; Ulmaceae: Wilde & Manchester 2003; Anacardiaceae: Manchester et al. 2007; Vitaceae: Chen & Manchester 2007, 2011; Cyclanthaceae: Smith et al. 2008; Cyperaceae: Smith et al. 2009; Malvales: Kvaček & Wilde 2010), they finally became the subject of a comprehensive treatise (Collinson et al. 2012). The study of flower remnants, for which Messel is particularly well-known, is still limited to a handful of taxa to date (Schaarschmidt 1984, 1986, 1988, 1992; Schaarschmidt & Wilde 1986; Harley 1997).

Remains of wood are only very rarely discovered in Messel; therefore, it is not surprising that only one sample has been subject to closer identification (Wilde & Süss 2001).

The state of preservation of plant remnants

The taphocenosis of plants from Messel is unusual in view of the fact that, apart from algal remains, it comprises a number of different plant parts side by side: leaves, fruits and seeds, entire flowers, woody twigs, stems and even occasional root remains. Articulated plant parts are an exception. Among these, flowering plants are dominant (Fig. 6.6). Remnants of conifers are a rarity; ferns are primarily represented by spores, and mosses and club mosses have only been documented by the presence of spores.

In most cases, the leaves entered the sediment more or less intact. Visible damage is usually due to feeding traces or fungal infection. Leaves that show increased signs of decomposition or skeletonization prior to being deposited are exceedingly rare (Wilde 1989). In almost all cases, the original leaf's organic substance is heavily compressed and, except for the cuticle, no longer shows any details of the leaf's internal structure. However, occasionally leaves are

Fig. 6.5: Highly resinous fruit of a flowering plant of unknown affinity, length including petiole: 9 mm.

Fig. 6.6: Flower of unknown affinity with translucent stamens, diameter: 3.5 mm.

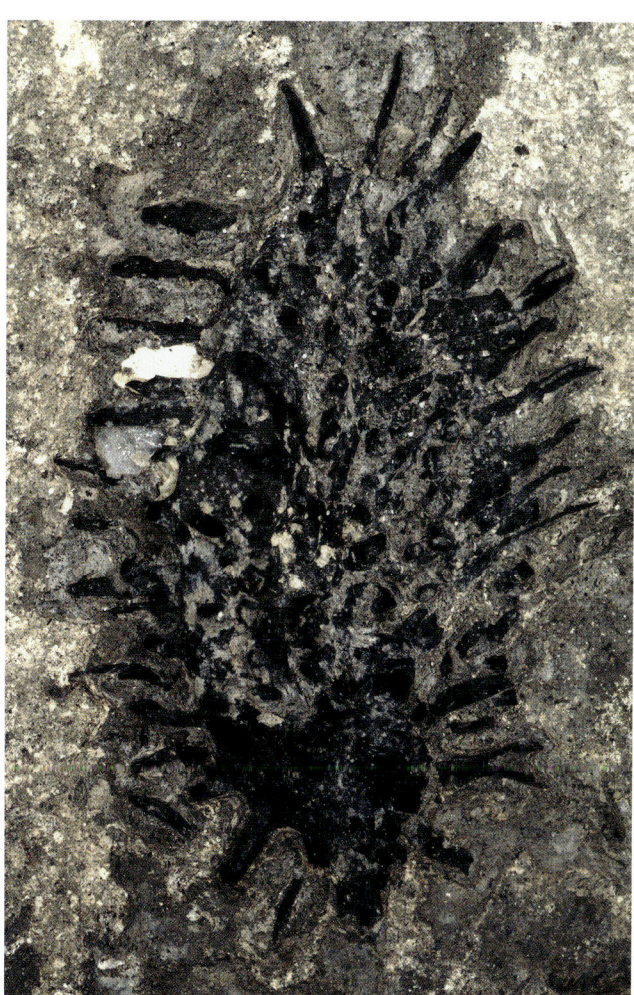

Fig. 6.7: Spiny fruit of a flowering plant of unknown affinity, length: 1.7 cm.

found where the cell structures have been preserved to some degree; in other cases, the leaf substance has been broken down entirely, leaving only the cuticle behind. A few leaves only appear as a powdery shadow of pyrite on the sediment. Interestingly, the state of preservation of leaves in general, and particularly that of the cuticle, appears to be specific for individual taxa (Wilde 1989). However, additional research is required in this area.

Fruits and seeds (Fig. 6.7) are also preserved in various stages of compaction (Collinson et al. 2012). The degree of compaction usually depends on the tissue's original consistency; fleshy tissues are affected more strongly than hard, sclerenchymatous integuments, which often still reveal some of the original cell structures. In rare cases, single fruits are also preserved in the form of a pyrite coating in the sediment. Since flowers were mostly rather delicate in their original state, they are without exception strongly compacted in Messel – in many cases with the pollen originally contained in the stamens still preserved *in situ* (Schaarschmidt 1984). Individual pollen grains, as well as spores and most algae, are only preserved in the form of their outer shells, which consist of a highly resilient organic material.

The original substance of the few wood remnants from Messel is usually heavily compressed or even homogenized, only occasionally revealing details of the wood's original structure (Wilde & Süss 2001). One single, as yet unprocessed, wood splinter from the pyroclastic debris contained in the research drilling at Messel in 2001 represents an absolute exception, since it shows an excellent preservation of the wood structure (Fig. 6.8).

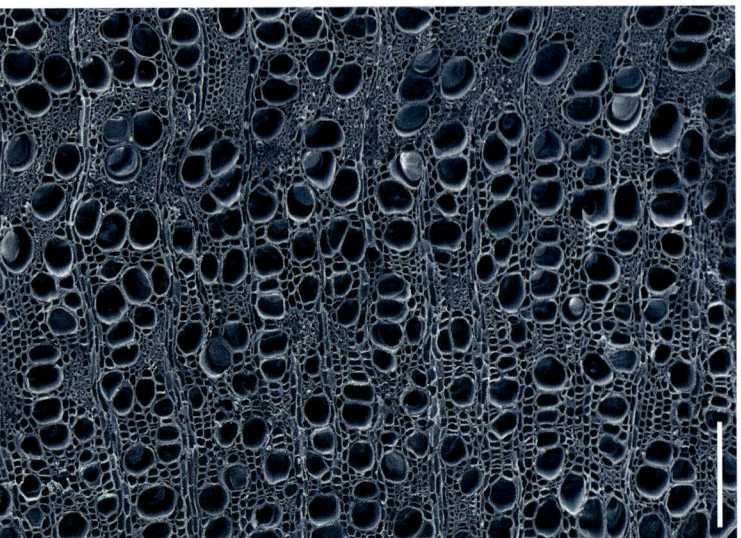

Fig. 6.8: Carbonized wood splinters (angiosperm wood) from a depth of 239 m in the Messel research drilling in 2001, scale 200 μm.

At this point, it should be mentioned that due to a low level of coalification, a process in which organic material is turned into coal over the course of millions of years, a large number of organic molecules can be found in the sediment that, as so-called biomarkers, may offer insights into their origin from particular groups of plants (e.g., Habermehl & Hundrieser 1983a, 1983b; Goth et al. 1988; Otto et al. 2002; Adam et al. 2006).

Systematics of the flora

The flora of Messel comprises a large number of individual species, genera and families. Since the affiliation of individual organs to particular biological species or even genera is only known or confirmed to a small extent, a comprehensive treatment is only possible at the family level for higher taxa such as conifers, ferns, club mosses, mosses and algae, as well as flowering plants (Table 6.1). Due to the large number of families represented (58 following the latest classification; APG IV 2016), it is impossible to cover them all in detail here. The true diversity hidden among the plant remnants from Messel is undoubtedly much higher, since many of these remains have not yet been systematically allocated, due to the lack of diagnostic characteristics.

Algae, mosses, ferns

Among the algae remnants in Messel, the cell walls of single-celled green alga *Tetraedron* are of particular importance (Chapter 3), since they make up a significant proportion of the organic material in the Messel oil shale, even forming almost pure layers in certain horizons (Goth 1990). Also widespread (and even quite common in certain layers) are the remains of colonies of the green alga *Botryococcus*, which is sometimes referred to as "oil alga" due to its possible role in crude oil formation. Another colonial green

Table 6.1: Number of flowering plant species in Messel that can be assigned to a family, broken down by their various organs. Amended according to Schaarschmidt (1988, 1992), Wilde (2004) and Collinson et al. (2012). Essential sources: **Leaves** – Wilde (1989), Wilde et al. (2005, Araceae), Kvaček (1988, Lauraceae), Kvaček & Wilde (2010, Malvales), Wilde & Frankenhäuser (2000, Myricaceae), Wilde & Manchester (2003, Ulmaceae); **Fruits and seeds** – Collinson et al. (2012); **Pollen** – Thiele-Pfeiffer (1988), Grimsson et al. (2016); **Flowers** – Schaarschmidt (1984, 1986), Wilde & Schaarschmidt (1993), Schaarschmidt & Wilde (1983a, b), Harley (1997, Arecaceae); **Wood** – Wilde & Süss (2001).

	Leaves	Fruits & seeds	Pollen	Flowers	Wood	Minimum number of species (Total number of taxa)
Primitive flowering plants (incl. magnoliids)	25	7	2	1	0	29 (35)
Monocots (Monocotyledons)	8	4	6	2	0	15 (20)
Higher dicots ("eudicots")	43	79	70	6	1	150 (199)
Total	76	90	78	9	1	194 (254)

Fig. 6.9: Pinnate leaf of a climbing fern (*Lygodium kaulfussi*, Schizaeaceae), length of the fossil: 3.5 cm.

alga (*Coelastrum*) was regularly observed by Richter et al. (2013). These are augmented by the organic shells of the permanent stages of freshwater dinoflagellates, which are particularly numerous near the beginning of oil shale deposition in the 2001 Messel drilling (Lenz et al. 2007). The presence of algae with siliceous shells, in particular diatoms, but also chrysophytes, is frequently evidenced by the hollow spaces left behind after the dissolution of the silica.

Contrary to the maar sediments from Eckfeld in the Eifel, which also date to the middle Eocene, no macroscopically visible remnants of mosses have been preserved in Messel. Only a few rarely observed spores may originate from mosses. With 23 types of spores, ferns are represented in Messel in a variety of forms (Thiele-Pfeiffer 1988). Due to their herbaceous nature, fern leaves are among the rarer plant fossils in Messel. Nevertheless, seven different species of fern fronds are known from Messel so far, which, except for *Thelypteris* sp., are widespread in the Paleogene (Wilde 1989; Frankenhäuser & Wilde 1993): *Acrostichum aureum*: Pteridaceae, "*Blechnum*" *dentatum* (Blechnaceae), *Lygodium kaulfussi* (Schizaeaceae; Fig. 6.9), *Osmunda lignitum* (Osmundaceae), *Ruffordia subcretacea* (Schizaeaceae) and

Fig. 6.10: Part of a fern frond (*Gleichenia*, Gleicheniaceae), length: 5.6 cm.

Rumohra recentior (Dryopteridaceae). A unique find is the remnant of a frond that can clearly be assigned to the genus *Gleichenia* (Gleicheniaceae; Fig. 6.10), which today often overgrows open areas in tropical and subtropical regions. Of particular interest among the ferns is the occurrence of the mangrove fern *Acrostichum*, which is always found under somewhat saline conditions today. Messel is one of the sites

Fig. 6.11: Remnants of twigs from a plum yew species (*Cephalotaxus messelensis*, Cephalotaxaceae), length of individual needles approc. 2 cm.

that demonstrate that during the Paleogene, this genus still existed in purely freshwater habitats (Wilde 1989; Frankenhäuser & Wilde 1993).

Gymnosperms

In Messel, the gymnosperms are only represented by conifers; the remnants of *Ginkgo* and cycads mentioned by Engelhardt (1922) could not be confirmed. Remains of needle-bearing conifer twigs are among the rarities in Messel, which may be partly due to the fact that they played a subdominant role in the vegetation around Lake Messel (Schaarschmidt 1988, 1992), but more significantly because these twigs only rarely separate from the mother plant. Most of the conifer twigs, along with the occasionally found cone scales, belong to *Doliostrobus taxiformis*, a widespread species in Europe during the Paleogene, which is characterized by numerous resin-filled ducts in its tissue. Its exact systematic position is still unclear, although the chemical analysis of the resin from Messel supports a relationship with the genus *Cunninghamia*, which today is restricted to China (Otto et al. 2002). According to the current view, it occupies a position among the cypress relatives *sensu lato* (Cupressaceae). The same applies for the genus *Sciadopitys*, only found in Japan today, which is also represented by pollen in Messel (Thiele-Pfeiffer 1988). Another lineage that is now restricted to China is the yew plum *Cephalotaxus* (Fig. 6.11), which is known from Messel by a few twig fragments with attached needles as well as a few individual needles (Wilde 1989). The isolated pollen grains regularly include grains that can be assigned to the cypress family *s.l.* On the other hand, the pollen of pinaceous trees, not represented by macro-remains, is found on a regular, albeit less frequent basis (Thiele-Pfeiffer 1988).

Systematics of the flora

Primitive flowering plants or basal angiosperms

Among the basal dicotyledonous flowering plants (such as magnoliids and Nymphaeales) known from Messel, one plant deserves special mention, which most likely belongs to the water lily-like plants *sensu lato* (Nymphaeales). This plant represents a rare case for Messel where the leaves (Fig. 6.1), flowers (Fig. 6.12), fruits and seeds as well as pollen grains can be clearly assigned to the same biological species. This association resembles a jigsaw puzzle. Typical water lily leaves were first described from Messel by Engelhardt (1922), followed by the mention of individual seeds (Collinson 1982, 1986) and the preliminary description of certain flowers (Schaarschmidt 1988, 1992), which were also associated with this family. Since the flowers always contained a specific, hitherto unknown pollen type (Thiele-Pfeiffer 1988) and some of the flowers in a more advanced state of development already contained respective seeds along with the pollen, the relationship became clear.

Due to the lack of an organic connection, the proof of relationship among leaves and flowers was more difficult. However, it was possible to document comparable support elements or sclereids in both the leaves and the petals, enabling an unambiguous association in this respect, as well. Based on the frequency of occurrence of the tender leaves and flower remnants, it can be assumed that these plants – similar to modern-day water lilies – grew along the edge of the maar lake in Messel.

The magnoliids include the family Lauraceae, or laurels. With at least 21 species of leaves and 3 species of fruits it is the family among the flowering plants represented in Messel by the highest number of taxa to date. On the one hand, this is due to the fact that many of the respective leaves have a rather characteristic appearance (Fig. 6.13), e.g., three-veined leaves that used to be considered a general trait of plants from the cinnamon tree group. On the other hand, the leaves of Lauraceae have characteristic stomata that become visible when examining the cuticle structure under the microscope. In addition, there is an inclusion of oil particles in the leaf's tissue, typical for Lauraceae, which are revealed during microscopic examination. The diversity of Lauraceae leaves in Messel, already noted by Sturm (1971), has caused some researchers to postulate the presence of a veritable laurel forest at Messel. According to our present knowledge, this is an erroneous assumption. However, it can certainly be assumed that plants from the laurel family played an important role in the vegetation around Messel in the form of shrubs and low trees (Schaarschmidt 1988, 1992) and that their leaves were thus available as a preferred food source for the early horses. Among the primitive angiosperms, the Magnoliaceae (magnolia family) with fructifications or seeds, the Myristicaceae (nutmeg family) with seeds (Collinson et al. 2012) and the Chloranthaceae with pollen (Thiele-Pfeiffer 1988) have been documented in Messel. In all cases, the respective leaves are lacking to date, either because they could not yet be identified despite specific characters (Magnoliaceae, Chloranthaceae) and are therefore missing completely, or because they are so similar to the leaves of other families that a clear distinction has not yet been possible (Myristicaceae).

Fig. 6.12: Flower of a water lily-like plant (Nymphaeales), length: 7 cm.

Monocotyledonous flowering plants or monocots

The remnants of monocotyledonous flowering plants or monocots are not particularly common in other fossil sites, either. Since these are herbaceous plants whose leaves rot while still attached to the mother plant, thereby rarely finding their way into the lake sediment, this is especially true for their leaves. However, a few families have been documented from Messel by leaf remnants or even entire leaves (Wilde 1989). These include, in particular, members of the arum family (Araceae; Fig. 6.14), which are represented by three types of leaves – two of which are even rather common (Wilde et al. 2005). It can there-

Fig. 6.13: Foliated twig of a plant from the laurel family (*Daphnogene crebrigranosa*, Lauraceae), length: 17 cm.

Fig. 6.14: Leaf of a plant from the arum family (*Araciphyllites tertiarius*, Araceae), length: 13.3 cm.

fore be assumed that they may have occurred in the marshy edges along Lake Messel, similar to modern-day arum relatives. Typical aquatic plants include the Hydrocharitaceae (frogbit family), which is represented in Messel with one leaf and pollen (Thiele-Pfeiffer 1988, Wilde 1989). The Smilacaceae (greenbriar family), today a family of climbing plants and vines primarily in warmer regions, is also regularly represented by leaves. Small to tiny leaf fragments with a characteristic cuticular structure are likely representatives of the Pandanaceae (screw palm or screw pine family), restricted today to moist and warm climates in Asia, Africa and Oceania.

Flowers and fruits of the Arecaceae (palm family) are among the most common macroscopic plant remains in Messel, and their pollen is regularly found in the sediment, along with associated leaf remnants (usually fragments of leaflets). The biological affiliation of the organs here is documented by numerous transitional stages between flowers and fruits, as well as pollen present in the flowers, while the affiliation of the leaf remnants remains a matter of conjecture (Schaarschmidt & Wilde 1986). The frequency of their remnants indicates that the palm trees grew in close vicinity to Lake Messel. A surprising discovery in Messel was the unambiguous documentation of the remains of fructifications of the Cyclanthaceae (Panama hat palm family) (Fig. 6.15), since they are currently restricted to a rather small area in tropical Central and South America (Smith et al. 2008). Of biogeographic interest for Messel is the occurrence of a pollen type that is found today in the Restionaceae (restio family), a family now only found in the south-

Fig. 6.16: Spike of a member of the sedge family (*Volkeria messelense*, Cyperaceae), length: 1.2 cm.

ern hemisphere (Thiele-Pfeiffer 1988, Schaarschmidt 1988, 1992). The remains of the fructifications of a sedge (Cyperaceae; Fig. 6.16) frequently found in Messel, partly with the associated pollen still attached, can also be compared to a genus with a current tropical distribution (*Mapania*; Smith et al. 2009). Another family found in the tropics today are the ginger relatives or Zingiberaceae (Fig. 6.17), which are represented in Messel by occasional leaf remnants.

Fig. 6.15: Disc of the fruit of *Cyclanthus messelensis* (Cyclanthaceae), diameter: 4.8 cm.

Fig. 6.17: Remnant of the leaf of a plant from the ginger family (Zingiberaceae), length: 6 cm.

Fig. 6.18: Fruit of a plant from the moon seed family (*Karinschmidtia rotulae*, Menispermaceae), diameter: 2.6 cm.

Higher flowering plants or eudicotyledons

Numerous plant families documented from Messel belong to the higher dicotyledonous flowering plants (eudicotyledons or "eudicots"). An important family for Messel in this regard is the Menispermaceae (moon seed family), of which 17 types of fruits or seeds have been described (Fig. 6.18) (Collinson et al. 2012). This is not least due to the fact that these, contrary to their leaves and pollen, are easily recognizable as such. On the other hand, based on their frequency it can be assumed that this family, which today includes many tropical and subtropical species of climbing plants and vines, played an important role in the vegetation surrounding Lake Messel. A family that is of particular interest due its modern-day parasitic lifestyle is the Loranthaceae (mistletoe family) from Messel. It is represented by two types of leaves that show the characteristic stomata design typical for this family (Wilde 1989). A speciality for Messel are flowers that were shown to belong to the bougainvilleas (Nyctaginaceae or four o'clock family), based on their very characteristic pollen (Wilde & Schaarschmidt 1993a, b). Today, these plants are popular shrub- or vine-like ornamentals in Mediterranean climates. The group of dogwood-like plants (Cornales) is particularly represented by fruits and seeds from the Alangiaceae (now considered part of the Cornaceae or dogwood family) and the Nyssaceae (tupelo family, including the former Mastixiaceae or mastixia family). In addition, there are several types of pollen and a type of leaves that also belong to this group. In this case, as well, it is not only the fact that these seeds and fruits and the pollen are easily recognizable (Thiele-Pfeiffer 1988; Collinson et al. 2012) but also the diversity of the respective plants with a modern tropical to subtropical distribution in the vicinity of Messel that plays an important role.

Not only members of the heather family or Ericaceae itself, but a number of related families, the Ericales, have been found at Messel. These include, among others, the Lecythidaceae (Brazil nut family) with a typical flower (Wilde & Schaarschmidt 1993a, b) as well as the Sapotaceae (soapberry family), the Symplocaceae (symplocos family) and the Theaceae or tea family, with pollen, fruits or seeds, and leaves (Thiele-Pfeiffer 1988; Wilde 1989; Collinson et al. 2012). Representatives of these families are frequently, if not primarily, found today as trees in tropical and subtropical forests. Today, the Icacinaceae (icacina family) is a family of mainly tropical trees, shrubs and vines, which is represented in Messel by not less than 11 types of easily recognizable fruits, along with leaves and pollen.

Among the rosids, the sweetgum family (Altingiaceae) is only documented by fruits (Fig. 6.19). The grapevine family or Vitaceae (Fig. 6.20, Fig. 6.21) is another family of plants with a predominantly vine-like growth pattern found in Messel, which has been frequently documented by leaves, several types of fruits and seeds as well as pollen. The same is true for the Rutaceae (rue or citrus family), which today is found as trees and shrubs and occasionally climbers in tropical and subtropical climates. In addition, there are winged fruits of a species of "tree of heaven" (Fig. 6.22). An interesting family in Messel is the spurge family or Euphorbiaceae (Fig. 6.23), which is represented by a large number of fruits of a single type (Collinson et al 2012). Although the correspond-

Fig. 6.19: Infructescence of a flowering plant related to the sweetgum (*Steinhauera subglobosa*, Altingiaceae), length: 7.3 cm.

Fig. 6.20: Seed of a plant from the grapevine family (Vitaceae), length: 7 mm.

Fig. 6.21: Leaf of a plant from the grapevine family (*Ampelopsis*, Vitaceae), length including petiole: 15 cm.

Fig. 6.22: Winged fruit of a "tree of heaven" (*Ailanthus confucii*, Simaroubaceae), length: 1.8 cm.

ing leaves can be expected, they have not yet been documented from Messel, likely due to the lack of sufficiently characteristic features and their great similarity to the leaves of other plant families. Today, numerous species of this family occur as herbs, shrubs and trees, mainly in the tropics, but also in temperate regions. In addition, there are fruits of *Sloanea* (Elaeocarpaceae, Collinson et al. 2012; Fig. 6.24), a genus

Systematics of the flora

Fig. 6.23: Fruit of a plant from the spurge family (*Euphorbiotheca gothii*, Euphorbiaceae, diameter: 1.1 cm.

Fig. 6.25: Remains of a large pinnate leaf (legume?), length: 9 cm.

Fig. 6.24: Fruit of a plant from the wood sorrel family (*Sloanea messelensis*, Elaeocarpacea), length including petiole: 6 cm.

that today represents tropical trees with characteristic buttress roots.

Another group of families that are important in Messel belongs to the beech-like plants (Fagales). The beech family itself (Fagaceae) as well as the birch family (Betulaceae) are only represented by pollen, albeit in many species and large numbers (Thiele-Pfeiffer 1988). The Myricaceae (bayberry family) has been documented by pollen (Thiele-Pfeiffer 1988) as well as leaf remnants, which in part belong to the genus *Comptonia* (sweetfern), a monotypic genus of low shrubs found across North America today (Wilde & Frankenhäuser 2000). Leaf remnants (Fig. 6.25) and pods (Fig. 6.26, Fig. 6.27) of several species of legumes (Leguminosae or Fabaceae) are rare but regular in Messel (Wilde 1989; Collinson et al. 2012). They represent a very species-rich family of herbs, shrubs and trees with a nearly worldwide distribution, including many tall trees in tropical forests. The rose family or Rosaceae is represented in Messel almost exclusively by rare leaf remains (Fig. 6.28).

Fig. 6.26: Pod of a legume (*Mimosites spiegeli*, Leguminosae or Fabaceae), length: 12 cm.

Fig. 6.27: Pod of a legume (*Leguminocarpon herendeenii*, Leguminosae or Fabaceae), length: 3 cm.

Fig. 6.28: Pinnate leaf of a plant from the rose family (Rosaceae), length including petiole: 8.2 cm.

The vegetation surrounding the maar lake

In order to correlate the association of plant remnants (taphocoenosis) known from the oil shale deposits of Messel with the vegetation around the maar lake, one has to consider how these remnants found their way into the lake and subsequently, the sediment. Since it may be assumed that the lake had a relatively limited catchment area, the vast majority of plant remains originated not too far from its banks and was likely transported by runoff from the surrounding areas (Storch & Schaarschmidt 1988, 1992; Wilde 1989; Ferguson 1993). In some cases, it is conceivable that fruits and seeds as well as pollen originated from the digestive products of animals, particularly birds, mammals and insects, that moved across greater distances (Ferguson 1993; Collinson et al. 2012). Transport by wind across large distances is likely, in particular, in the case of pollen grains, e.g., Betulaceae (birch family), Fagaceae (beech family) or Pinaceae (pine family) that are not represented by macro-remains (Ferguson 1993). The same applies for the various winged fruits (Fig. 6.30, Fig. 6.31), although they are only represented by a few species and usually in relatively

Fig. 6.29: Winged fruit of an engelhardioid plant from the walnut family (*Palaeocarya*, Juglandaceae), length including petiole: 4.4 cm.

A family of particular importance for Messel is the walnut family (Juglandaceae). Single leaflets or even parts of entire pinnate leaves from members of this family are by far the most frequently found leaves (Wilde 1989). At the same time, there are also numerous incidences of varied types of pollen with a high likelihood of belonging to this family (Thiele-Pfeiffer 1988). In addition, there are three types of winged fruits (Fig. 6.29, Fig. 6.30; Frankenhäuser & Wilde 1994; Manchester et al. 1994; Wilde & Frankenhäuser 1995; Collinson et al. 2012). Besides representatives of extinct genera or species among the fruits, pollen and leaves, there are also remains that can be associated with the engelhardioid Juglandaceae, which are found today in tropical Central America and Southeast Asia. In any case, it can be assumed that plants from the walnut family played a dominant role in the vegetation around Lake Messel.

Fig. 6.30: Several winged fruits of an extinct genus of the walnut family (*Cruciptera schaarschmidtii*, Juglandaceae), diameter of the individual fruits approc. 1.8 cm.

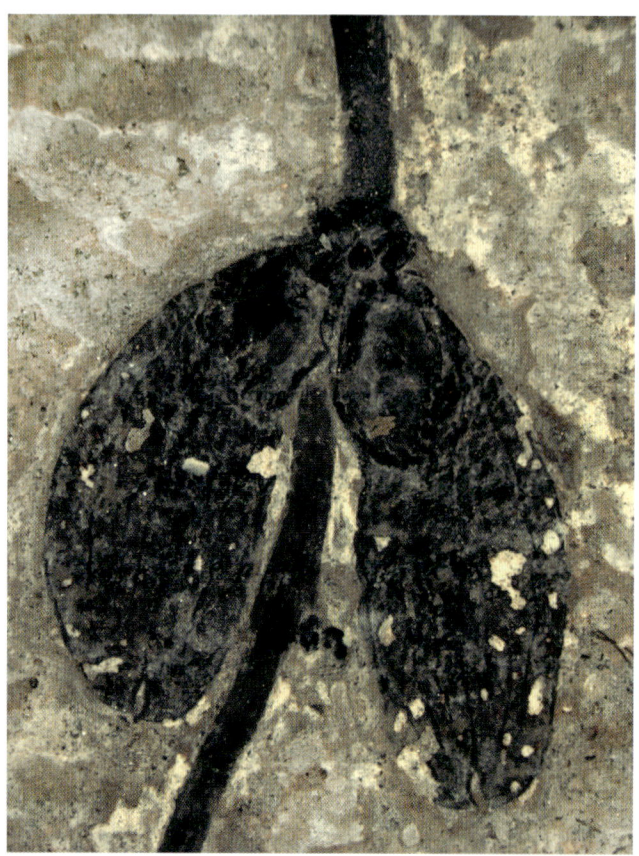

Fig. 6.31: Two winged fruits on a common axis (*Cedrelospermum leptospermum*, Ulmaceae), length of the individual fruits: 9 mm.

cated along the immediate edge of the lake and fell directly into the water.

The idea of a tropical forest (Chapter 3) surrounding Lake Messel was initially raised by Engelhardt (1922) and has been generally accepted by now (e.g., Wilde 1989; Schaal & Ziegler 1992; Gruber & Micklich 2007). Today, it is often referred to as a "paratropical forest" according to Wolfe (1977). Moreover, a certain zoning of the vegetation can be derived from the plant remains described to date (Wilde 1989; Collinson et al. 2012). Thus, it can be assumed that the

Fig. 6.32: Fruit of *Anacardium germanicum* (cashew nut, Anacardiaceae), length of the fossil: 2.5 cm.

small numbers (Collinson 1988). It is also conceivable that certain leaves near the lake were carried to the water's surface by air movements (Wilde 1989). A presorting of the plant material that ended up in the lake is most likely due to taphonomic filters (Chapter 4). Thus, based on the intense breakdown of organic materials under the humid and warm conditions to be assumed at Messel, any additional transport, in particular in the case of leaves, can be excluded (Wilde 1989). The rare occurrence of wood remains can be explained by the same reasons (Wilde & Süss 2001). Furthermore, an herbaceous vegetation along certain areas on the lake banks ensured that many plant remains from the vicinity could not even enter the lake (Wilde 1989, 2004). However, Collinson (1988) and Wilde (1989) work on the assumption that some of the leaves and seeds or fruits came from trees lowater lily-like plants were anchored in shallow water and their leaves and flowers occurred on the water's surface (Schaarschmidt 1988, 1992). Areas of shallow water were also home to rooting members of the Hydrocharitaceae (frogbit family) and the Lythraceae (loosestrife family), with the genus *Decodon* (water willow or swamp loosestrife) reported from Messel. It is possible that some of the ferns and monocots such as Araceae (arum family), Cyclanthaceae (Panama hat tree family), Cyperaceae (sedges) and Restionaceae (restio family) occurred in locally present swampy areas near the lake's edge. Moreover, the occurrence of swampy habitats near the lake is also indicated by the presence of conifers such as *Doliostrobus* and other coniferous trees formerly placed in the family Taxodiaceae (swamp cypress family) as well as *Nyssa* (tupelo tree) and representatives of the Arecaceae (palms), Pandanaceae (screw palm or screw pine family) and perhaps also Ericaceae (heather family) (Wilde 1989; Wilde & Micklich 2007a; Collinson et al. 2012). There are two plausible explanations for the large number of palm flowers and fruits: On the one hand, they were produced in large numbers by the plants, and on the other hand, they may have dropped directly into the lake (Collinson 1988; Collinson et al. 2012). It is noticeable that the plant families represented at Messel include an unusually high number of species with climbing or trailing habits, in particular, e.g., the Smilacaceae (greenbriar family), Menispermaceae (moon seed family), Icacinaceae (icacina family) and the Vitaceae (grapevine family). This points to the presence of a light-dependent "vine curtain" at the transition to the actual forest (Wilde 1989; Collinson et al. 2012). The species-rich forest itself was home to many angiosperm tree species, including the Juglandaceae (walnut family), Theaceae (tea family), Leguminosae (legume family) and possibly also the palms, while the layer of lower trees and shrubs was dominated by the Lauraceae (laurel family) (Wilde 1989; Wilde & Micklich 2007b; Collinson et al. 2012). Since they are only represented in Messel by their pollen, which is easily airborne, the Betulaceae (birch family), Fagaceae (beech family) and Pinaceae (pine family) probably occurred in somewhat drier locations farther away from the lake (Wilde 1989; Wilde & Micklich 2007b; Lenz et al. 2011).

When comparing the flora in Messel with recent floras (e.g., Wilde 1989; Storch & Schaarschmidt 1988, 1992; Collinson et al. 2012), only few taxa stand out that are exclusively found in purely temperate climates today; the vast majority has a tropical or subtropical distribution. These include many taxa with relations to modern-day plants that can be found in an area extending from India and East Asia to China and Oceania (Collinson et al. 2012). Some of the taxa known from Messel are found today across the Old and New World tropics, while *Anacardium* (cashew nut; Fig. 6.32) and the Cyclanthaceae are now restricted to the Neotropics (Central and South America) (Manchester et al. 2007; Smith et al. 2008). Thus, the example of Messel clearly demonstrates that climate changes since the middle Eocene have led to changes in the distribution of many groups of plants.

Chapter 7
Jewels in the Oil Shale – Insects and Other Invertebrates

Sonja Wedmann

In the past 20 years, our view of the relationships between different groups of animals has undergone significant changes. The traditional system of the animal kingdom was primarily based on morphological, anatomical and physiological features and on the comparison of the organisms' development (ontogeny). In the past decades, the use of genetic methods in the analysis of the phylogeny necessitated a review of the traditional system and led to changes in many areas. However, several of the traditional concepts were confirmed by the new methodology (e.g., Westheide & Rieger 2013).

The Metazoa (Fig. 7.2) encompass all of the multicellular animals. One of the most primitive groups are the sponges (Porifera), which do not yet possess any organs such as blood vessels or reproductive organs; moreover, they also lack a nervous system and a head. The head only develops in the Bilateria. As their name implies, the body of the Bilateria shows a bilaterally symmetrical design, i.e., it can be divided into a left and a right half. Large, traditionally recognized groups in the system of the Bilateria are the Protostomia (from Greek for "first mouth"), in which the blastopore ("original mouth") develops into the mouth during ontogeny, and the Deuterostomia (from Greek for "second mouth"), in which the blastopore develops into the anus during ontogeny. while the actual mouth opening is newly formed. According to the latest findings, the Protostomia can be divided into Ecdysozoa (molting or shedding animals) and the Lophotrochozoa (Spiralia) (e.g., Westheide & Rieger 2013, Dunn et al. 2014); Fig. 7.2 only illustrates the most important groups or the ones represented in Messel. The main distinction

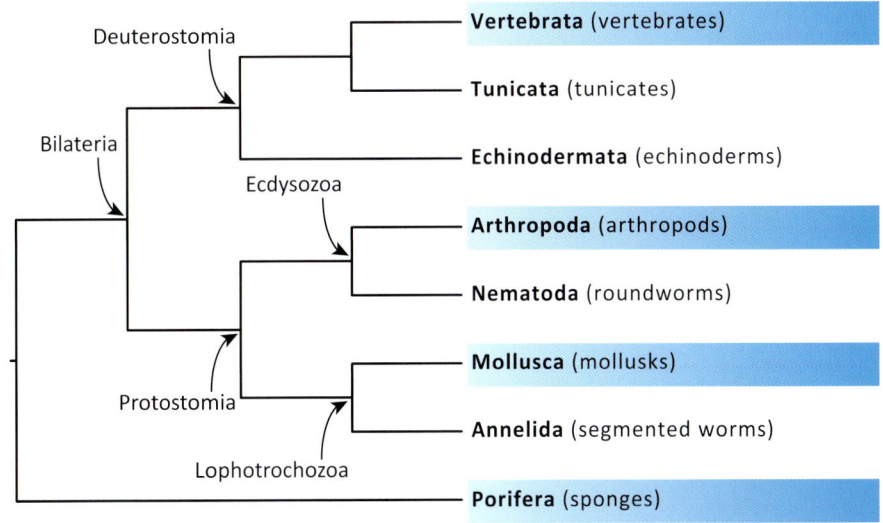

Fig. 7.2: Simplified phylogenetic tree of the animal kingdom (Metazoa). Groups recorded from Messel are marked blue.

Fig. 7.1: Fossilized male and recent female leaf insect (Phasmatodea).

among the Deuterostomia is between the Vertebrata (vertebrate animals) and the Echinodermata (echinoderms). Scientific research into the relationships between these animals continues to produce new insights (e.g., Satoh et al. 2014; Wanninger 2016).

Sponges (Porifera)

The majority of sponges lives in the ocean, but the group also includes denizens of freshwater habitats: the freshwater sponges or Spongillidae. The adult freshwater sponges are sessile (attached to a substrate) and live on a variety of hard surfaces such as wood and rocky bottoms, but they also attach to aquatic plants. They feed by filtering the finest organic particles from the water. The skeleton of freshwater sponges is composed of organic (collagen) and mineral elements. The latter are the spicules, also known as sclera, which can be divided into megasclera and microsclera, depending on size, position and function. Freshwater sponges are capable of sexual reproduction by producing larvae, but the most common method of reproduction is the production of gemmules. Gemmules are small, spherical permanent stages in the life cycle with a size of up to 1 mm that facilitate hibernation or survival during dry periods, for example. Under favorable environmental conditions, new sponge bodies develop from the gemmules. Their surface is densely covered with spicules, which represent a special type of needles called gemmosclera and which play an important role in the identification of sponges (Westheide & Rieger 2013).

Fig. 7.3: Freshwater sponge *Ephydatia gutenbergiana* (Spongillidae). Left: overview of the entire fossil, diameter c. 11.5 mm; right: greatly enlarged SEM image with impressions of spicules and a few heavily corroded spicules, scale: 0.1 mm.

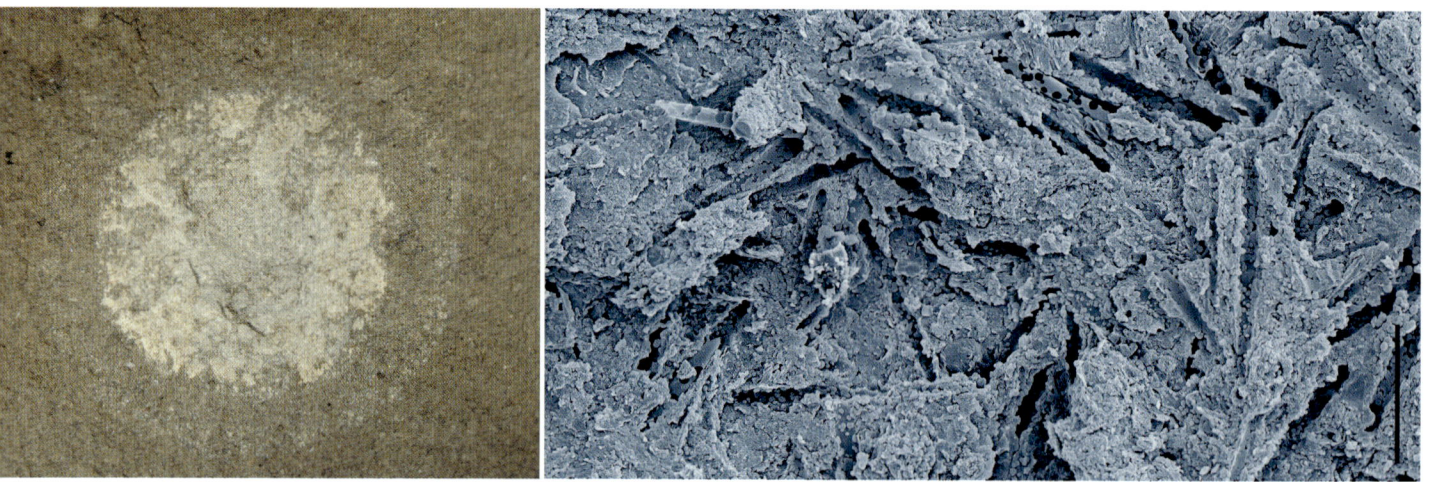

Fig. 7.4: Freshwater sponge *Lutetiospongilla heili,* holotype, (Spongillidae). Left: Overview of the entire fossil, fossil length approx. 8 mm; right: greatly enlarged SEM image of gemmosclera, scale: 0.1 mm.

Fig. 7.5: Shell of a mystery snail (Viviparidae) in siderite preservation, diameter c. 11 mm.

Fig. 7.6: Shell of a ram's horn snail (Planorbidae) in periostracum preservation, diameter c. 10 mm.

Two species of freshwater sponges (Spongillidae) are known from Messel, *Ephydatia gutenbergiana* and *Lutetiospongilla heili*. *E. gutenbergiana* (Fig. 7.3) was formerly assigned to the sponge genus *Spongilla*, but subsequent examination of the gemmosclera revealed its affinity to *Ephydatia* (Richter & Wuttke 1995, 1999).

The second sponge type, *Lutetiospongilla*, is known by megasclera and two types of gemmosclera. The holotype (Chapter 5) is an accumulation of numerous gemmules and remnants of the surrounding sponge tissue (Fig. 7.4).

Paleobiogeography and paleoenvironment

Spiculites, i.e., sediment deposits rich in spicules, occur primarily in the northeastern half of the Messel Pit, but individual spicules can be found in almost all of the layers (Richter & Baszio 2009).

With a thickness of 9 m, *Ephydatia* forms an extremely massive spiculite in the pit's southeastern part. This layer also contains larval cases of caddisflies that are composed of spicules (Richter & Wuttke 1999; Richter & Baszio 2009). *Lutetiospongilla* dominates in a layer of approximately 7 m thickness in the pit's eastern region. *Lutetiospongilla* presumably occurred in certain areas of Lake Messel close to the shore; thus, the site of discovery and the habitat broadly correspond with each other. *Ephydatia*, on the other hand, may have lived in slightly moving water, possibly at or near an inlet through which the specimens were transported into the lake (Richter & Wuttke 1999; Richter & Baszio 2009).

Mollusks (Mollusca)

Mollusks (Mollusca) are significantly underrepresented in Messel, which may be due to taphonomic reasons. To date, only shells of snails (Gastropoda) have been found, which occur in particular states of preservation, since the shell's calcium carbonates were dissolved (Fig. 7.5, Fig. 7.6). One preservation type shows whitish fossils that look as if they still contain remnants of lime. However, this substance does not contain calcium carbonate, but rather iron carbonate (siderite), which replaced the original shell substance. Other shells occur in the so-called periostracum preservation, where the fossil has a brownish coloration and only the periostracum, an organic membrane that covers the outside of shell, has been preserved (Wuttke 1988). Overall, representatives of three families of snails have been discovered in Messel. The mystery snails (Viviparidae) and the ramshorn snails (Planorbidae) are covered in the following paragraphs. There is only a single snail shell that may be associated with the Pleuroceridae. Although there is one mention of mud snails (Hydrobiidae) in the literature, no voucher specimens can be found. There are also descriptions of trace fossils that were initially interpreted as the feces of small snails; however, these may also represent fecal remains from other small invertebrates (Neubert 1999).

Mystery snails (Viviparidae)

Mystery snails are freshwater dwellers with a worldwide distribution. The name Viviparidae refers to the fact that they give birth to live young. They are grazers, i.e., they use their rasping tongue to feed on algae and bacterial lawns; however, they are also able to filter nutrients from the water. Mystery snails constitute by far the most frequently found snail shells in Messel. They occur in periostracum and siderite preservation (Fig. 7.5). Since the preservation does not reveal any exact details, the identification even at the family level is difficult, but the shell design indicates an affiliation with the Viviparidae (Neubert 1999).

Ramshorn snails (Planorbidae)

The freshwater ramshorn snails often show a disc-like shape. Many species are able to live in water with a low oxygen content, since they possess special adaptations such as an additional gill and the red blood pigment hemoglobin. Only a small number of specimens have been found in Messel (Fig. 7.6). They had originally been assigned to a species also known from Geiseltal in eastern Germany, but current studies show that due to missing characteristics an assignation to a genus would be premature (Neubert 1999).

Arthropods (Arthropoda)

Among others, the arthropods include the arachnids, which are represented in Messel only by the spiders and the harvestmen. In addition, there are crusta-

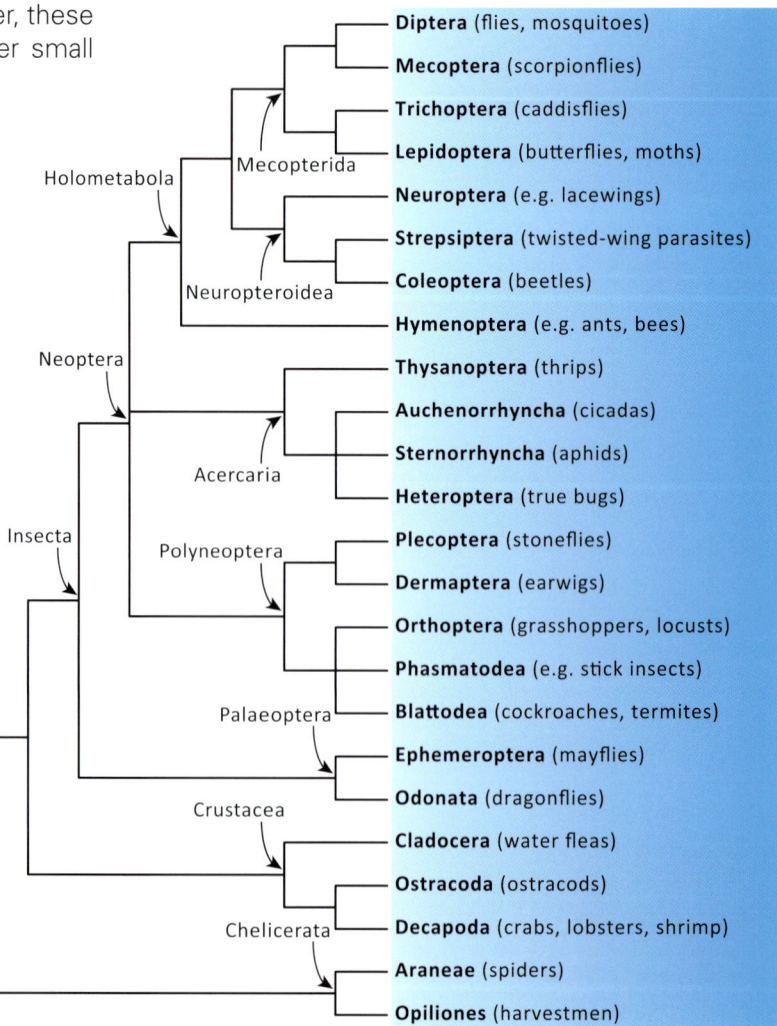

Fig. 7.7: Phylogenetic tree of the arthropods (Arthropoda). Groups known from Messel are marked blue.

ceans such as water fleas, seed shrimp and decapods. The species-rich insects are found in Messel in large numbers and in an exceptional state of preservation (Fig. 7.7; according to Westheide & Rieger 2013). As their name implies, one of the most important characteristics of the Arthropoda is the presence of segmented legs. They are composed of several tubes that are connected by joints and membranes. The entire skeleton is constructed of reinforced plates, which form a so-called exoskeleton. This external skeleton is similar to a knight's armor, with the muscles attached in the interior.

Spiders (Araneae)

Among the spider-like animals and their relatives (Chelicerata), the spiders are one of the most successful groups, with more than 43,000 species. They are typically terrestrial, occupying a wide range of habitats. The body is divided into two large parts, with four pairs of walking legs attached to the anterior part. The posterior body part bears spinnerets with spinning glands used for weaving the web. Insects constitute the main prey, but other animals are taken as well. Certain groups exclusively hunt other spiders. Due to their narrow mouth opening, spiders are incapable of eating solid food. They usually inject their prey with a digestive liquid and subsequently suck the liquefied nutrients from their victim (Westheide & Rieger 2013).

Currently, there are more than 160 fossils of spiders from Messel; however, not all of them are well preserved. There are several representatives of the orb-weaver spiders (Araneidae) (Fig. 7.8), which typically weave orb webs. They are presumed to be related to the recent genera *Singa* or *Zygiella* (Wunderlich 1986). One unusual specimen was identified as a representa-

Fig. 7.8: Male orb-weaver spider (Araneae, Araneidae), body length c. 4 mm.

Fig. 7.9: Female spider from the family Hersiliidae (Araneae), body length c. 6 mm.

Fig. 7.10: Harvestman (Opiliones), body length c. 4 mm.

Fig. 7.11: Resting eggs or ephippia of water fleas (Cladocera). Left: recent ephippium of *Daphnia pulex*, length c. 0.7 mm; right: fossil ephippium of the *Daphnia pulex* type, length c. 0.7 mm.

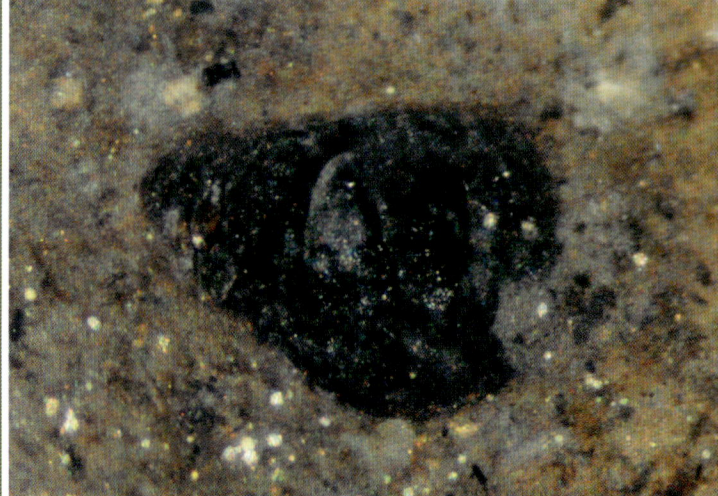

tive of the family Hersiliidae (Fig. 7.9) (Jörg Wunderlich, *pers. comm.* 1999).

Harvestmen (Opiliones)

With about 6,500 species, the harvestmen form a relatively small group. Their body is not divided into external segments. The long-legged harvestmen are universally known, but there are also species with short legs. Most are predatory (Westheide & Rieger 2013). In Messel, the long-legged harvestmen are represented by only six specimens, some of which (Fig. 7.10) may be members of the recent family Sclerosomatidae. Their slightly metallic structural colors possibly indicate an affinity with the group Gagrellinae, which today is mainly fond in the tropics (Jason Dunlop, *pers. comm.* 2012).

Crustaceans (Crustacea)

With 68,000 described species, the crustaceans may not be one of the most species-rich groups among the Arthropoda, yet they are one of the most disparate. There are elongated animals whose body consists of numerous uniform segments, while others have a short body, enclosed by a hard carapace. Some live in shells attached to a substrate; others are parasites that are barely recognizable as belonging to the crustaceans. Most species occur in the ocean, but some are found in freshwater and even on dry land. In Messel, the group is represented by water fleas, seed shrimp and decapods. Lake Messel must have been populated by many other small crustaceans, which are probably absent from the record due to taphonomic reasons, e.g. soft bodies that did not fossilize.

Fig. 7.12: Freshwater shrimp (Decapoda), fossil length c. 15 mm.

Water fleas (Cladocera)

Large numbers of water fleas in many forms and species are found in inland waters. They form an important link in the food chains. Water fleas have long been known from Messel in the form of their resting stage as so-called ephippia (Lutz 1991). These ephippia (Fig. 7.11) are occasionally found in extensive layers, but have also been documented as individual specimens from drill cores, thus giving evidence for the colonization of the early Lake Messel. Three different types of ephippia have been recorded (Richter & Wedmann 2005). The remnants of water flea bodies as well as juvenile clam shrimps (Conchostraca) have been found in fossilized fish feces (coproliths) (Richter & Baszio 2001).

Seed shrimp (Ostracoda)

Ostracods bear an outward similarity to tiny clams, since their body is enclosed by two shells. Most are aquatic, although a few species can be found in moist habitats on land. In Messel, their presence is evidenced by shells, although closer identification has not yet been possible (Goth 1990).

Decapods (Decapoda)

Among others, the decapods include lobsters, shrimp and spiny lobsters, along with many small, poorly known forms. They primarily occur in the ocean, but also in freshwater and even on dry land. Only a

Fig. 7.13: Exuviae of a freshwater shrimp (Decapoda), fossil length c. 16 mm.

small number of specimens of freshwater shrimp are known from Messel to date (Fig. 7.12, Fig. 7.13), their detailed systematic analysis is still pending.

Insects (Insecta, Hexapoda)

The name "Insecta" refers to the fact that their body is divided into three parts: head, thorax and abdomen. The three pairs of legs are attached to the thoracic segment. Winged insects carry an additional pair of wings each on the central and posterior segments of the thorax. This body plan has undergone numerous modifications among the insects, which largely contributes to their enormous success. An additional important factor is the inclusion of a pupal stage in the life cycle of the so-called holometabolous insects, which undergo a "complete metamorphosis." The Holometabola include the most species-rich groups of insects such as the beetles, butterflies, flies and hymenopterans, e.g., bees, wasps and ants (Fig. 7.7).

With about one million species described to date, insects make up more than half of all species known to science; they are the most successful group of animals in the world. And many additional species have yet to be discovered; estimates of the actual number of species range from 2 to 50 million. Based on their species-richness and their large biomass, it is apparent that insects play an important role in ecosystems.

Insects constitute by far the most numerous arthropods in the Messel Pit, as well as the most common animals overall. Senckenberg owns the most extensive collection of insects from Messel, which is housed in the Messel Pit Research Station and currently encompasses almost 17,000 specimens. An additional large collection with thousands of specimens

Box 7.1: Structural colors

Colors are either generated chemically by pigments, i.e., by various molecules that impart colors, or physically by the interaction of light with very small morphological structures, or by a combination of these two. The physically generated colors are called "structural colors," which often show a noticeable metallic iridescence. Structural colors can be caused by various physical processes, e.g., the scattering of light waves (coherent and incoherent scattering), wave propagation (interference) or bending (diffraction). As a prerequisite for the interaction with light, the morphological structures must be within the same size range as the light's wavelength, i.e., they must be minute. Especially among the insects, there is a large number of different nanostructures in the cuticle that are able to generate structural colors, e.g., multilayer reflectors, diffraction gratings or 3-D photonic crystals (e.g., Prum et al. 2006; Gebeshuber 2008). The top of the illustration shows a detail from the cuticle of a beetle, the bottom a detail of the shiny scales on a butterfly wing whose structural colors are based on multilayer systems (Parker & McKenzie 2003; McNamara et al. 2011).

is stored at the Hessian State Museum in Darmstadt. Smaller collections are located in various European museums, e.g., in the State Natural History Museum in Karlsruhe.

It is striking among the insects that structural colors (Box 7.1) are frequently preserved with their metallic sheen, even in the fossilized state (Lutz 1990). The most numerous and impressive examples can be found among the beetles. But structural colors have also been documented in certain true bugs, grasshoppers, dragonflies, hymenopterans and flies. Conspecific jewel beetles from Messel and Geiseltal show the same structural colors. Since these two fossil sites were subject to very different fossilization conditions, this offers proof of their originality (Hörnschemeyer & Wedmann 1994). Pigment colors are preserved in Messel as various shades of brown, and not in their original colors.

Abundance of the different insect groups in Messel

Meunier (1921) was the first to extensively study the fossil insects from Messel, and he described many new species. The second comprehensive treatment of Messel insects was undertaken by Lutz (1990), who for the first time reported and reviewed many of the insect groups named below. Wedmann (2005) compiled a list of all invertebrates known up to 2005.

Table 7.1: Abundance of the most common insect groups in the Messel collection of the Senckenberg Research Institute (as of Nov 2016)

Number of insects found in Messel	16,728
Beetles (Coleoptera)	10,066 (60.2 %)
thereof Weevils (Curculionoidea)	1932
thereof Click beetles (Elateridae)	1181
thereof Jewel beetles (Buprestidae)	1083
Hymenopterans (Hymenoptera)	2401 (14.4 %)
thereof Ants (Formicidae)	1406
thereof Giant ants Titanomyrma	654
True bugs (Heteroptera)	1933 (11.6 %)
Cicadas (Auchenorrhyncha)	373 (2.2 %)
Cockroaches and termites (Blattodea)	276 (1.6 %)
Flies (Diptera)	163 (1.0 %)

Literature consulted for the various sections in this chapter includes Westheide & Rieger (2013), Bellmann & Honomichl (2007), Dathe (2005), Günther et al. (1994). These books will not be cited again in the following, but served as the basis for the general information regarding each group.

To date, almost 100 insect species have been described from Messel; however, most groups have not yet been investigated in detail. The collections include hundreds of undescribed species. The most common insect fossils are terrestrial beetles, which also constitute the majority of specimens in the insect collections (Tab. 7.1). Second to beetles are the cases of aquatic caddisfly larvae. Field studies regarding the frequency of discovery of insects revealed that in some layers, such larval cases may constitute up to one third of all insect remains, while they are very rare in other layers. Ants (Formicidae) are also found frequently, particularly the extinct giant ants of the genus *Titanomyrma*, along with representatives of the true bugs (Heteroptera). All other insect groups occur at much lower frequencies, and some orders have only been documented by single specimens.

The abundance distribution of the different groups does not directly reflect the faunal composition in or around Lake Messel; rather, it is strongly influenced by taphonomic factors. In maar lakes such as Messel, beetles often dominate the spectrum of fossil discoveries, while flies or butterflies are only rarely found and are clearly underrepresented in relation to their abundance in the former habitat. In order to be preserved as a fossil in the sediments of a maar lake, several obstacles must be overcome. Once an insect drops onto the water's surface, it must first overcome the surface tension before it can sink. In this regard, certain morphological peculiarities are of taphonomic importance. For example, lepidopterans and flies have rather large wings in relation to their body weight, which, moreover, are resistant to wetting, so that they will often drift on the water's surface for extended periods of time. Beetles are much more compact and sink much more readily, at least as long as their wings are folded. But there are other factors as well, e.g., the insects' feeding status, environmental factors such as precipitation and many additional, in part only poorly studied factors that contribute to the fact that the preserved spectrum of fossils differs markedly from the former living communi-

Fig. 7.14: Larva of a mayfly (Ephemeroptera), fossil length c. 3.6 mm.

ty. Aquatic insects are also rather rarely found in maar lakes, since the fossil-bearing sediments formed in the lake's deeper regions, and not near the shores, where most of the aquatic insects occur.

Mayflies (Ephemeroptera)

Adult mayflies of many species have a very short lifespan. They spend most of their life as aquatic larvae, both in stagnant and flowing water. The larvae usually carry three (rarely two) long, multi-segmented terminal filaments at the tip of their abdomen. Most larvae feed on plant and animal detritus, but there also are filter-feeders and hunters.

Representatives of the mayflies are documented as larvae in Messel (Fig. 7.14). However, the delicate bodies of the larvae have not been preserved particularly well; in most cases, fragments of the abdomen and the long, terminal filaments are all that remains. It is possible that the specimens belong to the families Baetidae or Siphlonuridae (Richter & Krebs 1999).

Remnants of mayfly larvae have also been found in small coprolites of fishes, which offer special insights into the former ecosystem of Lake Messel (Chapter 13; Richter & Baszio 2001, 2002). To date, only one poorly preserved adult mayfly has been discovered.

Dragonflies and damselflies (Odonata)

Dragonflies are true aerial artists that can be found at many bodies of water. Both adults and larvae have a predatory lifestyle. Generally, dragonflies and damselflies are distinguished. The damselflies (Zygoptera) include the family Megapodagrionidae, which is documented in Messel by one isolated wing (Petrulevicius et al. 2008). There are fossil specimens of this group from several other sites in Europe and North America. Today, their closest living relatives occur worldwide across the tropics. Even a new family of odonates, Pseudostenolestidae, has been described from Messel (Fig. 7.15) (Garrouste & Nel 2015). It represents a very primitive group within the transition to the Anisoptera and is related to a number of odonate groups that otherwise became extinct in the Early Cretaceous. Dragonflies (Anisoptera) are represented in Messel by single individuals, including the hawkers (Aeshnidae) and clubtails (Gomphidae). In addition, there are a few records of larvae (Fig. 7.16).

Stoneflies (Plecoptera)

The name "stonefly" indicates that the adults are usually found along rocky or sandy riverbanks. The larvae are truly aquatic and usually develop in fast-flowing streams; they feed on algae or decomposing plant material as well as on small prey animals. Today, they

Fig. 7.15: Forewing of the dragonfly species *Pseudostenolestes bechlyi* (Odonata), paratype, wing length 34 mm.

Insects (Insecta, Hexapoda)

Fig. 7.16: Dragonfly larva (Odonata), fossil length c. 12 mm.

Fig. 7.17: Stonefly (Plecoptera), fossil length c. 9.5 mm.

Fig. 7.18: Earwig (Dermaptera), fossil length c. 16.5 mm.

are common and widespread in cool temperate zones. Tropical forms are found much more rarely. In Messel, there are very few records of stoneflies. There are several adult specimens that possibly belong to the family Perlidae, which today is found worldwide acorss the Northern and Southern Hemisphere (Fig. 7.17). An additional fossil of an adult animal is part of a group that has not yet been identified in detail (Yingying Cui, *pers. comm.* 2016). Larvae are very rarely found, and their fossils are only incompletely preserved.

Earwigs (Dermaptera)

About 2,000 species of earwigs are found around the globe, with the highest species diversity in the tropics and subtropics. They are characterized by their

Fig. 7.19: Katydid (Orthoptera, Tettigoniidae), length of forewing 62 mm.

Fig. 7.20: Pygmy grasshopper (Orthoptera, Tetrigidae), fossil length c. 12 mm.

shortened forewings and the pincers at the tip of their abdomen, which are mainly used for defense. Typical earwigs have a secretive terrestrial lifestyle, are frequently nocturnal, and their known food spectrum is very wide, from predatory to herbivorous. There are only few specimens from Messel (Fig. 7.18).

Grasshoppers, crickets and katydids (Orthoptera)

With approximately 22,500 species, the Orthoptera have a worldwide distribution, but most species occur in the tropics and subtropics. A characteristic trait of the Orthoptera is their hind-most pair of legs, which have developed into strong jumping legs. They are equally well known for their ability to produce sound, since they usually live in dense vegetation, where acoustic signals are much better suited to finding a partner than optical stimuli. Based on their antennae, the Orthoptera are divided into the Ensifera (with long antennae) and the Caelifera (with short antennae). Both groups occur in Messel. One specimen may belong to the Gryllacrididae (Lutz 1990), but katydids (Tettigoniidae) (Fig. 7.19) are also represented. Among the Caelifera, there are a few specimens of pygmy grasshoppers (Tetrigidae), which are characterized by a thorn-like process on their neck shield (Fig. 7.20).

Stick insects (Phasmatodea)

More than 3,000 species of stick insects have been described, primarily from the tropics. They have developed a variety of camouflage mechanisms, from imitating small twigs to mimicking leaves. There are a few specimens documented from Messel, including both camouflage mechanisms.

The world's first and to date only fossil record of the leaf insects or "walking leaves" (Phylliinae) is a specimen of *Eophyllium messelense* (Fig. 7.21) (Wedmann et al. 2007). The fossil is a male that was preserved in great detail and shows striking similarities to modern males of this insect group. As in recent males, the abdominal segments are laterally expanded in a leaf-like fashion, which leads to the conclusion that this advanced mimicking of deciduous leaves already developed in the Eocene. Leaf mimicking only leads to an optical deception of the predators (e.g., birds, primates) in conjunction with certain adaptive behaviors. Remaining immobile during the day or sudden jerky motions during disturbances imitate the movement of a leaf in the wind. Therefore, it can be assumed that the fossil species had already developed the same type of behavior. Currently, leaf insects are found in Southeast Asia and adjacent regions. The fossil from Messel shows that

Fig. 7.21: Leaf insects. Left: fossil male of the species *Eophyllium messelense* (Phasmatodea), holotype, body length 63 mm; right: recent male of the species *Phyllium celebicum*, body length 60 mm.

their distribution was much more widespread during the Eocene, which is most certainly connected to the globally much warmer climate at that time.

Cockroaches and termites (Blattodea)

Cockroaches ("Blattaria") and the termites (Isoptera), formerly placed in their own order, are closely related to each other, since termites developed from a subgroup of the cockroaches. Generally, cockroaches and especially termites prefer warm climates, and their highest species diversity is found in the tropics and subtropics. Cockroaches encompass around 4,600 species, and termites around 3,000 species. Most of the cockroaches have a solitary and nocturnal lifestyle, but there are species that exhibit brood care behavior, and the first stages of social behavior can be found among cockroaches most closely related to the termites. Termites have further developed the social system, with some species living in highly complex colonies that can include several hundred to a few million animals. Termite colonies involve several different castes: reproductive animals, workers, and soldiers. Reproductive animals are able to propagate; they are winged during their nuptial flight, but later shed their wings and found new colonies. Workers and soldiers never bear wings, and their reproductive organs are stunted. Genetically, there are both males and females. Often, the animals show a whitish coloration, which – in combination with their social habits – has earned them the nickname "white ants."

In Messel, the number of cockroaches clearly outnumbers the termites. Two cockroach species of the recent genus *Periplaneta* were described by Meunier (1921); today, this genus is widespread in human settlements. Recent studies show that the species diversity in Messel is much higher. Several recent families, subfamilies and genera can be distinguished among the cockroaches (Schmied 2009) (Fig. 7.22). Initial discoveries of termites were placed with the Hodotermitidae, albeit with reservations (Lutz 1990) (Fig. 7.23).

Thrips (Thysanoptera)

Thrips are also known as "thunderbugs," since the animals are often particularly active during humid thunderstorms in the early summer, when they can occur in huge numbers. They are usually only a few millimeters long and their wings are entirely covered by long fringes. More than a dozen records of thrips are known from Messel, but they are still pending further study.

Cicadas and "hoppers" (Auchenorrhyncha)

Auchenorrhynchan insects suck the sap of plants, using mouth parts that have been modified into a proboscis. At rest, they fold their wings over their back in a roof-like position. Species in Central Europe rarely exceed 5 mm in length, while representatives in the world's warmer regions may grow significantly larger.

The large singing cicadas (Cicadoidea) are the most striking group found in Messel (Fig. 7.24). Male cicadas are able to produce a species-specific song, which can be very loud. For sound-production, the males have tymbals, a drum-like organ, in their abdomen, which is lacking in females. Both males and females have special hearing organs (tympana), which are also located in the abdomen. It would be very exciting to examine these structures in the Messel fossils, but unfortunately, there are no fossil singing cicadas, especially males, that are sufficiently preserved to allow this. Many of the otherwise rather well-preserved specimens are missing the abdomen; this may be connected to the buoyancy of the tymbal, which is filled with large air sacs (Lutz 1990).

There are also records of about 300 small fossil cicadas from Messel to date, which are often not particularly well-preserved. Several families and genera of planthoppers (Fulgoromorpha) have been described (Fig. 7.25) (Szwedo & Wappler 2006). Apart from the singing cicadas, little is known to date about the remaining members of the Cicadomorpha.

Plant lice, scale insects and whiteflies (Sternorrhyncha)

The plant lice are a group of sap-sucking insects that tend to be rather small and encompass about 16,000 species worldwide. There are only a few records from

Fig. 7.23: Termite (Isoptera), body length 18 mm.

Fig. 7.22: Cockroach (Blattodea), wing length 50 mm.

Fig. 7.24: Female singing cicada (Cicadidae), body length 27 mm.

Messel; the aphids (Aphidina) are still pending further study. Most recent aphids occur in the temperate climate zones.

The scale insects (Coccina) are a group in which the males and females show a striking sexual dimorphism. Males only reach a length of a few millimeters, have wings and are easily recognizable as insects. The females, however, are larvae that have reached sexual maturity (neoteny); they are frequently covered by a scale-like shield that protects them from enemies and climatic influences. Leaves with numerous of these shields attached are also known from Messel (Fig. 7.26). Some of the representatives have been described as armored scale insects (Diaspididae: Aspidiotinae) (Wappler & Ben-Dov 2008).

True bugs (Heteroptera)

About 40,000 species of true bugs are known worldwide. Like cicadas and plant lice, they have a proboscis, which they not only use to suck sap from plants, but from animals as well. Most people are familiar with the bed bugs, which are among the few bloodsucking species. The forewing of true bugs is partly sclerotized and partly membranous.

Fig. 7.25: Planthopper of the species *Wedelphus dichopteroides* (Dictyopharidae), holotype, fossil length 21 mm.

Fig. 7.26: Scale insects (Coccina) on a leaf, length of leaf 57 mm; the inset has a width of 7.5 mm.

Fig. 7.28: Lace bug of the species *Oblongomorpha lutetia* (Tingidae), holotype, body length 5 mm.

In Messel, fossil bugs have been found in large numbers, but only few species are recognized (Lutz 1990). Aquatic bugs include the backswimmers or water boatmen (Notonectidae), which are predators and only rarely encountered in Messel (Fig. 7.27). Even less common are discoveries of giant water bugs (Belostomatidae), another aquatic, predatory group, whose recent representatives can reach a length of more than 10 cm. Water striders (Gerridae) have unusually long legs and are perfectly adapted to life on top of the water's surface, where they usually hunt other insects. The water strider *Cylindrobates messelensis* was described from Messel (Wappler & Andersen 2004).

Terrestrial bugs are much more commonly found in Messel than aquatic groups. Lace bugs (Tingidae)

Fig. 7.27: Water boatman (Notonectidae), body length 3 mm.

Fig. 7.29: Assassin bug (Reduviidae). Left: fossil representative of the Harpactorinae, body length 12.5 mm; right: recent species of the Harpacticorinae, *Eulyes pretiosa*, body length 30 mm.

are represented by more than 65 specimens. They can easily be identified to the family level, due to their bodies' characteristic reticulate surface structure (Fig. 7.28). Four species are known to date from Messel, which all belong to extinct genera (Wappler 2003a, 2006). While mirid bugs (Miridae) represent one of the largest true bug families today, they are rare in Messel.

Assassin bugs (Reduviidae) are represented by almost 90, partly very well-preserved, specimens (Fig. 7.29). They are predators that usually feed on other insects. The recent subfamilies Harpactorinae and Reduviinae can be distinguished in Messel (Christiane Weirauch, *pers. comm.* 2016). A species of Harpactorinae, *Amphibolus disponsi*, was already described by Kinzelbach (1970a).

The majority of the bugs from Messel belongs to the shield bug–like terrestrial bugs (Pentatomomorpha). Burrowing bugs (Cydnidae) (Fig. 7.30) and shield bugs (Pentatomidae) (Fig. 7.31) constitute the most common finds. Today, the burrowing bugs encompass only about 600 species, while there are 4,000 species of shield bugs. Surprisingly, the burrowing bugs are particularly common in Messel (Lutz 1990). Two species were assigned to the genus *Cydnopsis*, which is only known from fossils (Kinzelbach 1970a).

Several species of shield bugs are also known as "stink bugs," since they have defensive glands near the central part of their bodies that excrete a foul-smelling secretion when disturbed. These glandular areas even show up in the fossils as dark spots on the otherwise colorful body (Fig. 7.31). Milkweed or seed bugs (Lygaeidae) are rare. Flat bugs (Aradidae) (Fig. 7.32) are currently represented by a dozen specimens, which can be assigned to an astonishing six species (Wappler & Heiss 2006; Wappler et al. 2015).

Hymenopterans (Hymenoptera): Sawflies and parasites

Among others, the hymenopterans include bees, wasps and ants, along with a large number of generally lesser-known groups such as sawflies, ichneumon wasps, etc. As a rule, the hymenopterans can be divided into the very primitive groups, such as sawflies and wood wasps, in which the adults do not show the characteristically constricted "wasp waist," and the more highly developed hymenopterans that possess this trait. The latter encompass a large number of parasitic groups, comprising a combined total of more than 65,000 species. In additon, they also include the subclade of "stinging wasps" (Aculeata), in which the original ovipositor has been modified into a stinger.

Fig. 7.30: Burrowing bug (Cydnidae), body length 12 mm.

Fig. 7.31: Shield bug (Pentatomidae), body length 9 mm.

Fig. 7.32: Flat bugs (Aradidae). Left: fossil species *Neuroctenus kotejai*, holotype, body length 5 mm, right: recent species *Neuroctenus punctulatus*, body length 7.5 mm.

Fig. 7.33: Wood wasps (Siricidae). Top: fossil species *Xoanon? eocenicus*, holotype, wing length 9 mm; bottom: recent species *Urocerus gigas*, male, body length 27 mm.

This group includes the well-known bees, wasps and ants, in which the females usually possess a stinger.

Representatives of the primitive hymenopteran groups are very rare in Messel; there are single specimens of sawflies (Tenthredinidae), cimbicid sawflies with club-shaped antennae (Cimbicidae) and wood wasps (Siricidae) such as *Xoanon? eocenicus* (Fig. 7.33). The recent genus *Xoanon* is only found in eastern Asia today (Wedmann et al. 2014). Adult wood wasps are rather large, and the *Xoanon* species is very similar to our native wood wasp (Fig. 7.33 bottom).

Parasitic representatives of the hymenopterans are more commonly found in Messel than the sawflies; they include, e.g., the tiny chalcid wasps (Chalcidoidea) (Fig. 7.34), ensign wasps (Evaniidae), members of the Proctotrupoidea, the Stephanidae and the braconid wasps (Braconidae), as well as the "true" ichneumon wasps (Ichneumonidae) (Fig. 7.35). The true ichneumon wasps from Messel can often be assigned to recent subfamilies, and sometimes even recent genera (Seraina Klopfstein, *pers. comm.* 2016).

Fig. 7.34: Chalcid wasp (Chalcidoidea), body length 2.5 mm.

Hymenopterans (Hymenoptera): Bees and wasps

Bees (Apidae, also Apiformes or Anthophila), wasps and ants are representatives of the stinging wasps (Aculeata). All bees feed purely on a vegetarian diet, both as adults and larvae; there are also "parasitic" bee species whose larvae eat the food of their host's larvae. The best-known bees are the honey bees (*Apis*), which live in large, perennial colonies. However, within the group of bees, a large variety of lifestyles can be found, from solitary bees to primitive social forms and culminating in the highly developed social colonies of the honey bees. They play an important role as plant pollinators.

Several fossil bees are known from Messel (Wappler & Engel 2003; Wedmann et al. 2009). *Protobombus* (Fig. 7.36) and *Electrapis* (Fig. 7.37 top) belong to an extinct group of social bees, the Electrapini. The presence of pollen adhering to representatives of the Electrapini serves as evidence that bees already acted as pollinators in the Eocene. Many different types of pollen were found between the hairs on the bees' bodies (Wappler et al. 2015). Pollen grains emit a greenish glow when exposed to UV light, which renders them visible between the leg hairs, especially in a dark environment (Fig. 7.37 bottom).

Leafcutter bees (Megachilidae) are another group known from Messel. *Megachile* species use pieces of

Fig. 7.35: Ichneumon wasps (Ichneumonidae). Left top: fossil female ichneumon wasp, body length 5 mm; left bottom: recent female ichneumon wasp of the species *Pimpla spuria*, body length including ovipositor 13 mm; right top: fossil male ichneumon wasp, body length 5 mm; right bottom: male of the recent species *Ichneumon inops*, body length 13 mm.

Insects (Insecta, Hexapoda)

Fig. 7.36: Extinct bee of the species *Protobombus messelensis* (Apidae), holotype, body length 10 mm.

Fig. 7.37: Extinct bee, *Electrapis* sp. Top: complete fossil, wing length 7 mm; bottom: enlarged leg area, with pollen fluorescing green under UV light, basitarsus length 1.3 mm.

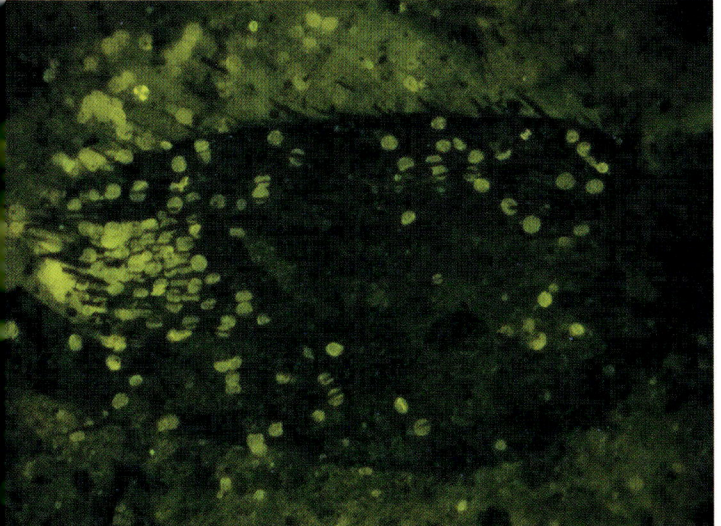

Fig. 7.38: Leaf of a walnut plant (Juglandaceae) with a segment excised by a leafcutter bee, leaf length 80 mm, the excised part of the leaf is about 10 mm long. Photo: copyright Torsten Wappler.

Fig. 7.39: Leafcutter bee of the species *Friccomelissa schopowi*, holotype; however, this species did not excise parts of a leaf, body length 7 mm.

plant leaves to pad the nests for their brood. The genus *Megachile* has only been documented indirectly for Messel, based on the characteristically shaped incisions along the edge of leaves (Fig. 7.38). However, there is also an actual specimen of the leafcutter bee *Friccomelissa schopowi*, although this species did not engage in the act of leaf-cutting (Wedmann et al. 2009) (Fig. 7.39).

Wasps (Vespoidea) are represented in Messel by several groups. These include the typical wasps (Vespidae) such as yellowjackets and hornets, several species of which are known to compete with us for our plum pastries in the late summer. The scoliid wasps (Scoliidae) and the Tiphiidae, both with a parasitic development, are also represented. The scoliid wasps could be identified as representatives of both an extinct and a recent subfamily. The Tiphiidae (Fig. 7.40) appear to belong to recent genera (Greschbach 2015).

Hymenopterans (Hymenoptera): Ants

Ants (Formicidae) are among the best-known colonial social insects. More than 14,000 species occur in large numbers in a wide variety of habitats. It is estimated that the biomass of ants in the Amazon rainforest is several times higher than that of all mammals, birds, reptiles and amphibians combined (Fittkau & Klinge 1973). In most ant species, three forms can be distinguished: males, queens, and the female, but sterile worker ants, which comprise the vast majority of animals in an ant colony. Only the males and the queens carry wings during the first part of their adult life. They use the wings to undertake nuptial flights; afterwards, the wings are shed at a predefined breaking point.

The fossil record of ants in Messel primarily involves winged males or queens. This has taphonomic reasons, since the wingless workers are often able to save themselves when they fall on the surface of a lake. When winged ants get onto the lake surface, e.g., by wind or rain, they are beyond hope of rescue; therefore, they tend to be overrepresented in the sediments at Messel.

The extinct giant ants (Formiciinae) are found most frequently; they currently constitute almost half

Fig. 7.40: Tiphiid wasp (Tiphiidae), body length 15 mm.

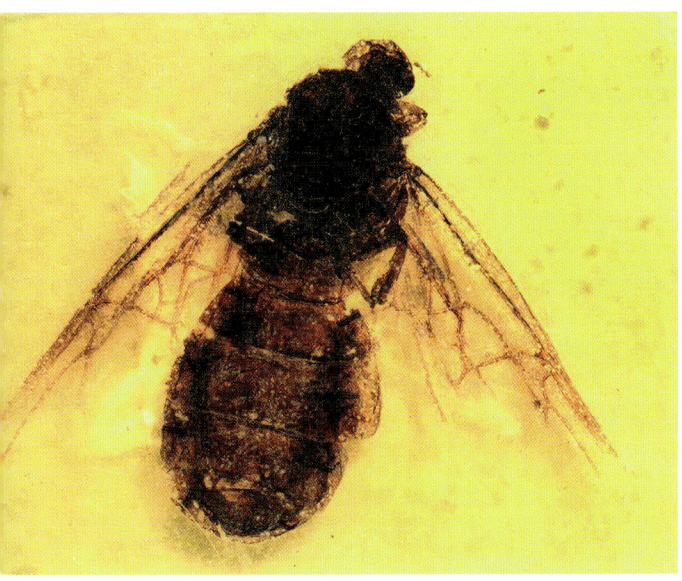

Fig. 7.41: Extinct giant ant *Titanomyrma giganteum*, holotype, female with a body length of 57 mm, transferred to synthetic resin.

Fig. 7.42: Extinct giant ant *Titanomyrma simillimum*, holotype, female with a body length of 53 mm.

Fig. 7.43: Extinct ponerine (stinging) ant of the species *Protopone germanica* (Formicidae), holotype, body length 14 mm.

Fig. 7.44: Forest-dwelling ant of the species *Gesomyrmex pulcher* (Formicidae), holotype, body length 5 mm.

of all ant specimens from Messel. Two species of giant ants are known from here, *Titanomyrma giganteum* (Fig. 7.41) and *T. simillimum* (Fig. 7.42) (Lutz 1986). With a wing span of up to 15 cm, they were the largest ants in the world. They occurred across Europe and North America and are assumed to have been thermophilic, since their fossil record is restricted to sites with an annual mean temperature above 20 °C (Archibald et al. 2011). Apart from the giant ants, all remaining ants from Messel can be assigned to recent subfamilies and, in part, even to recent genera.

Stinging ants (Ponerinae) are relatively primitive ants whose females possess a stinger and are able to deliver a sting. Stinging ants rarely occur in Central Europe, but the group has a worldwide distribution and was common in Messel. Twenty-two species have been described from Messel, five of which belong to extinct genera (e.g., Fig. 7.43) and one to a recent genus. Compared to ants from Baltic amber, several million years younger, the species diversity of ants in Messel is astonishingly high (Dlussky & Wedmann 2012).

Fig. 7.45: Isolated head of the ant *Gesomyrmex curiosus* (Formicidae), holotype, head length 3 mm.

Fig. 7.46: Weaver ant of the genus *Oecophylla* (Formicidae). Left: the fossil species *Oecophylla longiceps*, holotype, body length 15 mm; right: recent nest of weaver ants in South Africa, nest diameter c. 20 cm.

Box 7.2: "Zombie ants" – ants are being reprogrammed by fungi

Characteristic bite marks on a fossil leaf from Messel serve as evidence that "zombie ants" occurred as early as 48 million years ago (Fig. Box 7.2) (Hughes et al. 2011). Certain parasitic species of fungi are able to infest ants and manipulate their behavior in a way that aids the propagation of the fungi. These fungus-ant relationships are well-known today from fungi of the genus *Ophiocordyceps* from the tropics of Southeast Asia, which frequently infest carpenter ants (*Camponotus*). Once the fungal spores have germinated on the ant's body, the fungal hyphae enter the body and manipulate the ant's behavior. Like mindless zombies, the infected ants leave their normal habitat high in the tree tops and move near the forest floor into an area that offers optimal conditions for the fungus. Here, the ants attach themselves with their mouth parts to the veins of leaves, where they die soon afterwards. From the head of the ant zombie now grows the stalked fruiting body of the fungus, in which the spores develop that will ultimately infect new ants. The bite marks on the fossil leaf in Messel are specific enough to infer the special behavior described above. This documents the existence of this highly developed parasite-host relationship in the Eocene, and it is the first evidence of behavioral manipulation in the fossil record (Hughes et al. 2011).

"Zombie ants" in Messel. Left: fossil leaf with numerous bite marks, leaf length c. 100 mm; right top: detail of bite marks, inset width 6 mm; right bottom: recent dead "zombie ant" with fungal growth, image width 20 mm. Photo Photo right bottom © David Hughes.

Today, formicine ants (Formicinae) constitute one of the largest groups among the ants. Their stinging apparatus is greatly reduced; instead, they defend themselves by spraying formic acid at an attacker. The formicine ants include the recent genus *Gesomyrmex*, which has been described from Messel with three species (Fig. 7.44, Fig. 7.45) (Dlussky et al. 2009). Several additional fossil species are known from other Eocene fossil sites in Europe and far eastern Russia. Today, only six recent *Gesomyrmex* species survive. They are tree dwellers from tropical Asia, where they inhabit various types of forests.

Moreover, a fossil species of weaver ant (*Oecophylla*) is known from Messel (Dlussky et al. 2008) (Fig. 7.46 left). Modern weaver ants live high in the tree tops. Their larvae produce silk that is used to weave living leaves into nests (Fig. 7.46 right). There is no evidence that the ants in Messel already showed a similar behavior, but fossilized woven nests are known from the Miocene in Africa. Today, there are only two species of weaver ants; one occurs in the tropical and subtropical forests of Africa, while the other is found from India and Southeast Asia to northern Australia. However, fossil weaver ants are known from several Cenozoic sites in Europe.

A record of bulldog ants (Myrmeciinae) from Messel (Dlussky 2012) is of paleogeographic interest, since today, this group with fewer than 100 recent species is restricted to the Australian region. The fossil record of bulldog ants, however, shows a wide distribution during the Eocene, with additional finds in Europe, North and South America (Archibald et al. 2006).

Net-winged insects (Neuroptera)

Adult net-winged insects have wings with a reticulate venation. Their larvae are predators with a mostly terrestrial lifestyle. Some adults feed on pollen or plant sap, others are predatory. One species of mantidflies (Mantispidae), *Symphrasites eocenicus,* has been described from Messel (Fig. 7.47 top) (Wedmann & Makarkin 2007). At first glance, mantidflies (Fig. 7.47

Fig. 7.47: Mantidfly (Mantispidae). Top: forewing of the fossil mantidfly *Symphrasites eocenicus,* holotype, wing length 12 mm; bottom: recent mantidfly, entire animal with a body length of 21 mm.

Fig. 7.48: Parasitic twisted-wing (Strepsiptera) on an ant. Right: entire ant, body length 12 mm; left: detail of two male puparia of the twisted-wing parasite *Stichotema* sp.; the arrows indicate the puparia, length of the upper, larger puparium 1.4 mm.

bottom) resemble a praying mantis, with raptorial forelegs and a strongly elongated thorax. However, these traits are the result of independent (convergent) evolution, as mantises and mantidflies are not related. Their appearance merely reflects their similar lifestyle.

The Messel fossil represents an isolated forewing whose lower part is creased by multiple folds (Fig. 7.47 top). Nevertheless, it allows proper identification. The fossil belongs to the subfamily Symphrasinae, which today is exclusively restricted to South America and southern North America. Messel provides the only fossil record of this group outside of its current range.

Twisted-wing parasites (Strepsiptera)

Strepsiptera are parasites on other insects that show a complex development cycle in several regards. Adult males, which only reach a length of 1-8 mm, are able to fly and have large, foldable, thinly membranous hindwings. The females differ greatly from the males (sexual dimorphism) and are always entirely wingless. Primitive species have free-ranging females, while in most other species the females spend their entire life as parasites inside the abdomen of a host insect. Adult twisted-wing parasites do not eat; reproduction is their only purpose.

The infestation of insects by twisted-wing parasites is called being "stylopized." Lutz (1990) described a fossil ant from Messel which had been infested by two twisted-wing parasites. Its abdomen shows two puparia (modified pupa shells) that appear to be empty and from which presumably two males emerged (Fig. 7.48). The twisted-wing parasites from Messel are assigned to the highly developed family Myrmecolacidae, and within it, the genus *Stichotrema* (Kinzelbach & Pohl 1994). The females of this group remain inside the host's body.

Beetles (Coleoptera): Primitive groups

Beetles can easily be recognized by their elytra, which are modified forewings and usually form a protective cover over the abdomen. The hindwings are folded up under the elytra and are unfolded during flight. Beetles constitute by far the most commonly found insects in Messel, with weevils dominating among the identifiable forms.

Cupedidae is a very primitive family of beetles that no longer occurs in Europe today. Only a small handful of specimens is known from Messel, with one species described as *Tenomerga? messelense* (Fig. 7.49) (Tröster 1993). Generally, little is known about the lifestyle of recent cupedids. However, certain species are known to develop in fungus-infested wood.

The ground beetles (Carabidae) are a very old and large group that occur around the world with around 30,000 species. They are almost exclusively predatory. Currently, fewer than 200 fossils are known from Messel. Lutz (1990) already reported representatives of the Scaritinae. About 40 species can be distinguished, some of which can be identified more precisely due to their high–quality preservation. Half of the ground beetles in Messel are shore dwellers, the others flew in from various terrestrial habitats. The shore dwellers include groups that lived in the reeds (Lebiini) as well as animals burrowing in sandy or clayey sediments (Scaritini). So far, it has not been possible to assign the fossil species to a recent genus (Fig. 7.50). Compared to the extant fauna, the closest relatives of the Messel ground beetles appear to

Fig. 7.49: Beetle *Tenomerga? messelense* (Cupedidae), holotype, body length 5 mm.

Fig. 7.50: Ground beetle (Carabidae), body length 25 mm.

occur in the tropical and subtropical zones; however, there may also be a connection to the warm-temperate fauna (Joachim Schmidt, *pers. comm.* 2016).

Beetles (Coleoptera): Rove beetles, water dwellers and other handsome beetles

The usually rather small rove beetles (Staphylinidae) are easily recognized by their elongated bodies and the significantly shortened elytra, which leave the abdomen exposed. Nevertheless, they tend to be adept fliers whose large hindwings only become visible once they have been unfolded (Fig. 7.51). With more than 45,000 extant species, they represent a very species-rich group. To date, there are more than 100 fossil specimens from Messel, showing a high species diversity. Most rove beetles are predators that hunt for prey in dung, carrion or leaf litter.

Water scavenger beetles (Hydrophilidae) are represented in Messel by three recent genera (Fikácek et al. 2010), which all belong to the aquatic Hydrophilinae. The specimens represent the oldest fossil records (48 million years) of the genera *Hydrobiomorpha*, *Hydrochara* and *Hydrophilus*. A newly described species, *Hydrobiomorpha eopalpalis* (Fig. 7.52), reveals an unusual sexual dimorphism in the widened shape of the mouth parts' palps.

Fig. 7.51: Rove beetle (Staphylinidae), body length 8 mm.

Fig. 7.52: Water scavenger beetle (Hydrophilidae). Left: the fossil species *Hydrobiomorpha eopalpalis*, holotype, body length 20 mm; right: the recent great silver water beetle *Hydrophilus piceus* with a body length of 48.5 mm.

Fig. 7.53: Relative of the scarab beetles (Scarabaeoidea), body length 40 mm.

Fig. 7.54: Stag beetle (Lucanidae), body length 30 mm (including mandibles).

Among the rare aquatic insects, the Hydrophilidae are found rather frequently. From Messel, only adult specimens of hydrophilid beetles are known, which did not necessarily develop in Lake Messel, as it is known that adult water scavenger beetles are attracted to polarized, glossy surfaces such as lakes during their roaming flights.

Scarab beetles (Scarabaeidae), earth-boring dung beetles (Geotrupidae) and stag beetles (Lucanidae) are all part of the superfamily Scarabaeoidea, which is characterized by the leaf-shaped terminal segments of the beetles' antennae. This can be seen particularly well in the May bug or cockchafer. The antennae are covered with numerous olfactory sensors, giving the beetles a highly developed sense of smell. More than 500 fossils of Scarabaeoidea have been recorded in Messel, making Messel the site with the highest fossil incidence of this group of beetles in the world (Fig. 7.53) (Krell 2006). The vast majority of the specimens belongs to the scarab beetles, with stag beetles (Fig. 7.54) and earth-boring dung beetles represented in much lower numbers. Overall, at least 20 species can be distinguished (Krell 2006). The first scarab beetle was described from Messel by Meunier as early as 1921. A species of stag beetle, *Protognathinus spielbergi* Chalumeau & Brochier, 2001, was described on the basis of a published photograph. Many stag beetles immediately catch the eye due to the males' greatly enlarged upper mandible, which no longer serve the intake of food but are used during fights over females.

The beetle family Psephenidae has only been documented in Messel by larvae; no adults have been found to date. All of the currently more than 130 specimens from Messel appear to belong to a single species within the subfamily Eubrianacinae (Fig. 7.55); however, a more precise identification is not possible (Wedmann et al. 2011). The aquatic larvae of the Psephenidae, also known as water pennies, are rounded and strongly flattened, an adaptation to living in frequently turbulent waters, e.g., in rivers or in the rocky shore regions of large lakes, where they graze algae from the underside of rocks. The

Insects (Insecta, Hexapoda)

Fig. 7.55: Water-penny beetle (Psephenidae), aquatic larva, body length 6.2 mm.

high number of discoveries at various sites within the Messel Pit indicates that they occurred in the lake. This implies that Lake Messel had a rocky shoreline, at least in part, and was larger than previously assumed (Wedmann et al. 2011). Recent species of Eubrianacinae are found in Asia, Africa and North America. From Europe, a second fossil record of this group has been found in an Eocene site in France.

Jewel beetles (Buprestidae) live up to their vernacular name, since many species show a striking metallic iridescense. This is also true in Messel, where the structural colors have been preserved with their original gloss. About 16,000 recent species have been described. Many are thermophilic, which leads to a higher diversity of this group in the tropics and subtropics than in the temperate zones.

Many representatives of the subfamilies Buprestinae and Agrilinae are known from Messel. Agrilinae tend to be slender in appearance and often show a golden color, with characteristic wrinkled surface structures on their pronotum (neck shield) and elytra. At least 20 species of Agrilinae can be distinguished in Messel (Wedmann & Hörnschemeyer 1994). Among the Buprestinae, an even larger number of forms has been recorded in Messel (Fig. 7.56, Fig. 7.57), some of which could be identified to the species level. Two species, *Lampetis weigelti* and *L. transversovittata* (Fig. 7.57 left), are also known from Geiseltal (Hörnschemeyer et al. 1995). Since identical complex color

Fig. 7.56: Jewel beetle (Buprestidae) of the genus *Anthaxia*. Left: unidentified fossil *Anthaxia* species, body length 7 mm; right: recent species *Anthaxia concinna*, body length 6.5 mm.

Fig. 7.57: Jewel beetle (Buprestidae) of the genus *Lampetis*. Left: the fossil species *Lampetis transversovittata*, body length c. 22 mm; right: the recent species *Lampetis gorilla* with a body length of 21 mm.

Fig. 7.58: Extinct click beetle *Macropunctum meunieri*, holotype, body length 14.5 mm, transferred to synthetic resin.

patterns are found in the species from Geiseltal, from Messel, and in recent species, it can be assumed that the original colors and patterns are actually preserved here (Hörnschemeyer & Wedmann 1994).

Click beetles (Elateridae) possess the ability to suddenly snap back on their feet from a prone position on their back by means of a special click apparatus in their thoracic region. Among the many click beetles found in Messel, the extinct genus *Macropunctum* Tröster, 1991 is particularly common (Fig. 7.58). It owes its name to the numerous white dots distributed across the entire body. To date, nine species of *Macropunctum* have been described from Messel (Tröster 1991, 1999). It is of particular interest that *Macropunctum* has also been found in another fossil site, the Eckfelder Maar, which is several million years younger (Wappler 2003b). However, *Macropunctum* is not known from any other fossil site. The genus appears to have enjoyed a brief blossoming during the Eocene in Europe; it was probably never widespread and presumably became extinct near the end of the Eocene for climatic reasons. Representatives of recent click beetle genera are also known from Messel. These include *Agrypnus*, *Lacon* and *Lanelater* (Fig. 7.59), which, like *Macropunctum*, all belong to the Agrypnini and constitute approximately 70% of the click beetles in Messel (Tröster 1994). Agrypnini are recognized by their antennae, which can be concealed in special grooves on the underside of their thorax.

Darkling beetles (Tenebrionidae) are a large family with a worldwide distribution; some 18,000 species have been described to date. Their species diversity is highest in the arid regions of South America, Asia and Africa. Some of the species there show metallic

Fig. 7.59: Click beetle of the genus *Lanelater*. Left: the fossil species *Lanelater verae*, paratype, body length 21 mm; right: the recent species *Lanelater fuscipes*, body length 34 mm.

colors, but the majority is dark to black. Many species bear a striking resemblance to beetles from other families. Therefore, darkling beetles can be difficult to identify in the fossil record, since the finer structural details are not always very apparent. One species of darkling beetle, *Ceropria*? *messelense*, has been described from Messel, where it occurs regularly in many layers (Fig. 7.60). This species presumably lived in the near vicinity of Lake Messel. Recent relatives seem to develop in fungi that grow on trees; adult beetles have been found under the bark of trees (Hörnschemeyer 1994).

Beetles (Coleoptera): Various plant eaters

Comprising around 35,000 known species, the longhorn beetles (Cerambycidae) can be found on all continents. They have an elongated body shape and usually bear very long antennae. The larvae of longhorn

Fig. 7.60: Darkling beetle (Tenebrionidae) of the species *Ceropria messelense*, body length 20 mm.

beetles feed on wood; some adults do not eat at all, while others suck tree sap or feed on pollen. The intestinal contents of a longhorn beetle from Messel reveal that it had eaten pollen shortly before its death (Fig. 7.61). Insects that allow such ecological conclusions regarding food plants are rarely found. Next to the weevils, the leaf beetles (Chrysomelidae) are the most species-rich group of beetles. As their name implies, leaf beetles generally have a vegetarian diet. Their host species are almost exclusively higher plants (flowering plants), and the leaf beetles' evolution shows close connections to that of the plants. Similar to the weevils, this is a very diverse group with many species, especially in the tropics. They are often remarkably colorful, due to the presence of both pigment and structural colors. The specimens from Messel show impressive original structural colors, but they are often missing legs and antennae, which are required for a more detailed identification (Fig. 7.62). The subfamily Sagrinae, which today occurs mainly in Southeast Asia and across the Southern Hemisphere, has been documented from Messel. Its members can be recognized by their noticeably thickened hind legs (Fig. 7.63), which are not used for jumping, but rather in fights. The metallic-green body is also characteristic for this group.

The weevils and their relatives (Curculionoidea) comprise four families of beetles which usually have the head prolonged into a snout or beak. The mouthparts, which are often small and somewhat modified, are located at the front end of the snout. Most weevils are vegetarians and usually feed on plants, and rarely on decayed wood as well. More than 75,000 recent species have been described. With more than 57,000 species found around the globe, the true weevils (Curculionidae) are the largest family among the weevil-like beetles. Their antennae are usually geniculate (elbowed), i.e., similar to ants, a markedly elongated first segment is followed at an angle by several much shorter segments. The group can be divided into long-nosed weevils with a very long, thin, curved snout, and short-nosed weevils with a short, thick and flattened snout (Dathe 2005). Their body length can vary from less than 1 mm to about 60 mm. In addition to large species, many smaller weevils are also represented in Messel (Fig. 7.64), some of which only measure ½ mm in length. The larger specimens are up to 20 mm long. Only four species have been described from Messel, with one species distinguished by a metallic, light blue coloration (Rheinheimer 2007). Straight-snouted weevils in the narrow sense (Brentidae) are a heterogeneous group. The straight-snouted weevils are distinguished by their rather slender body. They are represented in Messel by just under 20 specimens, while the related Apionidae are found much more frequently.

There are a few additional groups of beetles in Messel, which are not covered in detail here. These include the throscid beetles (Throscidae), false click beetles (Eucnemidae), cylindrical bark beetles (Colydiidae) and hister beetles (Histeridae) as well as a single representative of the bark-gnawing beetles (Trogossitidae) described by Meunier (1921).

Fig. 7.61: Longhorn beetle (Cerambycidae) with preserved fossil pollen remains in the intestinal tract, body length 22.5 mm.

Fig. 7.62: Leaf beetle (Chrysomelidae). Left: fossil leaf beetle, body length c. 7 mm; right: the recent species *Chrysochloa tristis*, body length 11.5 mm.

Fig. 7.63: Leaf beetle (Chrysomelidae) from the subfamily Sagrinae. Left: fossil leaf beetle, body length 17 mm; right: the recent species *Sagra mutabilis* with a body length of 20 mm.

Caddisflies (Trichoptera)

The aquatic larvae of many caddisflies build a case using grains of sand. These cases protect the delicate body of the larva. Adult caddisflies resemble small moths, have hairs on their wings and special, modified mouthparts.

Caddisflies are almost exclusively represented by larval cases in Messel, where they are among the most common fossils. Their abundance varies from one excavation site to another, which was seen as an indication of transport via an inlet (Lutz 1990). Recently, it has become apparent that sand cases can be common not only in the northwest, but also in the eastern part of the pit, while they are entirely missing from the layers in the former center of the lake (Sonja Wedmann, *unpubl. data*).

The material used for the cases in Messel is mainly restricted to grains of sand (Fig. 7.65); in rare cases,

Fig. 7.64: Diversity of form among weevils in Messel (Curculionoidea). Body lengths: top left 11 mm, top right 8.5 mm, center left, body length 6 mm (without antennae), center right 7.5 mm (including head), bottom left 3 mm, bottom right 6 mm.

Fig. 7.65: Caddisfly case made of grains of sand (Trichoptera), case length 13 mm.

Fig. 7.66: Caddisfly larva in a silk case (Trichoptera), length of fossil 5 mm.

plant parts or spicules were used (Martini & Richter 1996). Contrary to the findings by Martini & Richter (1996), larval cases made exclusively of silk are found quite regularly, albeit nowhere near as commonly as cases made of sand. Rarely, accumulations of silk cases have been found (Lutz 1990). Only very few of the larvae that lived inside the cases have been preserved (Fig. 7.66).

Butterflies and moths (Lepidoptera)

Modern lepidopterans represent one of the most species-rich groups of insects. Although there are discoveries of basal lineages from the Mesozoic, the diversity of today's species arose primarily with the radiation of flowering plants and especially during in the Cenozoic (Grimaldi & Engel 2005). There are relatively few butterfly fossils, which is probably due to taphonomic reasons; butterfly wings are quite water-resistant and usually have a large surface area, which prevents them from easily sinking below the water's surface.

In Messel, lepidopterans are rare as well. A few specimens represent moth-like species with a glossy yellow-green color, which are documented by body fossils as well as fossil feces (coprolites). These are probably members of the forester moths (Zygaenidae) (Fig. 7.67) and belong to at least two different species. By means of ultra-structural analyses, it could be demonstrated that their structural colors underwent only slight changes during fossilization and originally were of a bright yellow-green hue (McNamara et al. 2011).

The occurrence of lepidopterans was further documented by the analysis of stomach contents, which included characteristic scales of moths (Micropterigidae) and perhaps also of macrolepidopterans (Richter 1993; Habersetzer et al. 1994). Traces of mines in a leaf and coprolites indirectly document the occurrence of two additional groups of butterflies (Kinzelbach 1970b; Schaarschmidt & Wilde 1986).

Fig. 7.67: Butterfly, probably a forester moth (Zygaenidae). Top: complete fossil, body length c. 9 mm; bottom left: greatly enlarged SEM image of the scales, showing the detailed exterior structures, image width 8 μm; bottom right: TEM image of the scales, showing the interior layering, image width 4 μm. Copyright for both bottom images: McNamara et al. 2011, PLOS Biology.

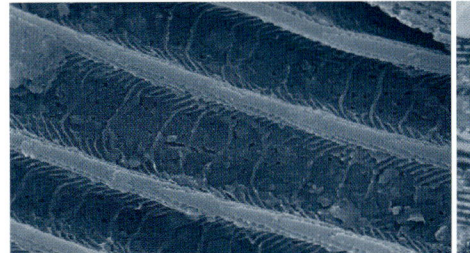
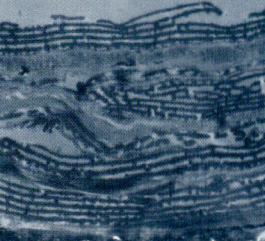

Flies (Diptera)

The "two-winged" insects or dipterans (Diptera) only possess a single pair of wings; the hindwings are reduced to so-called halteres, which serve sensory functions. A further characteristic trait is the presence of sucking mouthparts in many species, most notoriously exemplified by the mosquitoes (Culicidae). Modern-day flies belong to the mega-diverse insect orders with far more than 134,000 described species. Although the dipterans were presumably common and occurred with many species around and – as larvae – inside Lake Messel, they are clearly underrepresented in the fossil record. The collection includes barely over 150 specimens. This has taphonomic reasons, since the small, lightweight bodies of gnats and flies float on the water surface for a long time and only rarely sink below.

Analyses of fossilized fish feces (coprolites) have shown that a large number of aquatic midge larvae populated Lake Messel, serving as a main food source for young fishes. In particular, phantom midges (Chaoboridae) that inhabit the open water areas far from shore are very common. But near-shore dwellers such as the larvae of mosquitoes (Culicidae) and the usually bottom-dwelling chironomids (Chironomidae) were also eaten by fishes, as evidenced by fragments found in the coprolites (Richter & Baszio 2001, 2002; Richter & Wedmann 2005; Wedmann & Richter 2007). There are only few specimens of adult midges, including representatives of the craneflies (Tipuloidea), which have a slender body and remark-

Fig. 7.69: March fly (Bibionidae), female of the species *Plecia hoffeinsorum*, body length 4.5 mm.

Fig. 7.70: Dark-winged fungus gnat (Sciaridae), body length 1.5 mm.

Fig. 7.68: Crane fly (Tipuloidea), body length 10 mm.

ably long legs (Fig. 7.68). Three species of March flies (Bibionidae) are known (Fig. 7.69), of which one species is also documented in Baltic amber (Skartveit & Wedmann 2015). Very rarely found are fungus gnats (Mycetophilidae), biting midges ("no-see-ums") (Ceratopogonidae), dark-winged fungus gnats (Sciaridae) (Fig. 7.70) and other groups of midges (Victor Baranov, *pers. comm.* 2016).

The higher flies (Brachycera), which are distinguished, among other things, by antennae with ten or fewer segments, are represented in Messel by relatively few specimens but many species. There are representatives of the partly blood-sucking horse flies (Tabanidae), the snipe flies (Rhagionidae), the often colorful and glossy soldier flies (Stratiomyidae) (Fig. 7.71), the robber flies (Asilidae) and the hoverflies (Syrphidae). *Comptosia pria*, a species from the bee fly family (Bombyliidae), which have a parasitic development, has been described from Messel. Based on characteristic features in the wing venation, an assignment to the recent genus *Comptosia* could be confirmed (Fig. 7.72). This genus is represented today by more than 50 species in Australia, and closely related species occur in South America (Wedmann & Yeates 2008). One of only six Cenozoic fossil discoveries involves the record of a member of the family of tangle-veined flies (Nemestrinidae) (Wedmann 2007).

Scorpionflies (Mecoptera)

Scorpionflies are distinguished by elongated, beaklike mouthparts. The English name refers to the males of some groups which have genital appendages which resemble a scorpion's stinger. They are a small group with a worldwide distribution and many extinct families in the fossil record. Only a single fossil is known from Messel (Fig. 7.73).

Paleobiogeography of the insects in Messel

The insects in the Messel Pit document a wide range of paleobiogeographic relationships. Representatives of the mantidflies (Mantispidae, Neuroptera), leaf insects (Phasmatodea) and the bee flies (Bombyliidae, Diptera) constitute the first fossil records of these groups outside their current range. Many insects from Eocene or Paleogene fossil sites belong to re-

Fig. 7.71: Soldier fly (Stratiomyidae), wing length 5.5 mm.

Fig. 7.72: Bee fly of the genus *Comptosia* (Bombyliidae). Top: fossil species *Comptosia pria*, holotype, wing length 10 mm; bottom: pair of the recent species *Comptosia vittata*, perched on burnt eucalyptus seeds. Photo: copyright Jiri Lochman/Lochman LT; David Yeates identified the species.

Box 7.3: More than just chewed-up leaves – Plant-insect interactions

Plant-eating (herbivorous) insects leave behind a variety of feeding traces. These include traces such as hole feeding and chewed-up edges on leaves (margin feeding), foraging traces on the leaf surface, or skeletonization – leaving behind only the sturdier veins of the leaf. Characteristically shaped feeding tunnels (leaf mines) are created when the larvae of butterflies or beetles feed on the inner green leaf substance. Plant galls are another example of these interactions; they are an outgrowth of plant tissue frequently triggered by insects that offer developing larvae protection against the weather and predators.

The study of plant-insect interactions in the fossil record offers insights into the long-term development of these relationships and food webs. The quantification of the diversity of interactions at various fossil sites across different geological eras reveals information about the climate and climate change (Labandeira & Currano 2013).

For example, numerous small plant galls were found on fossilized laurel leaves in Messel (top right on *Laurophyllum lanigeroides*; scale: 10 mm), which were likely caused by gall midges (Wappler et al. 2010). Gall midges (Cecidomyiidae) are small midges with a current world-wide distribution that are capable of causing economic damage by infesting cereal grains.

The overall diversity of plant-insect interactions discovered in Messel is higher than expected. In addition to plant galls, various leaf mines and traces of hole feeding were found (bottom and top left; scales: 10 and 5 mm, respectively). In Messel, the proportion of plant-insect interactions is clearly higher on evergreens than on deciduous plants, which is explained by the atmospheric carbon dioxide content that was about 2.5 times higher than today (Wappler et al. 2012).

Fig. 7.73: Scorpionfly (Mecoptera), body length 20 mm (without beak).

cent genera or are closely related to genera that are currently found in subtropical and tropical latitudes. For example, the closest relatives of the mantidfly from Messel are today restricted to South America and southern North America. Leaf insects are now distributed in Southeast Asia and adjacent regions, and wood wasps of the genus *Xoanon* (Siricidae, Hymenoptera) are found in Eastern Asia today. Modern tree-dwelling *Gesomyrmex* ants occur in Southeast Asia, but they were recorded from Europe and Russia in the Eocene. Bulldog ants (Formicidae, Hymenoptera) are now restricted to Australia, but occurred widely across Europe and America during the Eocene. Bee flies of the genus *Comptosia* are exclusively found in Australia in modern times. *Comptosia* and its close relatives from South America were taken as a typical example for a Gondwana distribution. However, fossil records of representatives of *Comptosia* in Messel and in the Eocene fossil site Florissant in North America show that this taxon was formerly much more widespread than it is today, and that the apparently Gondwanan distribution of today represents a relic distribution (Wedmann & Yeates 2008).

In case of the mantidflies (Mantispidae), a paleobiogeographic scenario is being discussed that could serve as a possible explanation for the current distribution patterns of the subgroups of the Mantispidae. The concept of "ousted relicts," which postulates that formerly widespread groups of animals were replaced by more modern taxa that usually arose in the tropics, may best serve to explain the modern distribution of the subgroups among the mantidflies (Eskov 2002; Wedmann & Makarkin 2007). This concept may also be applicable to other groups.

The obviously wider distribution of many insect groups in the Paleogene is closely related to the globally much warmer climate (e.g., Janis 1993; Zachos et al. 2001; Mosbrugger et al. 2005). Moreover, during the Paleogene, the latitudes of what is now Europe were subject to limited seasonal temperature fluctuations, which not only led to much warmer but also more stable temperatures than today (e.g., Ivany et al 2000; Archibald & Farrel 2003). This warm climate may be one of the factors that favored a high species diversity during the Eocene. The new findings from the sediments of the Messel fossil site show that during the Eocene climate optimum, Europe was home to a large number of insect groups that have since become extinct in the Northern Hemisphere. On the other hand, there are also records of many groups (e.g., the fly *Hirmoneura)* that still occur in Europe today.

Considering the Messel insects investigated to date, there is no indication of close ties among these groups to any specific biogeographic region. However, the overall picture emerges that during the Eocene the species diversity in higher geographic latitudes was many times greater than it is today.

Chapter 8
Actinopterygians – the Fishes of the Messel Lake

Norbert Micklich

All fish species found in Messel belong to the ray-finned fishes (Actinopterygii), the largest and most successful group among the bony fishes. Among others, they are distinguished by the uniform design of their paired fins. There are almost 27,000 species known to date, which can be found both in the ocean and in freshwater habitats. They are distributed from the deep sea to the high mountains, and some species are able to withstand the most extreme conditions. The first actinopterygians are known from the Lower Devonian (Chapter 2). The most important main groups were already present in the Carboniferous.

While the Actinopterygii are considered to form a confirmed monophyletic group, the phylogenetic relations of several subgroups are controversial (Fig. 8.2, following Near et al. 2012). Especially among the higher-ranking forms such as the perch-like fishes (Percomorpha), different phylogenetic trees have been inferred, depending on the type of data they are based on.

The fish fauna of Messel is composed of both basal as well as more advanced actinopterygians (Fig. 8.1). Here, the Holostei (fishes with ganoid scales) count among the basal forms, along with the freshwater eel *Anguilla ignota* and members of the genus *Thaumaturus*. While the eels (order Anguilliformes) are assigned to the Elopomorpha (tarpon-like fishes), the position of *Thaumaturus* is still unresolved; it is probably part of the Clupeocephala ("herring heads").

A similar situation is found among the Messel perches. Newer phylogenetic analyses have led to a splitting of the previous order Perciformes (perch-like fishes), which renders an exact allocation more difficult. Representatives of the genus *Amphiperca* probably belong to the Percichthyidae (temperate perches), while *Palaeoperca* shows certain similarities with the Moronidae (temperate basses). On the other hand, *Rhenanoperca* contains characteristics reminiscent of the Centrarchidae (sunfish).

Range of species

Fishes constitute the majority of the vertebrates from Messel. In certain layers, they make up 80 % of all fossil specimens. To date, eight genera and species are known. The most primitive forms include the gars (Lepisosteidae) and the bowfins (Amiidae). The former are represented today by just two genera (*Atractosteus, Lepisosteus*) with several species, while the bowfins only include one single modern species, *Amia calva*. Both are considered "living fossils" that can be distinguished from other ray-finned fishes by a number of shared, presumably primitive traits.

Gars are represented in Messel by two species. The overwhelming majority of finds involves *Atractosteus messelensis* Grande, 2010. This species shows a close external resemblance to its living rela-

---

Fig. 8.1: Holostei from the Messel formation. Bottom left, gar *Atractosteus messelensis*; top right, "bowfin" *Cyclurus kehreri*. Scale: 5 cm.

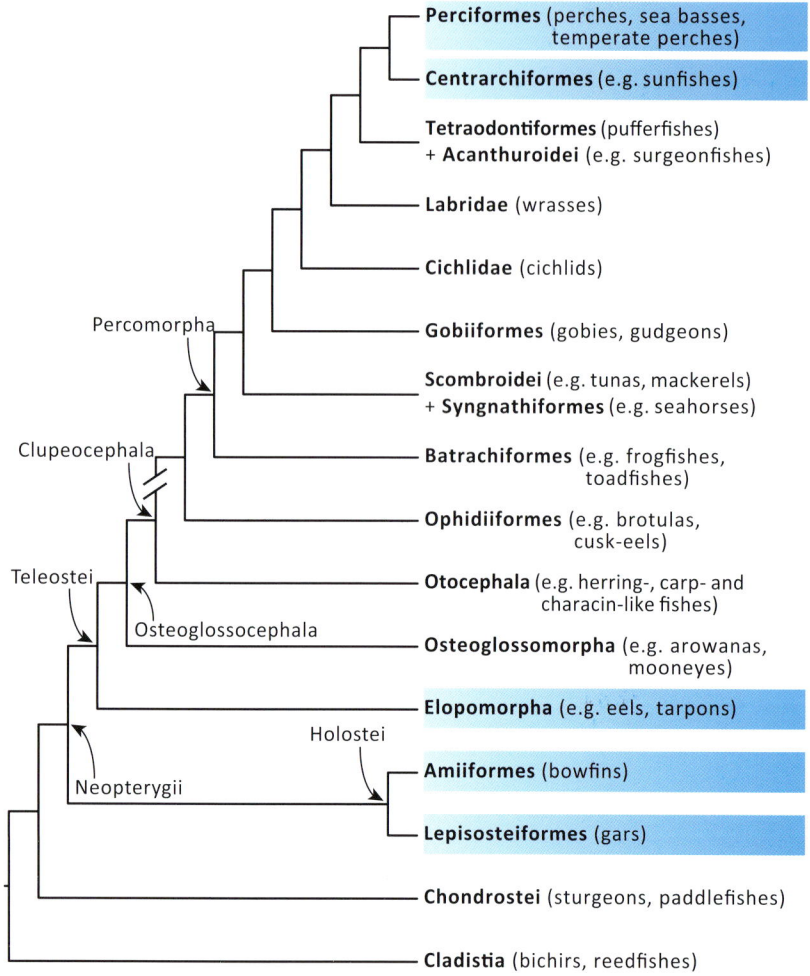

Fig. 8.2: Simplified phylogeny of the higher (top) and basal (bottom) ray-finned fishes (Actinopterygii). Groups known from Messel are marked in blue.

rare. It is characterized by dermal bones that are almost entirely covered with ganoine, an enamel-like substance, and especially by a very short snout whose jaws are primarily equipped with small and blunt rasping teeth (Fig. 8.4).

Cyclurus kehreri (Andreae, 1893) is one of the most common fish species in Messel. It is related to the modern-day species *Amia calva* and, like the latter, is characterized by a massive, cylindrical body and an elongated dorsal fin. Due to their massive skull that is enclosed by numerous sculpted membrane bones, the bowfins also have a rather primeval look. Their vertebral column is of an equally primitive design, as it already begins to bend upward in the anterior are of the tail fin, where it consists of "double vertebrae" with two ossification centers each (diplospondylia). However, bowfins also show more advanced traits that distinguish them from the gars.

The modern bony fishes (Teleostei) are distinguished by a number of shared characteristics that are considered advanced and that differentiate them from all other fishes. In Messel, they are represented by rather primitive as well as more highly developed forms.

The eel *Anguilla ignota* Micklich, 1985 is only known from a single specimen to date (Fig. 8.5). It likely represents a close relative of the modern freshwater eel *Anguilla anguilla* still found in Europe today, which shows a very similar morphology.

Thaumaturus intermedius Weitzel, 1933 (Fig. 8.6) is a species of fish with a more advanced skeletal structure than that of gars and bowfins. The scales and the ossifications of the skull are less massive, and the base of the tail fin shows a near-symmetri-

tives. The massive and strongly sculpted skull bones, the crocodile-like, elongated snout with a series of sharp teeth, and a body armored with chain mail–like rows of ganoid scales impart a particularly primeval look to these fishes. The second species, *Masillosteus kelleri* Micklich & Klappert 2001 (Fig. 8.3), is very

Fig. 8.3: Short-snouted gar, *Masillosteus kelleri*. Scale: 5 cm.

Fig. 8.4: Gars from Messel, details of dentition. *Atractosteus messelensis* (left), *Masillosteus kelleri* (right). Scale: left: 5 mm, right: 1 mm.

Fig. 8.5: Messel eel, *Anguilla ignota*. Scale: 5 cm.

cal design. To date, it is only known from young animals that reached an age of little more than one year (Micklich 2002).

The perch-like fishes are represented in Messel by three species: *Amphiperca multiformis* Weitzel, 1933, *Palaeoperca proxima* Micklich, 1978, and *Rhenanoperca minuta* Gaudant & Micklich, 1990 (Fig. 8.7). At one time, *Amphiperca multiformis* was placed in the genus *Properca* Sauvage, 1880 (Gaudant 2000). However, since the holotype of *Properca* clearly differs from *Amphiperca* in several morphological traits, we retain the original name here. Overall, perches are less common in Messel than the other species. *Amphiperca multiformis* can easily be recognized by its rather deep body and the uniform dorsal fin. The "double-finned" species *Palaeoperca proxima* has an elongated, spindle-shaped body. Its vernacular name refers to the dorsal fin, whose spiny and soft-rayed parts are separated from each other. Like *Thaumaturus intermedius*, *Rhenanoperca minuta* is also represented primarily by very young individuals. The majority of specimens shows massive "grinding teeth" in the pharyngeal jaws, which closely resemble those of the modern "shell-cracker" perches (Fig. 8.8). However, a few also have fine, pointy pharyngeal teeth that are usually characteristic

Fig. 8.6: Primitive bony fish, *Thaumaturus intermedius*. Scale: 5 mm.

Fig. 8.7: Perch-like fishes from the Messel formation: "predatory" perch *Amphiperca multiformis* (top); "double-finned" perch *Palaeoperca proxima* (middle); "shell-cracker" perch *Rhenanoperca minuta* (bottom). Scale: top and center: 1 cm, bottom: 5 mm.

A closer look reveals an exceptional diversity among almost all of the fishes in Messel. In many cases, this clearly exceeds the variability known from closely related fossil or living species (Fig. 8.9). The reason for this may be the fact that Lake Messel existed for an extended period of time. Many of today's long-term lakes are known for being inhabited by groups of closely related species that only show minor differences in their skeletal structure (e.g., Martens et al. 1994). It is possible that the fish fauna in Messel also consists of such groups (Micklich 1996, 2002, 2007c, 2012b, Micklich & Klappert 2004). Even in an isolated system such as a crater lake it may be possible for one or a few original species to evolve into new species within a relatively limited period of time (e.g., Schliewen et al. 2001). However, Lake Messel was a maar, which, at least in its initial stages, was isolated by a ring-wall. This should have prevented the immigration of a widespread stem species as a beginning for this type of speciation in the same location (sympatric). Moreover, the populations are mainly composed of juvenile fishes. In order to facilitate a sympatric origin of new species in the lake itself, several complete generation changes directly on site would have been required. Therefore, speciation in different bodies of water (allopatric) appears more plausible. Here, the different morphotypes developed separately from each other and only came together in the lake at a later date.

A closer look at the microstratigraphic distribution patterns of the different fish species reveals some particularities (Micklich 2012a). In certain layers, there are no records of fishes at all. In other areas, layers with only one or two species alternate with layers that held a more diverse fish fauna. Sometimes, certain species are only found in particular layers. On the one hand, this illustrates that there were numerous modifications in the fish fauna over time. On the other hand, it also shows that it is unlikely that Lake Messel was populated by means of passive mechanisms, e.g., the introduction of fertilized eggs by waterfowl. Apart from the rare occurrence of such findings or the fact that this could only allow the spread of fishes that attach their eggs to aquatic plants and similar structures (adhesive spawners), this would have meant that the birds introduced different species in different numbers and over varying time periods, which seems rather unlikely.

for insect and detritus eaters. Additional information regarding the composition of the fish fauna of Messel and the characteristics of the individual species can be found in Micklich (1988, 2007a, 2007b).

The fish fauna of Messel is clearly less species-rich than that of other fossil sites of a similar age, e.g., the Green River Formation in Wyoming (Grande 1980, 2013). However, this only refers to forms that have been properly described, differentiated and named.

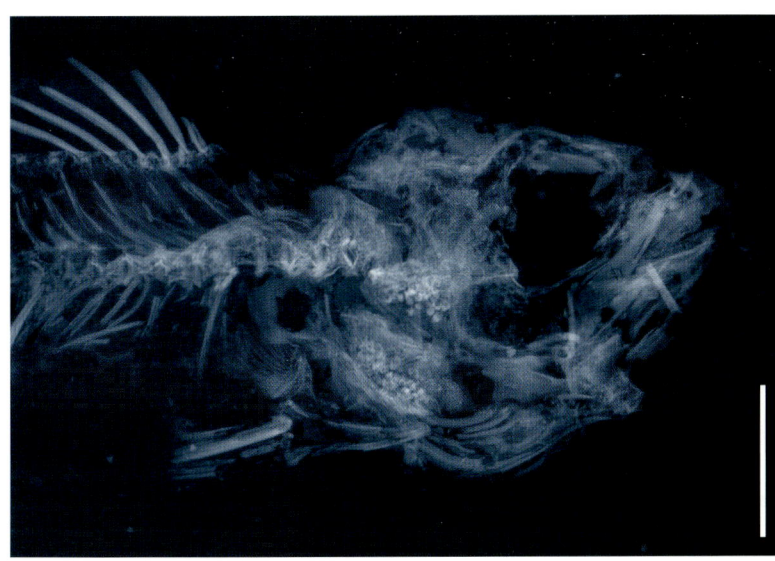

Fig. 8.8: X-ray image of the head and anterior body of *Rhenanoperca minuta*, with clearly visible "grinding teeth" in the pharyngeal area. Scale: 5 mm.

Paleobiology

As at least temporary inhabitants of the former maar lake, the fishes of Messel are further characterized by a number of additional peculiarities, which may offer important information toward the reconstruction of their former habitat. Remarkably, there is a predominance of "survival artists" such as the gars and bowfins. Their modern-day relatives are capable of surviving adverse conditions such as the lack of oxygen during extended hot periods. It can be assumed that their relatives in Messel possessed the same abilities. Even a rarity such as the freshwater eel *Anguilla ignota* fits this concept of intermittent hostile living conditions. It is known that modern eels are highly sensitive and will emigrate if their habitat becomes subject to adverse conditions, or that they will not even colonize areas with an insufficient water quality.

In addition, there are numerous other indications that the fishes of Messel were able to withstand a wide range of adverse external factors. In this context, it is of particular interest that, contrary to fishes in other localities and despite the often excellent preservation, the fish species at Messel are only rarely preserved together with remnants of their prey. Moreover, based on their jaw morphology and dentition, it appears that both the short-snouted gars and the bowfins as well as the majority of the small perch species of the genus *Rhenanoperca* were specialized on "hard-shelled" prey. Yet, prey of this type is only rarely found in Messel. This lack may also have been the reason that in the few cases where remains of prey could be identified, these came from a range of animals that would not have been expected based on the general characteristics of the respective predators. The small "shell-cracker" perches were often cannibalistic (Fig. 8.10), and both the gars and the bowfins include specimens that were in the process of devouring individuals of *Palaeoperca proxima*, a species of perch comparatively rarely found in Messel.

Fig. 8.9: Variability in *Amphiperca multiformis*. Normal, rounded shape of the tail fin (left); deviating, straight or slightly notched shape of the tail fin (right). Such differences are not known among modern perch species, neither as sexual dimorphism nor any other type of intraspecific variability. Scale: 2 cm.

A phenomenon that is particularly difficult to explain is the exceptionally high frequency of scale regeneration. In almost all of the Messel species, this significantly exceeds that reported from closely related or ecologically comparable extant forms (Fig. 8.11). This is also astonishing since this type of regeneration usually only occurs when the respective scales were previously lost. This may be due to various causes. However, at least at first glance, none of these are able to explain the heavy loss of scales (which had to be survived, in the first place) among the Messel fishes. The reasons are not yet fully clear, but one thing seems certain: Both in the fossils and in the recent species, it appears to follow the same patterns and may be due to similar causes. However, the Messel species were much more severely affected than their modern counterparts. For example, it is possible that the scales were lost during the initial immigration into the lake, and subsequently during later, possibly regular forays into obstacle-rich regions.

With few exceptions, the fishes at Messel are represented by both relatively small as well as larger individuals. It seems reasonable to interpret these as juveniles and adults, respectively. A look at the growth rings visible on some of the scales, however, reveals that the majority of fish skeletons found in Messel are indeed those of young animals. Even the largest specimens known to date appear not to be fully grown and had not reached sexual maturity. Truly sizeable specimens are only known from scanty remains, if any. This is likely due to the fact that they rarely lived in the lake itself, but mainly occurred in the wider vicinity. Similarly, there are age-gaps in the populations of certain species. For example, *Palaeoperca* and *Amphiperca* are represented almost exclusively by medium-sized specimens, whereas juveniles and very large individuals are only rarely documented. In case of *Rhenanoperca* it even appears that the population consists almost entirely of individuals that hatched in the past year. Additional information about the fishes of Messel can be found in Micklich (1983, 1985); Micklich & Klappert (2001); Micklich (2002, 2005, 2007c, 2012a); Micklich & Mentges (2012).

Paleogeography

The "overland" (i.e., by way of inland waters) immigration potential of the fish species in Messel and other European freshwater localities in the lower and middle Eocene was severely restricted toward the east (Gaudant 1993). However, similar to the mammals (von Koenigswald & Rust 2007; Chapter 12), individual taxa show clear connections to the Green River Formation in Wyoming. For example, the bowfin species *Cyclurus gurleyi* (Romer and Frixell, 1928) is closely related to *Cyclurus kehreri* (Grande and Bemis 1998), and among the gars, *Masillosteus janeae* Grande, 2010 represents a sister species to *Masillosteus kelleri*. Together, they are even placed in their own subfamily, Masillosteinae (Grande 2010). The gar species *Atractosteus simplex* (Leidy, 1873) and *Atractosteus atrox* (Leidy, 1873) also show close relations to *Atractosteus messelensis* (Grande 2010). At least in regard to its type of adaptation, *Eohiodon* appears to correspond with the Thaumaturi of Mes-

Fig. 8.10: *Rhenanoperca minuta*, specimen in the process of devouring a smaller conspecific. Scale: 5 mm.

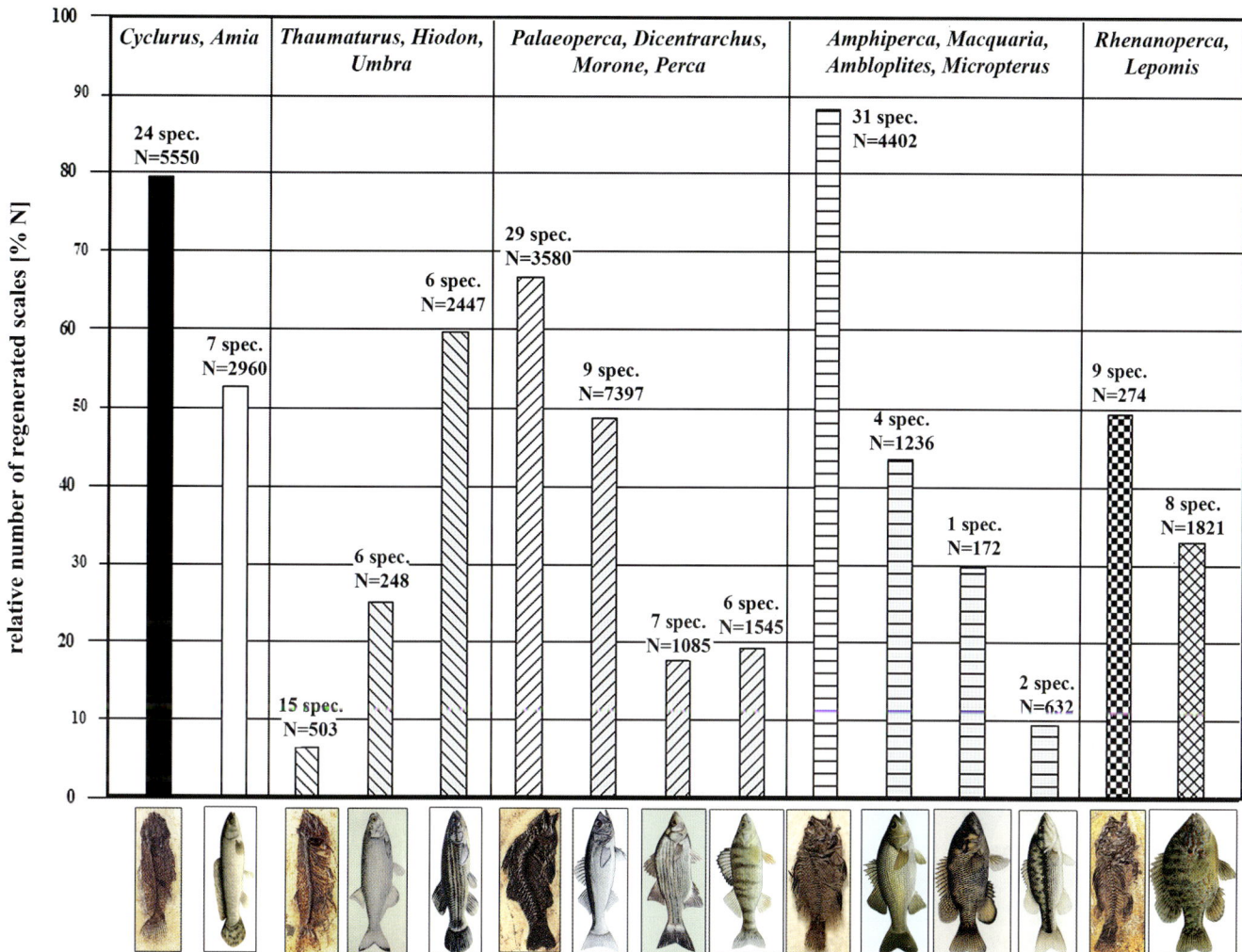

Fig. 8.11: Scale regeneration in Messel fishes and modern comparison species. The sequence of the bars (left to right) in the diagram corresponds with the sequence of the taxa listed in the headings. N = total number of the respective scales examined; spec. = number of specimens examined in the process.

sel, although this genus is usually assigned to the Osteoglossomorpha, and not the Clupeocephala, like *Thaumaturus*. Among the Percomorpha, the genera *Priscacara* and *Mioplosus* also represent adaptation types similar to those found in Messel (*Amphiperca*, *Palaeoperca*), even though the relationships are reversed (Whitlock 2010).

Numerous peculiarities in the fish fauna of Messel can only be explained due to external influences (Micklich 2012a). Thus, at certain times and under certain conditions, a varying number of different species from the presumably fairly homogenous range of species in the vicinity would have entered Lake Messel, e.g., during periods of flooding over already eroded parts of the ring-wall. Depending on the size and type of the respective catchment areas, this would have led to changes in the lake's fish fauna. Once they entered the lake, the fishes would have been subject to competitive pressure from conspecifics as well as to negative conditions such as a diminished water quality during subsequent phases of isolation.

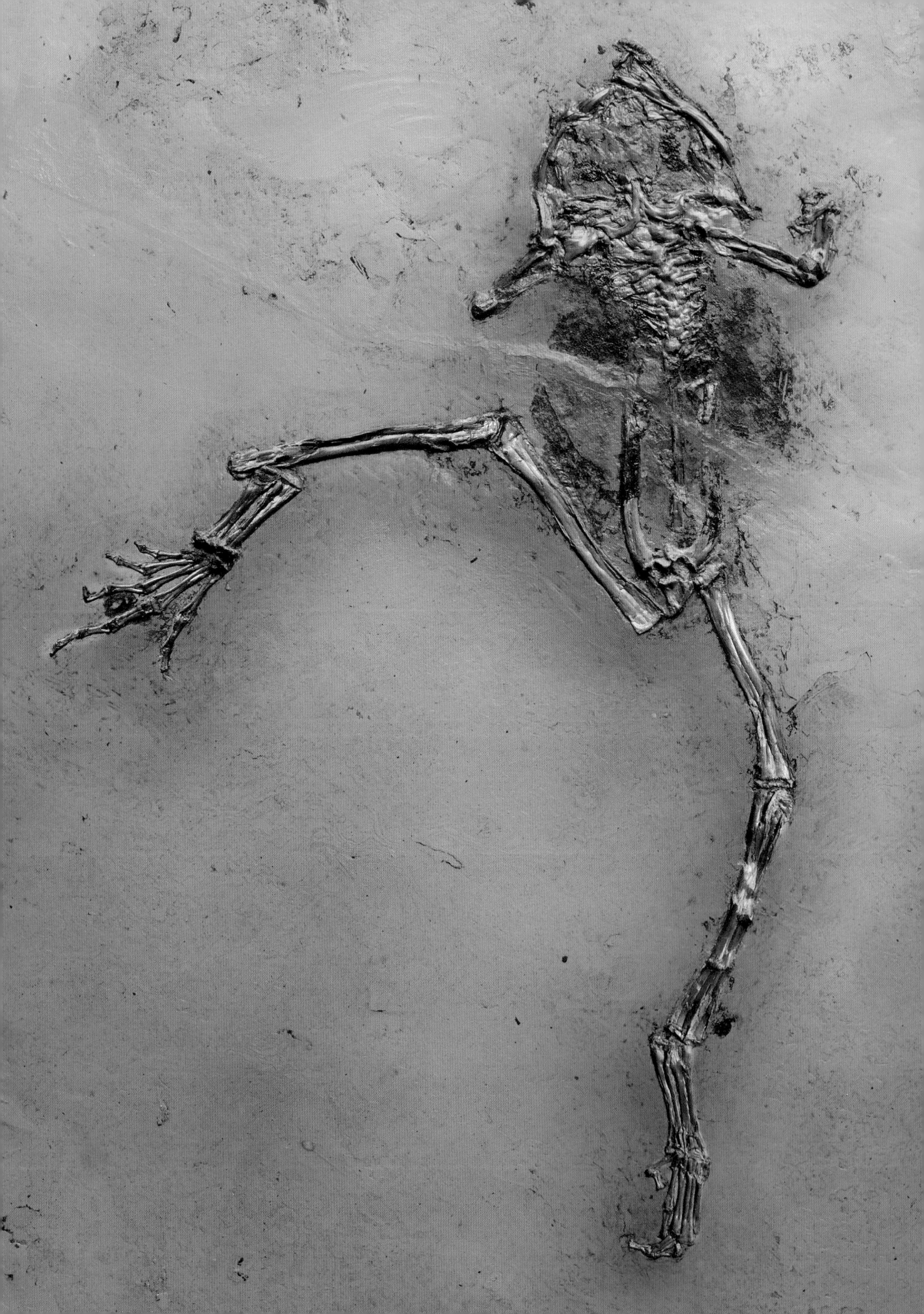

Chapter 9
Amphibians in Messel – in the Water and on Land

Michael Wuttke

Amphibians – Caecilia (legless, snake-like amphibians), Anura (frogs) and Urodela (salamanders) – form an important part of the world's vertebrate fauna (Fig. 9.2). They are distinguished from the other terrestrial vertebrates by their dependence on bodies of water for reproduction. More than 7,000 species have been described to date, and this number increases each year. Amphibians can be found in almost all terrestrial and freshwater habitats, with the exception of the coldest as well as the driest and most arid ones.

Frog fauna

With only three recorded species, the frog fauna of the Messel Lake is poor in species. There are specimens from only two families of frogs, from young animals to very old individuals. It is particularly noticeable that practically all early, post-embryonic development stages are missing from the fossil record. Only one specimen of a tadpole has been documented (Fig. 9.5), along with several larval stages of a water-dwelling species whose metamorphosis had not yet been fully completed.

Terrestrial: *Eopelobates wagneri*

Eopelobates wagneri, a member of the family Pelobatidae (spadefoot toads), is documented by almost 200 specimens, from young frogs to fully grown adults (Fig. 9.1). It can reach a head-body length of more than 10 cm (Wuttke 2012b). In the past, this extinct frog genus was not only widely distributed in Europe but also occurred in North America (Roček et al. 2014). It can be assumed that the two faunas were connected via a high-latitude land bridge (Chapter 2).

In order to reconstruct the former habits of *E. wagneri*, it is necessary to take a closer look at the skeleton, whose design can reveal characteristic adaptations to the preferred habitat, such as dry land or a body of water. Of note is the skull with heavy dentition in the upper jaw, whose individual parts are either fused or meet along wide contact areas that offer support against twisting. At the same time, the cranial roof, the upper jaw and the vertebrae are strongly sculptured by small grooves and bone ridges. The pelvis rests upon wing-shaped extensions of the pelvic vertebrae. Long spinous processes of the dorsal vertebrae as well as their surface texture indicate the presence of a strong dorsal musculature. Fully grown, the lower legs surpass the upper legs in length. In some specimens, soft-part preservation (Chapter 4) shows individual structures that point to the presence of a very horny skin.

Overall, the skeletal features as well as the dermal structure suggest a predominantly terrestrial lifestyle. The robust skull with its pronounced dentition made it possible to overpower even stronger prey animals such as small vertebrates. While the proportions of the hind legs indicate a well-developed ability

Fig. 9.1: *Eopelobates wagneri* with preserved soft parts in the torso and leg area. Scale: 5 cm.

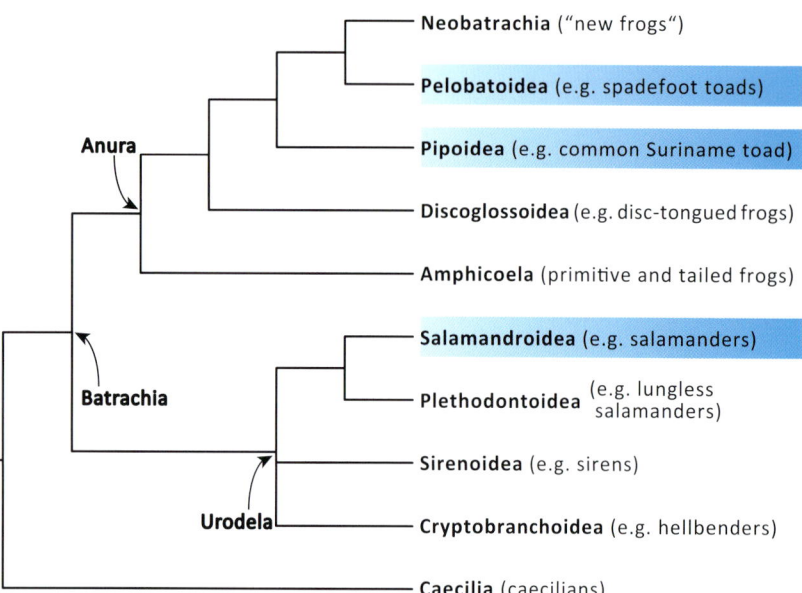

Fig. 9.2: Simplified phylogenetic tree of the amphibians. Groups found in Messel are marked in blue.

to jump, this was in fact restricted by the placement of the pelvis on the wide extensions of the pelvic vertebrae. Therefore, it is rather more likely that the animals moved in a toad-like, walking or hopping fashion.

In a small number of specimens, food remains have been preserved as well. As already indicated by the skull structure, the prey of *E. wagneri* included vertebrates (in this case, reptilians) (Keller & Wuttke 1997). In the first specimen, the preserved remnants could not be identified in further detail; the remaining two specimens have not yet been fully studied. In addition, several specimens contained preserved fragments of insects, such as the elytra of beetles (Fig. 9.3). In one case, an animal even defecated on the bottom of Lake Messel; the feces were like forced from the body by decomposition gases (Fig. 9.4) and, as far as visible under the binocular microscope, primarily contain insect remnants.

With preliminary reservations, the only tadpole documented to date can also be assigned to the species *E. wagneri*. This is mainly supported by the bones of the cranial roof, which even in this stage are already mostly fused with each other (Fig. 9.5). A more detailed examination is currently still pending.

Aquatic: *Palaeobatrachus tobieni*

Contrary to the terrestrial *Eopelobates*, the strictly aquatic frog species *Palaeobatrachus tobieni* (family Palaeobatrachidae, Wuttke et al. 2012) has been documented much more sparingly, with with fewer than 50 specimens known. This European frog family became extinct near the beginning of the Quaternary (Villa et al. 2016); this was likely due to the seasonal freezing of its habitat. It is closely related to the tongueless frogs (Pipidae) from Africa and South America and in particular with the African clawed frog *Xenopus* (Roček et al. 2014).

P. tobieni was a medium-sized frog (head-body length up to 7 cm, Fig. 9.6) with

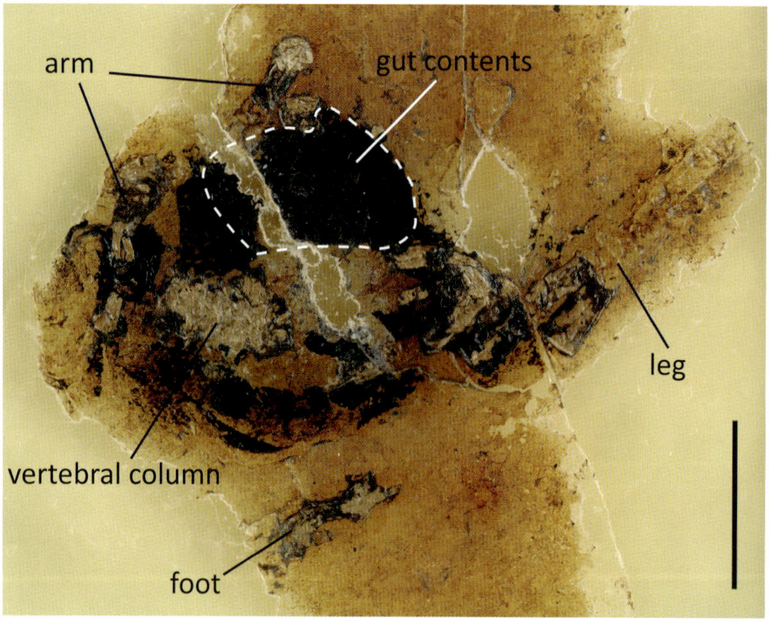

Fig. 9.3: Image of a headless, large *Eopelobates* with large soft part structures with a rounded outline in the torso area; intestinal area (dashed line) with insect parts. Scale: 2 cm.

Fig. 9.4: *Eopelobates wagneri*; inset: coprolite with insect remnants. Scale: 2 cm.

Then it propelled itself from its cover by a powerful backstroke of the hind legs and sucked the prey into its mouth, similar to the modern-day clawed frogs. Larger prey was pushed deeper into the mouth with the aid of the forelegs. During the back stroke, the significantly enlarged webbing between the long toes enabled a rapid forward acceleration. As documented by several discoveries from Messel, the prey of *P. tobieni* mainly included small crustaceans and aquatic insects. It probably also took young fish, although this is only documented for this genus by one specimen from the significantly younger Enspel Maar (Wuttke & Poschmann 2010).

The sole remnant of a frog larva, which already shows the incompletely ossified front and hind legs while still carrying the fully developed tail of a tadpole, probably represents this species (Fig. 9.7). Unfortunately, an exact assignment is not possible since diagnostically important parts of the skull bones were lost during the splitting of the fossil.

only a few sharp, backward-pointing teeth in its upper jaw. The skull bones are mostly reduced, and many of them were merely connected by tendons. Only the bones of the cranial roof are entirely fused, which is a characteristic trait of this frog family (Roček et al. 2015). Of special note are the extremely elongated fingers and toes, which on the hind feet were connected by webbing. In this frog genus, the extensions of the pelvic vertebrae are also widened in a wing-like shape, including the processes on the ultimate lumbar vertebrae. The predatory *P. tobieni* lurked among aquatic plants, waiting for a prey animal to come within reach.

Fig. 9.5: Tadpole, most likely *Eopelobates wagneri*. Scale: 5 mm.

Fig. 9.6: *Palaeobatrachus tobieni* with preserved soft parts in the torso and leg area, plates A and B (mirror image) were superimposed photographically. Scale: 1 cm.

Fig. 9.7: Frog larva, likely of *Palaeobatrachus tobieni*. Scale: 1 cm.

A new species of the genus *Palaeobatrachus* has been documented in Messel by only three specimens (Fig. 9.8). They show a much heavier bone structure, and their skull is not bent in a U-shape but shows straight and elongated upper jaws. This gives the skull a more triangular impression. However, two of the specimens were heavily compressed during the compaction of the mud on the lake bottom, making it difficult to detect all of the anatomical details during the ongoing study of this new species.

Lutetiobatrachus gracilis, an almost blank canvas

To date, it has not yet been possible to safely assign the third frog species from Messel, *Lutetiobatrachus gracilis* (Fig. 9.9), to one of the frog families (Wuttke 2012a). Recently, doubts have been raised whether the specimen might not in fact represent a young individual of the genus *Eopelobates* (Roček et al. 2014), but there are no comparison specimens available with an individual age similar to that of the present discovery. Other documented young of *E. wagneri* are slightly older and show different skull proportions than *Lutetiobatrachus*. Thus, this question will have to remain unanswered for the time being. It is also not yet possible to offer detailed conclusions regarding the habits of this new species.

As a rule, there are very few records of tadpoles from ancient bodies of water; in this regard, Messel is no exception. However, the exceptionally large number of tadpoles documented from the Enspel Maar in the Westerwald (Germany), dating from the Upper Oligocene (about 25 million years ago) (Maus & Wuttke 2004) and from the Miocene (15 million years ago) Bechlejovice Lake in the Czech Republic (Roček et al. 2012) can be given as counter-examples,

Fig. 9.8: *Palaeobatrachus* sp., a new, hitherto undescribed species for Messel. Scale: 2 cm.

which paint a rather different picture. But even in those two sites, not all frog families found there are documented by tadpoles. Therefore, it appears that other reasons play a role in preventing fossilization in certain lake sediments, provided they actually represented spawning grounds (Chapter 4).

Salamanders

Despite more than forty years of research history concerning the deposits in the Messel Maar, only two fully grown salamanders have been unearthed to date. This is even more astonishing in view of the fact that salamanders are normally exceptionally abundant in shore habitats (ecotone). For example, on Vancouver Island (Canada), a single species has

Fig. 9.9: *Lutetiobatrachus gracilis*, a new species, or a young frog of *E. wagneri*? Scale: 1 cm.

been documented in densities of up to 11,600 individuals/ha (1.1/m^2) (Ovasaka & Gregory 1989). Thus, salamanders significantly contribute to the biomass in many ecosystems, as well as to the transport of material between bodies of water and dry land (Davic & Welsh 2004). High numbers of individuals can also be found in habitats farther removed from open water; however, the fossilization potential in the sediments of a body of water outside the breeding season is extremely small in those cases.

A striking feature of both salamander specimens from Messel is an exceptionally heavily ossified skull, reminiscent of a turtle shell. This similarity was already duly noted by the scientist who named the genus *Chelotriton* (Pomel 1853), which translates to "turtle newt." The older specimen (Fig. 9.10) was found in the early 1970s and described as a new species, *Chelotriton robustus,* by Westphal (1988); "robustus" denotes the heavy ossification of the skull and the spinal column. Contrary to other species of this fossil genus, *C. robustus* has rather short ribs. Its final meal consisted of insects, whose fragmented remains are easily visible under the binocular microscope, although they defy further identification.

Fig. 9.10: *Chelotriton robustus* with insect remains in the intestinal tract (arrow). Scale: 2 cm.

Based on the heavily ossified skull, the latest, not yet scientifically described discovery from the year 2015 (Fig. 9.11) can likely also be assigned to the genus *Chelotriton*. In contrast to the former species, however, its ribs are strongly elongated, usually forked and/or provided with processes on their surface. They show a strong resemblance to the ribs of modern terrestrial crocodile newts of the East Asian genera *Echinotriton* and *Tylototriton*. These salamanders possess venom glands in their skin above the bone processes, which are emptied by the processes once a predator bites into the salamander's body, squeezing it in the process. It is likely that they served a similar function in the Messel salamanders, which constitute a new species for this fossil site.

However, it remains an enigma why the salamander spectrum is restricted to only two species from one genus, along with their extremely sparse record in the overall range of fossil discoveries, despite the presumedly high density of individuals in the shore area in the past. When comparing the frequency of discovery with that from lignite deposits in the Eocene locality of Geiseltal near Halle (Germany), it is noteworthy that there are only four individuals of the genus cf. *Tylototriton* (uncertain affinity) known from the latter location, as well. Only the olms, a group of aquatic salamander relatives, are abundant, with more than 300 individuals from Geiseltal; there is not a single record of these from Messel. However, the Geiseltal olms originate from sediment layers that are younger than the Messel sediments.

The only exception in the fossil record from localities formed in bodies of water are the sediments in Enspel Maar, where the species *Chelotriton paradoxus* (which shows strong anatomical similarities to the new species from Messel) is represented in large numbers. This may be explained by a predominantly terrestrial lifestyle of the Messel species following larval development, while the species from Enspel was likely bound to aquatic habitats throughout its life and only temporarily forayed onto land (Roček & Wuttke 2010; Schoch et al. 2015; Smith & Wuttke 2015). This begs the question whether Lake Messel even served as a spawning ground for these salamanders at all; other conceivable reasons for the rare fossil record may be related to so-called taphonomic filters (Chapter 4).

Fig. 9.11: First specimen of a new species from the genus *Chelotriton*. Scale: 1 cm.

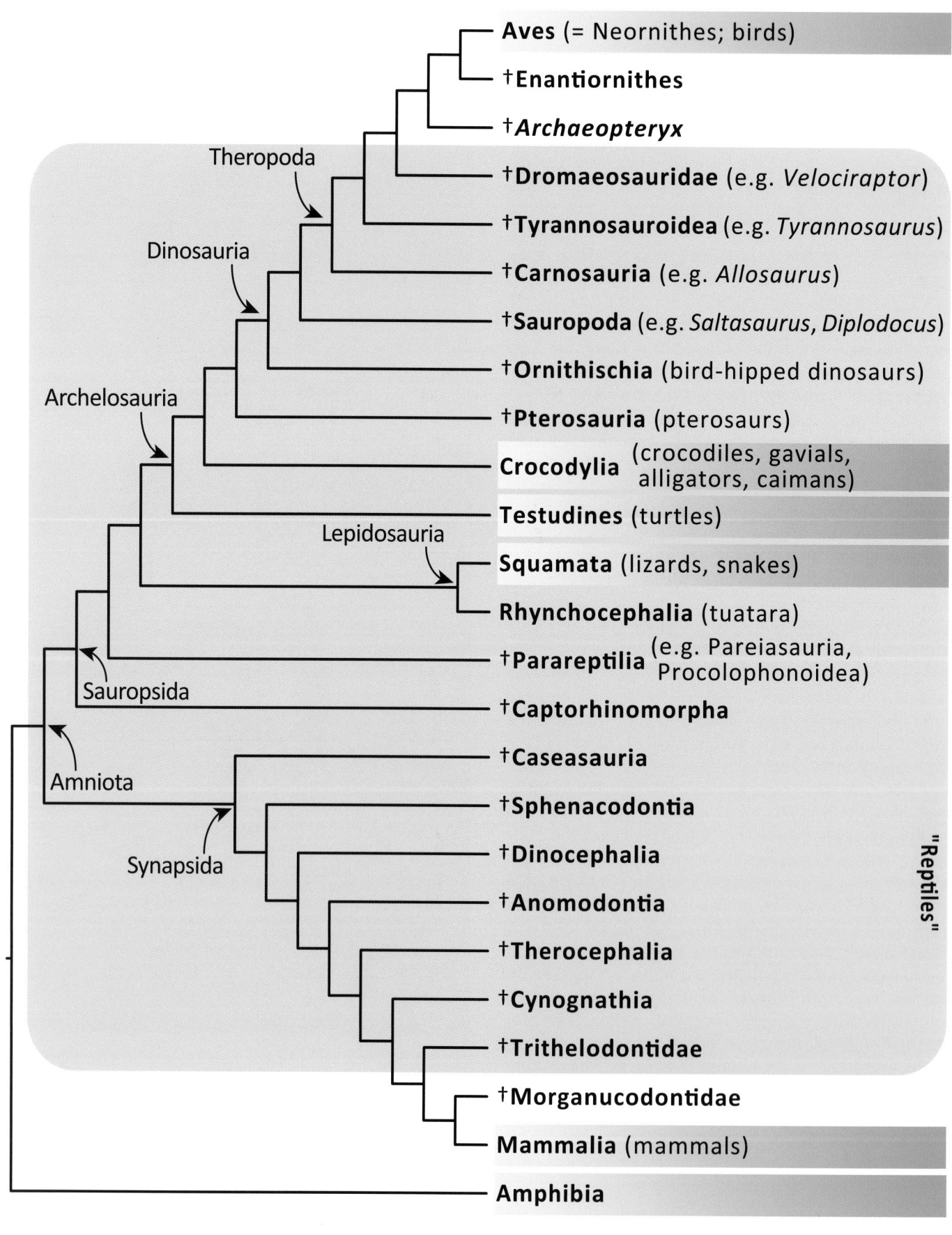

Chapter 10
Amniotes: Mammals, Birds and Reptiles

Editors

Vertebrates treated in the following chapters all belong to the clade Amniota, which includes animals typically called reptiles as well as birds and mammals.

While the transition from water to land in vertebrates required adaptations on a tremendous scale in critical anatomical systems – organs of olfaction, vision, and hearing, of locomotion, feeding and breathing (Clack 2002; Sumida & Martin 1997) – the earliest tetrapods (land vertebrates), like basal vascular plants (Chapter 6), were still dependent on water for reproduction. It is widely thought that amniotes overcame this challenge with two innovations (Carroll 1997). First, internal fertilization allowed sperm and egg cells to combine within the body of the mother. Second, a complex post-fertilization structure, the amniotic egg, evolved. It contains several additional membranes that facilitate gas exchange and waste disposal and allow the embryo to develop in considerable isolation from the environment for an extended time.

The earliest amniotes lived over 300 million years ago (Benton & Donoghue 2007), by which time the clade had split into its two major branches: synapsids and sauropsids. The basal members of both branches were historically classified as reptiles because they were cold-blooded and scaly-skinned and laid eggs – that is, they are amniotes lacking the warm-bloodedness and skin enhancements of mammals and birds. The absence of a trait is a primitive feature and no longer considered useful for defining groups of organisms, but the term "reptiles" remains in common, if informal, use.

The sauropsids include the living Lepidosauria (lizards and snakes, tuatara), Testudines (turtles), and Crocodylia (crocodiles) (Chapters 10) as well as Aves (birds; also called Neornithes) (Chapter 11). It was long thought that turtles are the most basal lineage of sauropsids, partly because they lack the temporal openings in the skull common to all other living sauropsids, but genetic studies strongly support a closer relation between birds, crocodiles and turtles (Roos et al. 2007). It appears that turtles may originally have possessed these openings and lost them at a later date (Bever et al. 2015).

The living sister-group of Lepidosauria is Archelosauria, which also includes crocodiles and dinosaurs. Dinosaurs diversified into a wide range of forms (Brusatte et al. 2010; Benson et al. 2014), and a profusion of fossils – most notably skeletons from Mongolia and feathered dinosaurs from northeastern China – clearly demonstrate that birds belong to this group (Norell & Xu 2005). Dinosaurs underwent a catastrophic extinction at the end of the Cretaceous, an event widely attributed to an asteroid impact (Brusatte et al. 2015), leaving birds as their only living members.

The synapsids include the recent mammals as well as a host of extinct relatives (Chapters 12). Although living mammals are separated by a vast anatomical gulf from the earliest amniotes, these extinct relatives document the stepwise accumulation of mammalian characters (Gauthier et al. 1988), a story that is increasingly supported by extraordinary fossils. Fossils also show a dramatic ecological radiation of stem-group mammals on land, interrupted most notably by the mass extinction at the end of the Permian, which decimated life on land (Fröbisch 2013) and in the oceans.

Fig. 10.1. Simplified phylogeny of the amniotes. Groups known from Messel are highlighted in blue. Groups traditionally referred to as reptiles are circumscribed.

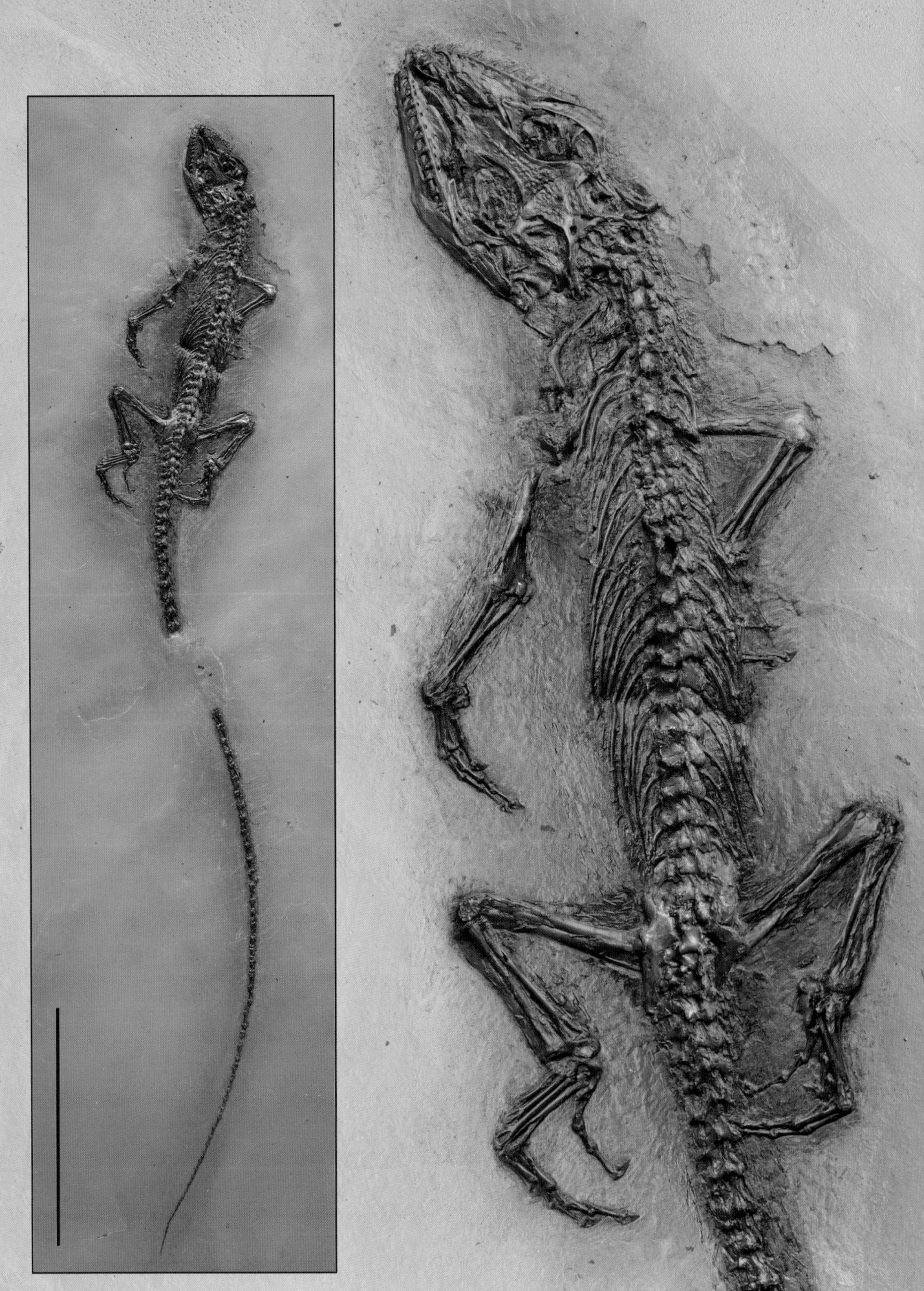

Chapter 10.1
Lizards and Snakes – Warmth-loving Sunbathers

Krister T. Smith, Andrej Čerňanský, Agustín Scanferla, Stephan F. K. Schaal

With over 10,000 living species, Squamata (lizards and snakes) is by far the most diverse group of living reptiles. It is thought that squamates originated in the Jurassic, over 190 million years ago, when dinosaurs also were diversifying (Jones et al. 2013). However, during their earliest history the squamates were overshadowed by their sister-group, the Rhynchocephalia, of which only the Tuatara (*Sphenodon*) has survived to the present. The oldest squamate fossils date to the Jurassic, around 160 million years ago, and only in the Cretaceous do they become diverse (Evans 2003).

That squamates are so species-rich may surprise readers from the middle latitudes, because the strongly seasonal present-day climes of the temperate latitudes have not fostered much diversity there. When we look back to the warm climates of the Eocene, however, we find that mid-latitude squamate communities were much more diverse. At Messel, 19 species of squamates have been found so far, and extrapolations show that many more will still be discovered.

The relationships of squamates to one another are controversial at present (Fig. 10.1.2). Trees based on genes diverge in several crucial respects from those based on morphology (see Chapter 5). First, the morphology strongly supports Iguania (including the clades Iguanidae and Acrodonta) as the basal-most clade in Squamata, whereas genetic data support Iguania as the sister-group to Anguimorpha and Gekkota as the first lineage to diverge. Second, numerous lineages of squamates have acquired a snake-like body-form: snakes are just the best and most successful example of this phenomenon. Because of convergent evolution of the skeleton, all snake-like forms cluster together in phylogenetic analyses of morphology. Only separate analysis of each group, or genes, shows where the snakelike groups actually belong (Gauthier et al. 2012). The phylogeny above is based on the most recent genetic studies. However, the reasons for the discrepancy between morphology and genes has yet to be explained (Gauthier et al. 2012). Fortunately, the interpretation of the relationships of most squamates from Messel is not greatly dependent on the broad-scale tree structure.

Lepidosaurs, and especially squamates, are distinguished by a number of evolutionary novelties (synapomorphies; Gauthier et al. 1988). In squamates, the skull joints are relatively loose, permitting some degree of movement in several places. An elaboration of this characteristic as well as changes to the mechanical properties of the ligaments binding the bones to one another facilitated the radiation of snakes. The exterior part of the skin consists of a two overlapping layers of keratin, which are shed in large chunks, rather than in small flakes. The male copulatory organs are not single, but rather the paired hemipenes; the vent (cloaca) is a transverse rather than longitudinal slit, unlike in mammals. Finally, the ancestral lepidosaur had a fragile tail, that is, a tail that could be shed and regenerated. Additionally, squamates retain many primitive features (symplesiomorphies) in common with fishes and amphibians. For instance, they are ectothermic ("cold-blooded," meaning that their body temperature is strongly dependent on the environment), and their teeth are replaced throughout their lifetime.

The Messel gecko

While the majority of vertebrate specimens from Messel are complete, articulated skeletons, even unspectacular fossils can be highly illuminating. Case in point is the small and unassuming collection of bones shown in Figure 10.1.3, which proved to be the first record of a gecko (Gekkota) from Messel. Geckos are an extremely diverse group of lizards, with at least 1,700 living species mainly distributed in the world's

---

Fig. 10.1.1: Juvenile *Necrosaurus feisti*, a predatory lizard with a powerful body. Scale: 10 cm.

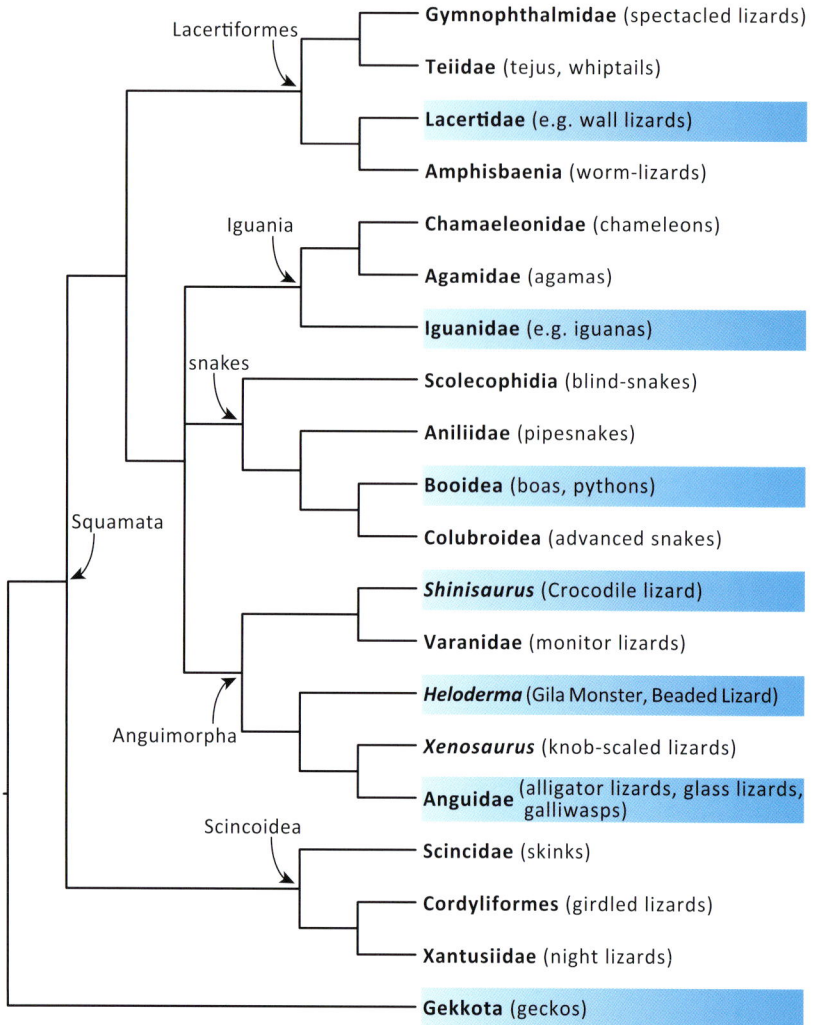

Fig. 10.1.2: Simplified phylogeny of Squamata. Groups known from Messel are highlighted in blue.

skeletons from the 150 million-year-old Solnhofen and correlative limestones of southern Germany and France. Given their age and present diversity, it is therefore surprising that geckos are so rare in the fossil record. In Europe, some Cenozoic fossils have been assigned to various geckos, including leaf-toed forms (Augé 2005; Bolet & Evans 2013; Čerňanský & Bauer 2010).

Like most geckos, the Messel gecko is small-bodied, perhaps 5 cm in snout-vent length (the distance from the tip of the snout to the vent, or cloaca). It shows a number of derived features (synapomorphies) of geckos, such as paired parietal bones, the absence of a parietal foramen (for the lizard "third eye"), and numerous small, simple teeth. Unlike modern geckos, however, it has a well-developed jugal bone and other bones behind the orbit as well as a well-developed supratemporal bar, similar to the Cretaceous stem-gecko *Norellius* (Conrad & Norell 2006).

The skeleton is partly disarticulated, bones of the skull separated, and the arms askew. The pelvis, hind-limbs and tail are missing, and the bone surfaces appear corroded. This suggests that the specimen might have been consumed by a predator, such as a raptor, and then regurgitated. It is the most nearly complete fossil gecko since the Jurassic.

tropical regions. Many of them are nocturnal, but whether nocturnal or diurnal, they have extremely acute vision. They vocalize, an unusual feature among squamates that is the onomatopoetic origin of the name *Gecko*. Special toe pads evolved numerous times in geckos (Gamble et al. 2012) and allow them to ascend smooth vertical surfaces with ease, and even to run across ceilings.

The oldest probable relatives of the group are *Eichstaettisaurus* (Estes 1983), known from several

Ornatocephalus

Ornatocephalus metzleri, named after a generous benefactor of Senckenberg, is one of the largest (>250 g) and most common lizards known from Messel (Fig. 10.1.4-5). It is also one of the most specialized and enigmatic (Weber 2004). Initial studies placed it near the skinks and relatives (Scincoidea), but to date, no consensus has been reached regarding its relationships.

One feature that has tended to obscure a deeper evolutionary understanding of *Ornatocephalus* is the head ornamentation for which the genus was named (*ornatus* - ornamented", *cephalon* - "head"). In contrast to most scincoids and lacertiforms (see Eolacertidae below), the head scales did not consist of broad plates but rather were broken up into many small scales. In addition, an osteoderm (bony plate in the skin) supplemented each of these scales above the skull bones, obscuring them. At the back of the skull, the osteoderms form a jagged crest, similar to many girdled lizards (Cordylidae). Osteoderms are present in the scales above the eyes and in the temple region, similar to Lacertidae, but are otherwise absent.

The limbs of *Ornatocephalus* are relatively short, and the forelimb and hind-limb are equal in length, features that are thought to indicate an arboreal, climbing lifestyle. The penultimate phalanges of the hand are relatively long and curved, reminiscent of primates, while the claws are robust and strongly curved, as in animals that climb and grasp (Weber 2004). The vertebrae have very tall and vertically oriented neural spines, similar to chameleons. The scales are exceedingly well preserved in one recently excavated specimen (Fig. 10.1.4, bottom). The scales of the tail form whorls.

Most noteworthy about the axial skeleton of *Ornatocephalus* is the long tail, which is more than thrice the snout-vent length. In the three complete specimens it is curled up toward its end (Fig. 10.1.4, top). The curling of the tail and the abrupt reduction in the length of the hind-most tail vertebrae, which would lend greater flexibility to the tail tip, suggest that *Ornatocephalus* had a prehensile tail. Modification of the tail into a grasping organ can occur in two ways: the tail can serve as an accessory extremity to facilitate locomotion in the vegetation by grasping branches, or it can be used to suspend the whole body from a branch. In the case of *Ornatocephalus*, its delicacy implies that it was only used to facilitate locomotion (Fig. 10.1.5; Weber 2004).

Large lizards (like other lower vertebrates) tend to be either carnivorous or herbivorous, or at least omnivorous, whereas small lizards are mostly insectivorous. The reasons for this are that, as the predator grows larger, the energetic expense of capturing small, agile prey increasingly outweighs the energy gained from their capture (Pough 1973). Thus, it becomes more efficient to eat plants (which do not run away) or larger vertebrate prey (which provide

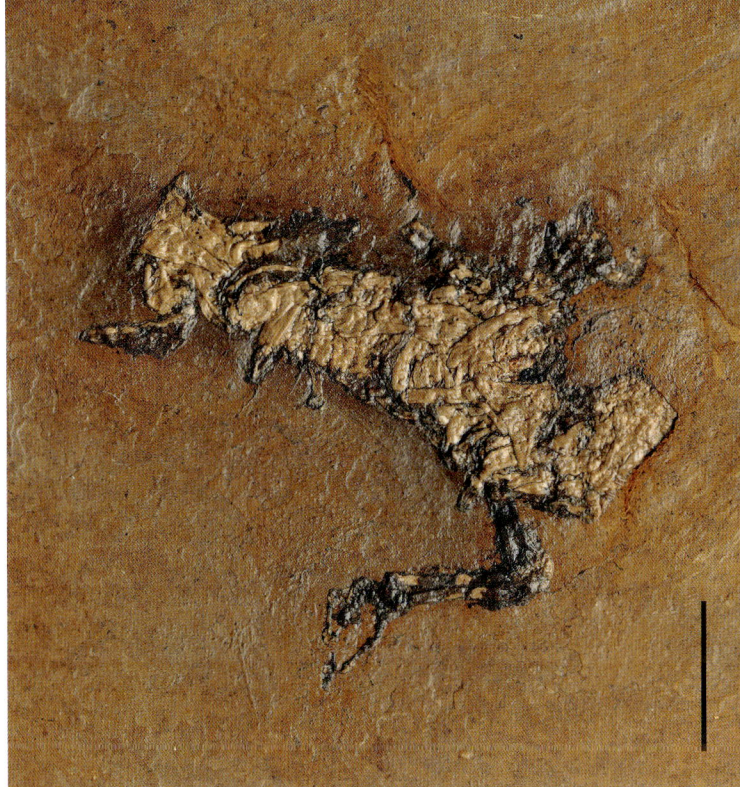

Fig. 10.1.3: First record of a gecko from Messel. This partially disarticulated specimen appears to have been regurgitated. Scale: 1 cm.

more energy per captured animal). In fact, in many squamates the diet changes during individual development from insectivory to carnivory or herbivory. In Messel, several specimens of *Ornatocephalus* include preserved gut contents, which showed that it was indeed omnivorous, taking insect prey and leaves (Weber 2004).

Lacertiformes: the early success

The most species-rich group of squamates known from Messel is related to the Lacertidae, the wall-lizards and allies of Europe, Asia and Africa. The Paleogene of Europe is brimming with lacertiform species known predominantly from isolated jaw bones and a few skull bones (Augé 2005), and one taxon, *Scincoideus*, migrated to North America in the early Eocene

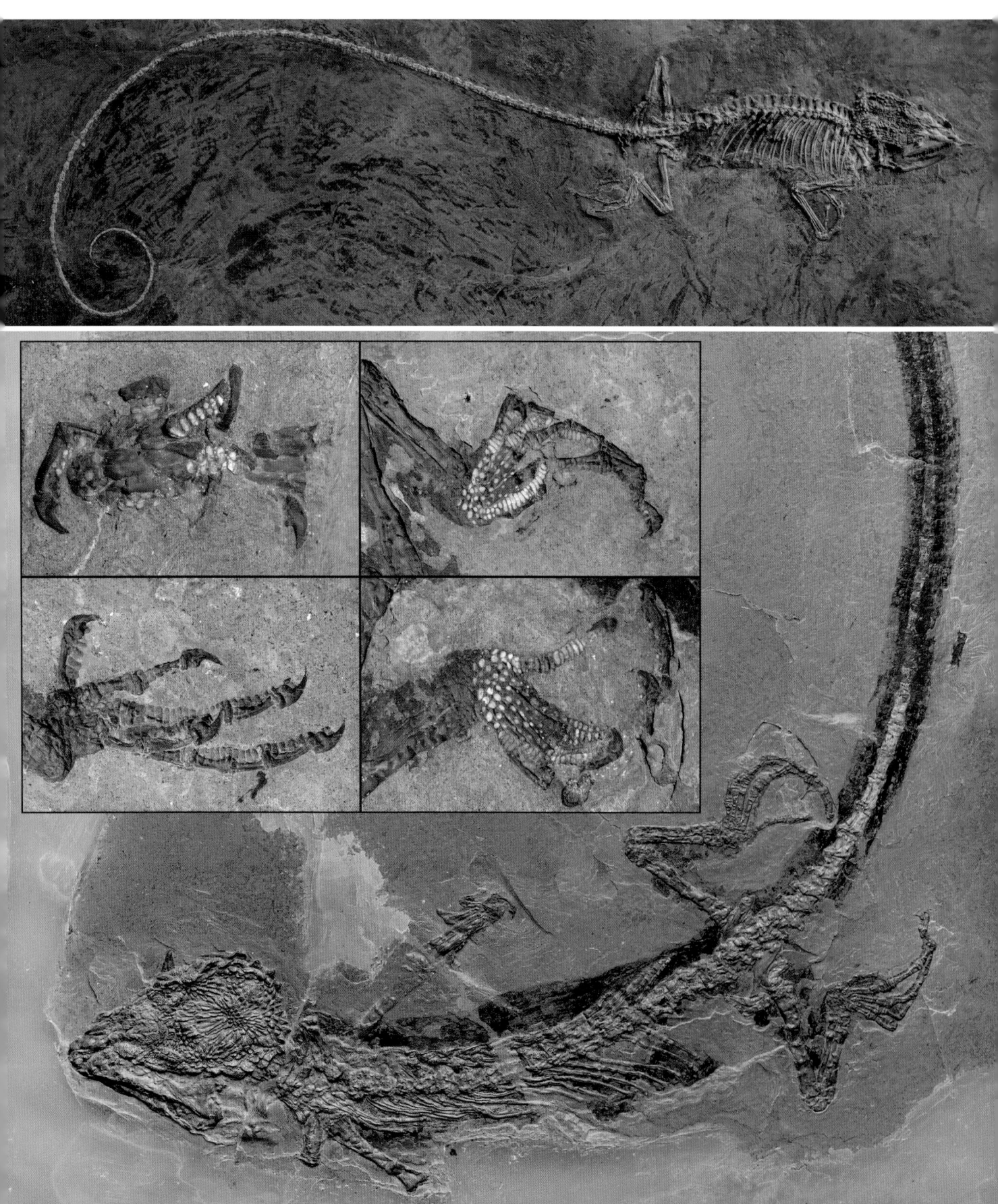

Fig. 10.1.5: Reconstruction of *Ornatocephalus metzleri*.

Fig. 10.1.4: *Ornatocephalus metzleri*, a large arboreal lizard with a prehensile tail. Top: the holotype with complete tail; bottom: specimen with exceptionally well-preserved scales on the limbs and tail. Insets: hands and feet on unprepared oil shale. Scale: 5 cm.

(Smith & Gauthier 2013). A species preserved in Baltic amber – of which only the exterior (scales) have been described – is close to or within the crown group Lacertidae (Borsuk-Białynicka et al. 1999). The complete skeletons preserved at Messel paint a more detailed portrait of these lizards and their place in the greenhouse ecosystem of the Eocene.

At least two of the Messel lacertiforms are closely related: *Eolacerta robusta* and a smaller form, *Stefanikia siderea*. *Eolacerta* was first described from the slightly younger lignite mines of Geiseltal in eastern Germany (Nöth 1940), but was later identified in Messel, based on nearly complete skeletons (Rieppel 1980; Müller 2001), and in France, based on isolated bones (Müller 2002). Studies since 1980 have emphasized the primitive features shown by *Eolacerta* in comparison with Lacertidae and thus cast doubt on the relationships of the fossil. The most recent work, however, concludes that *Eolacerta* is indeed closely related to Lacertidae, albeit outside of the crown group (Čerňanský & Smith 2017).

Eolacerta is the largest of the Messel lizards (Fig. 10.1.6), with a snout-vent length greater than 30 cm and a mass approaching 1 kg. The skull roof is highly ornamented, the ornamentation probably composed of osteoderms fused to the roofing bones. Osteoderms are also present in the scales above the eyes, two of which are predominant in size, a typical lacertid feature. Osteoderms are lacking on the rest of the body. The limbs are powerful, and the tail is very long. In the most complete specimen the tail had been shed and regenerated (Box 10.1.1). Given its body size and powerful proportions, it seems possible that *Eolacerta* included some vertebrate prey in its diet, although gut contents are not yet known.

Stefanikia is similar in skull structure (Fig. 10.1.7). However, it is much smaller and differs in a variety of relatively minor ways. It was also almost certainly terrestrial (Čerňanský & Smith 2017). At least one other terrestrial species related to Lacertidae is known from Messel. A further, quite distinctive species is rather common at Messel, but most specimens are poorly preserved. A recently excavated skeleton preserves

Fig. 10.1.6: *Eolacerta robusta*, a large terrestrial relative of Lacertidae. The tail had been shed and regrown. Scale: 5 cm.

not only details of the skull but also of the squamation (Fig. 10.1.8). The scales of the hands and feet are beautifully preserved, like those of the tail. The tail was probably prehensile, as in *Ornatocephalus*, implying that the species was arboreal.

The most enigmatic of the relatives of Lacertidae is *Cryptolacerta hassiaca*, a small lizard described on the basis of an adult specimen (Müller et al. 2011). The first studies of this highly distinctive animal suggested it was related to the worm lizards, Amphisbaenia, probably the sister-group to Lacertidae. More recent work has found *Cryptolacerta* to be more closely related to Lacertidae instead (Longrich et al. 2015), so we include the species here.

Cryptolacerta is a strange creature with a stoutly built skull (Fig. 10.1.9). The upper surface is highly ornamented. The scales did not leave deep grooves in most places, but a slightly arched furrow across the frontal bones suggests that some consolidation ("enlargement") of the scales had occurred. The frontals and the parietal bone have a strongly interdigitating mutual suture, as in Lacertidae, and the bones behind the orbits are greatly expanded backwards, nearly obliterating the supratemporal fenestra common to most amniotes. A third eye was present (Fig. 10.1.9, arrow). The jaws have a very low tooth count and may have been countersunk (Müller et al. 2011). The limbs are very short, particularly the forearm and lower leg,

Box 10.1.1: Autotomy

Autotomy ("self-slicing") refers to an animal's ability to shed a body part when an external force is applied, e.g., when it is attacked by a predator. Autotomy of the tail is a hallmark of squamates and their closest relatives, Rhynchocephalia. It is commonly assumed that lizard tails, when they are shed, break between the vertebrae, but this is not the case. To understand why, and how alternative patterns have arisen, we must look at the embryonic development, one of the more peculiar stories in vertebrate anatomy (Fig. 10.1.7).

The vertebrate body shows a partial segmentation, which is most clearly visible in the vertebrae. The ribs develop within the segment boundaries, but the development of the vertebrae is more complicated. In each body segment, there are areas of condensed tissue (step 1). Two successive condensations unite to form a single vertebra, but those two condensations are not from the same body segment. Rather, the condensations on either side of the segment boundaries fuse, so that the segmentation of the vertebrae does not correspond to segmentation of the body, a process called resegmentation (step 2). This process is common to all tetrapods (Williams 1959).

In lepidosaurs, an entirely new process takes place. Certain cells destroy the bone tissue along the original body segment boundary, creating a plane of weakness, the so-called autotomy plane (step 3, left). Therefore, when a lizard sheds its tail, the tail breaks through the middle of a vertebra along the autotomy plane. The tail as an organ regrows, but the bone does not. Instead, it is functionally replaced by a stiff tube (or, more precisely, a cone) of cartilage.

In some groups of squamates, such as *Basiliscus*, a further transformation takes place, often at a very early developmental stage. Other cells move into the autotomy plane and lay down new bone, thus sealing the plane of weakness so that breakage can no longer occur (step 4, left). This process usually begins at the tip of the tail and proceeds toward the base; ultimately, all autotomy planes may be sealed.

In a few squamates that had previously lost autotomy, a different transformation takes place. Unknown mechanisms change the relations of individual vertebrae to one another, making it relatively easy for the tail to be separated between the vertebrae (step 3, right). Because the tail is shed, but not in the manner typical for squamates, the ability is called pseudoautotomy. The selective pressures that lead to this result are unknown, but pseudoautotomy is well documented in the Messel basilisk, *Geiseltaliellus maarius* (see below).

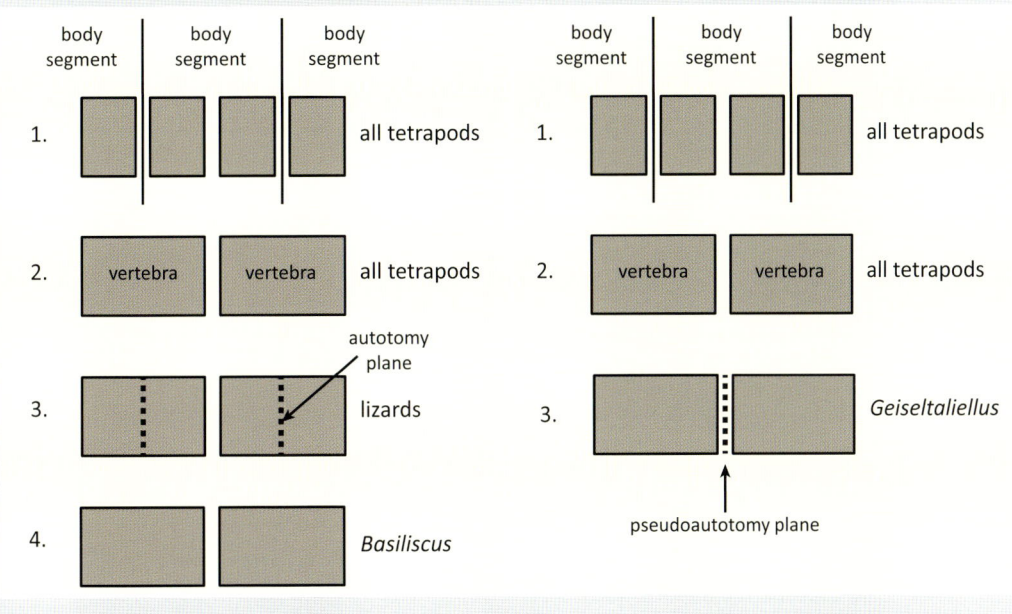

Fig. 10.1.7: *Stefanikia siderea*, a small terrestrial relative of *Eolacerta*. The tail had been shed. Scale: 5 cm.

and the hands and feet are reduced in size. The tail had been shed.

Quantitative studies of limb and body proportions suggested that *Cryptolacerta* was cryptozoic or semifossorial, that is, it spent much of its time hiding in the near-surface, probably burrowing in leaf litter or humus (Müller et al. 2011). Consistent with this conclusion is the consolidation of the head scales (which is commonly observed in burrowing squamates), the shape and orientation of the premaxilla, the generally stout build of the skull, and the reduced limbs.

The relationships of all of these animals have not yet been analyzed simultaneously, but a new picture of this interesting group is emerging. In the small ecotope of Messel there are at least five species, making it the most species-rich group of lizards at Messel. Furthermore, these species are disparate in terms of ecological specialization, including one large and at least two small terrestrial species, an arboreal species, and a cryptozoic species. Elsewhere in the Eocene of Europe there are species that are clearly closer to the crown group of Lacertidae than any of the species known from Messel (e.g., Borsuk-Bialynicka et al. 1999; Čerňanský and Augé 2013). These data have

Fig. 10.1.9: The holotype of *Cryptolacerta hassiaca*, a small cryptozoic lizard with reduced limbs and a stout skull. A "third eye" (arrow) is present. The tail had been shed. Scale: 1 cm.

Fig. 10.1.8: Undescribed arboreal relative of Lacertidae with a prehensile tail. The scales on the limbs and tail are exceptionally well-preserved. Insets: left hand and tail (mirrored) on unprepared oil shale. Scale: 1 cm.

suggested that the total clade of Lacertidae (i.e., the crown group and its stem) underwent an adaptive radiation in Europe, with a large number of species of various ecological types. Most of these had gone extinct by the end of the Eocene, leaving only one or a few lineages that radiated again in the Oligocene and Miocene – while colonizing Asia and Africa – to produce the diversity of lacertid lizards known today.

Iguanidae: Immigrants from the New World

Iguanidae (also called Pleurodonta; see Schulte et al. 2003) is a clade of lizards with over 1,100 living species. Today they are found predominantly in North and South America. The interrelationships of the major iguanid groups have proven exceedingly difficult to determine, possibly because the clade underwent an explosive radiation when it colonized the New World from Asia or after the extinction of dinosaurs at the end of the Cretaceous (Gauthier et al. 2012). Two species from Messel pertain to lineages that reached Europe from North America during the early Eocene.

Geiseltaliellus maarius (Fig. 10.1.10) is the most common lizard species from Messel, represented by over a dozen specimens. It was originally thought to be the same species that occurs in Geiseltal, *G. longicaudus*, but it has been shown to differ from it in minor ways (Smith 2009a). *G. maarius* is closely related to the extant basilisks (*Basiliscus*, *Corytophanes*, *Laemanctus*) of Central and northern South America, which are distinguished, among other things, by a pronounced bony crest projecting from the parietal bone on the top of the skull (Lang 1989), giving them the popular name "casque-heads."

The skull is moderately strongly built (Fig. 10.1.11). The jaw-closing muscles were attached on the top of the parietal bone, as in iguanid lizards generally. Like most iguanids, but unlike extant basilisks, the third eye is found at the boundary between the frontal and parietal bones. The front teeth are conical, but the cheek teeth show three cusps. Gut contents of *G. maarius* hint at a mostly insectivorous diet in which plant matter

Fig. 10.1.10: The Messel basilisk *Geiseltaliellus maarius*, a predominantly arboreal species with long limbs and a long tail. Scale bar: 5 cm.

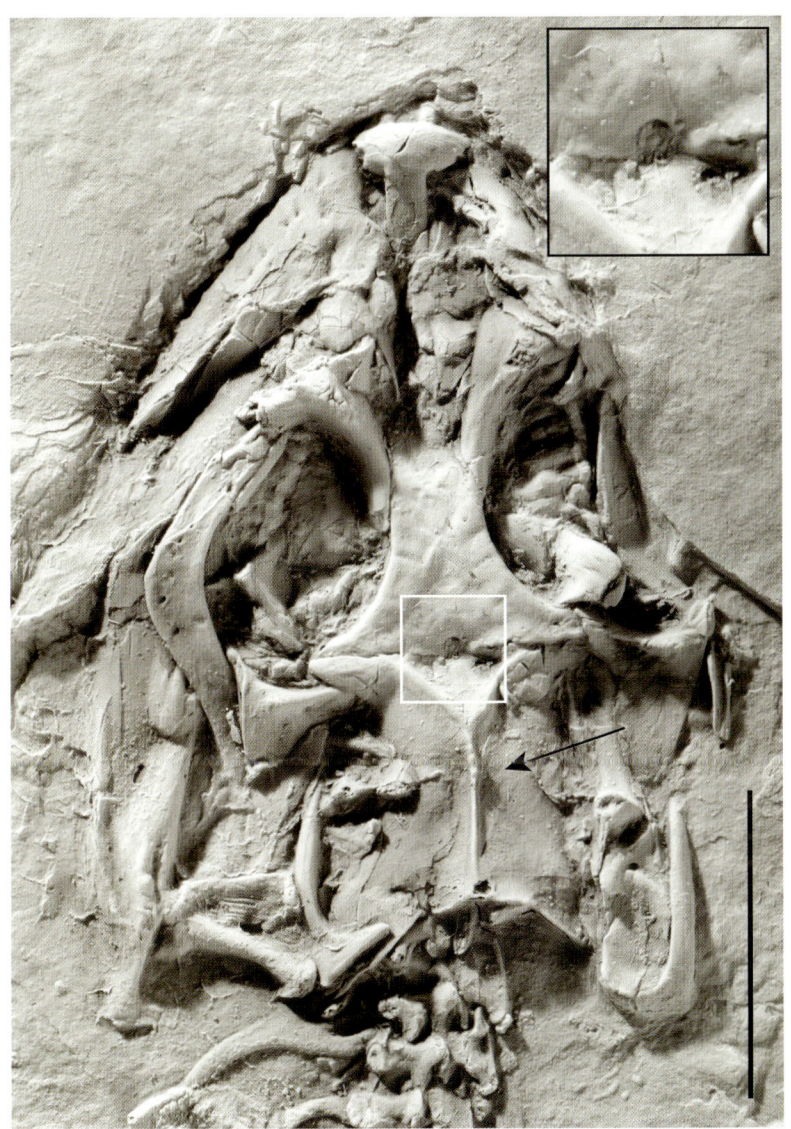

Fig. 10.1.11: Skull of the holotype of *Geiseltaliellus maarius* (inset: the "third eye"). The midline crest on the skull is expanded toward the front (arrow), reminiscent of the projecting crest of extant basilisks. Scale: 1 cm.

ly pronounced in males than females. (In *Corytophanes* and *Laemanctus* the crest is equally well developed in both sexes.) The parietal bone in *G. maarius* therefore suggests that this species was also sexually dimorphic, and it is possible that in the males, a crest of soft tissue projected from the top of the head (Fig. 10.1.12). Fossil evidence for such a crest has not yet been discovered.

The body of *G. maarius*, including the head, was covered by small, granular scales, 0.3–0.5 mm in diameter (Smith 2017a). These were somewhat smaller on the upper side of the body and somewhat larger on the head and the lower side (Fig. 10.1.12). No indications of keels or other ornamentations is present on the body scales, so it is presumed that they were smooth. The body, like that of living basilisks, has the proportions of an active lizard. The snout-vent length was up to 9 cm and the mass around 25 g. The hind-limbs are especially long, nearly two-thirds the snout-vent length. The unusual length of the tail was noted in 1931 by Johannes Weigelt in Geiseltal and is reflected in the name *G. longicaudus*. It is up to three times the snout-vent length: a large organ that must have played an important role in balance.

Most unusual about the tail of *G. maarius* is the fact that the individual tail vertebrae do not show breakage (autotomy) planes, even though the tail is missing in two-thirds of the specimens after the tail base, even in the smallest known specimen (Smith & Wuttke 2012). The last vertebra is always whole. *G. maarius* is therefore different from most lizards in shedding the tail *between* the vertebrae, a phenomenon called "pseudoautotomy" (Box 10.1.1).

The long legs and long tail are features that occur in two ecological groups of iguanid lizards today: tree-climbers and bipeds. *G. maarius* was probably a predominantly arboreal lizard, but very agile on the ground as well, where it laid its eggs and sometimes hunted. Extant *Basiliscus* have the special ability to run across water, which has earned them the com-

played a role in the adult (Smith & Scanferla 2016). *G. maarius* fell prey at times to the boa *Palaeopython fischeri* (Box 10.1.4).

In contrast to extant basilisks, the midline ridge on the parietal bone does not have a strongly projecting bony crest, but the front part of the ridge is expanded in three of six specimens (Fig. 10.1.11, arrow). In extant members of *Basiliscus*, the development of the crest is sexually dimorphic: more strong-

the middle latitudes associated with global cooling and became restricted to the tropics.

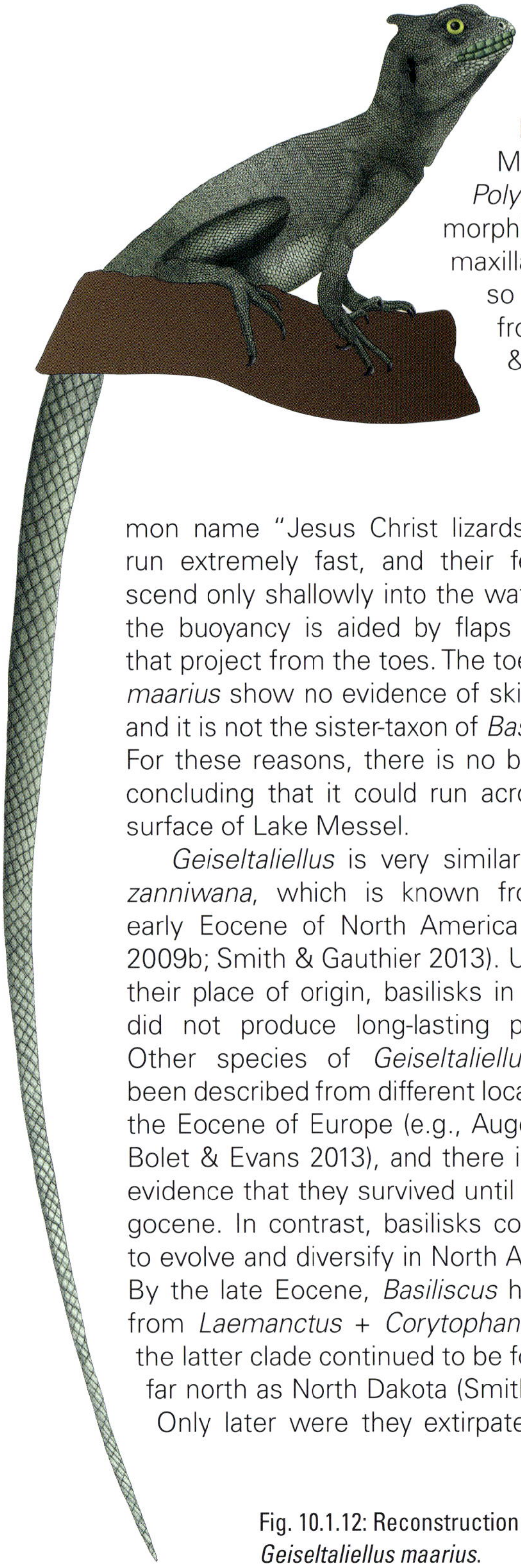

Another iguanid lizard from Messel (Fig. 10.1.13) belongs to Polychrotinae, which includes the genus *Anolis*, represented by hundreds of living species in the Neotropics. In particular, the Messel species seems to be related to the extant monkey lizards, *Polychrus*, of South America, as suggested by a number of synapomorphies, including a highly unusual relationship of the premaxilla to the maxilla. That a South American lineage should be found in Messel is not so surprising in view of the discovery of stem members of *Polychrus* from the early and late Eocene of North America (Smith 2011; Smith & Gauthier 2013). The species is known from a single specimen.

mon name "Jesus Christ lizards." They run extremely fast, and their feet descend only shallowly into the water, and the buoyancy is aided by flaps of skin that project from the toes. The toes of *G. maarius* show no evidence of skin flaps, and it is not the sister-taxon of *Basiliscus*. For these reasons, there is no basis for concluding that it could run across the surface of Lake Messel.

Geiseltaliellus is very similar to *Suzanniwana*, which is known from the early Eocene of North America (Smith 2009b; Smith & Gauthier 2013). Unlike in their place of origin, basilisks in Europe did not produce long-lasting progeny. Other species of *Geiseltaliellus* have been described from different localities in the Eocene of Europe (e.g., Augé 2005; Bolet & Evans 2013), and there is some evidence that they survived until the Oligocene. In contrast, basilisks continued to evolve and diversify in North America. By the late Eocene, *Basiliscus* had split from *Laemanctus* + *Corytophanes*, and the latter clade continued to be found as far north as North Dakota (Smith 2011). Only later were they extirpated from

Fig. 10.1.12: Reconstruction of *Geiseltaliellus maarius*.

Creepers in the underbrush

Anguidae is an almost exclusively New World clade, most diverse in North America, where its earliest relatives are also found. Two groups of anguid lizards, however, migrated to Eurasia during the Eocene: the glass lizards (Anguinae) and the glyptosaurs (Glyptosaurinae). Both are represented at Messel. Both emigrated from North America during the earliest Eocene. Anguinae eventually spread throughout Eurasia and North Africa. Two very distinct representatives are found in Europe today: *Anguis fragilis* and the much larger *Pseudopus apodus*. Glyptosaurinae also spread throughout Eurasia but did not survive the Paleogene there.

One of the Messel species is the large glass lizard *Ophisauriscus quadrupes* (Fig. 10.1.14). The largest known specimen is 49 cm long, and it is incomplete: the tail had been shed and only partly regrown (Sullivan et al. 1999). Even so, the tail as it is preserved is more than twice the snout-vent length! The body is moderately elongated as well, with 55-65 vertebrae in front of the vent, similar to living members of Anguinae (Wiens & Slingluff 2001). Thus, it was clearly a grass-swimmer (Box 10.1.2), and probably not as cryptic as *Anguis* today. Some scientists have concluded that it is closely related to *Anguis* (Sullivan et al. 1999), others to *Pseudopus* (Conrad 2008). The question is not yet settled.

The body was completely enclosed in an armor of osteoderms, which (like the scales that once covered them) are arranged in a regular grid of longitudinal and transverse rows. Individual osteoderms overlap one another from front to back and side to side. Like

Fig. 10.1.13: First record of a polychrotine from Messel. It is probably related to the extant monkey lizards, *Polychrus*. Scale: 1 cm.

all anguids, *Ophisauriscus* had a "lateral fold" (Sullivan et al. 1999), a longitudinal band on either side of the body where the osteoderms and scales are much smaller and the osteoderms do not overlap strongly, allowing the skin to stretch more easily. This serves important functions in these fully armored lizards, allowing them to expand the body for breath or to accommodate eggs or large meals.

Most remarkable about *Ophisauriscus quadrupes* are the limbs (Fig. 10.1.14), for which the species is named ("four-footed" little glass lizard). The limbs are strongly reduced in comparison to most lizards. In a specimen with a snout-vent length of 16 cm, the hind-limb is merely 5.3 mm long. However, the hind-limbs retain a pelvis, femur, lower leg bones, ankle bones and digits (Sullivan et al. 1999). Forelimbs are also present, though somewhat smaller than the hind-limbs. Because *Ophisauriscus* has already achieved the extent of body elongation shown by recent glass lizards, yet the small limbs still retain most of the bones, it is likely that factors other than developmental constraints are involved in limb reduction in this group.

The second anguid species at Messel is *Placosauriops abderhaldeni* (Fig. 10.1.15), which also was first described from Geiseltal (Kuhn 1940) and is known from about ten specimens from Messel (Keller 2009). They reached a total length of about 20 cm, half of which is the tail. The body is stout and completely covered by rectangular osteoderms. These osteoderms have a very peculiar texture consisting of small tubercles, a feature most prominently displayed by advanced Glyptosaurinae, in which they are also covered by a novel, enamel-like tissue. In one specimen, a number of lesions are observed in the osteoderms (Keller 2009), which might have resulted from necrotic infection by any number of microorganisms.

The teeth are chisel-shaped and blunt, but furnished with a strong carina from front to back and striations on the inner surface. The limbs are relatively short, but not strongly reduced as in *Ophisauriscus* (Keller 2009). *Placosauriops* probably fed on a variety of invertebrate prey, including some with a harder exoskeleton. Its lifestyle may have been similar to living galliwasps.

Box 10.1.2: Limblessness

It may seem strange that the limbs, which so splendidly adapted tetrapods to life on land, should be lost in the course of evolution, yet it has happened repeatedly. Nowhere is this phenomenon more abundantly displayed than in Squamata, where at least 62 independent instances of limb reduction are known (Greer 1991). Snakes just happen to be the most obvious and, with over 3,600 living species, the most successful of these lineages. The biologist Charles Camp (1923) first recognized that limbless squamates can be divided into two main groups: burrowers and grass-swimmers. The burrowers are generally small-bodied; the body is elongated, whereas the tail is short (top, *Natrix natrix*). The grass-swimmers, as the name implies, tend to live on the surface, although they may be cryptic. They are generally larger and are characterized by long bodies *and* long tails (bottom, the Slow Worm *Anguis fragilis*). The extent of limb reduction in the two groups – loss of digits, loss of forearm or lower leg bones, humerus or femur, shoulder or hip bones – varies among all the groups. Despite this fairly well-founded dichotomy, there remains uncertainty as to why these features evolved repeatedly and to what extent the reduction of the limbs is developmentally coupled with the elongation of the body (Wiens & Slingluff 2001).

Eurheloderma: an early Gila Monster

One of the rarest lizards known from Messel is a species of *Eurheloderma* (Fig. 10.1.16). This lineage was first described on the basis of some isolated bones from the late Eocene Phosphorites du Quercy of France. It is related to the large-bodied Gila Monsters and Beaded Lizards of North America (*Heloderma*), some of the most highly specialized lizards alive today. They prey almost exclusively on eggs and nestlings of birds and mammals. Since this is a highly seasonal food source, they store fat in the tail and remain inactive for much of the year, sticking to their shelters. Members of one population were found to be active above-ground for just 10 hours *per month* (Beck and Lowe 1991). They are venomous, with powerful jaws, and grooves in the teeth conduct the venom from the glands into the victim's flesh. Ritual wrestling – usually between males, lasting for hours and conducted in slow motion – can take place during the mating season when the males travel long distances to seek out females (Beck 2005).

The specimen from Messel is the first complete skeleton of *Eurheloderma* known, and its form is already highly similar to that of *Heloderma*. The head was covered by osteoderms, some of them embedded in the soft tissue of the skin and others fused to the underlying skull bones. The supratemporal arch is reduced, perhaps to create room for the large muscle masses that close the jaws and make the Gila Monster's bite so powerful. The shape of the lower jaws suggests that enlarged venom glands were present, and the teeth show incipient venom grooves.

The scales of the upper body surface were also covered with osteoderms. These are circular in outline, similar to extant *Heloderma*. However, the os-

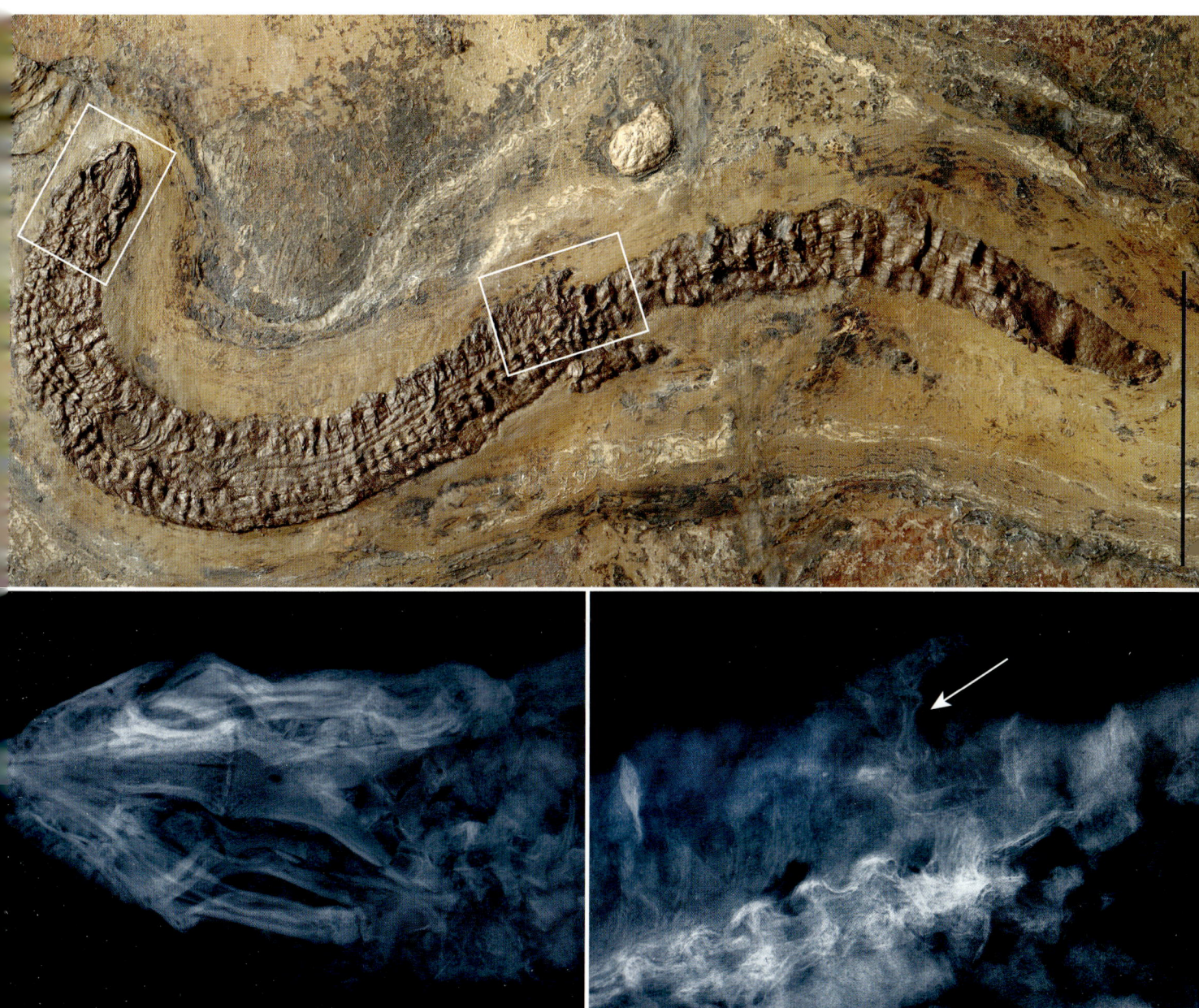

Fig. 10.1.14: The large, primitive glass lizard *Ophisauricus quadrupes* (top). The tail had been shed and partly regrown. Bottom left: X-ray image of the skull; bottom right: X-ray image of the pelvic region with reduced hind-limb (arrow). Scale: 5 cm.

teoderms are relatively flat. The first part of the name *Heloderma* refers to the resemblance of these lizards' scales to nail-studs or beads (Greek: *helos*), and it is unlikely that the skin of *Eurheloderma* from Messel would have the same beady appearance. Like extant *Heloderma*, the Messel *Eurheloderma* has an elongated body, with over 35 vertebrae in front of the vent.

Based on skull bones from the state of Utah (Nydam 2000), it is thought that helodermatids diverged from other lizards at least 100 million years ago. The

Fig. 10.1.15: The stout, primitive glyptosaur *Placosauriops abderhaldeni*. The animal was completely covered by osteoderms, which were ornamented with small tubercles. Scale: 2 cm.

Messel *Eurheloderma* specimen suggests that the typical body form of *Heloderma*, except for the beady scales, had already evolved by the early Eocene, and the lizards were probably already venomous.

The semi-aquatic shinisaurs

Another species from Messel is related to the living species *Shinisaurus crocodilurus*, the Chinese Crocodile Lizard (Smith 2017b). This rare, highly endangered species is confined to small populations in southern China and northern Vietnam, where it inhabits streambanks. The specific name, *crocodilurus*, refers to the superficially crocodile-like tail of these lizards, which bears paired rows of strongly keeled scales on its upper surface (Fig. 10.1.17). Taken together, these scales form a pair of ridges that make the tail a powerful organ of propulsion in the water. *Shinisaurus* feeds on insects and small fishes, waiting for them to pass by the branch on which it is resting.

The first described shinisaur from Messel is an isolated tail (Fig. 10.1.18), which had been shed naturally. The tail shows a puncture injury near its hind end, suggesting that it might have been shed during a predation event. The tail was completely covered in osteoderms, which clearly show the arrangement and shape of the individual scales and the tail as a whole. As in living shinisaurs, there were paired rows of strongly keeled scales. This is the earliest evidence of a crocodile-like tail in the lineage and a strong indication that the Messel shinisaur, like its living relative, was semi-aquatic.

Fig. 10.1.16: Complete skeleton of the Gila monster *Eurheloderma* sp. The tooth form suggests that a venom apparatus was present. Scale: 2 cm.

Necrosaurs: the "death lizards"

Necrosaurs are known by two species in Messel. They are closely related to *Necrosaurus cayluxi*, the so-called "death lizards" (*nekros-* "corpse") first described from the Phosphorites du Quercy of France. Many similarities were recognized between necrosaurs and extant monitor lizards, *Varanus*, to such an extent that a Hungarian noble and scientist even assigned the species *N. cayluxi* to that genus! Similarly, the first necrosaur described from Messel, *Necrosaurus feisti*, was originally assigned to the extinct monitor lizard genus *Saniwa*. Although there is no doubt that necrosaurs are anguimorph lizards, precisely how they are related to other anguimorphs remains an open question.

The larger and more common of the species is *Necrosaurus feisti*, named after one of the early private collectors at Messel. The holotype specimen has a snout-vent length of 20 cm, and the second described specimen is even larger (Stritzke 1983). The most complete specimen is the juvenile Senckenberg specimen (Fig. 10.1.1). The scales of the head were small but covered with osteoderms, especially over the skull bones. There is no tendency of the nasal openings to be shifted backward on the snout, unlike in living monitor lizards, where the shift is associated with elongated, U-shaped nasal passages. There were over 20 teeth in each of the upper and lower jaws, which taper toward sharp, backward-curved tips.

N. feisti had a powerful body with relatively long limbs. The form of the teeth and the powerful body suggest that it was a predatory, terrestrial lizard, a conclusion supported by the stomach contents of a specimen in private hands. Here, a specimen of the

Fig. 10.1.17: Extant *Shinisaurus crocodilurus* in its natural habitat in China.

small, semifossorial lizard *Cryptolacerta* (see above) is preserved, only the second of its kind known.

A second, much smaller species of necrosaur, yet undescribed, is also known from Messel (Fig. 10.1.19). It is only half the length of *Necrosaurus feisti*, and is distinguished from its larger relative by an extremely long, narrow, pointed snout. The palatal bones are long, strap-like, and covered by teeth. The numerous teeth of the upper jaw are relatively short but strongly curved backwards.

Small and large boas

Thus far, all of the Messel snakes that have been studied belong to Boidae, a clade of non-venomous snakes widely distributed in tropical and subtropical environments worldwide. In this respect, Messel is similar to other Eocene localities in Europe (Ivanov et al. 2000). Most recent phylogenetic studies suggest that boas and pythons represent two independent basal branches of Macrostomata (Box 10.1.3), the large-mouthed snakes (Reynolds et al. 2014). The clade Boidae comprises Boinae (anacondas, constrictors, rainbow boas), Erycinae (sand boas, rubber boas) and dwarf boas. They include minute forms (such as the fossil *Rieppelophis ermannorum*, less than 40 cm long), as well as giant species (such as *TItanoboa cerrejonensis*, perhaps reaching 13 m; Head et al. 2009). Boids show an extraordinary diversity in the anatomy of the skull, including large forms with the typical macrostomatan architecture with enlarged anterior teeth, and small species with short jaws and teeth of equal height. Also, many boids have infrared receptors in their lips that permit the extension of their "vision" into the infrared spectrum.

The unusual global distribution of living boids (Central and South America, Madagascar and some Pacific islands) has made them the focus of many studies regarding phylogenetic and biogeographic aspects, which have yielded important insights into evolutionary processes (Noonan & Chippindale 2006).

The minute boas *Messelophis variatus* and *Rieppelophis ermannorum* (Fig. 10.1.20) are the most common snakes recovered from Messel. These small boids (the largest specimens of *Rieppelophis* are 40 cm

Fig. 10.1.18: Shinisaur tail, showing a defect, probably an injury, near its tip; the tail had been shed. Scale: 2 cm.

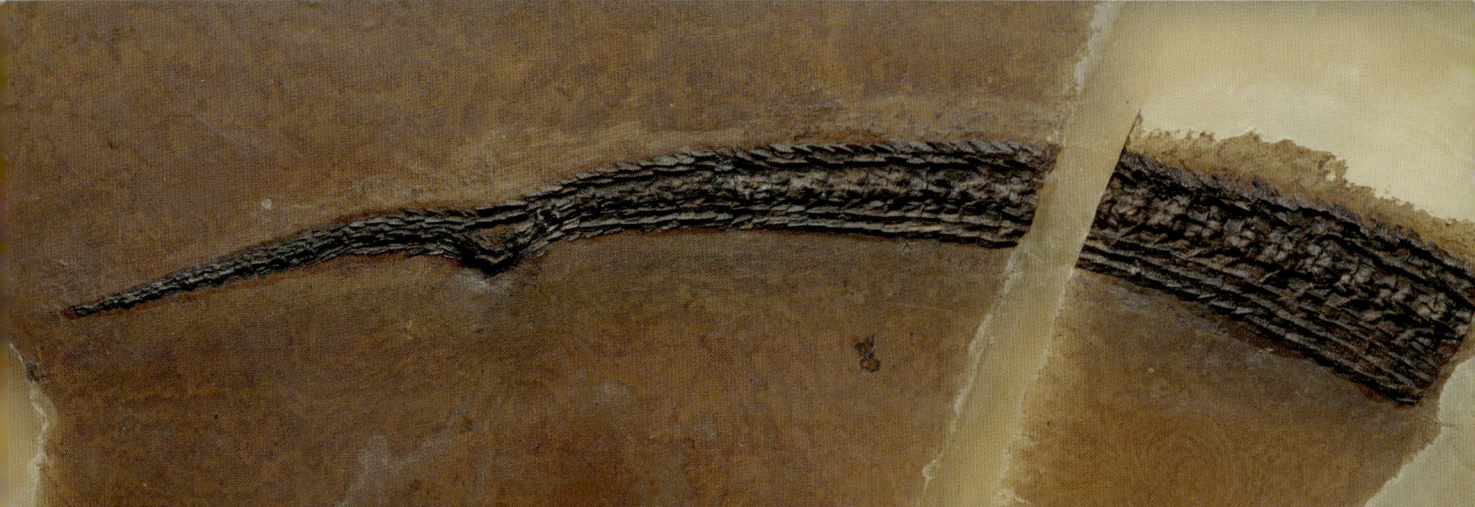

long) are among the smallest known representatives of the clade, extinct or extant. Both species were first placed in the genus *Messelophis* (Baszio 2004; Schaal & Baszio 2004), but further studies demonstrated that they represent divergent body plans and belong to different genera (Scanferla et al. 2016). Despite certain divergent anatomical traits (such as teeth on the premaxilla), the skulls of these snakes share short upper and lower jaws, as found in other small macrostomatans. Thus, the gape size was small and both had to take prey with a small cross-sectional area. Most traits differentiating these minute boas are in the postcranial skeleton. *Messelophis* has an elongated body with over 360 vertebrae, whereas *Rieppelophis* exhibits a shorter body with about 220 vertebrae. This disparate postcranial anatomy suggests that these species occupied different terrestrial microhabitats, *Messelophis* being a surface-dweller and *Rieppelophis* a cryptozoic snake that remained hidden in leaf litter or shallow burrows.

The numerous specimens of *Messelophis* recovered in Messel include one that contains skeletal elements of a tiny snake within its body cavity (Fig. 10.1.21). Skeletal remains include well-ossified toothed skull elements and some strings of vertebrae and ribs. These bones are positioned in the posterior third of the snake skeleton, near the vent and far from the presumed position of the stomach (see Smith & Scanferla 2016). The degree of development of the preserved bones and their position inside the host snake suggest that they represent an advanced embryo. It cannot be determined whether other embryos were present because the specimen is incomplete.

Development of the embryo inside the body of the female (viviparity, or live birth), is a widespread reproductive strategy present among lizards and snakes (Blackburn & Stewart 2011). Viviparity provides many advantages to protect the embryo from challenges presented by the environment, such as predation, temperature

Fig. 10.1.19: Small necrosaur with an elongate snout and short tail. Scale: 5 cm.

Box 10.1.3: Macrostomy: a big mouth

Most of living snake species belong to Macrostomata, a large group that currently contains more than 2,800 described living species around the world. Macrostomata means "big mouth" – a reference to the extraordinary ability of these snakes to swallow prey items whole that larger than their own head (right, the water snake *Nerodia*). To do so, macrostomatan snakes require complementary modifications in soft tissues (to allow greater stretching) and in the bones (lengthening the upper and lower jaws during early development) that contribute to a wide gape (left) (Cundall & Greene 2000; Scanferla 2016). Macrostomy is *the* principle modification that permitted macrostomatans to exploit an enormous diversity of prey types: a key adaptation of the group. Curiously, underground macrostomatan snakes have reversed the anatomical traits that permit macrostomy and returned to a diet based on prey items with a low cross-sectional area such as worms, insects or elongated vertebrates and thus resemble the condition present in basal snakes. Different episodes of occupation of underground and surface habitats coupled with changes in behavior and diet have shaped the skull morphology of macrostomatan snakes (Scanferla 2016).

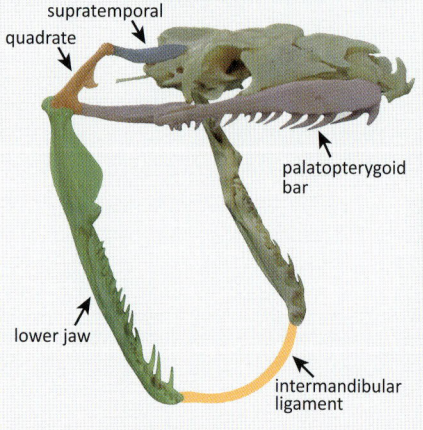

extremes, dehydration, and microbial infection. Despite its prevalence in living squamates, there are only two fossil records of viviparity in squamates, a terrestrial generalized lizard from the Lower Cretaceous of China (Wang & Evans 2011) and an aquatic mosasaurian from the Upper Cretaceous of Slovenia (Caldwell & Lee 2001). In consequence, this specimen of *Messelophis* represents the first fossil record of viviparity in snakes. Although interesting, the presence of this trait in *Messelophis* is not surprising because nearly all extant boids are viviparous.

Old and New World erycine snakes are found today in Eurasia, Africa and North America. This group of small to medium-sized, cryptozoic snakes exhibits a peculiar club-like structure in the tail, which is the result of the presence of additional bony extensions of the caudal vertebrae. This unique anatomy of the vertebrae represents a valuable tool to study the fossil history of erycines. Currently, only the genus *Eryx* is present in the southeast region of Europe, but erycines had a wider distribution in the past. However, this fossil record is mainly represented by isolated vertebrae, as is generally typical in snakes (Szyndlar & Rage 2003).

Among the numerous snakes recovered in Messel, a nearly complete specimen of a new, tiny boid snake was recently discovered (Fig. 10.1.22), which clearly displays the typical club-like tail skeleton present in living sand boas. The skull of this undescribed snake is also short, as in extant Erycinae.

Fig. 10.1.20: Top: the small boa *Messelophis variatus*. The head is located beneath a row of vertebrae in the middle of the specimen. Bottom: the tiny boa *Rieppelophis ermannorum*. Scale: 2 cm.

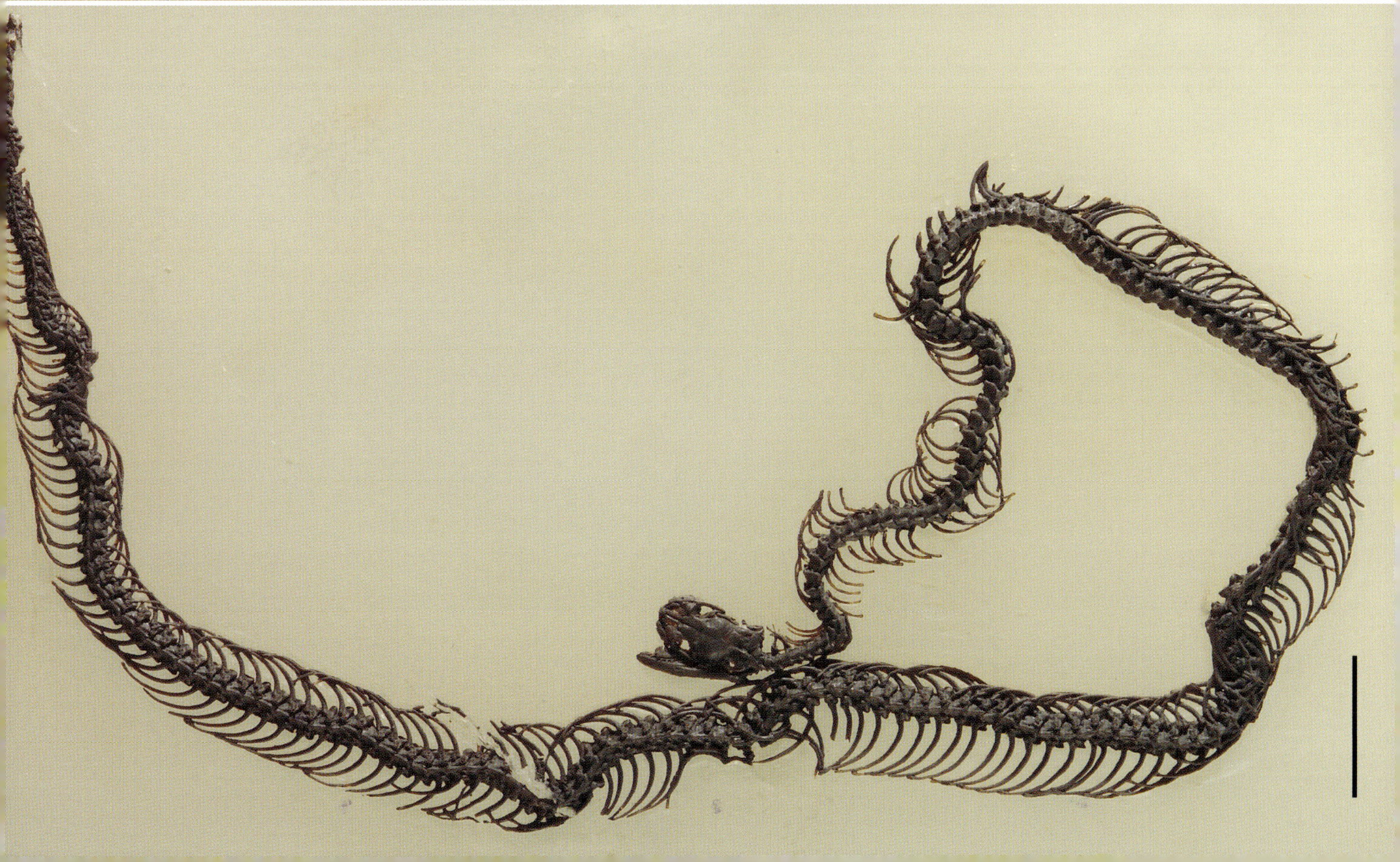

Small and large boas

Palaeopython

With adult specimens 2 m length, the boid *Palaeopython fischeri* is the largest snake known from Messel (Schaal 2004; Fig. 10.1.23). The relatively large number of specimens suggests it was an important component of the terrestrial community. *Palaeopython* displays some features of the skull shared with Neotropical boines, such as the anatomy of the sagittal crest and the morphology of the quadrate bone.

Extant boines are among the largest vertebrate predators in their ecosystems. Taking into account its large size and abundance, we can infer that *Palaeopython* was an important predator in the Messel ecosystem. Stomach contents are the most direct way to determine the diet of a fossil species, and some spectacular specimens of *Palaeopython* offer precise information about its diet. Greene (1983) described a large specimen with a crocodylian in its belly. More recently, a juvenile *Palaeopython* was described that had swallowed a specimen of the lizard *Geiseltaliellus maarius* (Box 10.1.4).

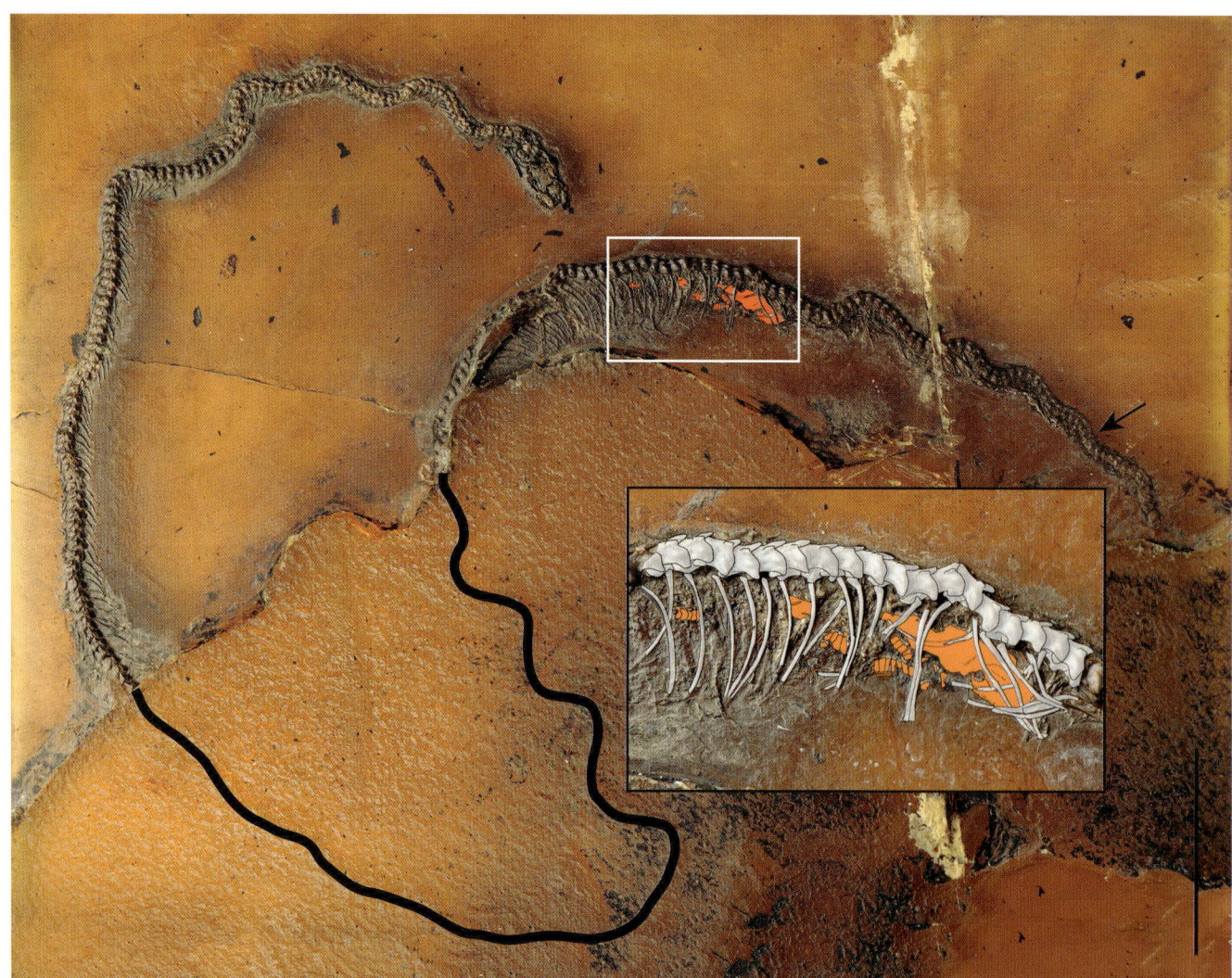

Fig. 10.1.21: A gravid female of *Messelophis variatus*. The extension of the body, where missing, is indicated by a thick black line. Inset: skull bones and vertebrae of the embryo. Scale: 5 cm.

Fig. 10.1.22: First record of a sand boa from Messel. Inset: club-like tail tip with accessory processes on the vertebrae, a unique feature of Erycinae. Scale: 2 cm.

Albeit indirectly, we can add birds to the diet of *Palaeopython* on the basis of an undescribed specimen that seems to have been regurgitated by a large snake. Regurgitation is a common behavioral response to stress (predation, low temperature), infection, or over-eating (in size or quantity of prey) in snakes, and the appearance of the regurgitated prey provides some clues about its predator. During prey constriction and swallowing, boid snakes alter the shape of their prey in order to reduce its cross-sectional area. For mammalian and avian prey, this process results in a straightening of the head and spine, which lengthens the carcass; the forelimbs are pressed against the side of the body, and the hind-limbs are extended caudally (Close & Cundall 2012). This is the typical shape of a tetrapod carcass regurgitated by a snake, and it is extraordinarily similar to the skeleton of an undescribed bird (Mayr & Schaal 2016). The skeleton is mostly intact (Fig. 10.1.25), indicating that little digestion had occurred prior to its regurgitation. The size of this bird (2.5 cm head length) points toward *Palaeopython* as the snake that swallowed it. The same is true of the regurgitated carcass of the carnivorous mammal *Lesmesodon* (Chapter 12.7).

The squamate community

As in birds and insects, the number of lizard and snake species known from Messel has undergone a most dramatic increase over the last few decades. Even so, extrapolations using new statistical methods (Chapter 5) suggest that a significant fraction of the species diversity is yet to be discovered. Additionally, comparisons with early and middle Eocene assemblages in Europe suggest that a number of species must have been present in the vicinity of Messel: monitor lizards, agamids, and worm lizards as well as additional glyptosaurs and snakes.

Fig. 10.1.23: Adult, 2-m-long specimen of *Palaeopython fischeri*. Scale: 5 cm.

The high temperatures of the early Paleogene climate exerted a decisive influence on squamate faunas in Europe. In particular, the Paleocene-Eocene Thermal Maximum (Chapter 3) had long-lasting effects on the squamates. As with mammals (Chapters 12), numerous species wandered from North America to Europe at this time. They are first recorded at sites such as Dormaal in Belgium (Augé 1990), and their relatives persisted until the time of Messel. In fact, nearly half of the known squamate species found at Messel are descended from North American clades whose representatives emigrated to Europe in the early Eocene: the basilisks and monkey lizards, the glass lizards and glyptosaurines, the Gila Monster, and perhaps the boids. In addition, the shinisaur and the erycine boid are probable immigrants, although their point of origin has not yet been settled.

It has been proposed that Europe – and Messel, in particular – served as a refugium, an isolated ecotope in which many ancient Mesozoic lineages survived, which were later replaced by modern groups that immigrated near the Eocene-Oligocene boundary. An alternative model is now emerging from the study of newly excavated and privately held specimens. Messel contains a large number of immigrant species from the New World that generally are closely related to extant lineages, and a species closely related to Old World shinisaurs. It also hosts a radiation of lizards related to Lacertidae that is remarkable for its species diversity and ecological breadth. While the relationships of a number of other species (geckos, necrosaurs and certain snakes) must still be clarified, these new data suggest that the European Eocene was a highly dynamic evolutionary stage, where various immigrants joined a vigorous endemic radiation.

Box 10.1.4: An ancient matryoshka

A magnificent juvenile specimen of *Palaeopython fischeri* (white) was discovered, whose stomach contains an arboreal lizard of the species *Geiseltaliellus maarius* (orange). The state of preservation of the lizard skeleton tells us that the snake died soon after its last meal, no more than one or two days later. But this newly discovered specimen held more surprises, since the lizard contains an insect (blue) in its belly. This extraordinary specimen reveals for the first time three levels of an ancient terrestrial food chain, and offers invaluable information about the diet of *Palaeopython*. The dietary preferences of living boid snakes change during their lifespan, from small cold-blooded prey (mostly lizards and frogs) with a small cross-sectional area as juveniles to bulky, warm-blooded prey as adults. Thus, the presence of a lizard as prey of a juvenile *Palaeopython* for first time provides solid evidence that such an ontogenetic (developmental) dietary shift has been present in boas for a long time (Smith & Scanferla 2016).

Fig. 10.1.24: Skeleton of a bird that was presumably regurgitated by *Palaeopython fischeri*. The body is straightened and the limbs closely pressed to the sides. Scale: 1 cm.

Chapter 10.2
Turtles – Armored Survivalists

Edwin Cadena, Walter G. Joyce, Krister T. Smith

Turtles (Testudines) are a successful group of vertebrates with nearly 350 living species. The most prominent feature of these animals is their protective shell. The shell is composed of two layers: an exterior layer of horny scales (scutes) formed by the outermost skin layer (epidermis), and an interior layer of bone (the carapace above and the plastron below), which is composed of numerous bony elements whose full evolutionary history has only recently been elucidated (Lyson et al. 2013a). The sutures between the scutes and between the bony elements of the shell generally do not correspond, such that the scutes, after they have decayed, leave a complex pattern of traces on the shell bones (Fig. 2, left). The plastron derives in part from the belly ribs (gastralia), which are ossifications of the dermis, the thick layer of the skin that underlies the epidermis, whereas most of the carapace develops in association with the ribs and vertebrae. The carapace usually has a row of marginal elements (called peripherals), which unite carapace and plastron at the bridge. Both carapace and plastron receive contributions from the shoulder girdle (Fig. 2, center, right).

Because many reptiles show ossifications of their skin in the form of osteoderms, paleontologists historically presumed that the turtle shell formed as the result of the successive accumulation and expansion of osteoderms in the thoracic region and the eventual fusion of this dermal shell with the underlying skeletal elements (e.g., Lee 1997). However, developmental biologists had long noted that this conclusion is not supported by embryology: the ossification of the dermis appears to be triggered during growth by the underlying skeletal elements (e.g., Burke 1989). The recent discovery or renewed study of fossil reptiles with intermediate morphologies – *Pappochelys* (Schoch & Sues 2015) and *Odontochelys* (Li et al. 2008) from the Triassic and *Eunotosaurus* from the Permian (Lyson et al. 2010) – contributed substantially to bridging the gap between primitive amniotes and early turtles and now corroborate the initial hypothesis by developmental biologists with paleontological data (Lyson et al. 2013a).

Although turtles have remained faithful to their shell, it is a commonly held misconception that the group underwent little change over the course of its evolutionary history. Among others, turtles have conquered a wide range of habitats, from tropical forests to subarctic oceans, evolved different ways to retract their heads and limbs within the protection of the shell, including the addition of hinges in the plastron, devised different ways to strengthen their jaws, and adapted to a wide range of diets, such as hard-shelled mollusks, terrestrial plants, and jellyfish (Ernst & Barbour 1989). Turtles rarely achieved high species-richness over the course of their evolutionary history, but fully shelled members of the stem or crown group have been persistently present since the Triassic in most tropical to temperate biomes.

Extant turtles are universally regarded to form two primary groups (Fig. 3), which are easily identified by the way they retract their necks: side-necked-turtles (Pleurodira), which double fold their elongate necks horizontally to hide below the shell, and hide-necked turtles (Cryptodira), which double fold their necks vertically to withdraw the head between the shoulder girdles inside the shell (Ernst & Barbour 1989). The ancestral crown turtle could not yet withdraw its neck in either fashion, but was nevertheless able to partially protect itself by laterally tucking its head below the shell (Werneburg et al. 2015). To strengthen their bite, turtles have evolved expanded temporal (jaw) muscles that are packed towards the back of the skull, but the ear region blocks the direct path to the mandible. The two primary groups of turtles also differ in their solution to this problem. In cryptodires, the ear capsule forms a trochlea (pulley) that allows the muscles to glide around the ear, whereas in pleurodires, a trochlea is formed instead by an expansion of the lateral process of the pterygoid (one of the palatal bones) that circumvents the ear capsule entirely (Gaffney 1975). Although this feature is useful

Fig. 10.2.1: Soft-shelled turtle, *Palaeoamyda messeliana*, seen from above. Scale: 5 cm.

150 Chapter 10.2 Turtles – Armored Survivalists

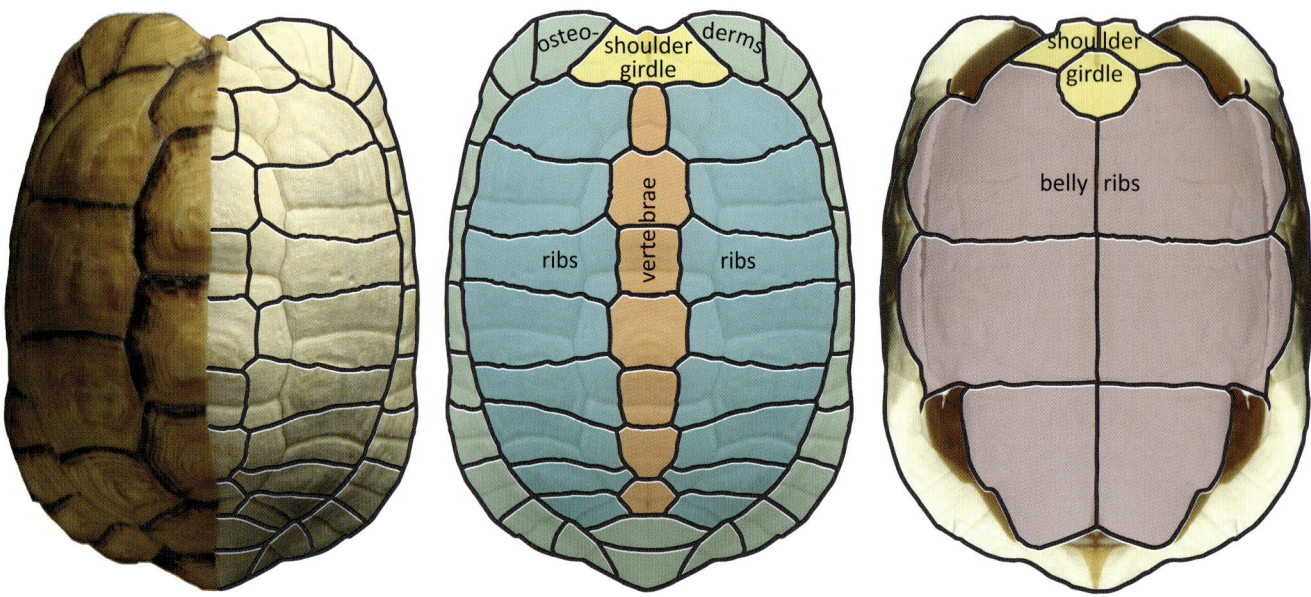

Fig. 10.2.2: Horny scales cover the carapace and jointly form the turtle shell. Left: the scale margins do not correspond to the bone sutures. Center and right: evolutionary origin of the carapace and plastron, respectively. Scale: 2 cm.

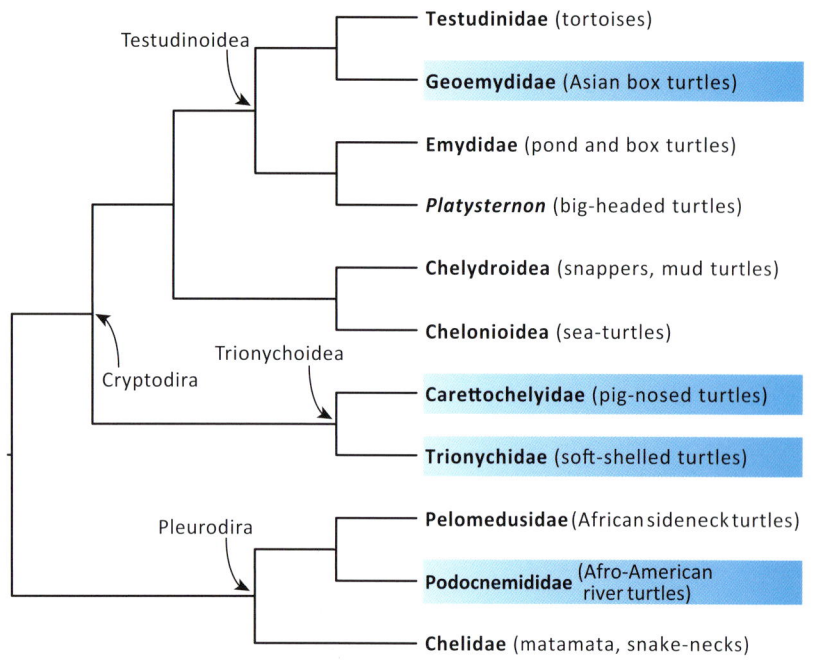

Fig. 10.2.3: Simplified phylogeny of turtles. Groups known from Messel are highlighted in blue.

in identifying modern representatives of these two groups, fossils strongly suggest that the cryptodire pattern is plesiomorphic for turtles (Joyce 2007), and the pleurodire pattern a later transformation.

Pleurodires originated on the southern continents in the Jurassic, some 170 million years ago. They colonized the oceans and the northern continents in the Cretaceous, before withdrawing once again to their original home range over the course of the Cenozoic. Cryptodires, on the other hand, originated around the same time in Asia, but have a nearly global distribution today (Joyce et al. 2016). The Messel fauna documents a time just after pleurodires from Africa and cryptodires from North American and Asia had fully displaced the original turtle fauna that had developed in Europe during the Mesozoic (Lapparent de Broin 2001; Joyce et al. 2016).

We recognize only four species of turtles at Messel, which represent four distantly related groups of freshwater turtles: the pleurodiran Podocnemididae (Afro-American river turtles), and the cryptodiran Carettochelyidae (pig-nosed turtles), Geoemydidae (Old World pond turtles and relatives), and Trionychidae (soft-shelled turtles). The only group of turtles that is notably absent from Messel is Testudinidae (tortoises), which are well documented from Geiseltal, Germany, a locality several million years younger (Hummel 1935). It is unclear to us whether tortoises were truly absent from the forests surrounding Lake Messel, or simply have not yet been discovered.

Palaeoemys messeliana

Geoemydids are a group of about 70 living species that currently live in tropical to temperate climate zones across the world. The group is ecologically diverse, ranging from terrestrial leaf turtles that live on the forest floor to highly aquatic river turtles. These turtles also have varied dietary preferences, ranging from entirely herbivorous to exclusively carnivorous. The smallest extant representatives have a shell length of just 15 cm, while the largest have shells up to 80 cm in length (Ernst & Barbour 1989). More than 100 geoemydid skeletons are known from Messel, including hatchlings with a shell length of only a few centimeters and large adults with a shell length of more than 30 cm. This material was historically assigned to the extant East Asian genus *Ocadia* (Reinach 1900; Staesche 1928), but it now is apparent that no close relationship exists (Hervet 2004). Some scientists have recognized two geoemydid species at Messel: a smaller one with a median keel, and a larger one lacking this keel (*Francellia messeliana* and

Fig. 10.2.4: Juvenile (left) and adult (right) pond turtles, *Palaeoemys messeliana*, seen from above. The adult was originally attributed to a different species. Scale: 5 cm.

Fig. 10.2.5: Hatchling turtle, probably *Palaeoemys messeliana*, seen from above (left) and below (right). The marginal elements of the bony shell have begun to ossify. Scale: 2 cm.

Euroemys kehreri, respectively) (Hervet 2009). However, since living geoemydids are known to have pronounced keels as juveniles that often disappear later in life, we agree with others (Claude & Tong 2004) that all geoemydids from Messel belong to a single species, *Palaeoemys messeliana*.

The limbs of this turtle are typical for geoemydids that live in ponds (Joyce and Gauthier 2004) by being intermediate in length and relatively sturdy. In life, the hands and feet of *Palaeoemys* would have been webbed, allowing these turtles to swim in water with ease, but hard scales would have covered the palms of the hands and feet, allowing them to venture onto land as well, particularly in search of new habitat. It is therefore plausible that this turtle may have reached the lake on foot. The broad and short jaws of this animal are typical of omnivorous geoemydids, whose diet consists of invertebrates, plants, fish, and carrion. At Lake Messel, this turtle likely searched for food near the edge of the lake and was frequently found basking near the shore on logs protruding from the water. The relatively frequent presence of juveniles attests to the quality living conditions that usually prevailed at the margins of the lake.

It appears all but certain that geoemydids originated in Asia at the beginning of the Cenozoic, but the group suddenly appeared in North America (Hay 1908; Hutchison 1998; Holroyd et al. 2001) and Europe (Lapparent de Broin 2001) at the beginning of the Eocene, perhaps as a result of high-latitude land bridges (Chapter 2) that became habitable during high temperatures that prevailed during the Paleocene-Eocene Thermal Maximum (Chapter 3). However, the taxonomy and phylogenetic relationships of Paleogene geoemydid turtles remain poorly understood at present, and it is therefore unclear if the European representatives immigrated directly from Asia, or via North America.

Neochelys franzeni

Although pleurodires are now restricted to the southern continents, the group used to be prevalent in the northern continents as well, particularly during the Eocene in Europe (Lapparent de Broin 2001). A handful of specimens documents the presence of the pleurodire *Neochelys franzeni* at Messel (Schleich 1993; Cadena 2015). At first sight, this turtle greatly resembles *Palaeoemys* in its size and the presence of a well-developed shell, a short skull, and robust limbs, but a number of details of the shell clearly distinguish them, particularly the number and distribution of scutes and certain midline elements in the shell. Additional differences are found in the skull (consistent with their respective affinity to Pleurodira and Cryptodira), the cervical vertebrae, and the internal aspects of the shell, but these are typically apparent in CT scans only (Cadena 2015). At Messel, *Neochelys* was ecologically closest to *Palaeoemys* in living in the shallow portions of the lake and being adapted to an omnivorous diet. It is likely that at times both turtles could be found basking together on the same log.

Neochelys is a representative of Podocnemididae, which are currently distributed in Madagascar and the tropical portions of South America (Ernst & Barbour 1989), but used to be common in Europe and Africa as well (Gaffney et al. 2011). Although unrelat-

Fig. 10.2.6: Juvenile side-necked turtle, *Neochelys franzeni*, seen from above (left) and below (right). Scale: 2 cm.

ed pleurodires had previously colonized Europe during the Mesozoic (Lapparent de Broin 2001; Cadena & Joyce 2015), the complete absence of *Neochelys* or related forms in the Paleocene of Europe reveals that this taxon entered Europe during the Eocene as well, but this time from Africa.

Allaeochelys crassesculpta

Fossils of the carettochelyid turtle *Allaeochelys crassesculpta* (Harrassowitz 1922) are certainly among the most unusual turtle finds made at Messel, not only because they represent the only known complete fossil carettochelyid skeletons in the world (Joyce 2014), but also because about a quarter of the individuals occur in pairs. A recent study confirmed the long-held suspicion that these pairs represent couples that perished while mating (Box 10.2.1).

The vast majority of carettochelyids found at Messel appears to be adult in size, but a small number of juvenile specimens clearly document that not all attempts at mating ended tragically, but rather that prospering populations existed in the lake. Historically, two species of carettochelyids were recognized at Messel, the larger *Anosteira* (now *Allaeochelys*) *crassesculpta* and the smaller *Anosteira gracilis* (Harrassowitz 1922). Although these two had long been considered synonymous (Gramann 1956), only the recent analysis of the mating pairs revealed that the larger morphotype represents the female, whereas the smaller is the male.

Carettochelyidae, much like its sister group, Trionychidae, originated in Asia during the Early Cretaceous, but spread across the northern hemisphere (North American and Europe) in the Paleogene, only to move southwards into Africa, southern Asia, and Australia during the Neogene (Joyce 2014). The only surviving species, *Carettochelys insculpta*, now lives in southern New Guinea and northern Australia (Ernst & Barbour 1989). Although the group displays little morphological diversity during its history, Asian fossils reveal the presence of at least two lineages during the early Paleogene, namely the smaller and in many aspects more primitive *Anosteira*, which is more common in northeastern Asia, and the larger and more derived *Allaeochelys*, which is closer to living *Carettochelys* and is more common in southern Asia, including India. *Allaeochelys crassesculpta* is clearly attributable to the *Allaeochelys* lineage in that it exhibits an expanded plastron and shares the loss of the carapacial scutes with *Carettochelys* that are plesiomorphically retained in the *Anosteira* lineage. It therefore seems reasonable to postulate that the European representatives of *Allaeochelys* arrived directly from Asia in the early Eocene (Joyce 2014).

Palaeoamyda messeliana

Trionychidae are unusual among turtles in that a tough, leathery skin replaces the horny scutes of the shell, and the bony elements along the periphery are absent. What remains of the shell of these cryptodires has a unique exterior texture consisting of pits and ridges, which makes even small fossil fragments readily identifiable as soft-shelled turtles, but complete skeletons are exceedingly rare. Messel has yielded over twenty soft-shelled turtles that are often exquisitely preserved and belong to a single species, *Palaeoamyda messeliana* (Reinach 1900; Cadena 2016). The bony shell of this turtle is more than 50 cm in length, and the full shell, including the leathery posterior margin, may have been up to 80 cm long. As a result, this turtle was clearly the largest in the Messel ecosystem. *Palaeoamyda* had elongated limbs that formed flexible, soft paddles well adapted to swimming, but poorly suited to walking on land. It is therefore highly unlikely that this turtle entered the lake on foot, but rather through a tributary that was temporarily or persistently connected with the lake. Extant soft-shelled turtles are primarily carnivorous and hide their flat bodies in soft sediments at the bottom of shallow lakes or rivers to ambush their prey (Ernst & Barbour 1989). The presence of soft-shelled turtles at Messel therefore clearly indicates that the shallows of the lake margin were habitable, with soft sediments in which these animals could hunt for their prey.

In a previously undocumented, unfortunately incomplete specimen, two individuals of *Palaeoamyda* are preserved together with their hind ends pressed together. It appears that this pair of soft-shelled turtles died while mating, thereby documenting that not only carettochelyids died while mating at Messel, but also trionychids. This is not entirely surprising, as numerous features that enabled the preservation of

Box 10.2.1: In flagrante delicto

All couples of *Allaeochelys* at Messel consist of one female and one male and are oriented with their rear ends facing each other (scale: 5 cm). Furthermore, the tails of some pairs are still aligned with each other in the mating position (Joyce et al. 2012). Mating vertebrates had never before been described as fossils, because few mechanisms exist that would allow couples to be preserved in this compromising position, especially large vertebrates such as turtles. The fact that carettochelyids died sporadically while mating at Messel is a result of the unique behavior and physiology of *Allaeochelys* in combination with the special circumstances that prevailed in Lake Messel. First, much like soft-shelled turtles (see below), carettochelyids lack hard, protective scutes; instead, a soft, permeable skin surrounds their bodies. This confers an adaptive advantage that allows these turtles to respire under water and remain submerged for prolonged periods of time. Second, the male *Allaeochelys* are smaller than the females, which is typical for living turtles that court and mate in the open water, as opposed to shallow water near the shore. Under normal circumstances, these two characteristics were likely unproblematic for these turtles, but Lake Messel was different from the river and pond habitats elsewhere in central Europe in being a deep, stratified volcanic lake with a thin layer of habitable surface waters overlying subsurface waters that had either been poisoned by hydrogen sulfide produced by decaying organic matter or volcanic carbon dioxide that percolated from the rocks below (Chapters 2, 13). It is probable that the pairs met in the open waters of the lake, but became unconscious and died when their skin absorbed poisons while sinking into toxic subsurface waters during their conjugal embrace.

Fig. 10.2.7: Carettochelyid turtle, *Allaeochelys crassesculpta*, seen from above. Scale: 5 cm.

Fig. 10.2.8: Soft-shelled turtles, *Palaeoamyda messeliana*, which died while copulating. Scale: 5 cm.

mating pairs in carettochelyids also occur in trionychids, most notably the presence of permeable skin. However, mating trionychid pairs appear to be rare by comparison to their small-bodied carettochelyid cousins, generally consistent with the lower abundance of isolated finds in recent collections.

Trionychids originated in Asia some 120 million years ago, but colonized all continents except Antarctica over the course of their evolutionary history: in particular, North America in the Late Cretaceous, Australia and India in the Paleocene, and Africa and South America in the Neogene (Joyce et al. 2016).

The group was long thought to have invaded Europe at the beginning of the Cenozoic, but the recent description of a fragmentary find in the Campanian of Sweden recently pushes this event back into the Late Cretaceous (Scheyer et al. 2012). Among Paleogene trionychids from Europe, at least one, the giant *Axestemys vittata*, shows a clear connection to coeval forms from North America (Georgalis & Joyce 2017). It remains unclear, however, if the remaining Paleogene trionychids, including the species from Messel, had an origin in North America as well, or arrived directly from Asia in an independent dispersal event.

Chapter 10.3
Crocodyliforms – Large-bodied Carnivores

Christopher Brochu & Jessica Miller-Camp

Crocodyliforms – members of the group including Crocodylia (alligators, gharials, and crocodiles) and their extinct relatives (Fig. 10.3.2) – are common at Messel. They ranged in total length (body and tail) from 1 to 4 m. Some resembled a modern alligator or crocodile, but others were small and had enlarged crushing teeth. Still others, with flattened, serrated teeth and toes that ended in blunt, hoof-like claws, may have spent less time in the water than their living counterparts.

No study of crocodyliform evolution can ignore Messel. Seven species have been found there, more than at any other Cenozoic site in the Northern Hemisphere. It is furthermore the only site preserving all of the primary lineages known from continental Europe during the Eocene. The species found at Messel are closely related to, if not conspecific with, some of those found at Geiseltal, a locality several million years younger in eastern Germany; because some are better preserved at Messel, and others at Geiseltal, the two sites complement each other. And some of these are known from exquisitely preserved, nearly complete skeletons.

Diplocynodon darwini

The most common Messel crocodyliform is the 2 to 2.5 m-long *Diplocynodon darwini* (Figs. 10.3.1, 10.3.3). It is part of a basal alligatoroid lineage found throughout Europe in rocks of Paleocene through middle Miocene age (e.g., Kuhn 1938; Aráez et al. 2016). Although not the oldest, some phylogenetic analyses suggest it is the basal-most species of *Diplocynodon* (Martin et al. 2014). It is also the best-known; dozens of nearly complete skeletons, from small juveniles (Fig. 10.3.1) to adults (Fig. 10.3.3), have been found. Indeed, *D. darwini* is one of the best-known fossil crocodyliforms in the world and gives us a glimpse of what early alligatoroids looked like.

Diplocynodon means "double dog tooth" and refers to the presence of two enlarged, canine-like teeth, or caniniforms, near the front of the lower jaw. In most crocodyliforms, only one lower tooth – usually the fourth – is enlarged, and when the jaws close it inserts either in a pit on the palate or in a notch on the side of the snout between the premaxilla and maxilla. In *Diplocynodon*, the third and fourth lower teeth are both enlarged – the "paired caniniform" condition (Fig. 10.3.4, top). *Diplocynodon* also has characteristic bipartite ventral osteoderms. That is, the bony plates in the skin of the belly region were composed of two elements sutured to one another front to back (Fig. 10.3.4, bottom).

What we now call *Diplocynodon darwini* used to be considered two separate species – *Alligator darwini* and *Crocodylus ebertsi*. Both were named by Rudolph Ludwig in 1877. The two forms were very similar, but differed primarily in how the teeth came together (occluded) when the jaws were closed. Like modern alligatorids, *Alligator darwini* had a complete overbite, with the paired caniniforms occluding in a pit on the palate. The caniniforms occluded in a notch on the side of the snout in *C. ebertsi* and could be seen when the jaws were closed, as they do in modern crocodylids. Ludwig was mistaken, however.

Two things led to the error. The first was an oversimplified view of how occlusal patterns evolved in crocodylians. Modern alligatorids have a complete overbite, whereas modern crocodylids have interlocking dentition in which the caniniforms and most of the lower teeth behind them are visible in closed jaws. Fossils were usually interpreted as if they had one of these two conditions, but in fact both are derived from an ancestral condition that no longer exists. The last common ancestor of alligators and crocodiles had a notch between the maxilla and premaxilla, allowing the fangs of the lower jaw to be exposed, but it had an overbite behind the notch. Alligatoroids lost the notch, and crocodyloids lost the overbite (Brochu 2003). Ludwig conflated the ancestral condition (exposure only of said fangs) with the derived crocodylid condition (exposure of most of the lower tooth row).

Fig. 10.3.1. Juvenile of the common alligatoroid *Diplocynodon darwini*. Scale: 2 cm.

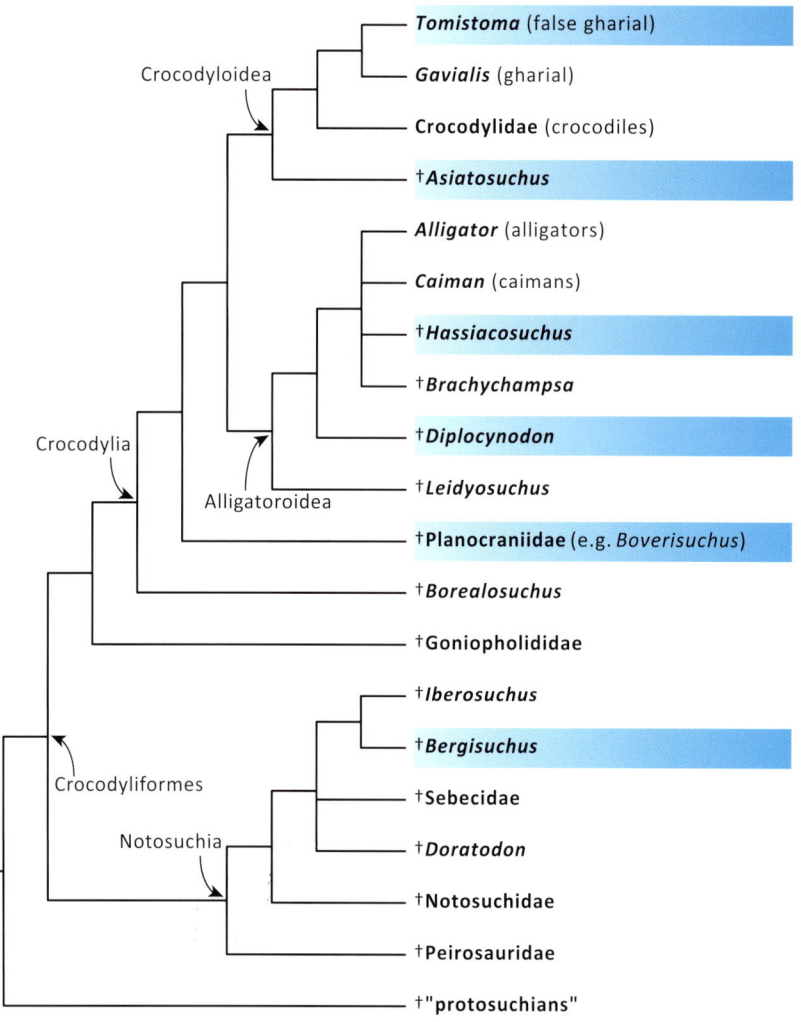

Fig. 10.3.2. Simplified phylogeny of Crocodyliformes. Taxa represented at Messel are highlighted in blue.

A close relationship between *Diplocynodon* and two North American lineages – *Leidyosuchus* and *Borealosuchus* – was once suspected (Rauhe & Rossmann 1995), but these genera do not appear to be closely related (Brochu 1997a, 1999; Wu et al. 2001; Brochu et al. 2012). Instead, phylogenetic analyses place *Diplocynodon* near the base of Alligatoroidea (Fig. 10.3.2).

Diplocynodon deponiae

The second species of *Diplocyndon* at Messel, *D. deponiae*, is less common. It was a much smaller animal – the largest known individuals are less than 1 m in length (Fig. 10.3.5). It was originally described as *Baryphracta deponiae* and not thought to be an alligatoroid (Frey et al. 1987). The first phylogenetic analyses to include it supported a close relationship with *Diplocynodon* (e.g., Brochu 1999), and later work, based on a more fully prepared complete skeleton, placed the species within *Diplocynodon* (Delfino & Smith 2012).

Juvenile specimens of *D. darwini* resemble *D. deponiae* in overall shape, but there are significant differences. For example, the large openings on top of the skull behind the eyes (supratemporal fenestrae) of juvenile *D. darwini* resemble those of other juvenile crocodylians – they are elliptical and shallow (Fig. 10.3.1). But in *D. deponiae*, they are constricted and have sharper margins resembling the supratemporal fenestrae of most caimans (Fig. 10.3.5, bottom). *D. deponiae* is also more heavily armored, and the tail is almost completely encased in osteoderms (Fig. 10.3.5, top).

Hassiacosuchus haupti

The third alligatoroid at Messel, *Hassiacosuchus haupti*, looks very different from *Diplocynodon* (Weitzel 1935). Its snout is deeper and shorter, and its

The second confounding factor was a change that frequently occurred in early forms of *Diplocynodon* during growth. The bone forming the roof and side of the occlusal pit for the fangs of the lower jaw was thin. It would commonly wear away as the animal grew, leaving the caniniforms exposed (Berg 1966). This happens frequently in modern caimans – the fang occludes in a pit in young individuals, but often in a notch in large animals (Kälin 1933). What Ludwig and others viewed as a taxonomically significant difference was a matter of ontogenetic (growth-related) variation (e.g., Rauhe & Rossmann 1995).

Fig. 10.3.3. Adults of the common alligatoroid *Diplocynodon darwini* seen from above (top) and below (bottom). The tail is almost devoid of osteoderms. Scale: 10 cm.

Fig. 10.3.4. Belly osteoderms (top) of *Diplocynodon darwini* are divided: the front part is connected to the back part by a suture (arrow). The third and fourth teeth in the lower jaw are enlarged and caniniform (bottom, here in *D. ungeri*). Scales: 2 cm.

posterior teeth were large and globular (Fig. 10.3.6). Though larger than *D. deponiae*, *Hassiacosuchus* was a small crocodylian; the largest known specimens are less than 2 m in length.

Hassiacosuchus is an alligatorine – an alligatorid more closely related to the living American alligator than to caimans. Unlike *Diplocynodon*, it belongs to a group found throughout the Northern Hemisphere during the Paleogene. It resembles several North American alligatorines, including *Allognathosuchus*, that also had bulbous back teeth. In fact, *Hassiacosuchus* was often considered a species of *Allognathosuchus* (e.g., Berg 1966; Wassersug & Hecht 1967; Rauhe 1990; Rauhe & Rossmann 1995), but phylogenetic studies suggest that the similarities between *Allognathosuchus* and *Hassiacosuchus* are plesiomorphic for alligatorines (Brochu 1999, 2004). They all had very short snouts and enlarged cheek teeth. Modern alligators secondarily lost the specializations found in their ancestors.

Historically, evolutionary biologists argued that animals with specializations for a specific habitat, diet, or function evolved from animals lacking these specializations, but that more generalized animals did not evolve from more specialized ancestors. This was called the "Law of the Unspecialized" (Cope 1896). Modern alligators are ecological generalists that will eat almost anything they can swallow, but early alligatorines such as *Hassiacosuchus* were much smaller and had cheek teeth suggesting a diet focused on hard-shelled prey (Abel 1928; Ősi 2014). Based on the Law of the Unspecialized, we would expect the ancestors of *Hassiacosuchus* to resemble modern *Alligator* in overall shape.

In fact, the opposite is true. The earliest alligatorids and close alligatorid relatives, such as *Brachychampsa*, resembled *Hassiacosuchus* in many ways. They were small and had short, blunt snouts with enlarged back teeth (Williamson 1996; Brochu 2004). They lived alongside other crocodyliforms lacking these specializations. Crocodyliform diversity in the Northern Hemisphere dropped sharply in the late Eocene (Hutchison 1982; Markwick 1998; Mannion et al. 2015). The only remaining lineage in North America after the Eocene was *Alligator*, which was descended from a form resembling *Hassiacosuchus*. In the absence of large generalized crocodylians, *Alligator* progressively grew less specialized (Camp 2016). In this case, at least, a generalist evolved from a specialist.

4Fig. 10.3.5. Holotype of *Diplocynodon deponiae* seen from below (top). The tail is completely covered with osteoderms. Scale: 5 cm. Skull of *D. deponiae* (bottom), showing the restricted supratemporal fenestrae (arrow). Scale: 1 cm.

Hassiacosuchus haupti

stf

Asiatosuchus germanicus

The largest Messel crocodyliform was a 4-m-long crocodyloid currently named *Asiatosuchus germanicus* (Fig. 10.3.7). It is neither as common nor as well known as *D. darwini*, but it was the largest predator in that ecosystem. It would have resembled a modern crocodile in general appearance and, like modern crocodiles, preyed on anything smaller than it.

Asiatosuchus is the most completely known of several early crocodyloids found throughout the Northern Hemisphere during the early and middle Eocene (e.g., Berg 1966; Delfino & Smith 2009; Wang et al. 2016). As such, it provides critical information on early conditions in the line that gave rise to the pantropical radiation that dominates modern crocodylian diversity today (e.g., Salisbury & Willis 1996). This is a group that crossed major marine barriers – including the Atlantic Ocean – to achieve its present distribution (Brochu 2000; Meredith et al. 2011; Oaks 2011) and became one of the most morphologically diverse of all crocodylian clades.

Tomistominae – Gharials in Europe

A second possible crocodyloid is represented by a fragment of lower jaw belonging to a slender-snouted crocodylian referred to Tomistominae – a group including the modern false gharial (*Tomistoma schlegelii*) and its extinct relatives – by Rossmann (2002). It would have resembled the tomistomines known from the early and middle Eocene elsewhere in Europe, such as *Kentisuchus* and *Megadontosuchus* (Mook 1955; Brochu 2007; Zvonok & Skutschas 2011; Jouve 2016).

The taxonomic status of the tomistomines is disputed. Morphological studies usually support trees in which tomistomines are crocodylids (Norell 1989; Salisbury & Willis 1996; Brochu 1997b). However, genetic studies support a close relationship between *T. schlegelii* and the Indian gharial, *Gavialis gangeticus* (e.g., Roos et al. 2007; Gold et al. 2014). Efforts to resolve this conflict continue.

Fig. 10.3.6. The short-snouted alligatorine *Hassiacosuchus haupti*, seen from the left. Inset: bulbous, crushing teeth of this species. Scale: 5 cm.

Boverisuchus magnifrons

One of the most unusual of the Messel crocodyliforms is the 2-m-long planocraniid *Boverisuchus magnifrons* (Fig. 10.3.8). Although the Messel material is fragmentary, *B. magnifrons* is known from well-preserved specimens at Geiseltal (Kuhn 1938; Berg 1966; Rossmann 1998, 2000a; Brochu 2013), and in lateral view its skull would have closely resembled that of its close North American relative, *B. vorax* (Fig. 10.3.8, bottom). The Messel and Geiseltal form was called *Pristichampsus rollinati* until recently.

Boverisuchus was a ziphodont crocodyliform – that is, it had flattened and serrated teeth resembling those of predatory dinosaurs (Langston 1975; Fig. 10.3.8, upper left). It can also be called a "hoofed crocodile"; it had the same number of fingers and toes as any other crocodyliform, but each ended in a short bone that would have supported a hoof-like claw. Its snout was compressed and deep. Claims that *Boverisuchus* could have walked or run bipedally (Rossmann 2000b) are doubtful, but it probably spent less time in the water than other Messel crocodyliforms and was one of the larger land predators (Rossmann 1999, 2000a, b; Hastings & Hellmund 2017).

Planocraniids belong to the crocodylian stem-group (Fig. 10.3.2). Their early history is poorly known; their lineage must have existed in the Late Cretaceous, but they do not appear in the fossil record until the Paleocene, and the earliest forms already show the flattened teeth and deep, compressed snout seen in *Boverisuchus* (Li 1984). That they were already highly modified when they first appear complicates our efforts to resolve their relationships to other crocodyliforms – characters shared with other lineages in the earliest planocraniids may have been overprinted by the specializations seen in the forms we know.

Fig. 10.3.7. Skull of the large crocodyloid *Asiatosuchus germanicus* seen from above (top) and obliquely (bottom). A notch receives the caniniforms of the lower jaw. Scales: 5 cm.

Fig. 10.3.8. The planocraniid *Boverisuchus*. Top left: an isolated tooth of *Boverisuchus magnifrons* from Messel. The sharp edge and serrations are not found in living crocodylians. Scale: 1 cm. Top center: claw-bearing finger bone from below. Scale: 17 mm. Top right: narrow snout and lower jaws of *B. magnifrons* from Messel. Bottom: complete skull of *B. vorax* from Wyoming, USA. Scale: 5 cm.

Bergisuchus dietrichbergi

The most enigmatic Messel crocodyliform is *Bergisuchus dietrichbergi*. Like *Boverisuchus*, it is known from very incomplete material at Messel (Fig. 10.3.9), and like *Boverisuchus*, it had a compressed snout and flattened, serrated teeth (Rossmann et al. 2000). But *Bergisuchus* is very distantly related to the other crocodyliforms at Messel. In fact, it does not even belong to Crocodylia (Fig. 10.3.2).

The precise relationships of *Bergisuchus* are unclear. A similar and probably closely related crocodyliform, *Iberosuchus,* is known from France and Spain. They have been compared with groups known otherwise from the southern continents, such as the sebecids and peirosaurids (e.g., Berg 1966; Buffetaut 1989; Rabi & Sebök 2015). These groups belong to Notosuchia, a radiation found in the Cretaceous and Cenozoic throughout the Southern Hemisphere, especially in South America. A few, such as *Doratodon*, are known from Eurasia during the Cretaceous (e.g., Dalla Vecchia & Cau 2011; Puértolas-Pascual et al. 2016). Most notosuchians were terrestrial predators; a few were semiaquatic, and some were herbivorous (Pol & Leardi 2015).

It is unclear where *Bergisuchus* and *Iberosuchus* came from. Apparently, intermittent connections between parts of Europe and some of the southern

Fig. 10.3.9. Incomplete lower jaw of the only specimen of *Bergisuchus dietrichbergi* from Messel. Scale bar: 1 cm.

land masses existed during the Cretaceous (Buffetaut 1989), which might explain the presence of Notosuchia in Europe.

The crocodyliform community

Modern crocodylians are semiaquatic ambush predators. They are opportunistic and eat anything they can catch and swallow in or near the water. Hatchlings and small juveniles eat arthropods and add small vertebrates to their diet as they grow. The larger they are, the larger the prey they can subdue. The crocodyliforms at Messel, *Boverisuchus* and *Bergisuchus* excepted, probably lived the same way.

The most striking aspect of the Messel crocodyliform fauna is its diversity, with respect to both species richness and disparity. Seven species are known. There is no modern analogue for this – the largest number of crocodylian species in the same region today is four, few places have more than two, and neither land-dwelling ziphodont forms nor crocodylians with enlarged crushing teeth currently exist (Grigg & Kirshner 2015).

In the strictest sense, they may not have actually co-existed very much. This is how different crocodylian species in the same area avoid competition today – they tend to segregate based on environmental factors such as water depth, salinity, lake or river size, nesting space, or substrate (e.g., Rebêlo & Lugli 2001; Villamarín et al. 2011). Moreover, because they are quite rare compared to *Diplocynodon darwini*, most of the crocodyliforms at Messel may have primarily lived elsewhere and only occasionally occurred in the lake (Micklich 2007).

Crocodyliforms living in the same region usually have different snout shapes, and these are often perceived as indicating different prey preferences. Those with long and slender snouts, for example, are said to be specialized on catching and eating fish. Modern *Gavialis* primarily eats fish, but other living slender-snouted crocodylians have a broader dietary range. This includes *Tomistoma*, which preys on monkeys (Galdikas 1985) and has been known to kill and eat adult humans (Rachmawan & Brend 2009). The same is true for extant crocodylians in general – skull shape can have a profound impact on function (e.g., Pierce et al. 2008; Erickson et al. 2012; Ösi 2014), but there is surprisingly little evidence linking snout shape differences to diet. Nevertheless, the disparity in skull shape seen at Messel and similar sites, like Geiseltal (Hastings & Hellmund 2016) may indicate dietary differences between some species.

The large, bulbous teeth in the back of the mouth of *Hassiacosuchus* may indicate a preference for hard-shelled prey (Abel 1928; Carpenter & Lindsey 1980; Ősi 2014). Crocodylians have extremely powerful jaw muscles and are capable of crushing turtle shells (Erickson et al. 2012). That *Hassiacosuchus* was relatively small suggests that smaller hard-shelled animals, such as mollusks, may have been a more important component of its diet. But this is speculative - there is nothing comparable to *Hassiacosuchus* alive today.

Bergisuchus and *Boverisuchus* are usually thought to have been more terrestrial than other Messel crocodyliforms. The blunt, hoof-like toe bones of *Boverisuchus* are certainly consistent with more terrestrial habits, and the flattened, serrated teeth of both resemble those of predatory dinosaurs. The limb bones of *Boverisuchus* are robust, and muscle attachment scars on these bones are prominent (Rossmann 2000b), suggesting more time spent walking rather than swimming. *Boverisuchus* was probably a land-based predator; the same is true for *Bergisuchus*, although we know much less about its anatomy.

Although small compared to most modern crocodylians, the Messel ziphodont crocodyliforms were among the largest land predators in the ecosystem. The Cenozoic is often called the "Age of Mammals," but the dominant land predators at Messel may have been crocodyliforms; in a sense, the Age of Reptiles had not yet ended.

Chapter 11
Birds – the Most Species-rich Vertebrate Group in Messel

Gerald Mayr

Among the terrestrial vertebrates in Messel, birds represent the majority of all fossil specimens. The collection of the Senckenberg Research Institute in Frankfurt alone comprises around 1,000 skeletal remains of birds (Mayr 2016a), and numerous additional ones are found in the other larger collections. Approximately 70 different species of birds from the oil shale deposits could be distinguished to date, although not all of them have been scientifically described yet. In regard to their species diversity, birds exceed all other vertebrate groups from Messel (Box 11.1). Therefore, a detailed knowledge of the avifauna of Messel is of great significance for our understanding of the ecosystem that encompassed Lake Messel.

In recent years, numerous avian fossils have been described from Eocene deposits, but no other fossil site offers a volume of specimens comparable to Messel, and only few others have yielded similarly complete remains. While most mammalian groups from Messel had been previously known from other fossil sites, many of the bird species were documented for the first time in Messel, and several are still only known from this site. Moreover, since the study of bird fossils began rather late, the knowledge gained in regard to the Messel avifauna over the last 20 years surpasses that of most other vertebrate groups from the site.

Many of the birds from Messel can be assigned to phylogenetic lineages still in existence today. However, some of them belong to groups that are probably unfamiliar to most readers. These will be introduced later in their respective places; in order to better understand the following text, we simply give a brief overview here of a few new insights regarding the relationships among extant birds (Neornithes) (Fig. 11.2). Recent birds can be divided into two groups that differ in regard to the structure of their bony palate. The Palaeognathae ("old jaws") comprise the ostrich-like birds (ratites) and their relatives, while the Neognathae ("new jaws") include all remaining species. Neognathous birds are further subdivided into two groups, the Galloanseres, comprising the gallinaceous birds and waterfowl, and Neoaves, which include the majority of all living birds today. The relationships among the Neoaves are insufficiently understood, but analyses of genetic data support three major groups that are termed Strisores, Aequornithes, and Telluraves (Fig. 11.2; Ericson et al. 2006; Hackett et al. 2008; Prum et al. 2015). The taxon Strisores contains nightjars and allies as well as swifts and hummingbirds. Aequornithes comprises the aquatic and semi-aquatic groups within the Neoaves, while Telluraves, in addition to owls, diurnal raptors, and parrots, includes the majority of all small, tree-dwelling bird species.

Almost certainly the split of the Neornithes into Palaeognathae and Neognathae already took place in the Mesozoic. However, the fossil record of neornithine birds from the Cretaceous is rather sparse, although remains of presumed representatives of the

Fig. 11.1: The Messel swift *Scaniacypselus szarskii*. Scale: 1 cm.

Galloanseres are known from the Late Cretaceous. In addition, fossils of several highly specialized avian groups from the Paleocene, including penguins and owls, suggest that the diversification of the Neoaves already began near the end of the Cretaceous (Mayr 2009a, 2016a). What is missing from Cretaceous deposits to date are the tree-dwelling taxa of the Telluraves. This is particularly noteworthy since fossils of the Enantiornithes – an extinct group of Mesozoic birds that were adapted to life in the trees – are not rare in some Cretaceous fossil sites. It therefore appears unlikely that the lack of remains of arboreal Telluraves in the Cretaceous is simply due to poor preservation; presumably, the radiation of these birds only occurred at the beginning of the Cenozoic, after the Enantiornithes had already become extinct (Mayr 2016a).

All birds found in Messel belong to the Neornithes, i.e., to the crown group that comprises the modern birds. Yet, as outlined below, the birds in Messel differed in many aspects from modern avifaunas, both in the tropics as well as in temperate zones.

Large ratites and other terrestrial species

During the early Eocene, Europe was largely isolated geographically from other continents, and one of the peculiarities of the fauna at that time is the lack of large predatory mammals. As a result, several birds lost their ability to fly, and the avifauna of Messel includes an unusually high number of flightless species. Today, the distribution of flightless birds continues to be restricted to predator-free oceanic islands and to continents such as South America and Australia, which during the Cenozoic were not connected by land bridges to the northern continents, where carnivorous mammals underwent several large radiations. An exception is the African ostrich; however, this fast and aggressive species is able to defend itself and escape from potential predators by running.

The ground-dwelling avian taxa from Messel discussed in the following belong to a number of only distantly related groups, several of which occupy a basal systematic position among modern birds. This applies, in particular, to the two representatives of the Palaeognathae known from Messel.

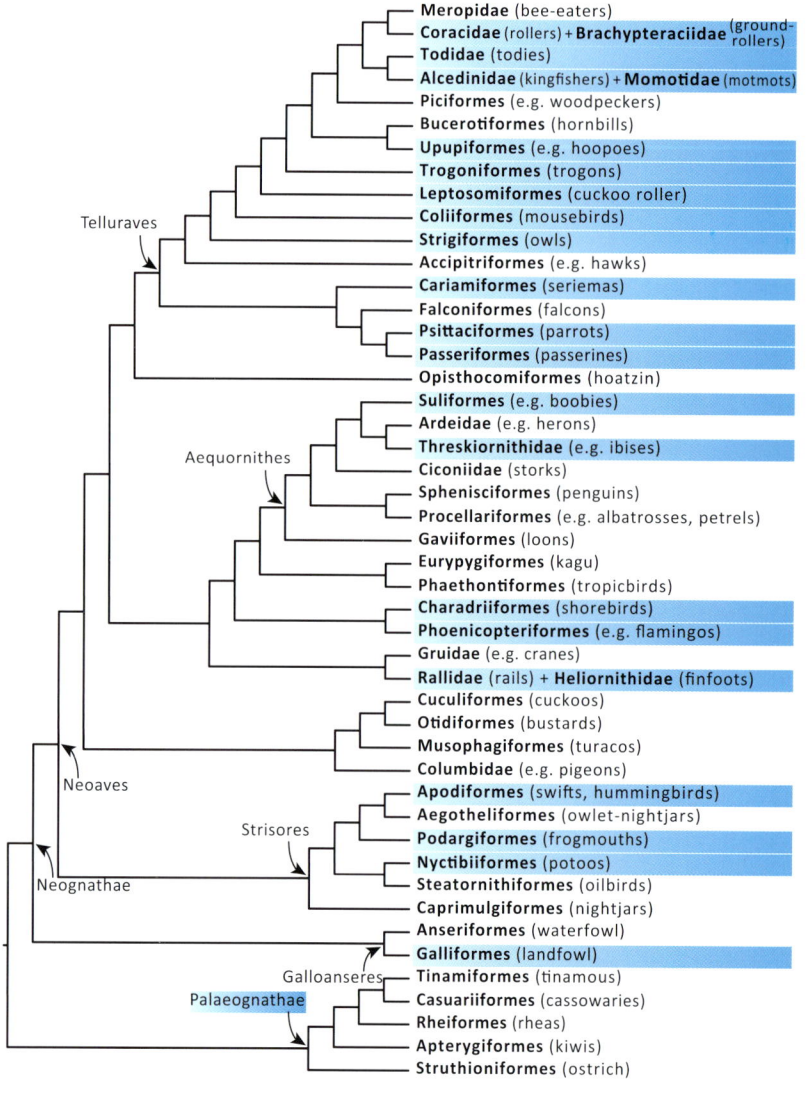

Fig. 11.2: Relationships among extant birds (Neornithes), based on analyses of genetic data. Taxa highlighted in blue have been documented from Messel.

Box 11.1: Why have so many birds been preserved in Messel?

The frequency of well-preserved remains of vertebrates in Messel is almost unparalleled, and birds and bats make up the largest part of these discoveries. This may be explained by the hypothesis that from time to time, toxic gases of volcanic origin exuded from the Messel Lake, killing a number of animals in the lake's surroundings (Chapter 4). Similar phenomena are known from recent maar lakes, where unfavorable conditions can lead to a catastrophic mass die-off near the shores (Franzen & Köster 1994). However, the overall frequency of vertebrates in relation to the Messel Lake's long period of existence is not particularly high, and to date, no layers have been discovered that show a concentrated occurrence of vertebrate skeletons. In regard to the birds in Messel, other causes of death appear more likely, and a striking number of skeletons show fractures of the long bones that cannot be explained by fossilization processes. It is therefore possible that a large part of the birds found in Messel constitute individuals that sustained injuries and flew over the open lake, where they dropped into the water and drowned.

The photos show examples of fractured long bones. Left: fractured leg bone of *Eurofluviovirida-vis*, right: fractured humerus of the Messel roller, *Eocoracias* (bone digitally cropped and background lightened). The arrows point to the respective bone ends. Scale: 1 cm.

The palaeognathous birds in the Messel forest

With the exception of the Neotropical tinamous (Tinamiformes), extant palaeognathous birds only comprise large, flightless species such as the African ostrich or the South American rheas. During the Paleocene and Eocene, however, there existed a widespread group of palaeognathous birds that were able to fly, which are known from fossil sites in North America and Europe. These birds are collectively known as Lithornithidae and were first described from the Lower Eocene of England and North America, where they are rather common in certain sites. The structure of the wings and shoulder girdle implies that the Lithornithidae possessed a well-developed ability to fly, which clearly surpassed that of the tinamous – the only flyers among extant palaeognaths. On the other hand, the long bill suggests that

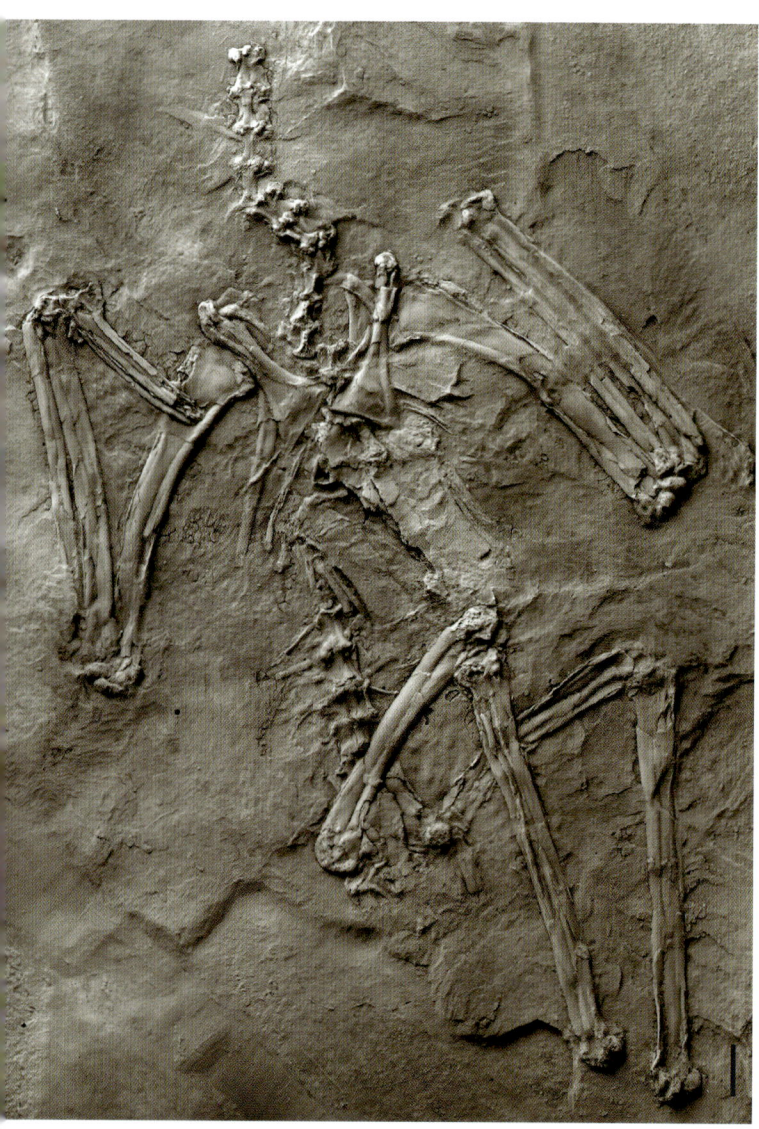

Fig. 11.3: Skeleton of an unidentified species of the Lithornithidae; the skull is not preserved. Scale: 1 cm.

Fig. 11.4: Skull of the lithornithid from Messel. Scale: 5 mm.

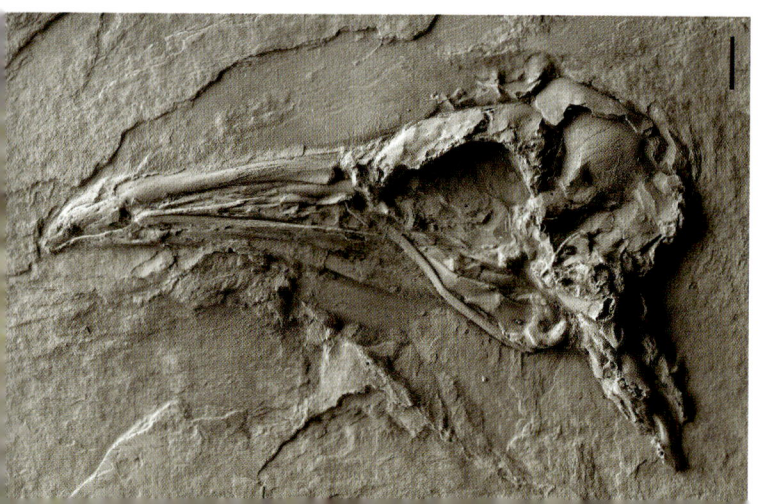

Fig. 11.5: Freestanding preparation of a skeleton of *Palaeotis weigelti*, a flightless palaeognathous bird. Scale: 5 cm.

these birds foraged on the ground. To date, one poorly preserved partial skeleton and one skull from Messel could be assigned to the Lithornithidae (Fig. 11.3, Fig. 11.4; Mayr 2008a, 2009b). Although the exact species affiliation is uncertain, these remains are among the geologically youngest discoveries of any members of the Lithornithidae.

Palaeotis weigelti was a flightless palaeognathous bird from the Lower Eocene of Europe. This long-legged species stood at a height of almost 1 meter and is only known from Messel by two skeletons and a few fragmentary remains (Fig. 11.5; Peters 1988a; Mayr 2015a). *P. weigelti* also occurs in the lignite deposits of Geiseltal, where the species was first described. The wings of *Palaeotis* are not as severely reduced as those of the cassowaries and emus, and compared to modern Palaeognathae, the skeleton most closely resembles the African ostrich and the South American rheas. However, the exact relationships to other groups have not yet been conclusively clarified. From a zoogeographic point of view, it appears most likely that *Palaeotis* is an early representative of the ostrich-like birds. Alternatively, a closer relationship to the South American rheas has been postulated, and at this point it is even possible that the skeletal traits of *Palaeotis* developed independent of ostriches and rheas.

The distribution of the modern palaeognathous birds is restricted to the southern continents, although the historical range of the ostrich extended from Africa all the way to China (Mayr 2016a). Until recently, it was assumed that the flightless representatives of the Palaeognathae were monophyletic, thereby suggesting that the ancestor of these birds was flightless as well. During the Mesozoic, the southern land masses formed the "supercontinent" Gondwana, and for a long time palaeognathous birds served as a textbook example for a southern distribution due to the breaking-apart of Gondwana in the course of plate-tectonic processes. The discovery of flightless ratites in the Eocene of the Northern Hemisphere does not fit this picture, since there existed no land bridges between Europe and most of the southern continents in the late Mesozoic and early Cenozoic. However, recent analyses of the relationships among palaeognathous birds (Prum et al. 2015) suggest that the flying tinamous occupy a phylogenetic position in the midst of the flightless species, which implies repeated, independent losses of flight capabilities in palaeognathous birds. This had already been suspected by several earlier authors and would render moot the assumption of land bridges in order to explain the geographic distribution of palaeognathous birds.

Fig. 11.6: Mold (with synthetic resin cast) of the femur of *Gastornis* cf. *geiselensis*. More completely preserved specimens of this giant flightless bird are also known from Geiseltal and several fossil sites in North America. Scale: 5 cm.

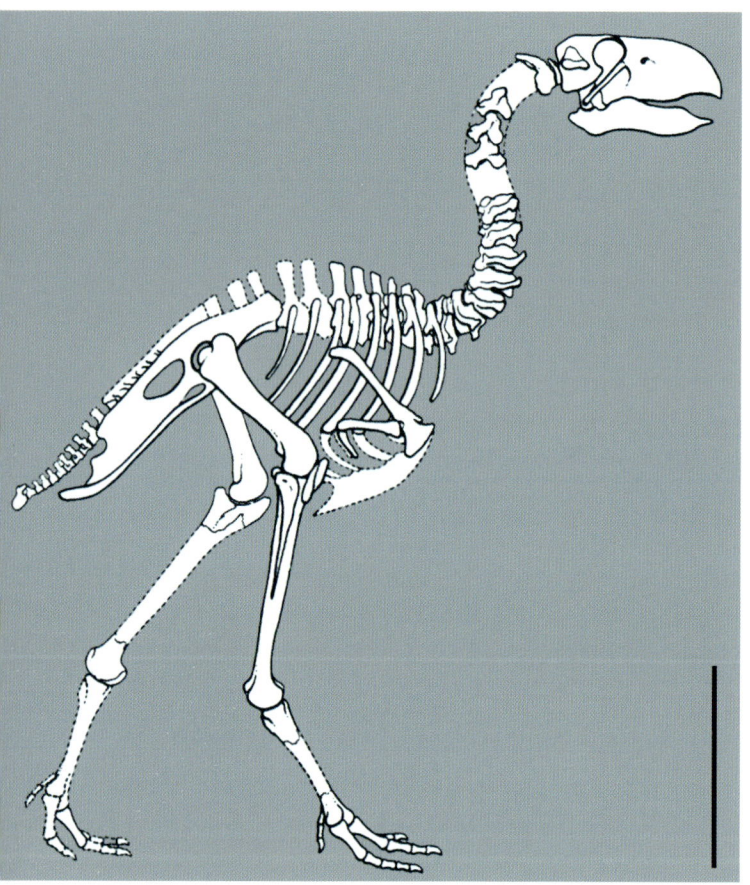

Fig. 11.7: Reconstructed skeleton of *Gastornis*. Scale: 50 cm.

Gastornithidae

In Messel, one of the most spectacular flightless birds, which occurred during the early Cenozoic in Europe, is only represented by the hollow mold of a single femur (Fig. 11.6). *Gastornis* – also known under its former name, *Diatryma* – is a representative of the Gastornithidae, skeletal remains of which have been described from numerous other fossil sites in Europe, North America and China (Mayr 2009a). The oldest fossils stem from the Paleocene in France, England and Germany, while North American discoveries are only known from the early Eocene. Substantial remains of these birds were also found in Geiseltal. All *Gastornis* species were giant, flightless birds, which stood up to 170 cm tall and had an exceptionally massive beak (Fig. 11.7). Several skeletal traits of the skull indicate a close relationship to the Galloanseres. The distribution of the Gastornithidae offers impressive proof of the existence of a land bridge between Europe and North America in the early Eocene, and remains of *Gastornis* from the Canadian Ellesmere Island show that the dispersal corridor was situated far to the north (Stidham & Eberle 2016). For a long time, it was believed that the enormous beak of *Gastornis* indicates a carnivorous lifestyle. Recent studies, however, concluded that these birds were specialized plant eaters instead (Andors 1992; Mayr 2009a; Angst et al. 2014). It is remarkable that a very similar group of birds, the Dromornithidae, evolved in the Cenozoic in Australia, independent of the Gastornithidae. While the demise of the Gastornithidae is probably due to the evolution of large mammalian predators on the northern continents, the Dromornithidae persisted on the geographically isolated Australian continent into the Pleistocene.

The gallinaceous bird *Paraortygoides*

The only gallinaceous bird described from Messel to date, *Paraortygoides messelensis*, belongs to the extinct taxon Gallinuloididae, which was first described from the Green River Formation. *Paraortygoides* reached about the size of a partridge (Fig. 11.8) und clearly differs from its modern relatives in the structure of the shoulder girdle and sternum. These differences suggest that *Paraortygoides* did not yet possess a large crop, which in extant gallinaceous birds serves to store and pre-digest the food that consists primarily of hard seeds. It is therefore likely that the species from Messel mainly fed on insects or fruits, and the transition to a grain-based diet in the evolution of gallinaceous birds may be connected to the spread of open grasslands in the latter part of the Cenozoic (Mayr 2006a). While *Paraortygoides* presumably foraged for food near the ground, the relatively low attachment of the hallux indicates that the taxon was better adapted to an arboreal life than most of the extant gallinaceous birds.

Seriemas

Several species apparently related to seriemas (Cariamiformes), which today comprise only two South American species, occur in Messel. Seriemas feed

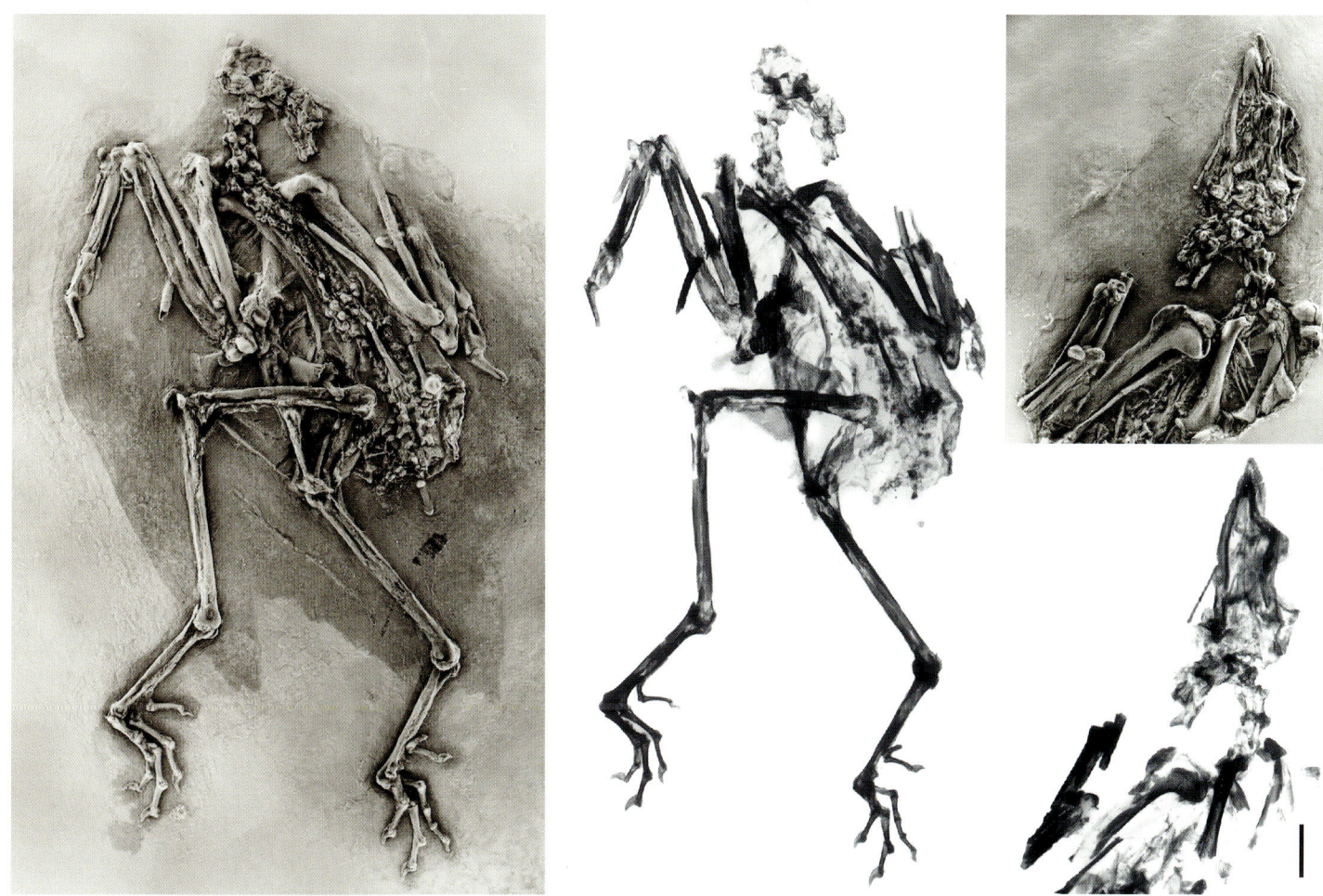

Fig. 11.8: The gallinaceous bird *Paraortygoides messelensis*. This species belongs to the taxon Gallinuloididae, which occupies a basal phylogenetic position within the Galliformes. The illustrated skeleton (with X-ray image) is preserved in two parts. Scale: 1 cm.

on insects and small vertebrates, and they are poor flyers. According to recent genetic analyses, seriemas are not related to the crane-like birds, with which they had been traditionally placed, but rather show close ties to falcons, parrots, and passerines (Ericson et al. 2006; Hackett et al. 2008; Prum et al. 2015); however, to date no morphological traits have been found that would support this hypothesis. Cariamiform birds showed a high diversity in the Cenozoic and were widely distributed across the Northern Hemisphere. Three definitive representatives of this group have been found in Messel, and several other species are most likely related to the Cariamiformes as well.

The unambiguously identified Cariamiformes from Messel belong to the Idiornithidae, a group of birds first described from late Eocene and Oligocene fossil sites in France. Their skeletal morphology is reminiscent of modern seriemas, although most species were noticeably smaller, and many anatomical traits, including the shape of the bill, remain unknown to date. The specimens from Messel are among the oldest representatives of the Idiornithidae. Two species of the taxon *Dynamopterus* (formerly known as *Idiornis*) have been documented by one foot each (Fig. 11.9). They show clear differences in size and the robustness of their foot bones, which indicate different modes of locomotion and possibly different

Fig. 11.9: Tarsometatarsi of two *Dynamopterus* species, which clearly differ in regard to size and bone proportions. Scale: 1 cm.

habitat preferences. An additional species, *Dynamopterus tuberculatus*, is one of the few large birds in Messel represented by a complete skeleton. An unusual trait of *D. tuberculatus* is the fact that many vertebrae and several other bones are covered with small bony tubercles (Peters 1995; Fig. 11.10). Similar structures are unknown in modern birds, and their function remains a mystery. The rather short wings of *D. tuberculatus* suggest greatly reduced flight capabilities.

Strigogyps

Currently, the roughly turkey-sized *Strigogyps* (formerly *Aenigmavis*) *sapea* (Fig. 11.11) can only be assigned to the Cariamiformes with reservation. Based on the wing skeleton, this species was also flightless, or nearly so (Peters 1987; Mayr 2005a). In addition to *S. sapea*, a smaller species of *Strigogyps* occurred at Messel, but it is only documented by the remains of a wing and has not yet been named. Another species of *Strigogyps* is known from Geiseltal, where two halves of a skeleton were found that were initially identified as the remains of a New World vulture (the half with legs) and those of a hornbill (the half with wings), respectively. None of the skeletons described to date includes a skull, which limits any inferences regarding the birds' possible lifestyle. While the unusually strong feet are equipped with large, raptor-like claws, the stomach content of one individual contains plant remains (Mayr & Richter 2011).

The relationships of *Strigogyps* are uncertain, but except for the powerful claws and certain characteristics related to their flightlessness, the taxon resembles another species, *Salmila robusta*, of which three skeletons have been found to date (Fig. 11.12; Mayr 2002a). Due to the unusual mosaic of traits characteristic of different avian groups, a phylogenetic assignment of *S. robusta* is difficult, even though the species shows similarities to the Cariamiformes in certain features, especially in the structure of sternum and wing bones. However, contrary to cariamiform birds, *S. robusta* had relatively long tail feathers, short legs and a slender bill, which indicate a very different habitat use. It is possible that the species was better adapted to an arboreal lifestyle.

Fig. 11.10: Skeleton of *Dynamopterus tuberculatus*, one of the few large birds represented in Messel by a complete skeleton. The image on the right shows a detail of the cervical vertebrae, which are covered with bony tubercles. Scale: 5 cm.

The Messel rail

There exists no doubt about the terrestrial habits of the most common species of bird known from Messel, the so-called Messel rail, *Messelornis cristata*. This species represents about two thirds of all avian specimens from Messel and is documented by several hundred skeletal remains (Fig. 11.13). Messel rails were about the size of the extant Corncrake (*Crex crex*). Obvious size differences between individuals indicate a pronounced sexual dimorphism (Hesse 1990). Messel rails were initially considered relatives of the South American sunbitterns (Eurypygidae), but many skeletal traits instead demonstrate a close relationship to the finfoots (Heliornithidae) and rails (Rallidae) (Mayr 2004a). It was also assumed that in one

Fig. 11.12: *Salmila robusta*, another presumed relative of the seriemas. Scale: 1 cm.

Fig. 11.11: *Strigogyps sapea*, a flightless bird that may be related to the seriemas (Cariamiformes). Scale: 5 cm.

specimen of *Messelornis* a skin flap is preserved on the head, which led to the Latin species name "*cristata.*" However, it is more likely that the presumed soft tissue remains actually represent plant material that found its way underneath the bird's head. Despite the large number of *Messelornis* specimens, only very few of them show plumage remnants, and those are usually very poorly preserved. Since the preservation of feathers in fossil birds from Messel is due to the fossilization of melanosomes (Box 11.2), this may indicate that the feathers of *Messelornis* only contained a small number of these melanin-producing cell organelles and therefore either had a very light coloration, or their colors were based on other pigments. Very few specimens of *Messelornis* are preserved with stomach contents, and if so these consist of single fruit seeds. Therefore, the species presumably fed mainly on invertebrates whose re-

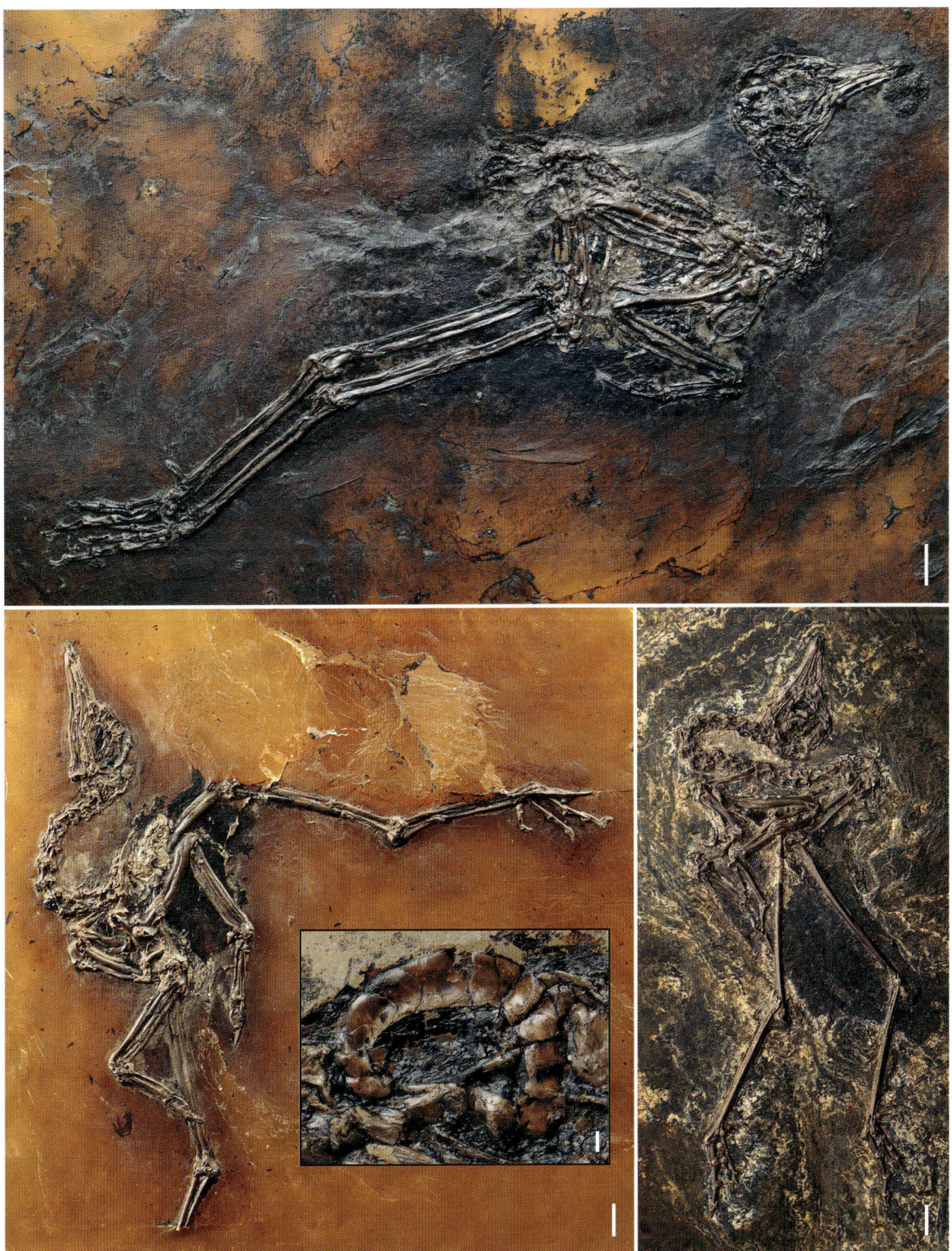

Fig. 11.13: The Messel rail, *Messelornis cristata,* is by far the most frequently found bird in Messel. Insert bottom left: Detail of the sclerotic ring (bony plates surrounding the eye). Scale: 1 cm.

Box 11.2: The oldest nectarivorous bird

The unique preservation conditions at the Messel fossil site frequently bear witness to the ecological interactions in a bygone ecosystem. A particularly impressive example for this is the specimen of the small bird *Pumiliornis tessellatus* illustrated here (left, scale: 5 mm), whose stomach contains a large amount of pollen grains (top right, scale: 5 mm) that emit a greenish glow in the image taken under a fluorescence microscope and are easily recognizable (bottom right, scale: 200 µm) (Mayr & Wilde 2014). The long beak of *Pumiliornis* is similar to that of certain extant sunbirds, suggesting that this may be the first example of a specialized bird-flower interrelationship. *Pumiliornis* almost certainly did not feed on the rather nutrient-poor pollen itself, but ingested it while foraging for nectar. Unfortunately, it has not been possible to definitively identify the plant that these pollen grains originated from, since they belong to a widespread type that is found in several, not closely related groups of plants. However, the large size of the pollen grains suggests that they belong to a species specialized in animal pollination, since the pollen grains of wind-pollinating plants are usually very small. A nectarivorous diet is only encountered today among hummingbirds, a few parrot species, and some passerines (e.g., sunbirds). Modern representatives of these groups are only known from significantly younger fossil sites, and the discovery from Messel constitutes the oldest record of a flower-visiting bird.

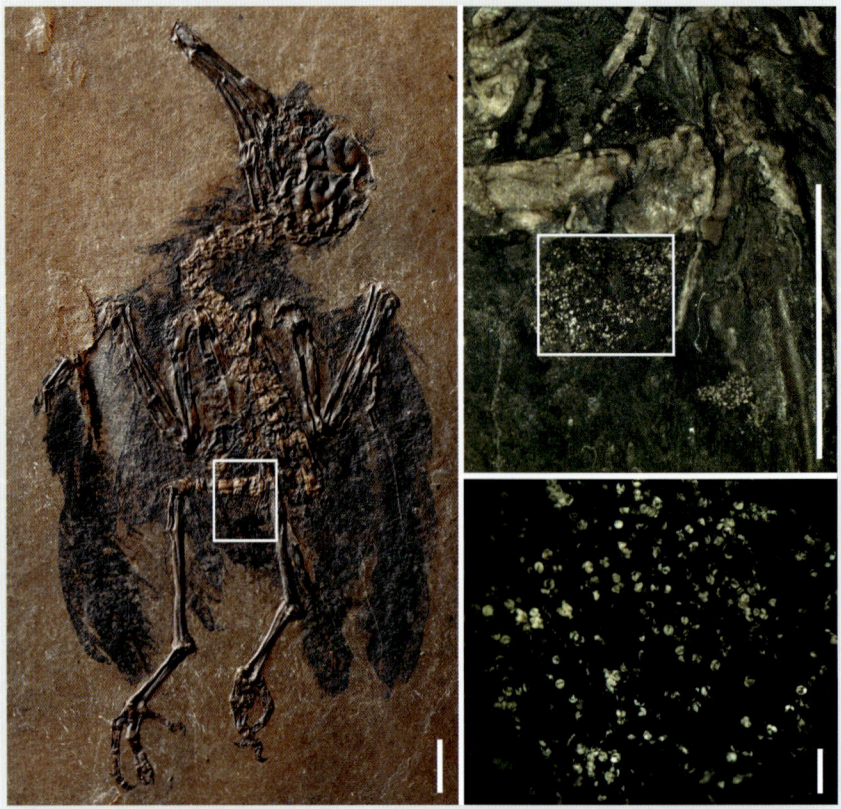

mains left no fossilized traces. The rather short toes suggest that *M. cristata* was a forest dweller, similar to the extant African Nkulengu rail (*Himantornis haematopus*). A terrestrial lifestyle near the shores could explain the exceptional frequency of *Messelornis* remains, and it appears highly likely that Messel rails also foraged in the immediate vicinity of Lake Messel.

Bird life at water's edge

It has long been known that true water birds are extremely rare in Messel (Peters 1988b, 2006). One of the few aquatic species was described as *Juncitarsus merkeli* and is only known from a single skeleton, which is housed in a private collection (Fig. 11.14). *J. merkeli* is among the largest birds discovered in Messel and stood approximately 60 cm tall. The species is characterized by unusually long legs; the length of the tarsometatarsus (a bone of the lower leg) corresponds to that in extant flamingoes. The taxon *Juncitarsus* was initially described on the basis of fossils from the Eocene of North America and was classified as a flamingo (Phoenicopteriformes). However, all of the North American fossils are rather fragmentary, and it was only the specimen from Messel that offered insights into the bill shape of *Juncitarsus* (Peters 1988b). This clearly differs from the characteristic flamingo bill and more closely resembles the bill of cranes or shorebirds (Charadriiformes). In fact, the shorebird-like skeletal traits of *Juncitarsus* were considered an indication of a close relationship between the flamingoes and Charadriiformes. However, several recent phylogenetic studies revealed that the grebes (Podicipediformes) actually constitute the closest relatives of the flamingoes, which is certainly one of the most surprising results of modern phylogenetic analyses. Externally, flamingoes and grebes differ greatly in appearance, and while the extremely long-legged flamingoes filter tiny food particles from the water, grebes use their rather short legs for forward propulsion when diving. Several skeletal traits suggest that *Juncitarsus* is not part of the direct evolutionary line of the flamingoes, but rather constitutes a sister taxon of flamingoes and grebes (Mayr 2014). The closest extant relatives of flamingoes and grebes are uncertain, but at least one newer study indicates that they may indeed be related to the Charadriiformes (Prum et al. 2015). The similarities between *Juncitarsus* and certain Charadriiformes could therefore be explained by the retention of primitive traits (plesiomorphies) of the last shared ancestor of Charadriiformes, flamingoes and grebes.

Another bird that may have been attracted to the open surface of the Messel Lake is only documented by a single skull (Fig. 11.15). *Masillastega rectirostris* was described as a probable relative of the gannets and boobies (Suliformes) (Mayr 2002b), which would make it the oldest fossil record of a group that is exclusively restricted to marine habitats today. However, a definitive identification of the only available specimen is not possible, and we can only hope that more complete skeletal remains will be found in the future that elucidate the systematic relationships of this relatively large bird.

Much more common are records of *Rhynchaeites messelensis*, a very basal representative of the ibis family (Threskiornithidae). This species (Fig. 11.16) was among the first birds described from Messel – in the year 1898. Not only is *Rhynchaeites* clearly smaller than any of the extant ibis species, it also differs from its modern relatives in numerous primitive traits. It has noticeably short legs, which indicate that *Rhynchaeites* most likely foraged on the forest floor, rather than along the shores of the Messel Lake. The long, down-curved shape of its bill is similar to that of its modern relatives (Peters 1983), although the tip of the bill – contrary to extant ibises – only shows few openings for the sensory nerves that lead to the tactile corpuscles. This suggests that *Rhynchaeites* did not use its bill to probe for food in the soil, but likely picked it off the surface "on sight."

An incomplete skeleton of a plover-like bird (Charadriiformes) constitutes one of the oldest records of this group (Mayr 2000). While the classification into the Charadriiformes appears well corroborated, an allocation to one of the subgroups of this rather diverse taxon, which includes gulls, auks, sandpipers and snipes, among others, is not possible, due to the fragmentary state of the only known skeleton. Although many extant Charadriiformes occur in aquatic habitats, this is not true for all of the species, and the specimen from Messel does not allow any conclusions regarding this species' lifestyle. A small, long-toed bird recently described as *Vanolimicola longihallucis* may possibly be affiliated with the jacanas (Mayr 2017a). These representatives of the Charadriiformes, which occur throughout all tropical regions, have the proportionally longest toes among all birds – an adaptation to their foraging strategy on the surface of water bodies overgrown by aquatic plants, in particular the leaves of water lilies.

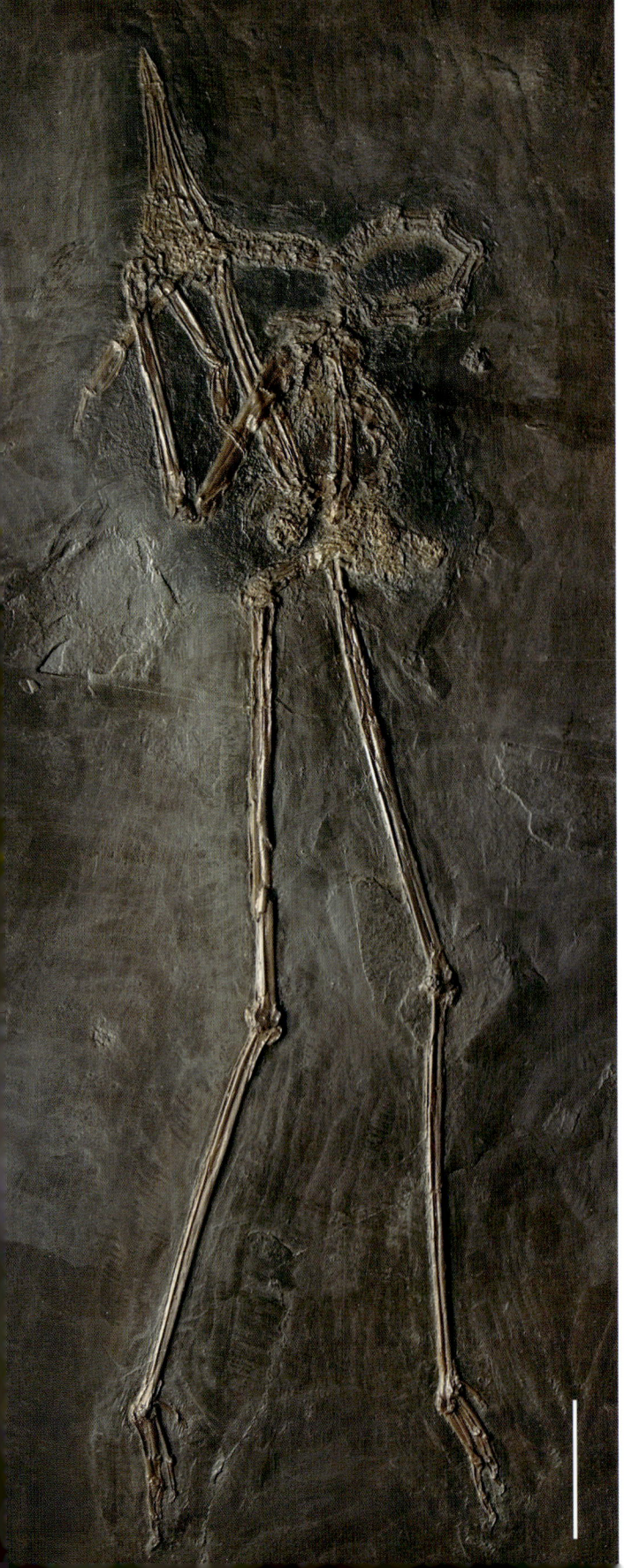

The aerial insect hunters

Birds with very broad bills that were specialized insect hunters are relatively common in Messel. Most of these species belong to the Strisores, which include nightjars and allies, swifts and hummingbirds. While most readers should be familiar with the latter two groups, the nightjar-like birds primarily comprise taxa that only include few species today and are restricted to the Southern Hemisphere. The only widespread group that also occurs in Central Europe are the nightjars proper (Caprimulgiformes). Surprisingly, records of this particular taxon are still absent from Messel or any other lower Eocene fossil sites in Europe to date.

The abundance of insect-hunting birds in Messel is striking and coincides with the large number of bats. On the one hand, the open surface of Lake Messel may have offered a suitable area in the forest for birds and bats to hunt for flying insects. On the other hand, it is also conceivable that the evolution of certain groups of insects in the early Cenozoic favored the occurrence of flying insect hunters among the vertebrates. In this context, the moths may deserve particular consideration, as they include a number of species-rich groups, such as the owlet moths (Noctuidae), that underwent a radiation in the early Cenozoic (Grimaldi & Engel 2005).

Nightjars and allies

The birds combined here as "nightjars and allies" do not constitute a monophyletic group, but are successive sister taxa of of swifts and hummingbirds. Together with the latter, they comprise the Strisores. Four of the five taxa of nightjar-like birds are relict groups that today only occur in the Southern Hemisphere, and all species are crepuscular or nocturnal. Since swifts and hummingbirds (Apodiformes), which are phylogenetically derived from an (owlet-) nightjar-like ancestor, are diurnal, the (as yet unanswered) question arises whether nocturnal habits evolved several

Fig. 11.14: Skeleton (cast) of *Juncitarsus merkeli*, a relative of flamingoes and grebes. Scale: 5 cm.

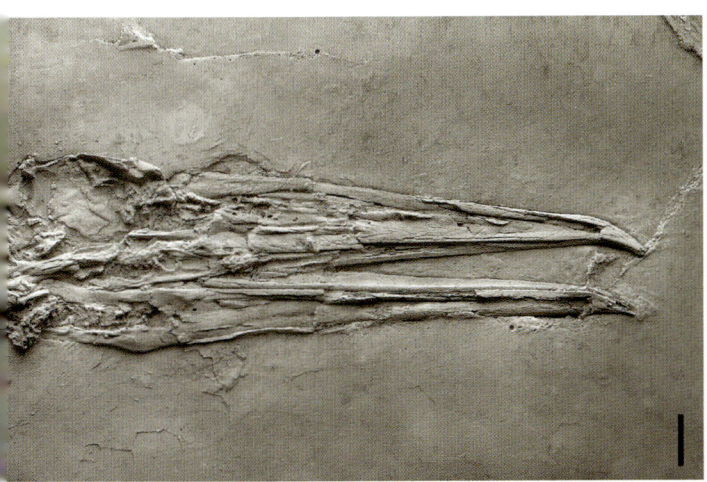

Fig. 11.15: Skull of *Masillastega rectirostris*, a presumed relative of the sulids. Scale: 1 cm.

Fig. 11.16: *Rhynchaeites messelensis*, one of the oldest known ibises. Scale: 2 cm.

times independently among the Strisores or whether their last common ancestor was already nocturnal. In the latter case, the ancestors of swifts and hummingbirds would also have been nocturnal, which means that these birds underwent a secondary transition to a diurnal lifestyle (Mayr 2010a).

The closest modern relatives of the Strisores are uncertain. A fossil taxon that was possibly close to the stem group of these birds is *Palaeopsittacus*, which from Messel is only known by a single skeleton without a skull (Fig. 11.17; Mayr 2003a). *Palaeopsittacus* was a roughly thrush-sized bird with skeletal traits reminiscent of the South American oilbirds (Steatornithiformes) – an archaic group among the Strisores. The taxon was initially described from the London Clay, where it is only documented by very fragmentary remains (Harrison 1982).

Today, the frogmouths (Podargiformes) are restricted to Southeast Asia and the Australian region. Contrary to many other Strisores, they are not aerial insect hunters but ambush their prey – large insects

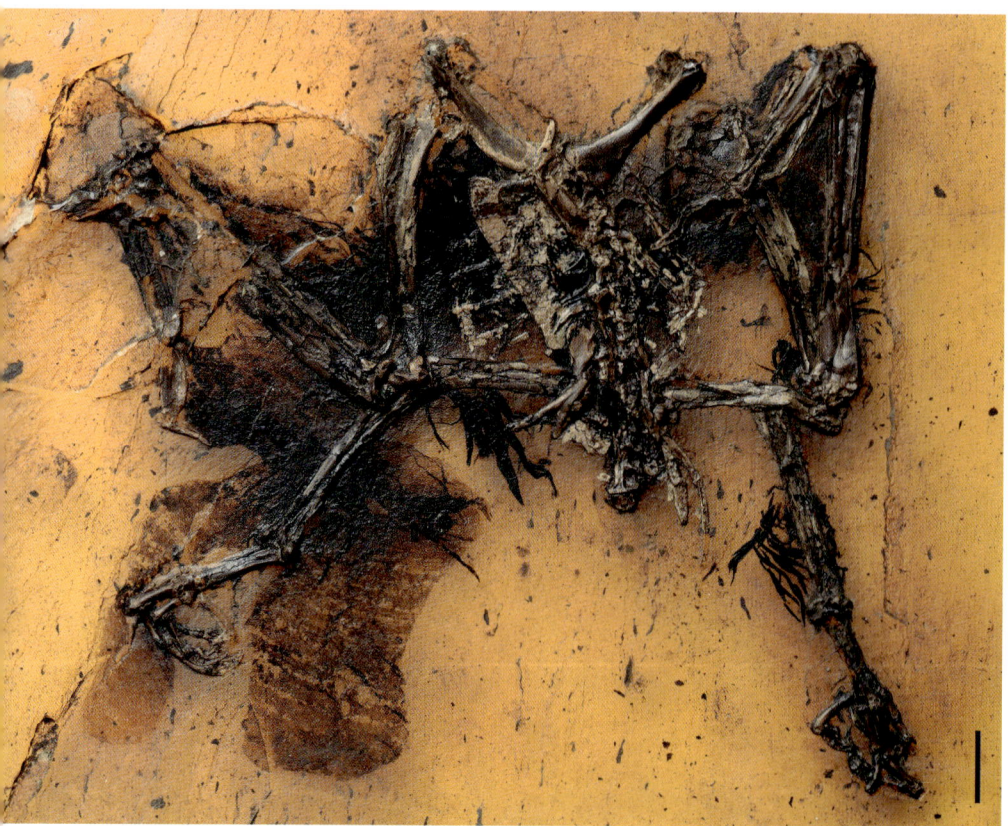

Fig. 11.17: The only skeleton of *Palaeopsittacus* sp. known from Messel. Scale: 1 cm.

and small vertebrates – from perches. In Messel, these birds are represented by *Masillapodargus longipes* (Fig. 11.18; Mayr 2015b). Like its extant relatives, the species is characterized by a broad, strong beak. However, *Masillapodargus* had much longer legs than extant frogmouths, which – together with the fossil taxon's smaller size – indicates differences in lifestyle and hunting behavior.

The aerial insect hunters among the Strisores have a broad beak as well, which furthermore tends to be very short. *Protocypselomorphus manfredkelleri* (Fig. 11.19; Mayr 2005b) may be one of phylogenetically most archaic representatives of these birds. Very short legs, long wings and a short, swift-like bill suggest that this species hunted flying insects in the air.

A further species of the Strisores, whose systematic relationship has not yet finally been clarified, is *Hassiavis laticauda* (Mayr 2004b). It shows similarities in its skeletal morphology to the Archaeotrogonidae, an extinct group of nightjar-like birds that is documented by many specimens from late Eocene fissure fillings in France. *H. laticauda* was a very small bird with relatively long tail feathers. In two specimens, these still show a barred pattern, similar to many of the extant nightjar-like birds (Fig. 11.20). The bill shape of *Hassiavis* resembles that of the extant Australian owlet-nightjars (Aegotheliformes), which, however, have much longer legs. Based on the bill shape, the short wings and long tail, *Hassiavis* was probably an ambush hunter that – similar to the extant owlet-nightjars – went after flying insects from an elevated perch.

Two species of *Paraprefica*, a representative of the potoos (Nyctibiiformes), have been described from Messel. The range of the seven extant potoo species is restricted to the Neotropics. Potoos are characterized by a rather unique skeletal structure, which enables an unambiguous identification; like its modern relatives, *Paraprefica* has a short but very wide, triangular bill and a grotesquely shortened tarsometatarsus (Fig. 11.21, Fig. 11.22). Furthermore like its extant relatives, this bird was an insect hunter that chased its flying prey with a wide-open beak. The two species of *Paraprefica* from Messel, *P. major* and *P. kelleri*, constitute the oldest records of potoos, which are not known from any New World fossil sites (Mayr 1999, 2005c, 2009a). One specimen with well-preserved plumage remnants shows that at least the outer primaries had a banded pattern similar to that of extant potoos (Fig. 11.22). This may be an indication that the plumage of *Paraprefica*, similar to its modern relatives, had a bark-like camouflage pattern that protected the bird from predators while it spent the daylight hours perched immobile on dead branches or tree stumps. This, in turn, would suggest that *Paraprefica* was crepuscular or nocturnal, like its extant relatives.

Fig. 11.18: *Masillapodargus longipes* has the bill shape characteristic of frogmouths. Scale: 1 cm.

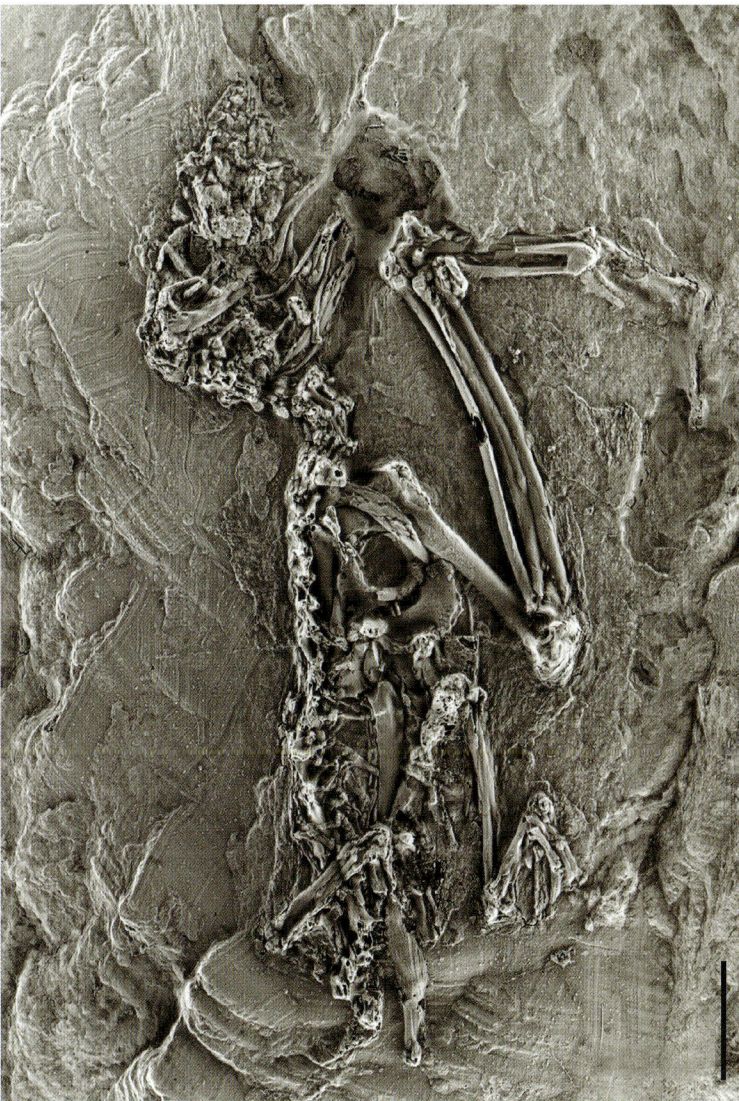

Fig. 11.19: *Protocypselomorphus manfredkelleri*, a bird whose extremities show the proportions of a typical aerial insect hunter. Scale: 1 cm.

Swifts and early relatives of the hummingbirds

Contrary to their closest relatives, swifts and hummingbirds have a diurnal lifestyle. One shared trait among these two externally very different groups is an extremely shortened humerus. In addition, several swifts and all hummingbirds have very short legs, which gave rise to the scientific name Apodiformes ("without feet") for the higher-level taxon that contains both of these groups.

The recently described *Cypseloramphus dimidius* may be a basal representative of the Apodiformes; it is only known from one incomplete skeleton (Fig. 11.23; Mayr 2016b). The assignment of this bird to the Apodiformes was based on the short and compact humerus and the swift-like bill. However, compared with indisputable Apodiformes, the hand section of the wing is not elongated in *Cypseloramphus*.

Scaniacypselus

The tiny swift *Scaniacypselus szarskii* is known from some 15 skeletons in Messel (Fig. 11.24). Modern swifts are divided into two groups, the tree swifts (Hemiprocnidae), which are only found in Southern Asia and New Guinea, and the true swifts (Apodidae) with a worldwide distribution. There are four species of tree swifts, as opposed to almost 100 species of true swifts. *S. szarskii* is clearly smaller than any extant swift, but its skeletal morphology shows many traits characteristic for the Apodidae. True swifts exhibit a specialized nesting behavior, which is probably at least partially responsible for their evolutionary success. While tree swifts attach their nests to branches, true swifts tend to choose almost inaccessible vertical surfaces, where they attach the nests with their saliva. As a possible adaptation to this behavior, the tarsometatarsus of many true swifts is relatively long, while their short foretoes all have the same length, making them well-suited for clinging to the substrate. Scani-

Fig. 11.20: A specimen of the presumed archaeotrogon *Hassiavis laticauda;* the fossil shows the preserved barring of the tail feathers. Scale: 1 cm.

Fig. 11.21: *Paraprefica kelleri*, a representative of the potoos. These birds are characterized by a short, very broad beak as well as a grotesquely shortened tarsometatarsus (arrows). Right: coated with ammonium chloride. Scale: 1 cm.

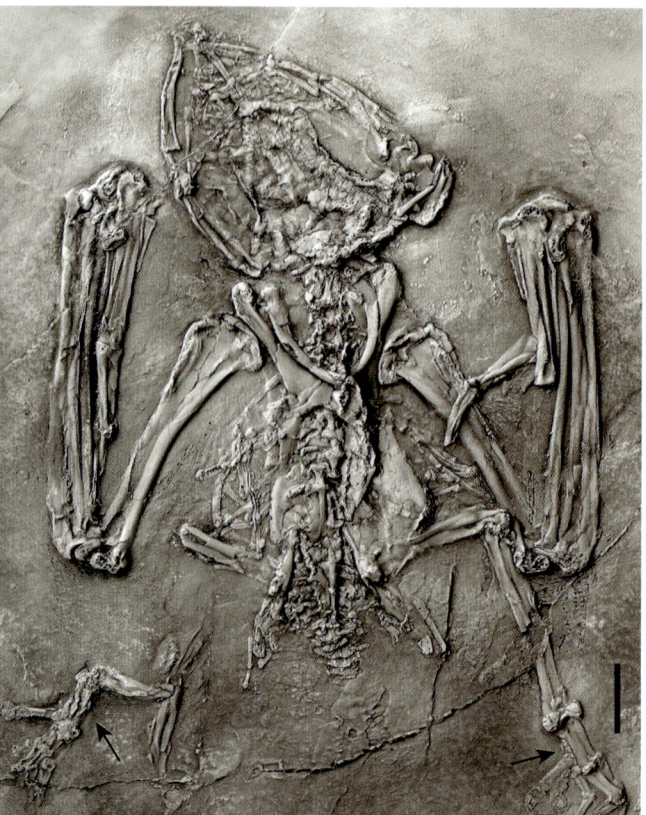

Fig. 11.23: *Cypseloramphus dimidius* is presumed to be a primitive relative of the swifts and hummingbirds, in which the wing's hand section is not as strongly elongated. Scale: 1 cm.

Fig. 11.22: An additional specimen of the potoo *Paraprefica*. Note the banding on the outer primaries. Scale: 5 cm.

acypselus, by contrast, had a significantly shortened tarsometatarsus, suggesting that the Messel swift nested in trees, like the extant tree swifts (Mayr 2015c). *S. szarskii* was also described from a lower Eocene fossil site in France, and a closely related species is known from the middle Eocene of Denmark.

Parargornis

A highly unusual apodiform bird is *Parargornis messelensis* (Mayr 2003b). While this equally tiny species is only known from a single fossil to date, the specimen shows extremely well-preserved plumage details (Fig. 11.25). *Parargornis* has a significantly short- ened humerus, a swift-like bill, short, broad wings and a long tail. In regard to its skeletal structure, this bird shows many similarities to the taxon *Argornis* from the late Eocene of the Caucasus, which represents a very early stage in the evolution of hummingbirds. Today, hummingbirds are restricted to the New World, but fossils from the late Eocene and the early Oligocene of Europe document that the evolution of these birds partly occurred in the Old World. The taxon *Eurotrochilus* from the lower Oligocene already bore a close resemblance to its modern relatives and probably used its long bill to sip nectar, like extant hummingbirds do (Mayr 2009a, 2016a). *Parargornis*, on the other hand, differs markedly from extant hum-

Fig. 11.24: Skeletons of the Messel swift *Scaniacypselus szarskii*. The specimen illustrated on the right with preserved feathers was photographed prior to its preparation. Scale: 1 cm.

mingbirds in some morphological traits and can only be identified as a hummingbird relative by comparison with *Argornis* and other fossils. All extant apodiform birds have narrow wings, and the combination of short, broad wings and a shortened humerus found in *Parargornis* is not known from any other bird. Modern hummingbirds are characterized by a hovering flight with rapid wingbeats that enables them to stand still in front of a flower for several minutes. The unusual wing morphology of *Parargornis* may possibly be an early adaptation to this highly specialized flight style, and the swift-like bill suggests that *Parargornis* did not feed on nectar but used hovering flight to collect insects from flowers and leaves. This lifestyle can be expected from an early representative of the hummingbirds, since insects still constitute part of the diet of several extant hummingbird species, and the phylogenetic position of these birds among specialized insect hunters suggests an insectivorous lifestyle of their ancestors.

The diversity of apodiform birds in Messel stands in stark contrast to the fact that in the only slightly older (by a few million years) fossil sites of the London Clay and the Green River Formation, Apodiformes are represented by only a single taxon, *Eocypselus*, which is not known from Messel. *Eocypselus* is a very archaic representative of the Apodiformes, and it is possible that the split of the Apodiformes into swifts and hummingbirds only occurred in the lowest Eocene, shortly before the Messel oil shale was deposited.

The arboreal birds of the Messel forest

As noted above, the current fossil record suggests that the true arboreal birds (Telluraves) within Neornithes – the clade including modern birds – only developed after the extinction events near the end of the Cretaceous. Prior to that time, the ecological niches for arboreal birds were occupied by the Enantiornithes, an avian group that was widely distributed throughout the Cretaceous (Chiappe & Meng 2016; Mayr 2016a).

Eocene avifaunas are characterized by the absence of passerines (Passeriformes), which today make up more than half of all avian species and are the dominant group of birds in many terrestrial ecosystems in regard to both the number of species and individuals. Most of the remaining groups of the tree-dwelling Telluraves already existed at the time of the Messel Lake, and their radiation must therefore have taken place during the approximately 17 million years between the extinction of the Enantiornithes and the deposition of the Messel oil shale.

Phylogenetic analyses based on genetic data suggest that the Telluraves also include the only distantly related owls (Strigiformes) and diurnal raptors (Falconiformes and Accipitriformes). The fossil record of owls goes back to the Paleocene. In Messel, only two skeletons of a small species have been found to date, which was described as *Palaeoglaux artophoron*; both specimens lack a skull (Fig. 11.26; Peters 1992). In addition to these specimens, there is also a small pellet from Messel, which may have been produced by *Palaeoglaux* (Box 11.3).

There are no definitive records of diurnal raptors from Messel, but for one species a relationship with falcons (Falconiformes) is being considered. The species in question is *Masillaraptor parvunguis* (Fig. 11.27; Mayr 2006b, 2009c), which is only documented by a small number of poorly preserved specimens. This bird had a relatively long beak, similar to that of certain modern South American falcons. The structure of the feet likewise shows similarities to diurnal raptors, although *M. parvunguis* did not have strongly developed raptorial claws (a fact reflected by its scientific species name). The long legs suggest that *Masillaraptor* spent most of its time on the ground, in which it would have resembled several of the phylogenetically basal South American falcons. An as yet undescribed skull of a related species was found in the Green River Formation (Grande 2013).

Fig. 11.25: *Parargornis messelensis*, a very archaic hummingbird. The only known skeleton of this species shows the exceptionally preserved plumage remains (right: coated with ammonium chloride). Scale: 1 cm.

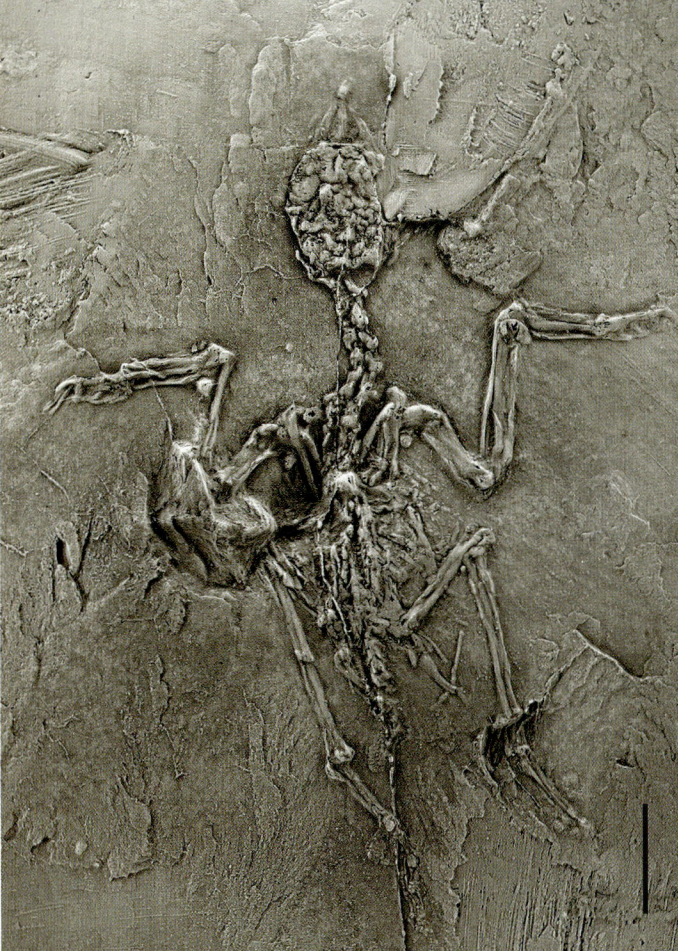

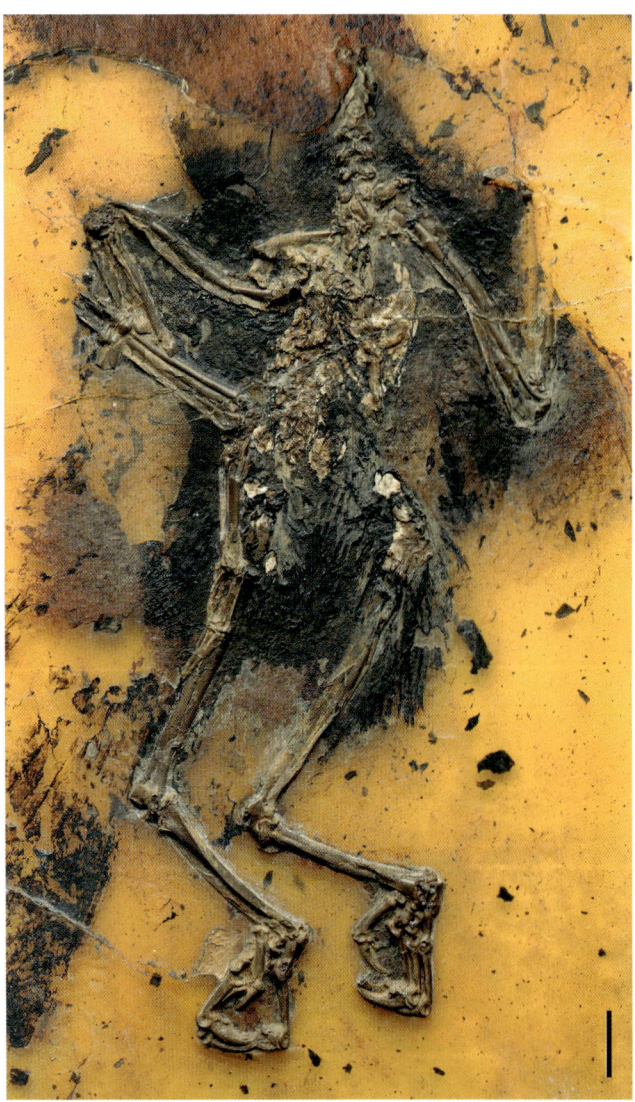

Fig. 11.26: The holotype of the small owl *Palaeoglaux artophoron*. Scale: 1 cm.

> **Box 11.3: Pellets**
> A few compacted masses of avian bones known from Messel are examples of pellets, i.e., indigestible food remnants that were regurgitated by predators and found their way into the lake (Mayr & Schaal 2016). The fossil illustrated here (scale: 1 cm) is similar in shape to the pellet of modern owls and may have been produced by the Messel owl, *Palaeoglaux artophoron*. Pellets can offer insights into the trophic relationships within the food web around Lake Messel, but their study is still in its beginning stages. In most cases, it is not yet possible to clearly identify the contained prey animals nor the species that produced the pellets.
>
>

Mousebird diversity

Extant mousebirds (Coliiformes) constitute a typical relict group that once had a much more widespread distribution. The six very similar recent species of these small, grayish-brown birds are only found in Africa south of the Sahara. All species have long tail feathers and a small, finch-like beak and feed on fruits, buds and other plant parts. Fossil mousebirds are known from the Eocene of North America and Europe, and these early forms showed highly diverse morphological adaptations. In North America, mousebirds became extinct by the end of the Eocene, whereas in Europe, they occurred well into the Miocene. Mousebird records from Africa go back to the early Miocene.

The phylogenetically most basal fossil representatives of mousebirds are the Sandcoleidae, documented in Messel by the taxon *Eoglaucidium* (Fig. 11.28; Mayr & Peters 1998). Sandcoleidae are also known from North America and other European fossil sites, and *Eoglaucidium* was initially described from Geiseltal – albeit erroneously as a small owl. *Eoglaucidium* is among the largest tree-dwelling birds known from Messel; it had a rather unspecialized, thrush-like bill. The skeletal morphology is reminiscent of the African turacos (Musophagiformes); however, *Eoglaucidium* shows a very different bill shape. In some of the specimens from Messel, stomach contents are preserved that include fruit seeds, suggesting that the sandcoleid species were most certainly arboreal birds that foraged in the Messel forest. Several specimens show well-preserved plumage remains, and the short, rounded wings and extremely long tail are characteristic for birds adapted to maneuvering in dense vegetation.

Chascacocolius cacicirostris, a species that is more closely related to extant mousebirds, is known from a skull (Fig. 11.29) and a complete skeleton in a private collection (Mayr 2005d). *Chascacocolius* was first described from the lower Eocene of North America. The ends of the lower jaw bones of this remarkable mousebird taxon bear long processes that increase the leverage of the muscles that open the beak. The beak proportions of *C. cacicirostris* strongly resemble those of certain New World icterids (Icteridae). These passerines are not closely related to mousebirds and use their beaks to forage in the ground or in bark crevices; some species also feed on the pulp of large fruits.

Another rather unusual mousebird species, *Masillacolius brevidactylus*, is characterized by a strongly elongated tarsometatarsus and short foretoes approximately equal in length (Fig. 11.30). This particular foot structure is not found among extant birds, leaving its function open to speculation. Similarities to the feet of certain swifts suggest that *Masillacolius* may have used its long legs to cling to vertical surfaces. However, it is not known whether this indicates a par-

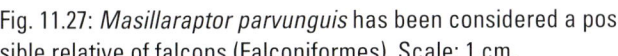

Fig. 11.27: *Masillaraptor parvunguis* has been considered a possible relative of falcons (Falconiformes). Scale: 1 cm.

Fig. 11.28: Three skeletons of the mousebird taxon *Eoglaucidium* sp. Scale: 2 cm.

Fig. 11.29: Skull of *Chascacocolius cacicirostris*, a mousebird with unusually long bony processes at the posterior end of its lower jaw. Scale: 1 cm.

Fig. 11.30: The mousebird *Masillacolius brevidactylus* is characterized by a strongly elongated tarsometatarsus and shortened foretoes. Scale: 1 cm.

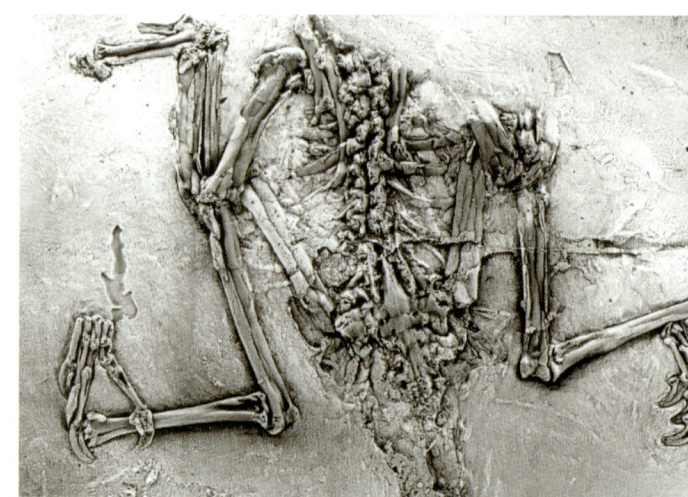

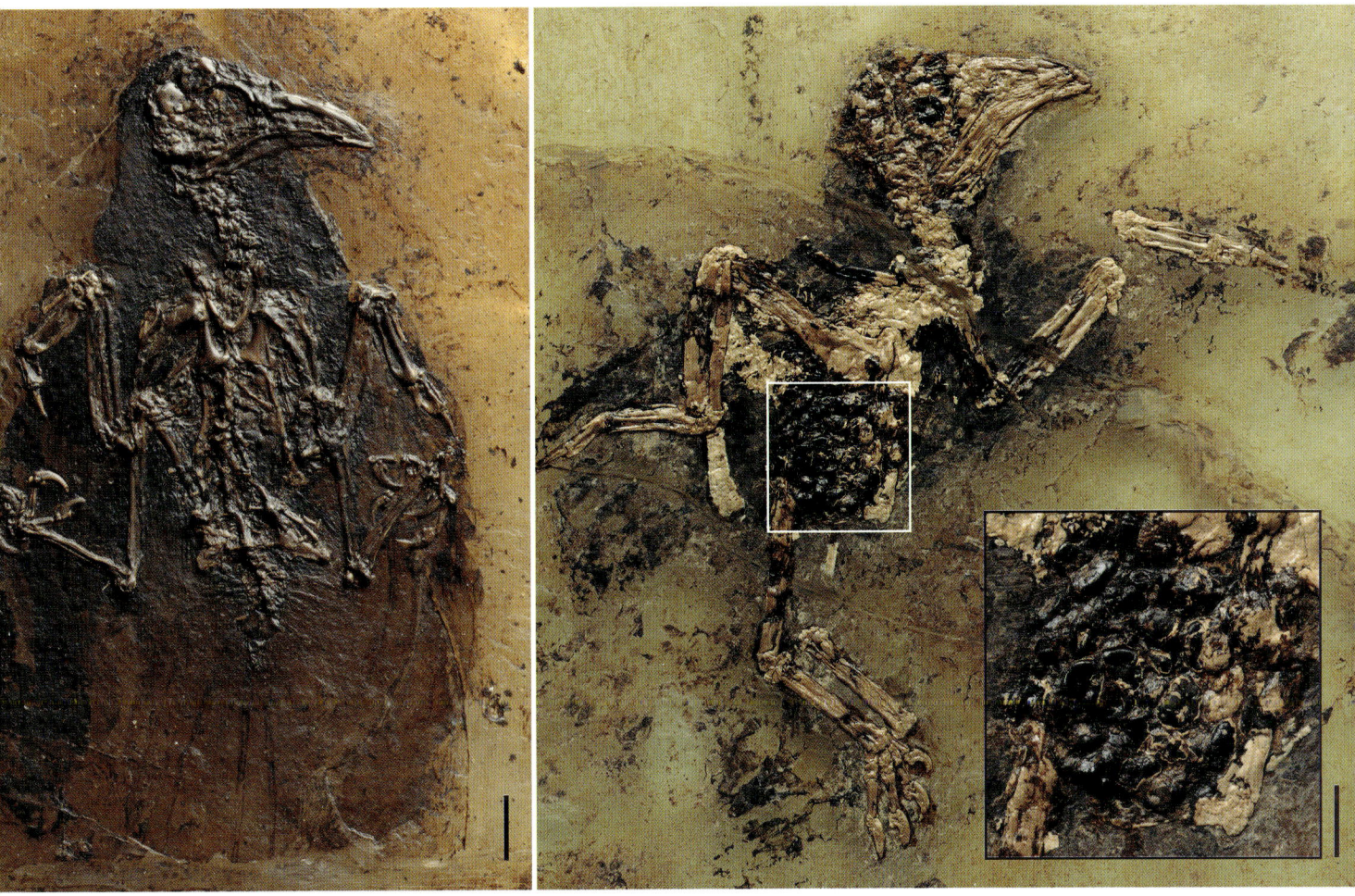

Fig. 11.31: *Selmes absurdipes* is the species of mousebird from Messel that most closely resembles the extant mousebirds in regard to its skeletal structure. The skeleton on the right is preserved with a dense cluster of fruit seeds as stomach contents (details in the inset). Scale: 1 cm.

ticular foraging mode or perhaps a specialized breeding behavior. In this species, as well, one specimen contains fruit seeds in its stomach (Mayr 2015d).

The taxon among the mousebirds from Messel most closely resembling modern Coliiformes is *Selmes absurdipes* (Fig. 11.31), a species also known from Geiseltal (Mayr 2001). Its skeletal morphology corresponds closely to that of extant mousebirds; however, contrary to its modern relatives, *Selmes* had a relatively long bill. The stomach content of one specimen of *S. absurdipes* included a large number of densely packed fruit seeds (Fig. 11.31).

The great diversity among the mousebirds from Messel in regard to their foot structure and beak shape indicates that these birds occupied numerous ecological niches in the local forest. It is not known what ultimately caused the extinction of all of these species, or why mousebirds are now restricted to Africa. These short-winged birds exhibit no migratory behavior, and as fruit eaters, they were obviously unable to cope with the cooling trend during the Cenozoic and the emergence of the cold winters in the Northern Hemisphere. This explains their disappearance from Central Europe in the course of the Miocene. However, it remains a mystery why mousebirds did not survive in the tropical regions of Asia, and non-climatic factors must have played a role in their extinction as well.

Parrots and passerines

Among other traits, parrots (Psittaciformes) are characterized by a particular arrangement of their toes, where, in addition to the hallux, the fourth toe is also turned backwards. Among other modern birds, this so-called zygodactyl foot is otherwise only found in the Piciformes and the cuckoos, and it is generally interpreted as a perching adaptation. In Messel, zygodactyl birds are quite common; one of the groups that possess this type of foot are the Halcyornithidae, which are known from Messel by the two species *Pseudasturides macrocephalus* (Fig. 11.32) and *Serudaptus pohli* (Fig. 11.33). These small birds have a strong bill, and, in addition to Messel, they have also been reported from other Eocene fossil sites, including the London Clay, Geiseltal, and the North American Green River Formation (Mayr 2009a). There is no doubt that the Halcyornithidae were highly specialized arboreal birds. However, details of their lifestyle are still poorly known, and none of the specimens holds clearly identifiable stomach contents. Especially the structure of the legs, in particular certain details of the tarsometatarsus, indicates a closer relationship to the parrots. However, in regard to other traits, e.g., the shape of the humerus and the coracoid as well as the presence of long bony processes at the anterior border of the eye socket, the Halcyornithidae clearly differ from other fossil psittaciform birds and show closer similarities to falcons and owls.

For some time, another species with presumed affinities to parrots, *Messelastur gratulator*, was only known by two skulls, which were initially considered to belong to raptors. Only recently did it become possible to identify skeletons of this species (Fig. 11.34). *Messelastur* has very strong claws that resemble those of owls and diurnal raptors, but in the overall skeletal structure, close similarities to the Halcyornithidae prevail (Mayr 2011a). Similar to the latter, *Messelastur* also displays an unusual mosaic of characters, combining skeletal features of parrots, owls, and falcons. A closely related taxon is known from the Green River Formation und the London Clay.

Fig. 11.32: *Pseudasturides macrocephalus*, a representative of the parrot-like Halcyornithidae. Scale: 1 cm.

For several other birds from Messel, at least a distant relationship with parrots has been postulated. For example, the tarsometatarsus of an isolated foot shows close similarities to the taxon *Vastanavis*, which was described from the lower Eocene (Fig. 11.51; Mayr 2016c; Mayr et al. 2013). Birds similar to *Vasta-*

Fig. 11.33: *Serudaptus pohli*, another species of the Halcyornithidae. Scale: 1 cm.

Fig. 11.34: An unusually well-preserved, post-cranial skeleton of *Messelastur gratulator*, a relative of parrots and passerines. Scale: 1 cm.

navis are also known from the Green River Formation and the London Clay. Another bird from Messel with possible ties to these parrot-like forms is *Eurofluvioviridavis robustipes*, which is only represented by one single skeleton (Fig. 11.35; Mayr 2015e).

Surprising relationships

Based on the results of recent phylogenetic analyses involving genetic data, it is now assumed that parrots are the closest relatives of the passerines (Passeriformes) – a hypothesis that was not considered by any morphologist before. In regard to this hypothesis, it is noteworthy that close relatives of passerines have been found in Messel which exhibit parrot-like traits. These features likely represent basal traits of the common ancestors of parrots and passerines and are in agreement with the hypotheses regarding genetic relationships.

One of these avian groups is the taxon Zygodactylidae, which was initially described from a Miocene fossil site in Germany. For a long time, these birds were only known from several leg bones, which showed such an unusual mosaic of characteristics that an unambiguopus phylogenetic assignment was not possible. Six species of the taxon *Primozygodac-*

Fig. 11.35: *Eurofluvioviridavis robustipes*, another supposed relative of parrots and passerines. Scale: 1 cm.

Fig. 11.36: Two skeletons of *Primozygodactylus* (Zygodactylidae), a zygodactyl relative of the passerines (left: *Primozygodactylus quintus*, right: *Primozygodactylus eunjooae*). Scale: 1 cm.

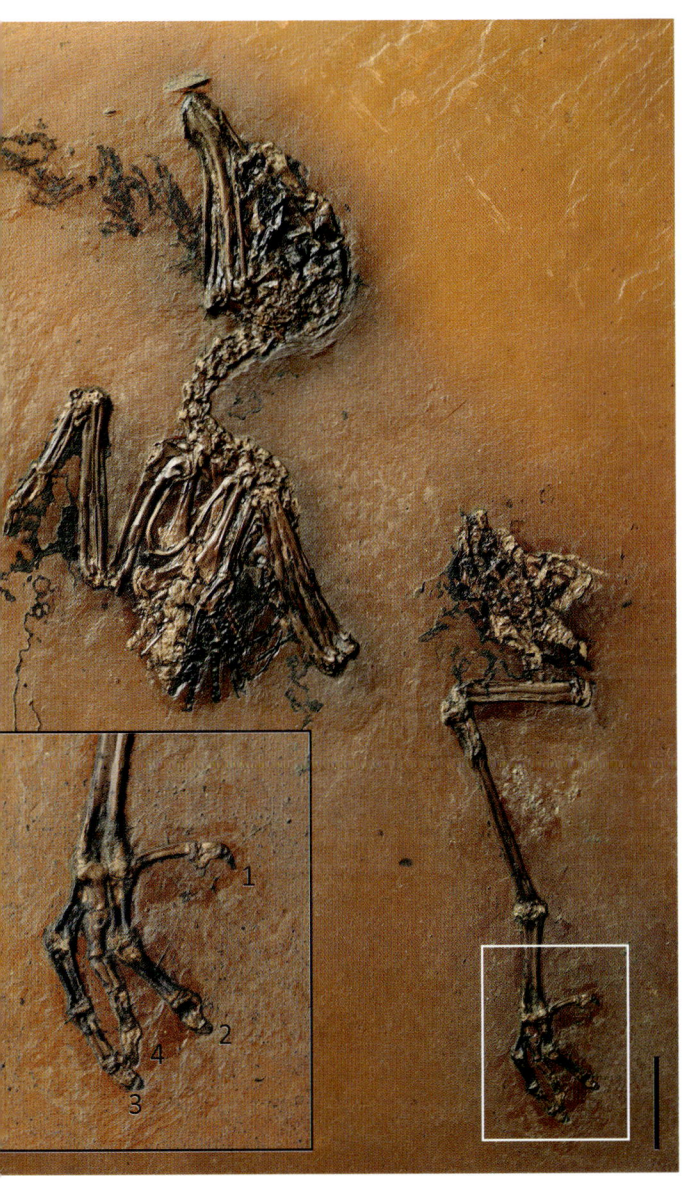

Fig. 11.37: *Psittacopes lepidus*, a parrot-like representative of the basal group of passerines with a zygodactyl foot. Inset: Detail of the toes, numbered. Scale: 1 cm.

in the structure of their tarsometatarsus, zygodactylids show strong similarities to passerines in their skeletal traits. Some of these traits are, however, also found among piciform birds, which were long considered close relatives of the Passeriformes. Only after recent analyses of genetic data shed a new light on the relationship between Piciformes and passerines was it possible to set these traits into a phylogenetically meaningful context (Mayr 2008b). One skeletal characteristic of zygodactylids is the unusually long tarsometatarsus, and all species have long tail feathers (Fig. 11.36; Mayr & Zelenkov 2009). The bill of *Primozygodactylus* shows proportions similar to that of reed warblers, although it has unusually long nostrils. One specimen of the largest *Primozygodactylus* species, *P. major*, is preserved with its stomach contents, which consist of a densely packed cluster of grape seeds (Fig. 11.52).

Another species of potential significance for our understanding of the evolution of passerines is *Psittacopes lepidus*. This little bird is only known by a few skeletons from Messel (Fig. 11.37). Additional specimens from the London Clay are housed in a private collection. Initially, *Psittacopes* was regarded as a representative of the parrots, but recent studies suggest that the taxon is more closely related to the passerines and that the similarities to parrots are limited to primitive traits (Mayr 2015e). A presumed close relative of *Psittacopes* is the long-billed *Pumiliornis tessellatus*, another zygodactyl species whose systematic relationships had not been clarified until recently. Preserved stomach contents of one specimen of *Pumiliornis* include pollen, indicating a nectarivorous lifestyle (Box 11.4; Mayr & Wilde 2014; Mayr 2015e).

Both parrots (as the closest living relatives of passerines) as well as *Psittacopes* and the Zygodactylidae have zygodactyl feet with a fourth toe that is rotated backwards. It can therefore be assumed that the ancestral species of the Passeriformes also possessed such a foot, and that the zygodactyl condition of their feet was secondarily lost in the evolution of passerines.

tylus (Fig. 11.36) were described from Messel, and fossil remains of these birds are also known from the Green River Formation and the London Clay. Based on their zygodactyl feet, the specimens from Messel were initially considered relatives of the Piciformes (Mayr 1998). However, apart from the backward-pointing fourth toe and the associated modifications

Box 11.4: Exceptional preservation: Foot scales and uropygial glands

The most striking trait of many avian fossils from Messel is the exquisite preservation of the plumage. As explained in Chapter 4, this fact is due to the fossilization of melanosomes – cell organelles that produce the dark pigment melanin (Vinther et al. 2010). This also explains why some feathers show noticeable barring, and it is meanwhile believed that a large percentage of all soft-part preservation in Messel is due to the preservation of melanin.

In one specimen of the potoo *Paraprefica*, parts of the foot scales are visible (Peters 1988b: Fig. 208). In this case, it is most likely not the keratin of the horny scales that has been preserved, which, as a protein, has a very low fossilization potential. Rather, the preservation of the scales is probably due to embedded calcium phosphate salts, which would also explain that only those parts on the bottom of the feet were preserved that are more heavily strained in the living birds.

Highly unusual – and only found in Messel to date – are records of the secretions from the uropygial gland (Mayr 2006c, 2016a). The uropygial gland (wax gland) is located just above the base of the tail and produces a secretion consisting of long-chained lipids that plays an important role in feather maintenance. This gland often shows a characteristic heart shape, which can be clearly seen in the corresponding position in several skeletons from Messel. During the fossilization process, the secretions from the uropygial gland were probably converted into adipocere ("corpse wax"), which – contrary to its name – consists of hardened stearic acids and is not a wax. The formation of adipocere was recently identified as a possible reason for the unusually high level of preservation of many mammalian remnants from Messel (Smith & Wuttke 2012).

The image on the bottom left shows an unidentified small bird (cf. *Messelirrisor*) after being transferred to synthetic resin; the two illustrations on the right show the uropygial gland prior to preparation. Scale: 1 cm.

One group barely known among non-ornithologists is represented in Messel by the species *Plesiocathartes kelleri* (Fig. 11.38). This approximately jay-sized bird is a primitive representative of the cuckoo-rollers (Leptosomiformes), which today are restricted to a single species, the courol (*Leptosomus discolor*) of Madagascar and the Comoro Islands. Courols feed on insects and small vertebrates, and according to recent genetic analyses, they are the closest relatives of the trogons, Coraciiformes and Piciformes (Prum et al. 2015). In Messel, *P. kelleri* is only documented by two specimens (Mayr 2002c), but fragmentary remains of similar birds were found in the London Clay and Geiseltal, and species of *Plesiocathartes* were also recorded in the Green River Formation (Mayr 2009a). In fact, the skeletal structure of *Plesiocathartes* and the extant courol is so similar that the latter could be called a "living fossil" among the modern birds (Mayr 2008c).

Trogons

The trogon *Masillatrogon pumilio* (Trogoniformes) is likewise known from only two skeletons (Fig. 11.39; Mayr 2009d). Trogons are characterized by a heterodactyl foot, in which, in addition to the first toe, the second (inner) toe is also rotated backwards. These very colorful, small to medium-sized arboreal birds feed on insects and fruits and are today restricted to the tropical regions of the Americas, Africa and Asia. Eocene trogon fossils are also known from the Danish Fur Formation and the London Clay (Mayr 2009a). *M. pumilio* was smaller than any extant trogon, but the heterodactyl design of the feet is clearly visible in the fossils from Messel, and the rest of the skeleton also shows great similarities to extant trogons. One specimen with well-preserved plumage remains shows that the tail feathers were very long, as is the case in modern trogons. Trogons have short, broad wings and do not exhibit migratory behavior. In Europe, these birds existed until the early Miocene, approximately 20 million years ago. Their extinction in the Northern Hemisphere can therefore easily be explained by the onset of the pronounced climatic seasonality with cold winters during the Miocene, which offered the trogons an insufficient food supply and limited their distribution to the tropics and subtropics (Mayr 2011b).

Fig. 11.38: *Plesiocathartes kelleri*, an early representative of the cuckoo rollers (courols). Today, this group of birds is restricted to Madagascar and the Comoro Islands. Scale: 1 cm.

Trogons and Coraciiformes

Among the tree-dwelling birds, the trogons, Coraciiformes and Piciformes constitute a monophyletic group. These birds are cavity nesters that usually excavate their own burrows in trees or in the soil. Many species are adorned with a colorful plumage, and most feed on insects or small vertebrates, although the clade also includes several specialized fruit eaters (certain trogons as well as toucans and hornbills).

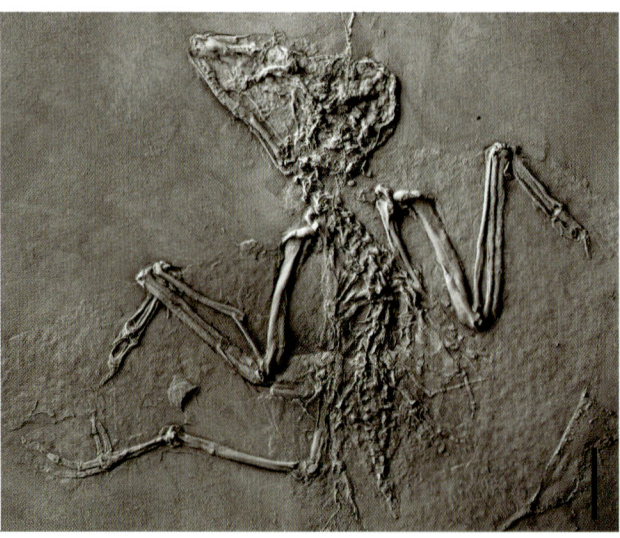

Fig. 11.39: The Messel trogon, *Masillatrogon pumilio,* represents one of the oldest fossil records of the trogons, a group of birds whose current distribution is restricted to the tropical and subtropical regions. The fossil in the image on the right has been coated with ammonium chloride. Scale: 1 cm.

The Messel hoopoes

Among the most common arboreal birds in Messel are the hoopoe-like Messelirrisoridae, of which three species have been described (Fig. 11.40). Extant representatives of the hoopoe-like birds (Upupiformes) are the hoopoes (Upupidae) from Eurasia and Africa and the wood hoopoes (Phoeniculidae), a family of nine extant species that occur in Africa south of the Sahara. Like its living relatives, *Messelirrisor* had a very long bill, which – analogous to extant hoopoes – was probably noticeably longer in the male than in the female (Mayr 1998, 2006c). Moreover, similar to extant hoopoes, the posterior ends of the lower jaw bones carry long processes, which increase the leverage of the jaw muscles attached at that point and are characteristic of birds that forage for food in the soil and in bark crevices. Modern hoopoes are either terrestrial (hoopoe) or trunk climbers (wood hoopoes), whereas the Messelirrisoridae presumably foraged among the branches in the canopy – most likely in search of insects and other invertebrates. The species of the Messelirrisoridae are smaller than any of the extant Upupiformes, and some of the *Messelirrisor* species only reached the size of a small hummingbird. In addition to Messel, fossils of these birds have been found in the London Clay and Geiseltal, although they are very rare in these localities.

Rollers

The Coraciiformes include the ground rollers (Brachypteraciidae) from Madagascar and the true rollers (Coraciidae), which are widespread throughout the Old World. All extant species of this group feed exclusively on small animals. In Messel, two groups of roller-like bird are represented. The phylogenetically most basal group is the Primobucconidae, which in Messel is documented by two species, *Primobucco perneri* and *P. frugilegus* (Fig. 11.41; Mayr et al. 2004). A clearly larger species, *Eocoracias brachyptera*, is more closely related to extant rollers and is known from specimens with well-preserved plumage details (Fig. 11.42; Mayr & Mourer-Chauviré 2000). These show that *E. brachyptera* had short, rounded wings and a long tail, similar to modern ground rollers, which identifies the species as a forest dweller. Stomach contents are preserved in both the Primobucconidae and *Eocoracias*, which – surprisingly – consists of fruit seeds. Apparently, therefore, the dietary range of these birds was significantly broader than that of their modern relatives. Primobucconidae as well as an *Eocoracias*-like form (*Paracoracias*) are also known from the Green River Formation. Among the extant Coraciiformes, the ground rollers and true rollers differ significantly from each other in their lifestyle and the associated anatomical characteristics. Four of

Fig. 11.40: Three skeletons of the Messelirrisoridae (top: *Messelirrisor grandis*, bottom: *M. halcyrostris*). The Messel hoopoe *Messelirrisor* is among the most common small arboreal birds in Messel. Scale: 1 cm.

Fig. 11.41: A skeleton of a representative of the roller-like Primobucconidae, showing an exceptional level of feather preservation. Note the feather crest on the head. Scale: 1 cm.

Fig. 11.42: The Messel roller *Eocoracias brachyptera*. Scale: 1 cm.

the five species of ground rollers from Madagascar are denizens of tropical forests, and all ground rollers forage primarily on the ground. In adaptation to this lifestyle, ground rollers have relatively long legs and short remiges and rectrices. The twelve species of true rollers, on the other hand, occupy open habitats and are better adapted to a life in the air. Contrary to the ground rollers, they have short legs and relatively long flight feathers. *Eocoracias* is a representative of the stem group of the Coraciiformes, and a comparison with this fossil taxon reveals that the ground rollers' wing and tail proportions are ancestral traits of the Coraciiformes, whose Eocene representatives also inhabited forested habitats, where short, rounded wings and a long tail improved their maneuverability in dense vegetation. However, like modern true rollers, *Eocoracias* had short legs, suggesting that the long legs of ground rollers are a secondary adaptation to these birds' terrestrial habits.

A kingfisher relative

The small *Quasisyndactylus longibrachis* (Fig. 11.43) is a representative of the kingfisher-like birds (Alcediniformes). A characteristic trait of these birds is the so-called syndactyl foot, in which the basal phalanges of foretoes are attached to each other by a connective tissue. The resulting rigid foot is an adaptation to the nesting behavior – the eggs are laid in tunnels that the birds excavate themselves in the soil. In all specimens of *Quasisyndactylus,* the toes are tightly adjoined to each other, which indicates the presence of a syndactyl foot, since during the decomposition process, the ligaments between the basal phalanges of the toes do not decompose any faster than those between the remaining skeleton elements. In addition to the globally distributed and species-rich kingfishers, the extant Alcediniformes also include the motmots (Momotidae) and todis (Todidae) of the New World; the phylogenetic affinities of the bee-eaters (Meropidae), which are traditionally also placed with the Alcediniformes, are uncertain. Compared to these extant groups, *Quasisyndactylus* shares most similarities with the todis, which today only occur on the Greater Antilles, a group of islands in the Caribbean. Todis have been documented in the Oligocene fossil record from Europe (Mayr 2009a). However, it is not yet clear whether *Quasisyndactylus* actually represented an Eocene todi or rather an archaic member of the Alcediniformes, which only resembles todis in certain primitive traits (Mayr 1998, 2004b). The long, wide and rather flat bill of *Quasisyndactylus* indicates that the taxon primarily fed on insects, which it hunted from elevated perches.

There are no unambiguous representatives of the Piciformes from Messel. In addition to the true woodpeckers, this taxon also includes the toucans and a few other groups that are not adapted to a tree-climbing lifestyle. While a possible affiliation with the Piciformes has been postulated for *Gracilitarsus mirabilis* (Fig. 11.44), there are only a few, mainly uncharacteristic traits that support this hypothesis (Mayr 2009a).

Fig. 11.43: *Quasisyndactylus longibrachis*, a relative of kingfishers, which resembles the extant todis in regard to its bill shape. Scale: 1 cm.

The skeletal morphology of *Gracilitarsus* is reminiscent of the Zygodactylidae, suggesting the possibility that it constitutes another stem group representative of passerines. Unfortunately, certain traits of significance for a systematic placement of this bird are only poorly preserved or entirely lacking in the known specimens. The unusually short and strong humerus

Fig. 11.44: *Gracilitarsus mirabilis*, an unusual bird with short humeri and long legs, whose affinities are unknown. Scale: 1 cm.

Fig. 11.45: *Perplexicervix microcephalon*, a species possibly related to the extant screamers. The spinal column is covered with numerous small bony tubercles. Enlargement: Detail of the cervical vertebrae of another individual. Scale: 2 cm.

of *Gracilitarsus* indicates a specialized flight style, while the structure of the feet – especially the long tarsometatarsus and the short toes – resembles the aforementioned mousebird taxon *Masillacolius*.

A skeleton contained in a snake pellet (Box 11.3) shows similarities to the North American taxon *Neanis*, which is believed to be related to the puffbirds and jacamars (Bucconidae and Galbulidae) (Mayr & Schaal 2016). These two Neotropical taxa are among the most primitive extant representatives of the Piciformes. However, this is another case where the remains from Messel are simply too incomplete to allow a definitive classification of this bird.

Several mystery birds

Owing to the great age of the site, it is not unexpected that many species from Messel cannot be assigned to any extant avian group. Several specimens whose systematic placement remains difficult have already been mentioned above. In the following, additional species are introduced that have defied all classification efforts to date.

One of these is *Perplexicervix microcephalon*, a species only known from poorly preserved skeletal remains (Fig. 11.45). This bird was about as big as a

Fig. 11.46: *Eopachypteryx praeterita*, another bird of unknown affinities. Scale: 1 cm.

crow. The skeleton resembles that of extant screamers (Anhimidae), a Neotropical group of primitive Galloanseres. However, the shared characteristics mainly concern the overall shape and the proportions of the wing and leg bones and are thus of little phylogenetic value. Similar to the cariamiform species *Dynamopterus tuberculatus* discussed above, the vertebrae in *Perplexicervix* are covered by numerous small bony tubercles (Mayr 2010b). This unusual characteristic is not found in any modern birds, although a few vertebrae of screamers show rudiments of small tubercles.

The systematic relationships of the recently described taxon *Eopachypteryx* (Fig. 11.46) are equally uncertain. The two known species are so clearly differentiated from all known groups of birds that their classification in a separate family Eopachypterygidae seemed appropriate (Mayr 2015f). *Eopachypteryx* has a relatively large skull with a rather short, massive beak, as well as strong wing bones and short legs. Not least due to the poor state of preservation, it is not even possible to reliably assign the taxon to one of the major extant groups such as the Strisores or the Telluraves.

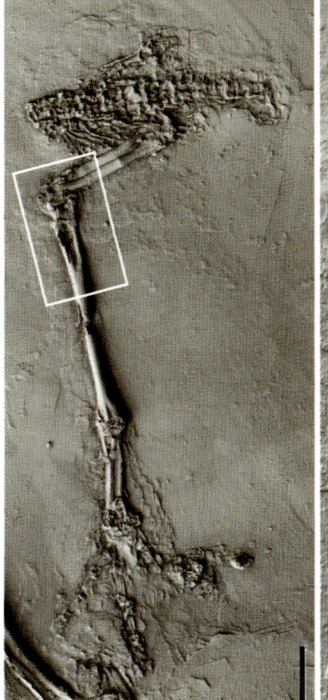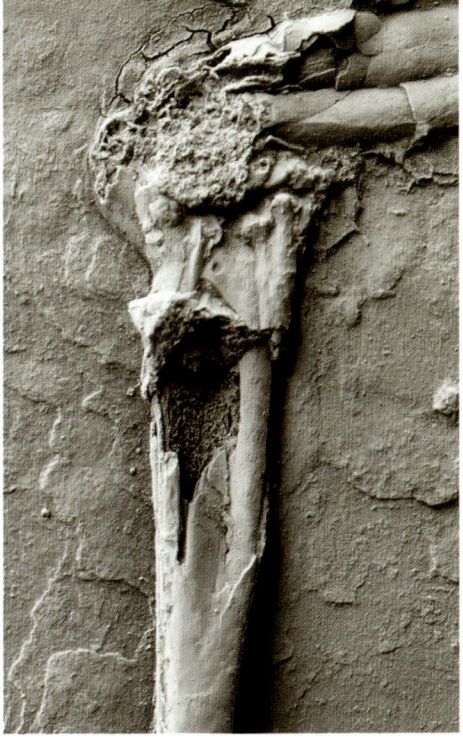

Fig. 11.47: Left leg of *Lapillavis incubarens*, a species only known by fragmentary skeletal remains. The image on the right shows details (marked with a rectangle on the left) of the tibiotarsus. Medullary bone is visible at the break, which offers proof that the specimen was a breeding female. Scale: 1 cm.

The same applies to the species *Lapillavis incubarens*, which is only known from a single fragmentary skeleton. Certain skeletal features of this relatively small bird are reminiscent of the cuckoo rollers (courols), but there are also similarities to the basal representatives of the Strisores. The skeleton of *Lapillavis* is remarkable, since some of the fractured long bones allow a view inside the bones' interior. For the first time in a fossil from Messel it was possible to document the so-called medullary bone (Fig. 11.47; Mayr 2016c). This new bone formation occurs in female birds shortly before they lay eggs, in order to support the increased calcium requirement during egg shell formation. The documentation of medullary bones in *Lapillavis* therefore enabled the first ever unequivocal determination of the sex in a bird from Messel.

Another bird is only documented by one isolated skull, which – apart from its clearly smaller size – bears a strong resemblance to the skull of *Foro panarium*, a species described from the Green River Formation (Fig. 11.48; Olson 1992; Mayr 2016b). The systematic placement of *Foro* is uncertain, although a connection to the African turacos (Musophagiformes) and the South American hoatzins (Opisthocomiformes) has been suspected.

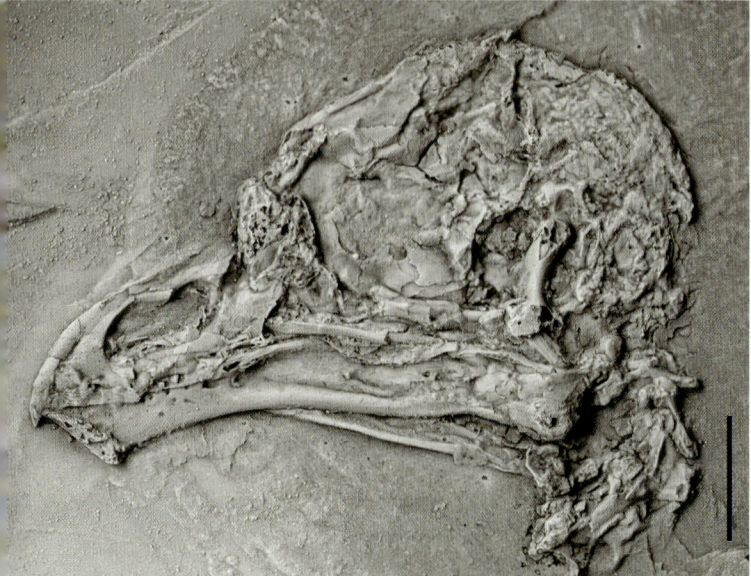

Fig. 11.48: Isolated skull of a bird resembling the North American taxon *Foro* (the image on the bottom shows the fossil coated with ammonium chloride). Scale: 5 mm.

Biogeographic connections

Compared to other Eocene fossil sites in Europe, Messel shows obvious similarities to the avifaunas of the London Clay in England, Geiseltal in eastern Germany, and the Danish Fur Formation. Numerous avian groups from Messel are also known from North American fossil sites, in particular the Green River Formation in Wyoming, whose geologic strata are a little older than the oil shale deposits in Messel. Examples of avian taxa from Messel that are represented by closely related species in North America include the aforementioned Lithornithidae, Gastornithidae, Gallinuloididae (Fig. 11.49), Juncitarsidae, Sandcoleidae, Halcyornithidae, Messelasturidae, Zygodactylidae, Leptosomiformes (*Plesiocathartes*, Fig. 11.50), and Primobucconidae (Mayr 2009a, 2016a). In fact, the similarities between the avifaunas of the lower Eocene of Europe and North America are so extensive that it may be assumed that the paratropical forests that formerly covered both continents were home to very similar groups of birds. Land bridges that existed during the lower Eocene (Chapter 2) enabled even the flightless *Gastornis*, along with numerous group of squamates and mammals, to spread across both continents.

Missing from the Eocene of North America are records of the palaeognath *Palaeotis* as well as potoos (*Paraprefica*), swifts (*Scaniacypselus*), and humming-

Fig. 11.49: *Paraortygoides messelensis* from Messel (left) and *Gallinuloides wyomingensis* (right) from the Green River Formation. Scale: 1 cm.

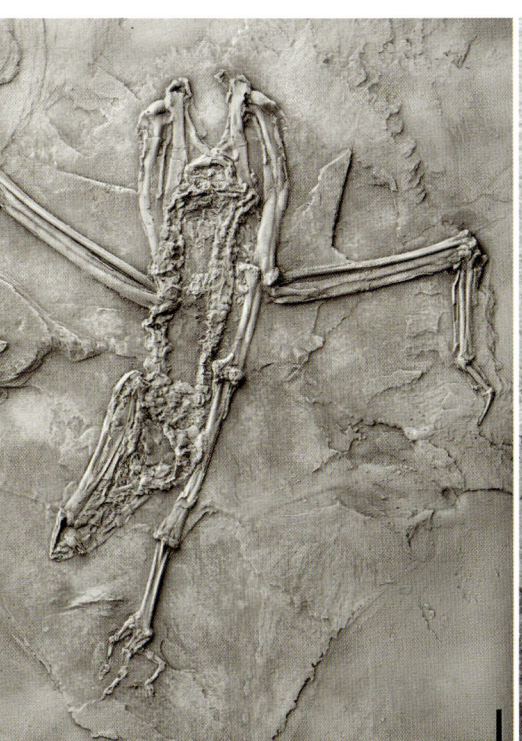 

Fig. 11.50: *Plesiocathartes kelleri* from Messel (left) and *P. wyomingensis* (right) from the Green River Formation. Scale: 1 cm.

birds (*Parargornis*). As detailed above for the latter two apodiform birds, these differences may partially be explained by the younger geologic age of the Messel site. On the other hand, there are no specimens of the Presbyornithidae from Messel, an extinct group of Galloanseres that is found rather commonly in North America. The same is true for oilbirds, which have been documented by a skeleton from the Green River Formation (Mayr 2009a; Grande 2013).

Unfortunately, little is known about Eocene birds from the southern continents. Only in recent years have avian remains from the lower Eocene of Africa been described, which – although they show certain similarities to European taxa of the same age – in part belong to groups that have not yet been discovered in Messel and other Eocene fossil sites in Europe (Mourer-Chauviré et al. 2016). Of particular interest is recently discovered avifauna from the lower Eocene of India. Here, the taxon *Vastanavis* is particularly common, which represents a very archaic parrot-like bird (Mayr et al. 2013; Mayr 2015e). A similar species was discovered in the London Clay, and a fragmentary foot of a *Vastanavis*-like bird is also known from Messel (Fig. 11.51; Mayr 2016c). India is part of the former supercontinent Gondwana and had barely reached Asia in the early Eocene as the result of plate-tectonic processes. The similarities to the European avifauna are in accord with the rather uniform global climate conditions at that time, but overall we do not yet have sufficient information about the avifaunas of the Southern Hemisphere in the Lower Eocene to be able to make far-reaching inferences.

Only a few avian groups from Messel can still be found in Central Europe today; these include swifts, owls, hoopoes and rollers. Apart from owls, however, these birds are currently represented in the Northern Hemisphere by only a small handful of species, with their centers of distribution being in the tropical and subtropical regions of the southern continents. The differences are partly due to the climate conditions in Central Europe during the Eocene. At the time of the Messel Lake, Central Europe had a subtropical climate (Chapter 3). By the end of the Eocene, and continuing throughout the Oligocene, a global cooling trend set in, accompanied by the development of pronounced seasonal temperature fluctuations. The emergence of cold winters of the Northern Hemisphere is certainly one of the reasons why insectivorous and frugivorous birds without the capabilities of long-distance migration to warmer climates are no longer found in Central Europe. One group affected by this are the trogons, which today are restricted to the subtropics and tropics of the New World, Africa and Asia.

Undoubtedly, the complicated historical biogeography of birds cannot be explained solely by climatic events. Of note is the high proportion of avian species in Messel whose closest living relatives today are restricted to the Neotropics (South and Central America). This includes the seriema-like *Dynamopterus*, the potoo *Paraprefica,* and relatives of hummingbirds (*Parargornis*). If the extinction of these groups

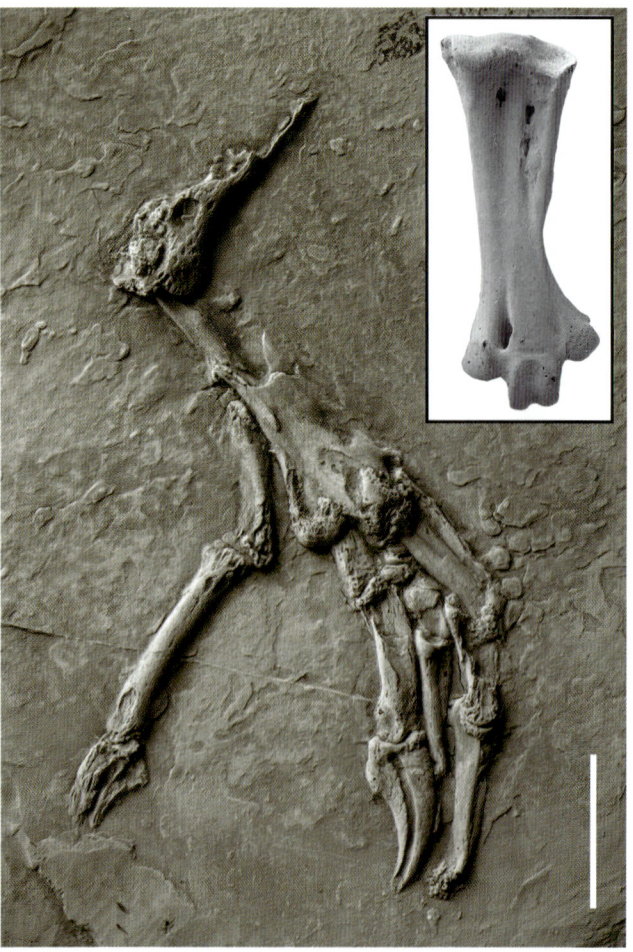

Fig. 11.51: The foot of an unidentified bird whose tarsometatarsus resembles the Indian taxon *Vastanavis* (inset). Scale: 1 cm.

in Europe were solely due to climatic factors, it could be expected that they would occur in the Old World tropics today; in particular, since – according to our current knowledge – these birds, which nowadays occur in the Neotropics, disappeared from Europe prior to the Miocene, i.e., at a time when the climatic gradient between the tropics and the Northern Hemisphere was not yet strongly developed (Mayr 2011b). Cuckoo rollers, represented in Messel by the taxon *Plesiocathartes,* are only found around Madagascar today, and in their case, as well, the question arises why they do not occur in the tropical regions of Central Africa and Asia. Mousebirds, formerly common in North America, are currently only found in Africa south of the Sahara, and the modern distribution of frogmouths is restricted to Southern and Southeast Asia, Australia and New Guinea.

Compared with extant tropical ecosystems, it is notable that the closest relatives of the aforementioned Southern Hemispheric avian groups occur in regions that were geographically isolated during the Cenozoic. Messel also shares certain ecological peculiarities with South America, Madagascar and New Guinea, such as a high number of flightless or terrestrial birds (as compared with Africa or Asia) and many representatives of the Strisores. This suggests that the survival of many groups that were widespread in the Northern Hemisphere during the Eocene in these areas is related to the geographic isolation over large periods of time during the Cenozoic (Mayr 2017b).

The first thing that comes to mind here are profound changes in the ecosystems in the Northern Hemisphere due to the diversification of placental mammals during the Cenozoic. In the early Eocene, marine straits separated Europe from Asia and Africa. The strait between Asia and Europe only closed in the early Oligocene, which led to the immigration of several mammalian groups that originated in Asia and did not occur in Europe prior to this time. These include felids and mustelids, that is, cat- and weasel-like carnivorans. Apart from the immediate threat posed by predators, avian populations are sensitive to nest predation and competitors for breeding sites. It is therefore conceivable that some of the "Southern Hemispheric" avian groups became extinct as a result of increasing competition for suitable nesting sites during the Cenozoic. In addition to arboreal snakes and other bird species, many mammalian species, in particular the tree-dwelling squirrels, constitute potential nest predators or compete with birds for suitable nesting sites in tree cavities. In this context, it is interesting to note that today, cavity-nesting parrots are much more common in South America and Australia than in Africa and Asia, the latter forming the centers of extant squirrel distribution (Corlett & Primack 2011).

Although the ecological interrelations between birds and other animals are often poorly understood, it can be assumed that the evolution of birds was closely correlated to that of potential predators. In the Eocene, many of the mammalian groups that now act as potential predators or nest-site competitors had not yet evolved. Comparisons with the ecosystem in Messel might therefore lead to a better understanding of modern tropical ecosystems.

Messel birds and tropical avifaunas

Fossil discoveries are generally very sparse in any of the areas that are covered by tropical rainforests today, and birds, in particular, are only rarely found. Therefore, it is difficult to place the evolution of the avifaunas in tropical ecosystems into a temporal perspective, which is particularly regrettable since birds are among the most extensively studied groups in extant tropical ecosystems.

Meanwile, however, there exists an extensive fossil record in Europe and North America from paratropical ecosystems of the early Cenozoic. At the time of formation of the Messel oil shale, the pronounced temperature gradient between the poles and the equator had not yet developed, and a tropical or subtropical climate prevailed up to much higher latitudes (Chapter 3). Therefore, a comparison with Messel and other Eocene fossil sites may lead to a better understanding of the evolution of extant tropical ecosystems, and birds were the first group of animals from Messel that were studied in this regard (Mayr 2017b).

To date, approximately 70 species of birds have been distinguished in Messel, although not all of them have yet been scientifically described, and some are only represented by fragments – usually isolated skulls or legs (Mayr 2016a, c, 2017b). However, the actual number of avian species in the ecosystem around Lake Messel was certainly much higher.

On the one hand, the various collections still contain several undescribed species, and on the other hand, only a small proportion of the oil shale has been excavated for fossils, so that numerous new discoveries can be expected in the future. Moreover, it is unlikely that all of the species were preserved in the lake sediments, and the crushed skeletons of many birds from Messel make it difficult to distinguish closely related species from each other, especially since many extant bird species are mainly distinguished by plumage characteristics. It is not possible to determine the actual number of bird species that occurred in the Messel ecosystem during the early Eocene, but an estimate of 100 is certainly on the conservative side. Even the 70 species identified to date surpass the average number of avian species encountered in Central European forest ecosystems. The number of bird species in modern tropical ecosystems varies greatly, and while the Amazon Basin holds the largest number of species – up to 400 in certain areas – the number of avian species in the rainforests of New Guinea barely reaches 100 (Mayr 2017b).

It has long been known that the tropical regions today are home to a much higher species diversity than the temperate latitudes, and many hypotheses were developed in order to explain this species gradient along the latitudes (Mittelbach et al. 2007). In analyses of species diversity and speciation in tropical ecosystems, birds often play a significant role, not least since they tend to be easily observable and their species diversity is well known. According to one theory, the species diversity of tropical ecosystems can be explained by the fact that the existence of stable climatic conditions over a long period of time favored the development of a high species diversity (Mittelbach et al. 2007). However, this hypothesis is countered by the fact that those birds from Messel that can be assigned to modern taxa are often members of relict groups with only a small number of extant species. On the other hand, modern-type crown group representatives of taxa that are particularly numerous in tropical ecosystems today, e.g., passerines, parrots, pigeons and hummingbirds, are not found in Messel and other fossil sites of a similar age. Therefore, the fossil record contradicts the assumption that today's most species-rich avian groups, many of which occur in the tropics, are also the oldest groups. It is therefore more likely that the large number of different ecological niches provided by tropical ecosystems played a major role in the evolution of their high species diversity (Mayr 2017b).

The ecosystem at Messel during the Eocene was characterized by relatively high temperatures throughout the year. It can therefore be expected that

Fig. 11.52: A fossil of *Primozygodactylus major* (Zygodactylidae) with numerous grape seeds in its stomach content (enlarged detail). Scale: 1 cm.

many birds fed on insects and fruits, and for some of the smaller species, stomach contents indicate at least a partially fruit-based diet (Fig. 11.52). However, the lack of larger frugivorous tree-dwelling birds is conspicuous. While all modern tropical ecosystems include specialized large fruit-eaters such as toucans in the Neotropics, hornbills in Africa and Asia, or large pigeons in the rainforests of New Guinea, none of the arboreal species identified from Messel to date even reached the size of a crow. That this is not just an artifact due to lack of preservation is evidenced by the fact that large arboreal birds are also unknown from other fossil sites such as the London Clay and the Green River Formation. It can thus be assumed that the coevolution of plants and birds that led to the development of large fruits primarily eaten by birds did not occur until later (Eriksson 2016).

What remains to be discovered

The lack of passerines was mentioned earlier as one of the peculiarities of the Eocene avifaunas in Central Europe. This significant difference compared to the modern avifaunas can be explained by the evolutionary history of the Passeriformes, which, according to our current knowledge, originated in the Australian region, whence they only began to spread to the Northern Hemisphere during the early Oligocene (Mayr 2009a, 2016a).

In addition, numerous other birds that are a familiar part of the avifaunas of the Northern Hemisphere today are missing from Messel and other similar-aged fossil sites. These include widespread arboreal groups such as woodpeckers, pigeons and cuckoos as well as aquatic and semi-aquatic birds, e.g.,

Box 11.5: The feeding remains of crocodiles

Remains of large birds are very rare in Messel, and largely complete skeletons are only available from the species *Palaeotis weigelti*, *Juncitarsus merkeli*, *Dynamopterus tuberculatus*, and *Strigogyps sapea*. More common are isolated feet of larger birds (scale: 1 cm), which normally show fracture points at the ends of the bones and apparently are the feeding remains of predators. Based on similar discoveries from younger fossil sites, these probably constitute leftovers from the meals of crocodiles (Mayr 2016c), which were very numerous in Messel and are represented by several species (Chapter 10.3). The majority of these isolated feet belongs to ground-dwelling bird species. It is likely that these animals were caught near the lakeshore, and the legs were separated from the remaining skeleton during the feeding process.

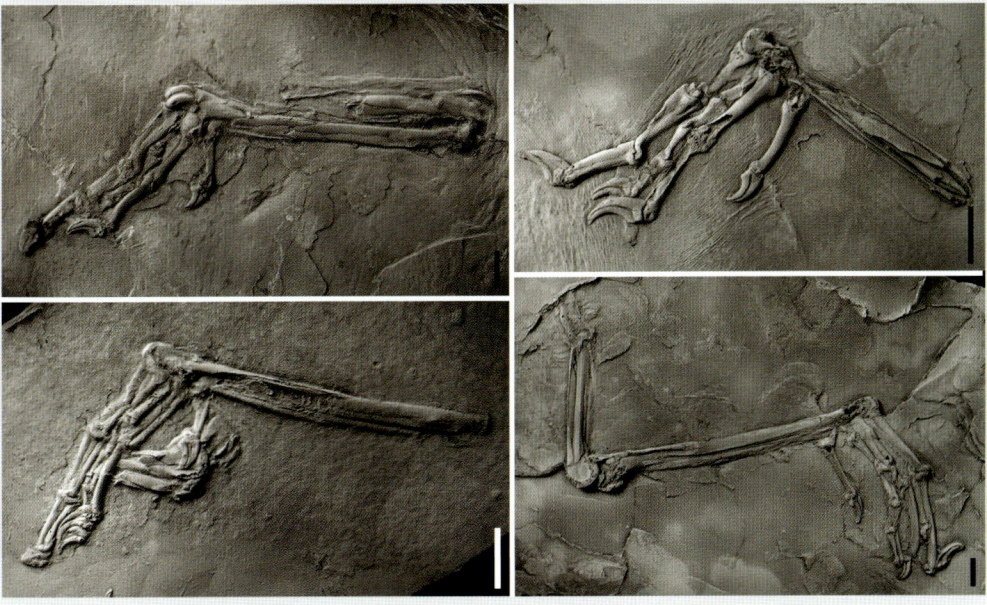

Box 11.6: Fragile fossils: Isolated feathers

Besides skeletons, a significant number of individual avian feathers has been found in Messel (scale: 1 cm). These specimens have not yet been studied, which is mainly due to the fact that a reliable phylogenetic assignment of isolated Eocene bird feathers appears almost impossible. However, the relative frequency of large feathers from the body plumage is noticeable and contrasts with the rarity of large bird skeletons.

storks, herons and cormorants. These differences are probably also based to a large extent on the fact that these groups, at least in their modern form, did not yet exist during the Eocene, or that they originated in regions in the Southern Hemisphere and only later spread to Europe.

On the other hand, the rarity of skeletons of larger terrestrial birds in Messel is most certainly a taphonomic artifact, since several isolated feet of larger avian species have been found, which presumably constitute the feeding remains of crocodiles (Box 11.5). Here, we can hope that future excavations will unearth additional fossil materials. There are also a few isolated feathers of larger birds that have not yet been studied in detail and whose systematic affinities are still uncertain (Box 11.6).

Despite the exceptionally rich fossil inventory, a number of avian groups known from other lower Eocene sites in Central Europe and North America have not yet been found in Messel. As mentioned above, several taxa known from the North American Green River Formation are missing, e.g., the oilbird *Prefica* (Steatornithiformes) or the apodiform Eocypselidae. The lack of remains of these birds may be an artifact of the incomplete fossil record, or it may indicate different ecological conditions at the fossil site. At least in the case of the Eocypselidae, which are well known from other fossil sites, it is also conceivable

Box 11.7: A particular rarity: Chicks

Skeletons of juvenile birds are extremely rare in Messel; to date, only two have been found. One is about the size of a large young chicken and probably belongs to a ground-dwelling species (scale: 1 cm). Unfortunately, the skeleton represents a very early developmental stage, which makes it difficult to firmly assign it to one of the avian groups known from Messel. It is certainly a member of a larger species that was primarily terrestrial and whose chicks were precocious. No nests or eggs have been found in Messel, and it is unlikely that eggs would be preserved, given the lake's acidic, calcium-dissolving properties.

that the slightly younger age of Messel compared to the Green River Formation or the London Clay played a role, and that some of these groups had already become extinct at the time of the formation of the Messel oil shale. Despite the presumably different paleoecological conditions, the avifauna of Messel appears to show noticeable similarities to that of Geiseltal. It is therefore remarkable that the apodiform Aegialornithidae known from Geiseltal have not been reported from Messel.

Moreover, many birds from Messel are only documented by incomplete remains. This not only applies to the aforementioned "mystery birds," but also to several species whose systematic affinities are known, or at least less controversial. For example, neither of the two known specimens of the owl *Palaeoglaux* preserves a skull, while a skull is the only evidence of *Masillastega* found to date. Further studies will also be required to clarify the relationships of certain problematic fossils, i.e., the small number of avian chicks known from Messel (Box 11.7).

Even in the case of the better-documented species, each additional discovery can offer insights into hitherto unknown aspects of the morphology or lifestyle, whether by means of preserved stomach contents or a particularly well-preserved plumage or skeleton. This even applies to the extremely numerous Messel rail, for which only very few specimens with identifiable stomach contents are known, and none with well-preserved feathers. And finally, the discovery of a skeleton of the giant flightless bird *Gastornis* would certainly constitute a highlight of the long-term excavation activities. Although complete skeletons of this bird are known from other fossil sites, the unique level of preservation of fossils in Messel might offer new insights into the still controversial diet of this bird or its plumage.

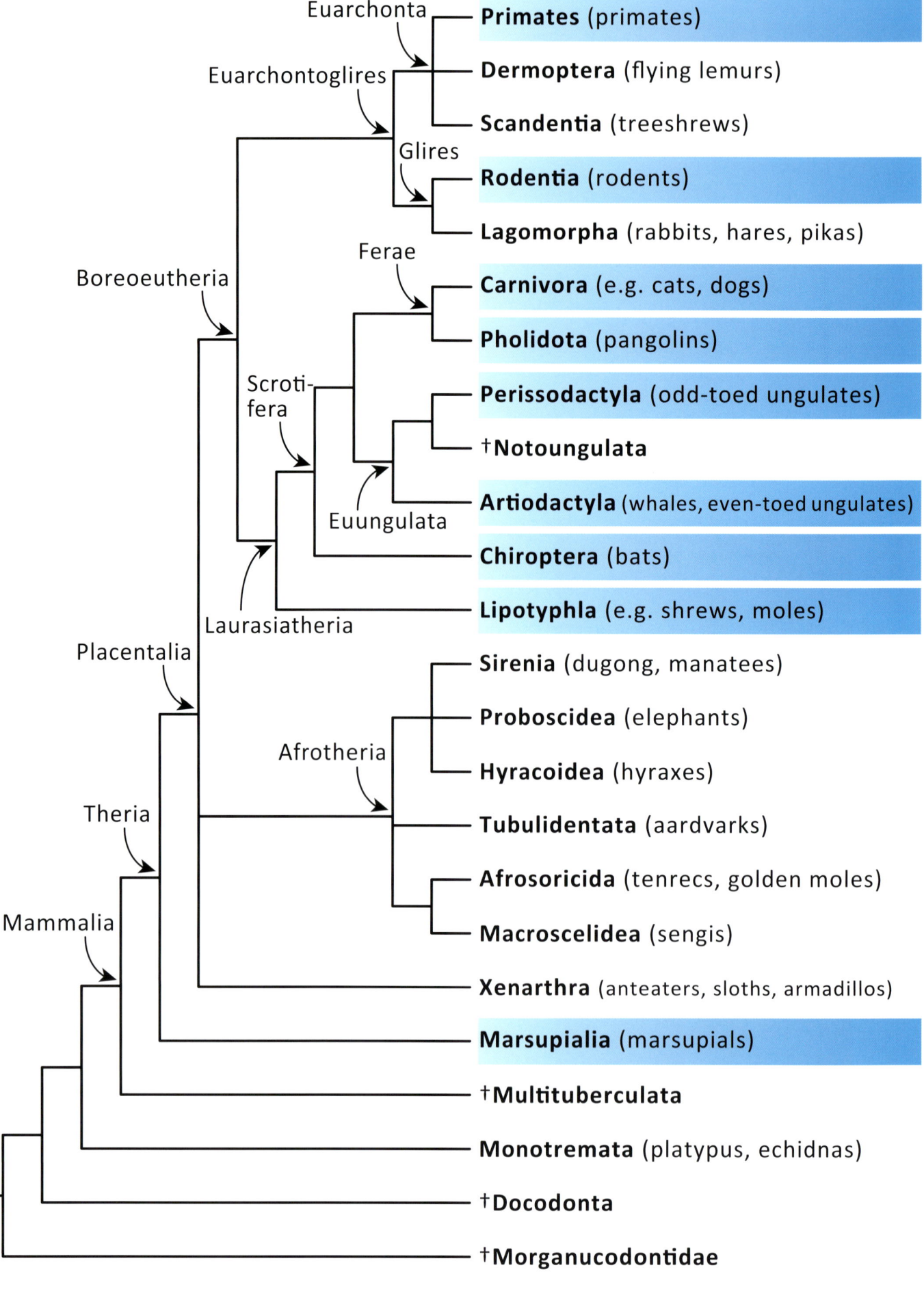

Chapter 12
Mammalia – Another Success Story

Editors

Three middle-ear bones, warm-bloodedness, hair, and mammary glands for the nourishment of hatchlings and newborns are unmistakable features of all living mammals (Mammalia). It was initially assumed that most early mammals were small, insectivorous, and nocturnal, and became specialized only after the extinction of the dinosaurs at the end of the Cretaceous. However, a series of spectacular new fossils has demonstrated a remarkable disparity of Mesozoic mammals, including swimming, gliding, climbing and digging forms. Some features previously thought of as unassailable synapomorphies may have arisen more than once (Luo 2007). Thus, our understanding of early mammal evolution is changing rapidly.

Many mammalian characteristics are related to their high metabolism and the ability to sustain vigorous activity: branching lungs with a high surface area, a four-chambered heart, and an upright posture (Carrier 1987). Fur, although it may have evolved for different reasons, insulates and helps maintain the high and constant body temperature. It might have been particularly useful to the closest relatives of mammals, which were probably crepuscular (active during twilight) or nocturnal (Angielczyk & Schmitz 2014).

Additionally, the mammalian sensory apparatus and associated neural capacity to process the sensations are considerably advanced in mammals. Olfactory sensitivity increased substantially (Rowe & Shepherd 2016). The shift of two jaw bones into the middle ear and the elaboration of the inner ear led to improved hearing, especially at high frequencies (Manley 2010).

Reproduction evolved dramatically within the clade. Monotremata lay eggs, a primitive feature (Chapter 10). In Theria (Marsupialia and Placentalia), instead, the embryo receives nutrition and gas exchange directly from the mother via a placenta. In Placentalia, gestation is prolonged. Placental mammals form the bulk of mammal diversity today, over 90% of the roughly 5,500 living species (and those from Messel). Placentals clearly radiated very rapidly after the extinction of the dinosaurs at the end of the Cretaceous (O'Leary et al. 2013), and Messel documents the early evolution of many of the lineages that arose.

Debates about the interrelationships of these lineages and their time of origin are not yet resolved, but genetic studies have produced a well-supported framework with four major placental clades – Afrotheria (including elephants, sea cows, hyraxes, tenrecs, and aardvarks), Xenarthra (including sloths, anteaters, and armadillos), Laurasiatheria (including carnivores, ungulates, bats, and shrews), and Euarchontoglires (including rodents, rabbits, and primates) – that is now generally accepted. Laurasiatheria and Euarchontoglires are generally supported as sister groups (comprising the Boreoeutheria). Xenarthra and Afrotheria might be sister groups (e.g., Tarver et al. 2016). More challenging is the inclusion of certain extinct species into a phylogeny, sometimes even when the fossils are of exceptional quality. The future might lie in the use of fossil proteins (e.g., Welker et al. 2015).

Fig. 12.1: Relationships among mammals under the current consensus. Groups known from Messel are highlighted in blue.

Chapter 12.1
Marsupials – a Surprise in Messel

Cornelia Kurz, Wighart von Koenigswald

At first glance, it may appear unusual for marsupials to be among the finds described from a European fossil site, since this group does not occur in modern-day Europe. However, various species of marsupials are found in Cenozoic deposits across large parts of Europe. Among the mammals, the marsupials (Marsupialia) form a sister group to the placentals (Placentalia); they date back to the late Early Cretaceous (Case et al. 2005).

Today, representatives of the marsupials are restricted to the Australian faunistic region and South America, whence they colonized North America only in the recent geological past. However, the fossil record shows that marsupials formerly occurred in Europe, Asia, Antarctica and North Africa as well.

The modern opossums (Didelphidae) from America are among the most primitive marsupials, the Ameridelphia (Fig. 12.1.2). However, these Didelphidae originate from a lineage dating to the Miocene; therefore, they are only remotely related to the much older opossum-like marsupials from Europe. Nevertheless, both groups are generally referred to as "opossums."

The European fossil marsupials are small, mouse- to rat-sized animals. They are placed in two large systematic groups, the Peradectia and the Didelphimorphia. Both groups have their origin in North America.

European marsupials have been known to science for more than 200 years. In 1804, George de Cuvier described the first specimen from the upper Eocene gypsum deposits at Montmartre in Paris. This involved an almost complete skeleton, including a skull. To date, complete skeletons with skulls have only been unearthed at two other European sites: in Messel and in Geiseltal, which is a few million years younger. Numerous other sites have merely produced teeth, remnants of jaws and isolated bones, as well as a small number of intact skulls.

So far, the Messel Pit has yielded a total of five almost complete specimens plus one isolated lower jaw. The first of these skeletons was discovered in 1980. Although it involved a juvenile, the characteristic dental formula and the epipubic bones were recognizable (Koenigswald 1982). While an additional specimen had already been found around the end of the 1970s, it was initially not recognized as a marsupial (Kurz 2007).

Anatomy and morphology

The primary differences between placentals and marsupials are found in the soft parts, in particular those related to the reproductive biology.

However, the identification of a fossil marsupial requires the presence of osteological and dental features, since only those are usually preserved. Formerly, the epipubic bones, in particular, were considered a useful indicator for the distinction between marsupials and placentals. At least in the European Cenozoic, the

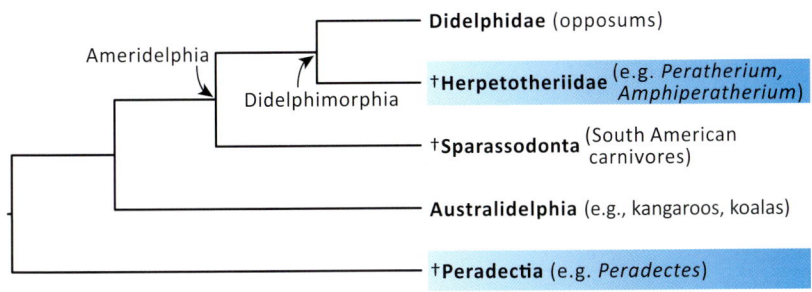

Fig. 12.1.2: Simplified phylogeny of the Marsupialia. Groups known from Messel are highlighted in blue.

Fig. 12.1.1: The climbing opossum "*Peradectes*" with a prehensile tail. Scale: 1 cm.

presence of epipubic bones offers reasonable proof that an actual marsupial has been found. The epipubic bones ("pouch bones") are not anatomically connected to the pouch, since they are present in males, as well. Another feature is a projection of the angle of the lower jaw (Processus angularis) that is bent toward the midline. However, certain insectivores show a similar structure, and it is difficult to distinguish between the two groups if this feature is not fully preserved. The dental formula, on the other hand, serves as a relatively reliable indicator for primitive marsupials. They have a larger number of incisors: five in the upper jaw and four in the lower, as opposed to three in the primitive Placentalia. In addition, the opossums' four molars exceed the three usually found in placentals.

Particularly important indicators in terms of systematics and biology can be found in the morphology of the molars' occlusal surfaces. Many of the fossil vertebrates from Messel were found with their mouths closed. This complicates the analysis of the dental features of some of the marsupial specimens. However, the use of a special micro-X-ray technique makes it possible to reveal the dentition of these specimens (Kurz & Habersetzer 2004). In combination with the findings from classic dental morphology, these studies indicate three different forms among the marsupials from Messel.

Form 1: Peradectidae: "*Peradectes*"
Form 2: Herpetotheriidae: *Amphiperatherium goethei*, *Amphiperatherium* sp.
Form 3: an additional herpetotheriid marsupial

Form 1 includes the three "*Peradectes*" skeletons (Fig. 12.1.1, 12.1.3). The affiliation with the "*Peradectes*" group is based on the morphology of the occlusal surfaces of the upper molars, which show three main cusps of equal size. These specimens show shallower crowns and more rounded cusps on their molars and incisors, which are inserted in the lower jaw at a relatively steep angle (Fig. 12.1.4). This form is characterized by a rather short snout, and – measured across the zygomatic arches – the skull is relatively wide. The skull protrusions above the eye sockets and on the cheekbones are well-developed. In addi-

Fig. 12.1.3: Right-side view of the opossum "*Peradectes.*" Scale: 1 cm.

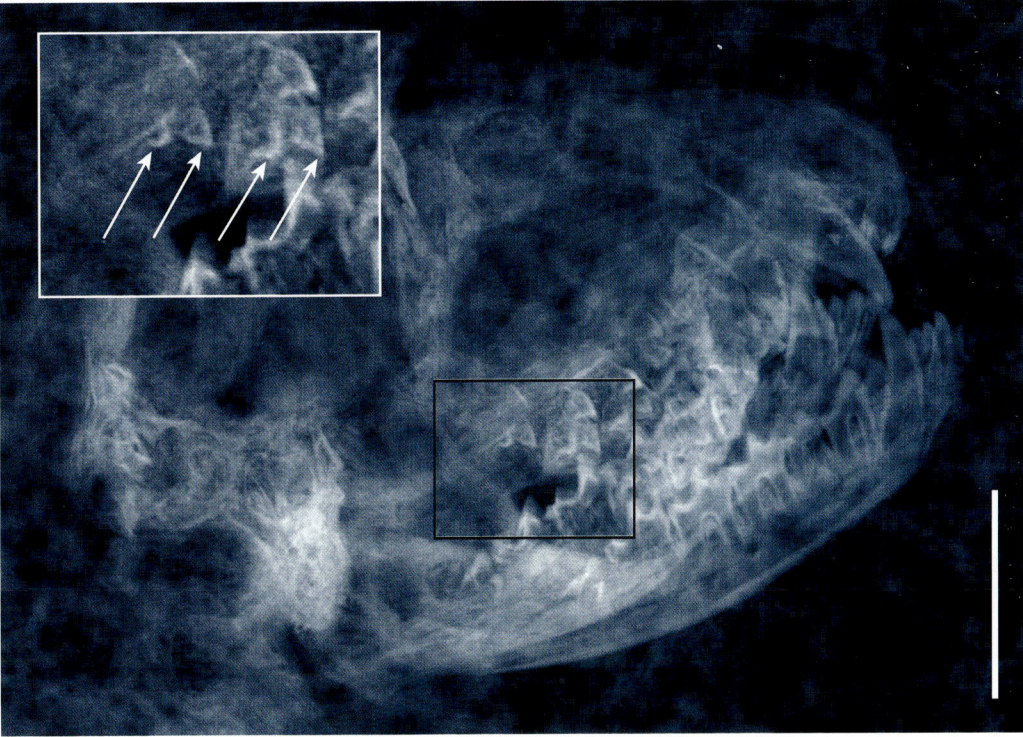

Fig. 12.1.4: Micro-X-ray image of the skull of "*Peradectes*." The upper and lower jaws are clearly visible, as well as the rounded snout with steeply protruding lower incisors. Diagnostic features include the almost equal cusps (see arrows in the inset image) of the upper molars. Scale: 5 mm.

tion, the spinal column is inserted near the edge of the skull base, rather than pointing straight backward from the occiput (position of the Foramen magnum).

The skeleton clearly shows the lack of transverse processes on the lumbar vertebrae. Moreover, the specimens (where preserved) show an extremely long tail in relation to their head and torso length. The metatarsals, however, are more or less of the same length as the first phalanges.

The skeletons of the Herpetotheriidae belong to the forms 2 and 3. In this group, the two main cusps of the upper molars are arranged at an angle toward each other. In addition, the position of the small cusps along the molars' edge can be used to determine the species. The herpetotheriids found in Messel can be distinguished primarily on the basis of the position of their lower incisors. In form 2 (Fig. 12.1.5), these protrude more strongly forward and at a shallower angle than in form 3. The skulls of both forms are characterized by elongated snouts, combined with relatively narrow skulls and small protrusions above the eye sockets and on the cheekbones. The two relatively complete specimens also show a spinal column attachment closer to the occiput.

As opposed to form 1, the lumbar vertebrae have rather expansive transverse processes. The tail length (where preserved) roughly equals the head to torso length, and the metatarsals in one preserved specimen are significantly longer than the first phalanges.

Paleoecology

The six marsupial specimens from Messel indicate that the European marsupials did not represent generalized forms. The Messel skeletons offer a much more detailed insight into the animals' habits than previously assumed. The morphological features and proportions of the skeletons document different levels of adaptation to the habitats of the systematically distinguishable taxa. In addition, the dental morphology also allows an analysis of the animals' dietary habits. This makes it possible to reconstruct habitat associations for the marsupials from Messel.

In regard to Messel, this means that the specimens of form 1 ("*Peradectes*") with their shallow occlusal surfaces and vertically protruding lower incisors were likely omnivores, with a large proportion of fruits

Fig. 12.1.5: Ventral view of the terrestrial opossum *Amphiperatherium* sp. The skull is partially obscured by the elongated lower jaw. Note the forward-protruding lower incisors in the inset (arrow). Scale: 5 cm.

in their diet. Due to the missing transverse processes on their vertebrae, the almost equal length ratio of metatarsals and first phalanges and the longer tail, this first form had a flexible pelvis perfectly suited for climbing in tree branches, a suitable gripping foot and a prehensile tail. In addition, this form – in adaptation to a climbing lifestyle – shows a different positioning of the skull and eyes: In climbing mammals, e.g., the primates, the spinal column is attached closer to the underside of the skull. The skull of form 1 is wide and carries well-developed protrusions, which serve as attachment points for the so-called Ligamentum orbitale. This offers support behind the eye and separates it from the chewing muscles. Thus, the eyes are pointed more directly forward, facilitating an improved spatial perception.

Form 2, with its strongly forward-protruding lower incisors and the longer snout, indicates that *Amphiperatherium* was more likely an insectivore. The mobility of *Amphiperatherium* was restricted by the large transverse processes on the lumbar vertebrae in the pelvic region. The longer metatarsals in relation to the first phalanges are a strong indication for an animal that moves on the ground. However, the resulting stability and the long foot enabled the herpetotheriid marsupial to transfer a significant amount of power to the ground while walking, resulting in a relatively fast gait. The rather short tail only served to keep the balance. The narrow skull with poorly developed protrusions also indicates a terrestrial lifestyle, since spatial vision was not well developed.

With the less strongly forward-protruding lower incisors, form 3 suggests that the unidentified herpetotheriid marsupial was more likely an omnivore with a low proportion of insects in its diet. As far as discernible, the features of the skull and skeleton are the same as found in *Amphiperatherium*, i.e., the second form.

Evolution and biogeography of the marsupials from Messel

The ancestors of the European marsupials likely reached Europe from North America via the Arctic during the Late Cretaceous or the early Paleogene. During this time, solid land bridges in the far north (Chapter 2) facilitated at least two instances of intensive faunal exchange by way of northern Greenland.

Independent of each other, two marsupial lineages immigrated to Europe: Peradectidae and Herpetotheriidae. The oldest documented occurrence of marsupials in Europe is a single tooth from the Late Cretaceous in the Netherlands, which was discovered south of Maastricht and described under the name *Maastrichtidelphys meurismeti* (Martin et al. 2005). The discoveries from Messel show that the descendants of those lineages adapted to different habitat types in the primeval forest of Messel. *Amphiperatherium* was a large omnivore with a high percentage of insects (if not exclusively) in its diet and strongly adapted to terrestrial movement. The small herpetotheriid marsupial pursued a similar lifestyle in the same forest and was also an omnivore, with a significantly lower proportion of insects in its food. "*Peradectes*" represents a rather typical marsupial for the primeval forest, small of size and adept at climbing in the tree branches, where it likely preferred a diet of fruits.

Due to the preservation of complete skeletons in Messel, it was possible for the first time to document this diversity in such detail. The same diversity presumably applies to other occurrences of marsupials in the Paleogene in Central Europe. However, the species diversity decreases over time, with only one single species, *Amphiperatherium frequens*, persisting into the Miocene and finally disappearing in the middle Miocene (Ziegler 2006).

Chapter 12.2
Four Archaic Yet Highly Specialized Mammals

Wighart v. Koenigswald, Gregg F. Gunnell †, Thomas Lehmann, Kenneth D. Rose, Irina Ruf

The species of the families Apatemyidae, Pantolestidae, Paroxyclaenidae and Pseudorhyncocyonidae found in Messel represent four extinct lineages that originated during the first major radiation of the Placentalia. However, these animals cannot really be labeled primitive, since their representatives already show highly specialized adaptations to a variety of lifestyles (Rose 2006a). They include terrestrial (ground-dwelling), arboreal (tree-climbing) and even semi-aquatic forms. Correspondingly, they also showed significant differences in their preferred diet.

The adaptations of these taxa can only be reconstructed by comparing their fossil skeletons to those of recent species that show similar adaptations. In this context, the actual relationships only play a subordinate role. Therefore, the four archaic placental species from Messel represent striking examples of convergent evolution and illustrate the adaptive radiation of the placental mammals and the occupation of ecological niches as early as the beginning of the Paleocene.

Systematics

The phylogenetic placement of the four mammals presented here, which belong to four different families, is still a matter of conjecture. Complete skeletons, as available from Messel, are extremely rare, but they contribute greatly to our understanding of the relationships and the lifestyles of these groups.

For a long time, the systematic position of the Apatemyidae was controversial, and they were placed with a number of different groups until their unusual specialization became known through the discoveries from Messel. Based on a phylogenetic analysis (Silcox et al. 2010), the Apatemyidae should best be placed as stem group of the Euarchontoglires (Fig. 12.2.2).

The position of the other three families – the Pantolestidae with *Buxolestes*, the Paroxyclaenidae with *Kopidodon* and the Pseudorhyncocyonidae with *Leptictidium* – remains questionable. They are either viewed as representatives of the stem group of Placentalia (Fig. 12.2.2), or on a somewhat higher level as stem representatives of the Laurasiatheria, although where they split from other mammals cannot be clarified any further with confidence.

According to a recent study, the Pseudorhyncocyonidae are monophyletic and are closely related to the Leptictidae, a species-rich North American family (formerly their sister group within the Leptictidae (Gunnell et al. 2008)), as well as the Pantolesta and Palaeanodonta (Hooker 2013). Should the latter actu-

Fig. 12.2.2: Simplified and rather conservative phylogenetic tree of the archaic mammals presented in this chapter. Groups known from Messel are highlighted in blue.

- Euarchontoglires (e.g. rodents, primates)
- †**Apatemyidae** (e.g. *Heterohyus*)
- Laurasiatheria (e.g. carnivores, ungulates)
- Afrotheria (e.g. elephants, hyraxes)
- Xenarthra (e.g. anteaters, sloths)
- †**Pantolestidae** (e.g. *Buxolestes*)
- †**Paroxyclaenidae** (e.g. *Kopidodon*)
- †**Pseudorhyncocyonidae** (e.g. *Leptictidium*)

Fig. 12.2.1: *Kopidodon macrognathus*. The remarkably well-preserved soft-tissue remains of the specimen, embedded in the typical lateral position, shows the bushy tail and a rounded outer ear. Scale: 10 cm.

ally be related to the Pholidota (pangolins) (Chapter 12.7), the Pseudorhyncocyonidae, Leptictidae, Pantolesta, and Palaeanodonta would all belong to the Laurasiatheria. However, according to O'Leary et al. (2013), it is also possible that *Leptictis* represents a sister group of the elephant shrews (Macroscelidea) within the Afrotheria.

The genus *Buxolestes* belongs to the Pantolestidae and shows convincing similarities to the North American genera *Pantolestes* and *Palaeosinopa*. The Pantolestidae probably belong to an only weakly supported group, the "Cimolesta" which includes the Paroxyclaenidae and may possess affinities to Laurasiatheria via the Palaeanodonta.

The family Paroxyclaenidae was established on the basis of a compressed skull of *Kopidodon macrognathus* (Weitzel 1933). The first jaw fragment of this genus had been originally assigned to primates (Wittich 1902). The Paroxyclaenidae then ended up with the "Creodonta," an archaic group of ancient predators. Later, the Paroxyclaenidae were variously placed with the Condylarthra (ancient ungulates), the "Cimolesta," and even the Ptolemaiidae (basal Afrotheria) (Russell & McKenna 1961; Tobien 1969; Koenigswald 1983a, 1988; Bown & Simons 1987; Russell & Godinot 1988). It has not yet been possible to unequivocally determine their systematic position (Fig. 12.2.2).

The remarkable adaptations of *Leptictidium*

The Pseudorhyncocyonidae were small to medium-sized mammals with an elongated snout, archaic dentition, and remarkably elongated hindlimbs (Fig. 12.2.3). The genus *Leptictidium* is particularly species-rich, with at least six species described from Germany, France, and England (Hooker 2013). First described by Tobien (1962) from the Messel Pit, the genus is meanwhile known by nearly 20 complete skeletons from Messel, whereas only fragmentary remains have been found in the other sites. Three species can be distinguished in Messel: *L. tobieni* (Fig. 12.2.4), *L. nasutum,* and *L. auderiense* (Fig. 12.2.5). Most of our knowledge regarding the paleobiology of the Pseudorhyncocyonidae is almost exclusively based on the skeletons from Messel.

The skull of *Leptictidium* is characterized by a long and narrow snout, whose reconstructed muscle insertion surfaces suggest the presence of a short proboscis (Storch & Lister 1985). The smooth skull surface and the lack of a sagittal crest imply a moderately developed jaw musculature and a rather low bite force. This indicates that its prey animals must have been rather small. The dentition of *Leptictidium* clearly shows adaptations to a carnivorous to insectivorous diet, which is also confirmed by analyses of the gut contents of three individuals of *L. nasutum* (Maier et al. 1986). Besides the digested bones of lizards and small mammals, they also contained cuticles and antennae of insects.

Fig. 12.2.3: Life reconstruction of *Leptictidium,* based on the skeletons from Messel. The animal in the drawing is about to jump.

The remarkable adaptations of *Leptictidium*   **225**

Fig. 12.2.4: *Leptictidium tobieni*, the largest species of the genus. The great difference in the length of the fore- and hindlimbs is striking. Scale: 10 cm.

Fig. 12.2.5: The complete skeleton of *Leptictidium auderiense* gives the impression that it was frozen in mid-jump. Scale: 5 cm.

Box 12.2.1: *Leptictidium auderiense* – an agile biped

The inner ear of mammals houses the sense of hearing in the snail-shaped cochlea and – within the remaining parts – the sense of balance. In particular, the three semicircular canals detect the angular acceleration of the head and are therefore very interesting for tackling the morpho-functional questions about an animal's adaptation to a certain mode of locomotion. Although in fossils the soft tissue structures of the inner ear are no longer preserved, the bony labyrinth – the corresponding cavity in the surrounding bone (petrosal bone) – reflects their shape and size.

Based on micro-CT data of the head of a new specimen of *Leptictidium auderiense* (right image, scale: 2 cm), a virtual 3D model of the bony labyrinth was created (left image, scale: 2 mm). Since the size of the semicircular canals is correlated both with the agility and the locomotion, several parameters of the semicircular canals were examined to characterize the specialized locomotion of *Leptictidium* (Ruf et al. 2016). Compared to recent terrestrial mammals, *L. auderiense* was a very agile animal, similar to the bipedal, saltatory rodents (e.g., desert jerboas), but also the quadrupedal elephant shrews (Macroscelidea). Conversely, the North American members of the Leptictidae, *Leptictis* and *Palaeictops*, show significantly lower agility values, which resemble those of recent generalists. The results of this study clearly demonstrate that *L. auderiense* was a locomotion specialist, as already postulated on the basis of the derived skeletal traits. Yet, the question whether *Leptictidium* was a bipedal jumper or a runner remains open.

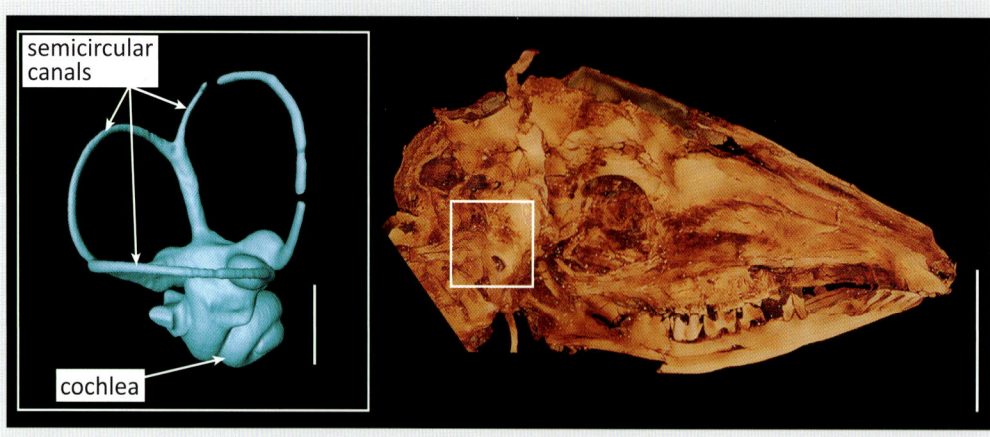

Soft-tissue remains indicate that the fur of *Leptictidium* was rather short, slightly longer on the thighs, and probably sparse on the tail (Fig. 12.2.3). The unusual body build, with very long hindlimbs in relation to its body length and markedly short forelimbs, immediately catch the eye (Fig. 12.2.4). The humerus (upper arm bone), the radius and the robust ulna are short but show strong muscle attachments. This suggests that *Leptictidium* was able to use its forelimbs efficiently for digging (Rose 1999, 2006b). The long tail consists of more than 40 vertebrae, is more than twice as long as the trunk, and thus counts among the longest tails known for mammals. Another remarkable character is the slender pelvis, which is only connected to the vertebral column by one of the three sacral vertebrae (Storch & Lister 1985). The hands are proportionally short, whereas the feet are very long and slender. Their first and fifth digits are reduced, while the third one is elongated. In the feet especially, the third metatarsal is almost half as long as the femur (thigh bone).

Among recent mammals, such asymmetrical proportions of the extremities are only found in bipedal, saltatory (hindlimb-dominated jumping) forms. Based on its body size, *Leptictidium* most closely resembles the African springhare (*Pedetes*) (Rose 2006b). This suggests that *Leptictidium* was an obligate biped and

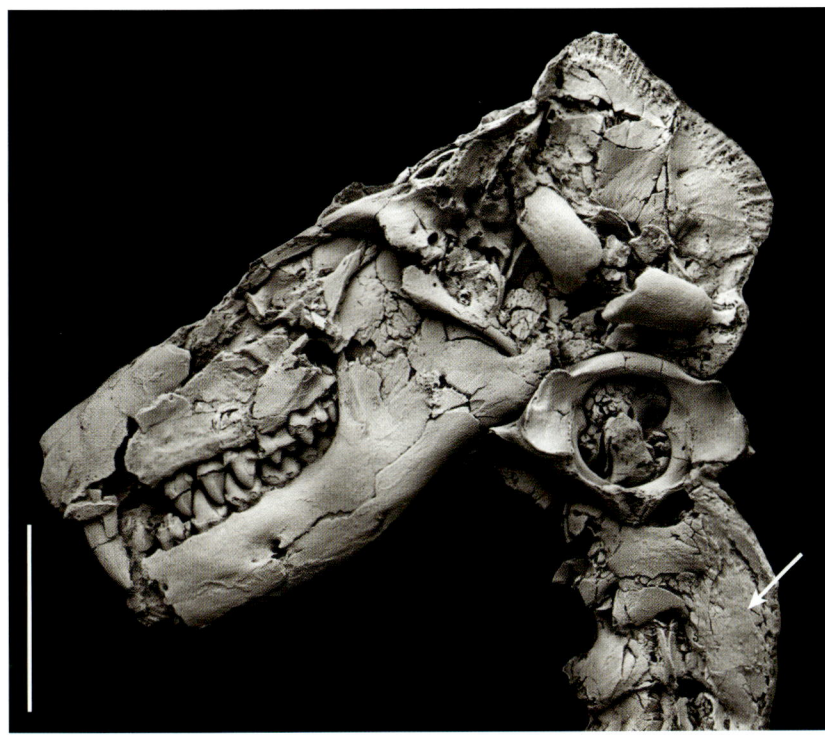

Fig. 12.2.6: Skull of *Buxolestes piscator*, with a flaring neck shield and the enlarged spinous process on the second cervical vertebra (axis or epistropheus) (arrow). Scale: 2 cm.

Therefore, we assume here the validity of the three species, although a detailed comparative description of their postcranial skeleton is still pending. Regardless, it seems remarkable that three closely related species, exhibiting the same highly specialized bipedal locomotion and sharing the same dietary preferences, would live sympatrically in the primeval forest of Messel.

The piscivore *Buxolestes*

The family Pantolestidae occurred in the Eocene across North America and Europe. However, most fossil sites have yielded only single bones or jaw fragments. The four skeletons found in Messel convey a rather complete picture of the lifestyle pursued by this group (Koenigswald 1980, 1987b; Pfretzschner 1999).

therefore dwelled on the floor of the primeval Messel forest. However, the exact mode of bipedal locomotion (walking or jumping) remains uncertain to date (Maier et al. 1986; Frey et al. 1993; Christian 1999).

An entirely different approach for reconstructing its locomotion is provided by the inner ear, which houses both the sense of hearing as well as the sense of balance (Box 12.2.1).

The validity of the three *Leptictidium* species from Messel has been recently questioned, since they show similarities in several traits and adaptations (Hooker 2013). An essential criterion used for their distinction is the body size. The head-body length of *L. auderiense* is 22 cm, *L. nasutum* measures 31 cm (about 40 % longer), and *L. tobieni*, with 37 cm, is again 20 % longer (Koenigswald & Storch 1987). Since adult specimens are known for all three species, these size differences are unlikely to be explained by different age classes. Further differences can be found in the dental morphology. For example, the upper canine and the first lower premolar of *L. nasutum* and *L. tobieni* have two roots, and the outer edge of the upper molars of *L. tobieni* bears an enlarged central cuspule (Koenigswald & Storch 1987).

The skeletons of *Buxolestes piscator* show a head-body length of 45 cm and a tail with a length of 35 cm (Fig. 12.2.7). The body weight probably reached no more than 2 kg. With regard to its physical proportions, *B. piscator* most likely resembled a small carnivore. However, despite the relatively large canines, *Buxolestes* does not belong to the Carnivora, since it lacks the shearing mechanism of the dentition used for cutting meat that is characteristic for this order (Fig. 12.2.6). The molars' chewing surfaces show a rather primitive pattern. The tips of the teeth are worn down to a shallow surface and show no adaptation to a specialized diet.

The suspicion that the Pantolestidae (and thus, *Buxolestes*) had an aquatic lifestyle (Matthew 1909) was confirmed by the discoveries from Messel. The heavily muscled upper arm further indicates burrowing habits. This is borne out by the enormously enlarged upper end of the ulna (olecranon), which serves as a lever and intensifies the powerful exten-

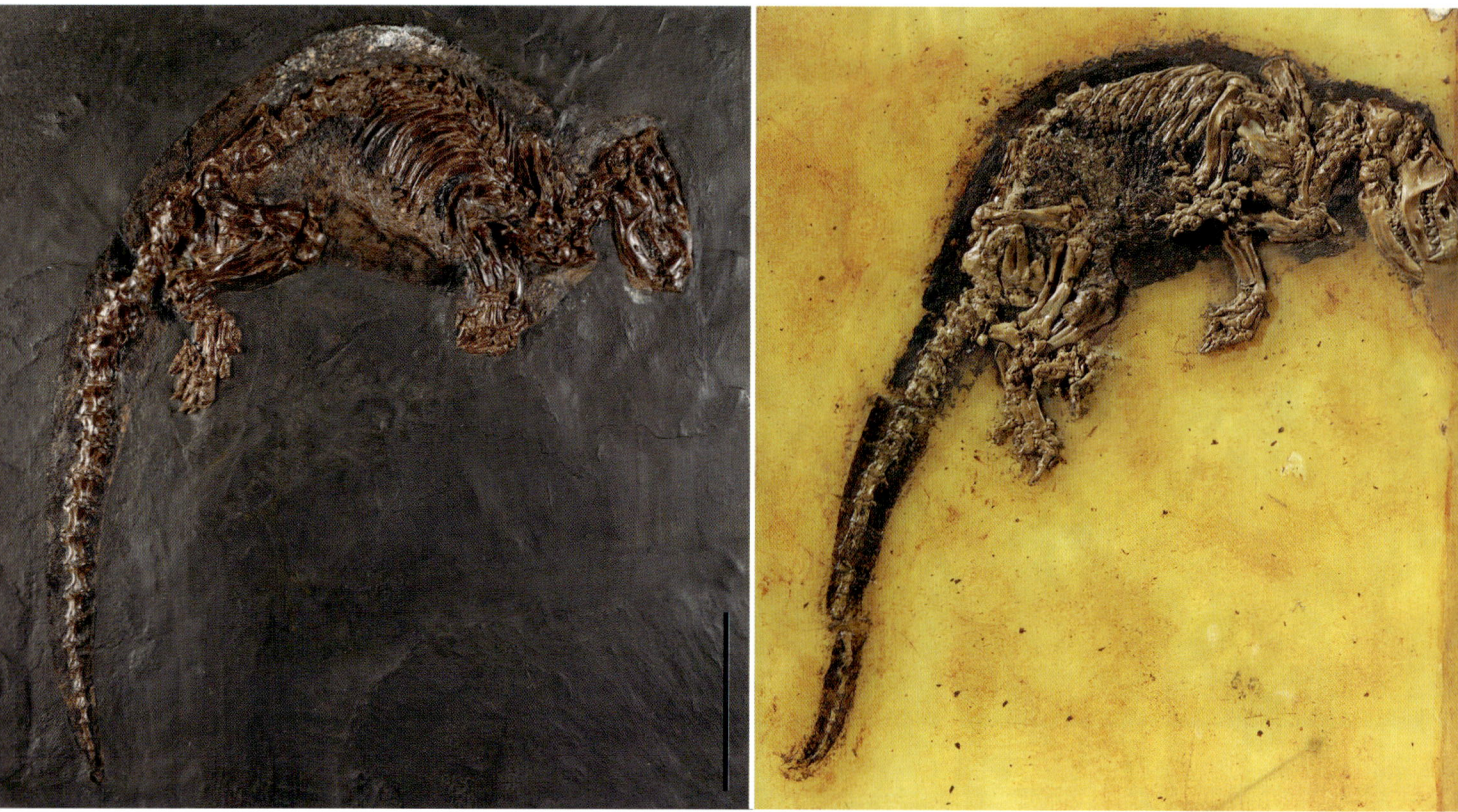

Fig. 12.2.7: Skeletons of *Buxolestes piscator* (left) and the smaller *Buxolestes minor* (right), both in the characteristic lateral position. Scales: 10 cm.

sion of the forearm. It is possible that *Buxolestes* used its strong claws to dig an underground lair.

At first glance, the skeleton does not show any typical adaptations to an aquatic lifestyle, since there is no evidence of fins, or a flattened tail such as found in the beaver (*Castor fiber*) (Fig. 12.2.7). However, even the first specimen showed some notable characteristics (Koenigswald 1980). For example, the rear of the skull bears a laterally flaring neck shield, and the spinous process of the second cervical vertebra is unusually enlarged (Fig. 12.2.6). Therefore, *Buxolestes* possessed a strong neck musculature, which on the one hand helped keep the tip of its nose above water, but mainly enabled fast and agile movements against the water resistance while hunting for fish (Koenigswald 1980; Pfretzschner 1993). A further peculiarity can be observed on the first caudal vertebrae. They carry noticeably large transverse processes, similar to those found in the genus *Lutra* (Old World otters), but which are lacking in terrestrial animals such as foxes (*Vulpes*) and martens (*Martes*).

The semi-aquatic and piscivorous lifestyle of *Buxolestes* can also be confirmed by a different line of evidence. Already in the first specimen unearthed in Messel, fragments of fish could be identified in the gut, leading to the specific epithet "*piscator*" for this fish predator. However, the second specimen of *Buxolestes* only contained plant remains (Richter 1987). The third specimen, as well as a related form from Wyoming, again contained fish remains (Rose & Koenigswald 2005).

One of the *Buxolestes* skeletons from Messel is about 20 % smaller and was therefore described as a second species, *Buxolestes minor* (Pfretzschner 1999) (Fig. 12.2.7). Since it shows neither an enlarged neck shield nor any transverse processes on the caudal vertebrae, it appears to be lacking any clear adaptations to a semi-aquatic lifestyle. Therefore, *B. minor*

Fig. 12.2.8: Complete skeleton of *Kopidodon macrognathus* in lateral-supine position. The enlarged view of the skull from below shows the closed rows of teeth and the large canines. The detailed views of the hands (bottom left) and feet (bottom right) show the strongly curved fingers and toes. The claws are laterally compressed and well suited for climbing. Scale: 5 cm.

was interpreted as a terrestrial scratch-digger. However, the question remains whether this skeleton could represent a not yet fully grown individual of the larger species (Rose et al. 2014). Although replacement of the milk dentition by permanent teeth had already been completed in the only specimen of *B. minor* known to date, the last molar is not yet fully erupted. Therefore, it remains questionable whether *B. minor* represents a separate species. It would not be surprising if the characteristic traits of the skull and tail were not yet fully developed in an immature individual.

The tree-climbing *Kopidodon macrognathus*

Kopidodon belongs to the Paroxyclaenidae and had a head-body length of almost 60 cm with a bushy, 53- cm-long tail. Its body weight can be estimated around 3–6 kg (Clemens & Koenigswald 1993) (Fig. 12.2.1, Fig. 12.2.8). Thus, it had the approximate size of a small raccoon, although it was in no way related to the latter. Yet the skeleton hints at a similar lifestyle among the branches of trees.

Meanwhile, *Kopidodon* has been documented by 24 more or less complete skeletons from Mes-

Fig. 12.2.9: The complete skeleton of *Heterohyus nanus* shows the large head and the long tail of this delicate animal. The hands (right image) show elongated bones in the metacarpus and the fingers. In contrast, the claws are very small. Scales: left 2 cm, right 5 mm.

sel, which reveal the specialized adaptations of these animals (Koenigswald 1983a). The massive skull bore a slight sagittal crest, indicating a strong chewing musculature. Despite its large canine teeth, *Kopidodon* was not a predator (Fig. 12.2.8). The dentition is characterized by teeth with low rounded cusps suitable for crushing food. Shearing edges for the cutting of meat are absent (Koenigswald 1983a). Shallowly worn teeth suggest an herbivorous diet that may have included tough-skinned fruits.

The number of thoracic and lumbar vertebrae varies among the different individuals. The long tail consists of up to 29 vertebrae. In several specimens, remains of the bushy hair has been preserved along the entire length of the tail. The arms and legs were very well developed. The humerus shaft bears well-developed muscle attachment surfaces. The radius had a substantial range of rotation at the elbow, allowing it to cross and uncross the ulna at the distal end (motions called pronation and supination), which is required for burrowing as well as climbing animals. The hands were extremely flexible, and the bony tips (distal phalanges) of the digits are claw-shaped and laterally compressed (Fig. 12.2.11). They were covered by sharp, horny claws that enabled an excellent grip on the bark of trees when climbing (Clemens & Koenig-

Fig. 12.2.10: Partial skeleton of *Heterohyus nanus,* with bushy hair only preserved at the tip of the tail. Scale: 1 cm.

Fig. 12.2.11: Drawing of the massive skull of *Heterohyus nanus,* with firmly embedded large incisors and premolars. Scale: 1 cm.

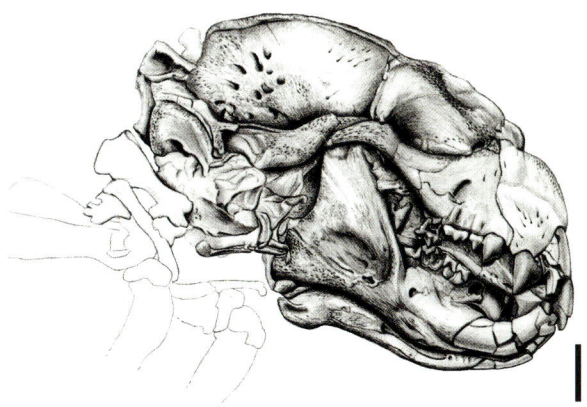

swald 1993). Such laterally compressed claws are found in many specialized climbers, while claws used for digging are usually broadened. It is unlikely that the hindlimbs were as flexible as the forelimbs, since the neck below the femoral head is relatively short (Stefen & Lehmann 2011). Tibia and fibula are rather strong and allow a good mobility of the foot. Similar to the hands, the feet possess toes that are curled inward, with laterally compressed distal phalanges. Thus, the claws of the feet served as anchors while climbing, whereas the soft sole could be precisely adapted to the substrate. Both hands and feet thus show a high level of specialization to a life in trees, where ample leaves and fruits are available for consumption.

The long-fingered *Heterohyus nanus*

The special lifestyle of the Apatemyidae was first recognized thanks to the skeletons of *Heterohyus nanus* from Messel (Koenigswald & Schierning 1987; Koenigswald 1987a, 1990). These animals lived in trees and reached the size of a squirrel, but they are not related to the rodents. Their similarities are actually the result of convergence.

At least four specimens of *Heterohyus* from Messel are housed in public and private collections (Fig. 12.2.9). The animal's head-body length was about 14 cm and its tail reached a length of 15–17 cm (Koenigswald 1990), which was probably covered with a dense layer of hair. At least one bushy tail tip has been preserved (Kalthoff et al. 2004) (Fig. 12.2.10). The delicate animal had a surprisingly robust head (Fig. 12.2.11). This may be linked with the two enlarged incisors, which provided the animal with a forceful bite (Koenigswald 1990; Koenigswald et al. 2005). The molars show multiple cusps, as evidenced by specimens from other fossil sites.

The animal had strong hindlimbs (Koenigswald & Schierning 1987; Koenigswald 1987a, 1990; Kalthoff et al. 2004); tibia and fibula appear to be fused at their lower end, which theoretically may have limited the ability to rotate the foot. However, the articular facets at the ankle allow free movement of the large feet. In most specimens, the five toes lie more or less parallel, so that the feet rested on a broad sole when set on the ground. The tips of the five toes carry small, laterally compressed claws that added extra grip when climbing.

The forelimbs are even more highly specialized than the hindlimbs. The well-developed clavicles enabled far-reaching lateral movement of the arms. The humerus shows a simple shape and lacks enlarged muscle attachment surfaces (Koenigswald 1990). The high degree of specialization of the Apatemyidae is shown by the extreme elongation of certain fingers (Fig. 12.2.9, Fig. 12.2.12). The metacarpals of the first, fourth and fifth finger are noticeably shorter than those of the second and third. The first and second phalanges (finger bones) further increase this disparity by being extremely elongated. The phalanges of the central digits are more than twice as long as those of the outer digits.

In the absence of comparable specializations in the recent fauna, it would be quite difficult to interpret such modified structure of the hands. But a very similar elongation of individual finger is found in two recent mammals: in a species of lemur, the aye-aye (*Daubentonia*), and in the striped possums or trioks (*Dactylopsila*). However, there are significant differences in the position of the elongated digits (Koenigswald 1990) (Fig. 12.1.12). This unusual elongation of individual fingers evolved several times in unrelated groups of animals (convergent evolution). Still, both recent genera use their elongated fingers in a rather similar way: they tear into the narrow cracks of tree bark, or even into the holes made by wood-boring insects, in search of insect larvae. Remarkably also, these recent forms bear strong incisors that can be used to tear open the bark and widen cracks. Thereby, the enlarged incisors, firmly rooted in the large skull of *Heterohyus*, further extend the number of similarities with the recent forms. This type of foraging strategy only makes sense with an arboreal lifestyle, which can therefore be postulated for the Apatemyidae as well.

Paleobiogeography

The Pseudorhyncocyonidae, which are represented in Messel by *Leptictidium*, were endemic to Europe and are generally rare in the fossil record. According to a recently published revision of the family (Hooker 2013), their temporal distribution extends from the Paleocene to the Eocene. The specimens from Messel therefore count among the last representatives.

In their dentition, skull anatomy as well as in the specializations of their appendicular skeleton, the Pseudorhyncocyonidae show certain similarities to the Leptictidae, which were essentially restricted to the North American continent. In Asia, a few scanty specimens are known from the middle Eocene to the early Oligocene, while discoveries from Europe are dubious. If the Leptictidae shared a common ancestor with the Pseudorhycocyonidae, they must have dispersed from North America into Europe during the Paleocene, since the Leptictidae are known in North America from the early Paleocene, and perhaps even from the Cretaceous (Rose 2006a).

The Pantolestidae, represented at Messel by *Buxolestes*, are known in North America by the mainly Eocene related forms *Pantolestes* and, in particular, *Palaeosinopa*, although the family was present there throughout the entire Paleocene and Eocene (Fig. 12.2.13). In Europe, the Pantolestidae first appeared in the middle Paleocene (Russell 1964). Accordingly, a faunal exchange must have taken place during the Paleocene between North America and Europe, probably from west to east. However, the many similarities between the skeletons of the much younger genera *Buxolestes* and *Palaeosinopa* strongly suggest a second faunal exchange during the early Eocene.

The Paroxyclaenidae are endemic to Europe. The oldest representative of this family, *Merialus*, is documented by a fragment of mandible from the lower Eocene of France (Russell & Godinot 1988). The remaining Paroxyclaenidae species were also found in Eocene deposits, for example in Geiseltal (Koenigswald 1983b), but most are documented by meager remains. Only *Kopidodon* from Messel is truly well known.

The Apatemyidae were already represented in North America by several genera during the Paleocene (Bloch & Boyer 2001). In the early Eocene, the genus *Apatemys* appeared in North America and Europe. Thus, the Apatemyidae only occurred in Europe at a later time, and they probably originated in North America (Russell et al. 1979; Strait 2001). An exquisitely preserved skeleton of *Apatemys*, which bears a close resemblance to *Heterohyus*, was found in the limestone ("plattenkalk") deposits at Fossil Lake in Wyoming (Koenigswald et al. 2005) (Fig. 12.2.13). The last Apatemyidae occurred in Europe at the end of the Eocene (Koenigswald et al. 2009).

The great similarities between the mammals from Messel and the forms from North America point to a

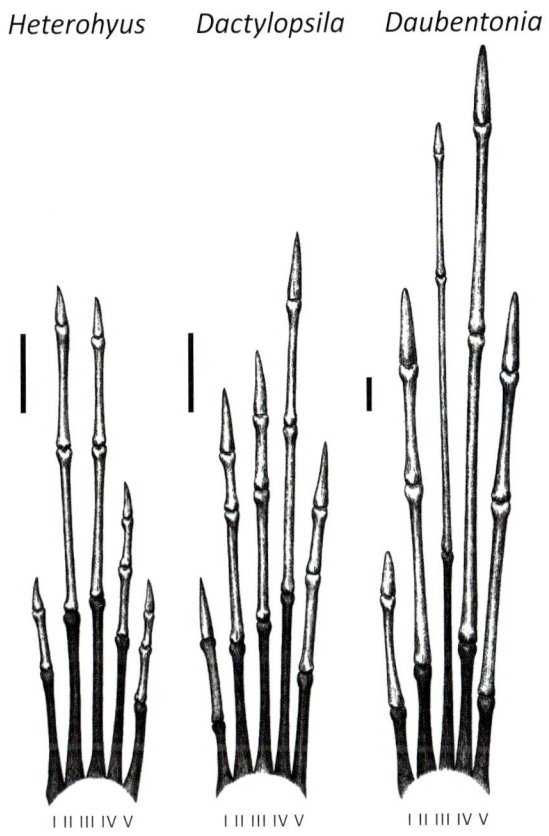

Heterohyus *Dactylopsila* *Daubentonia*

I II III IV V   I II III IV V   I II III IV V

Fig. 12.2.12: In addition to *Heterohyus,* selectively elongated fingers are also found in several other mammals, e.g., in *Dactylopsila* and *Daubentonia,* but in each case a different set of digits has been elongated. Scales: 5 mm.

close relationship, which can only be explained by a faunal exchange between the continents via temporary land bridges (Chapter 2). During the Paleocene, the early Leptictidae and the first Pantolestidae presumably spread via the older De Geer-route, which was located farther to the north. The remarkable similarities between *Palaeosinopa* and *Apatemys* from Wyoming (Fig. 12.2.13) and *Buxolestes* and *Heterohyus* from Messel suggest the likelihood of a second exchange, which probably occurred during the early Eocene via the Thule land bridge.

Specimens of a leptictid (*Prodiacodon*) and a pantolestid (*Palaeosinopa*) from lower Eocene layers on Ellesmere Island (Eberle & McKenna 2002) indicate that these groups already lived in the area of the land bridges before the related forms occurred in Messel. Moreover, these discoveries demonstrate that these animals were able to survive far north of the Arctic Circle. At that time, the region was covered by forests, enabling even climatically restricted animals such as crocodiles to live that far north; however, it continues to remain a mystery how they coped with the long periods of darkness during the polar night (McKenna 1980; Smith et al. 2006).

Contrary to the connection between Europe and North America, the connection to Asia was restricted by the Turgai Strait, which divided the Asian continent from north to south (Koenigswald & Rust 2011).

Fig. 12.2.13: In many details, the skeletons of *Apatemys* sp. (left) and *Palaeosinopa* sp. (right) from the Fossil Lake site in Wyoming show the same adaptations as *Heterohyus* and *Buxolestes* from Messel, thus providing evidence of the close relationship between certain groups of mammals from North America and Europe. Scales: left 2 cm, right 10 cm.

Chapter 12.3
With and Without Spines: the Hedgehog Kindred from Messel

Thomas Lehmann

Long considered to represent the most primitive order of placental mammals, the "Insectivora" was thought to incorporate almost all insectivorous placental mammals, but lacked clear unifying features. A growing number of studies has shown that "Insectivora" does not constitute a monophyletic group (e.g., Novacek 1986). Today, hedgehogs, moles, shrews and solenodons are the exclusive living members of the monophyletic order Lipotyphla (also referred to as Eulipotyphla), whereas certain superficially similar taxa such as tenrecs and golden-moles are now included in the clade Afrotheria (e.g. Stanhope et al. 1998), while others even belong to the placental stem. Lipotyphla seems now securely established within the clade Laurasiatheria, where it is probably in a basal position.

The Lipotyphla lineage could go back as far as the Cretaceous, but the oldest ascertained representatives are found in the Paleocene of Mongolia (Lopatin 2006). With almost 400 living species, the Lipotyphla is one of the most species-rich orders of placentals, mostly thanks to the high diversity of shrews. But in the past, the hedgehog-like Erinaceomorpha were more numerous and quite diverse, with representatives in Europe, Asia and North America. Erinaceomorpha is today entirely absent from the American continent.

Lipotyphla was long considered to be divided in two major clades: the Erinaceomorpha (hedgehogs, gymnures, and their fossil relatives) and the Soricomorpha (shrews, moles, solenodons, and their fossil relatives) (Symonds 2005). In the phylogenetic tree presented here (Fig. 12.3.2), the Soricomorpha are disbanded, following the most recent genetic studies (Roca et al. 2004; Brace et al. 2016), and the Talpidae are clearly separated from the Erinaceomorpha.

The oldest fossil lipotyphlans belong to the Erinaceomorpha (Gould 1995). The Amphilemuridae for instance, which were successively considered primates, and then "insectivores," is a common family in Paleogene sites (Maître et al. 2008). However, a recent study (Hooker & Russell 2012) suggests that Amphilemuridae were stem macroscelideans (i.e., elephant shrews and fossil allies), and thus members of the clade Afrotheria instead. This hypothesis has not always been confirmed in subsequent studies (e.g., Hooker 2014; Manz & Bloch 2015).

In Messel, only two lipotyphlan genera are known, both referred to Amphilemuridae: *Macrocranion* (Fig. 12.3.1) and *Pholidocercus* (Fig. 12.3.6). Although they are represented by complete skeletons in Messel, most fossil amphilemurids are known from teeth and jaw fragments. Thus, *Macrocranion* and *Pholidocercus* have been assigned to the family based on dental features. For instance, all amphilemurids retain a primitive placental dentition, with three incisors, one canine, four premolars and three

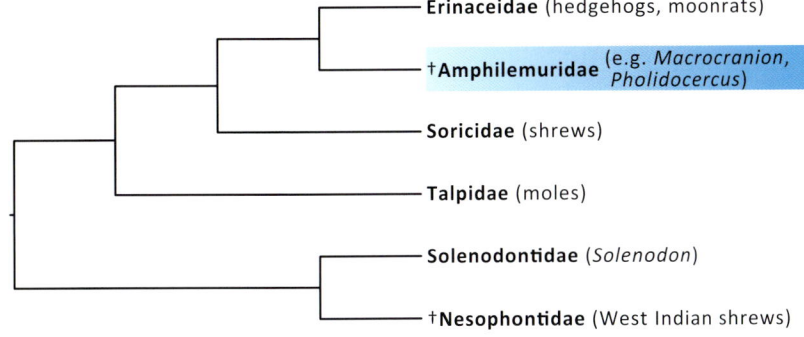

Fig. 12.3.2: Simplified phylogenetic tree of the Lipotyphla. Taxa known from Messel are highlighted in blue.

Fig. 12.3.1: *Macrocranion tupaiodon,* with delicate hair preservation, including numerous vibrissae on the snout (inset). Photos were taken before transfer. Scale: 1 cm.

Fig. 12.3.3: Details of the head of *Macrocranion tupaiodon*, showing the large last lower premolar and the inclined lower incisors. Scale: 1 cm.

molars (Fig. 12.3.3). Their lower teeth in front of the last premolar are in a tight series (no diastemata), have only one root, and become increasingly forward-inclined (procumbent) toward the spatulate incisors (Koenigswald & Storch 1983).

A fish-loving hedgehog

Macrocranion tupaiodon was one of the first mammals described from Messel (Weitzel 1949), and with over 50 individuals counts among the most abundant. *M. tupaiodon* was a small, lightly built animal about 16 cm in length (head and trunk) with a relatively long tail up to 15 cm (Storch 1996). Soft tissue preservation shows large external ears, long and numerous vibrissae (Fig. 12.3.1, inset), thick fur, and a rather naked tail (Maier 1979). The gut contents of ten specimens included animal and vegetal matter. But as three of these individuals ate almost exclusively (and a fourth partially) fish before death, *M. tupaiodon* can be de-

Fig. 12.3.4: Life reconstruction of *Macrocranion tupaiodon,* feeding on a bowfin (*Cyclurus*) washed up on the shore of Lake Messel.

Fig. 12.3.5: Plate A and B of the best specimen of the rare *Macrocranion tenerum*, showing short hair. The tail is missing. Scale: 2 cm.

scribed as an omnivorous mammal with a predilection for fish (Storch & Richter 1994). The lack of postcranial adaptations for swimming suggests that *M. tupaiodon* did not actively hunt for fish in the water, but could have rather scavenged for fish carcasses (Fig. 12.3.4). This would suggest that *M. tupaiodon* regularly went foraging on the shore of the Messel Lake.

Anatomical features of the humerus, in particular, suggest a terrestrial lifestyle for *M. tupaiodon* with a capacity for running. The tibia is fused at both ends with the fibula, which would be disadvantageous for an arboreal climber (Maier 1979). The claws are short and stout, recalling hooves, and certainly not adapted for digging or climbing. Altogether, *M. tupaiodon* has been reconstructed as a quadrupedal forest floor dweller with adaptations for running. The brain was large and the snout slender (Maier 1979). The large hole in front of the orbit (infraorbital foramen) that transmits the infraorbital nerve suggests a highly sensitive nose and vibrissae, which, in view of the small orbits, indicate reliance on the sense of smell and touch when foraging.

Macrocranion tenerum: the smallest lipotyphlan from Messel

Macrocranion tenerum is a small and gracile animal of only 9 cm head-body length (Fig. 12.3.5). It shows elongated hindlimbs (Storch 1993). However, the proportions of the limbs are similar to those of *M. tupaiodon*. The feet of *M. tenerum* are reported to be particularly long, as in saltatorial mammals (Storch 1993). Accordingly, *M. tenerum* is described as a quadrupedally jumping mammal, similar to modern Checkered Sengis (*Rhynchocyon cirnei*), with sporadic bipedal hopping (Storch 1996). However, since the proportions of *M. tenerum* fall within the range of variation of *M. tupaiodon*, further studies are required to confirm the bipedal jumping possibilities in *M. tenerum*.

The skull of *M. tenerum* has roughly the same proportions as that of *M. tupaiodon*. The large hole for the infraorbital nerve, the small orbit, and the reported depressions on the side of the snout, which may serve as insertions for nasal muscles linked to a mobile proboscis, suggest that olfaction was the predominant sense, backed up by a fine tactile sense (Storch 1993). The antemolar dentition of *M. tenerum* is less complex than in *M. tupaiodon* (Storch 1993). The dentition suggests an omnivorous diet for *M. tenerum* (Storch 1996). This is confirmed by the gut contents of one specimen, which consists overwhelmingly of the cuticles of social insects such as ants, but no insects with an aquatic larval phase, and very little plant material (Storch & Richter 1994). *M. tenerum* probably relied more heavily on insects than *M. tupaiodon* and was a less frequent visitor of the shores of Lake Messel. The fantastic preservation of the second specimen of *M. tenerum* (Fig. 12.3.5) reveals a body covered with thick bristles or spines, somewhat similar to extant spiny mice (*Acomys*).

A spiny, strong-headed, and scaly-tailed hedgehog

Pholidocercus hassiacus (meaning "Hessian scaly-tail") is definitely one of the most striking fossils from Messel. This robust animal of about 19 cm (head and body) had a 16 to 20 cm long tail encased in a sheath of imbricate bony scales (Koenigswald & Storch 1983) (Fig. 12.3.6). In the skull, fine grooves and furrows indicate an enhanced blood supply, which could suggest the presence of a horny pad (Fig. 12.3.7). Possibly defensive in function, this feature is also somewhat similar to structures found in living mammals that dig with their forehead and nose (Storch 1996). Moreover, several of the ten known specimens preserve the outline of their fur, which seems to consist of long and possibly stiff bristles, oriented backwards (Fig. 12.3.8).

The postcranial skeleton of *Pholidocercus* is more generalized in proportions and shows fewer speciali-

Fig. 12.3.7: Cranium of *Pholidocercus hassiacus,* showing the sculpted bones of the forehead, which might have borne a horny pad. Scale: 1 cm.

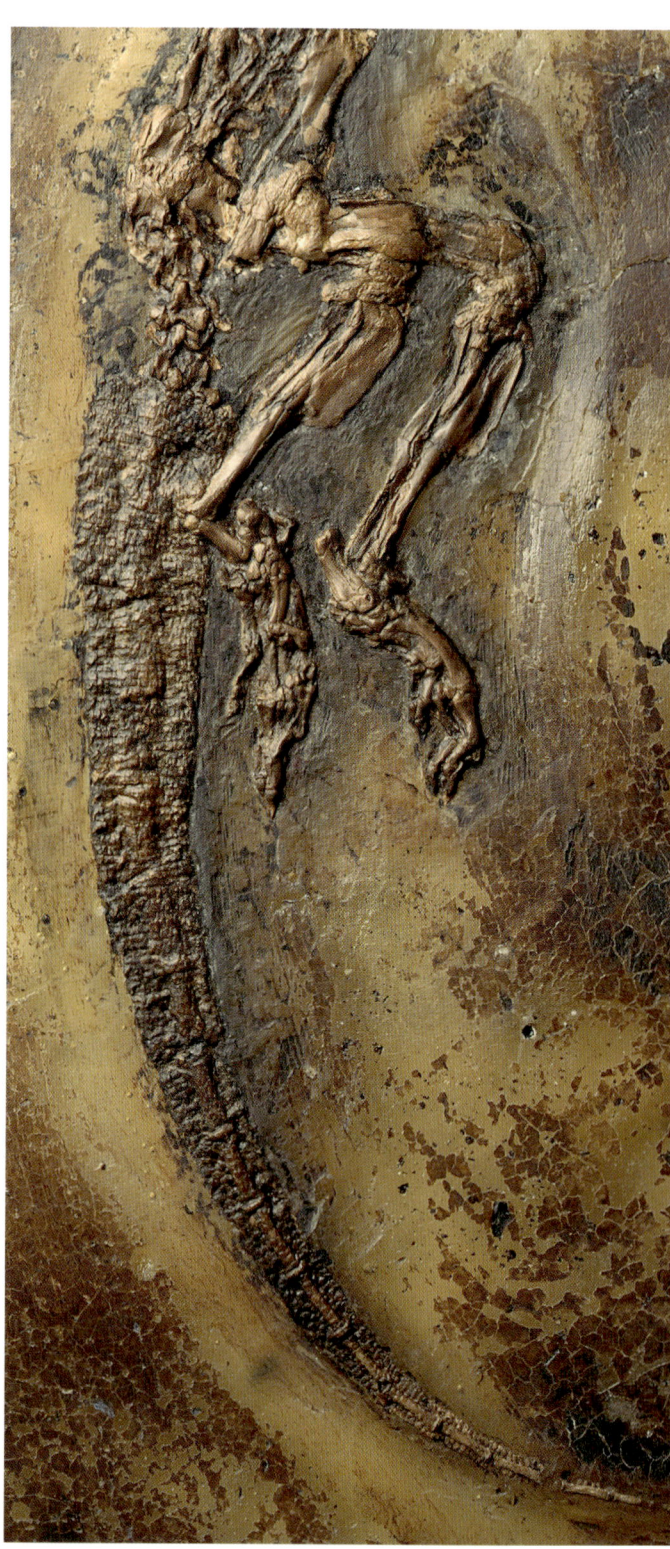

Fig. 12.3.6: *Pholidocercus hassiacus,* showing its massive scaly tail. Scale: 2 cm.

Paleobiogeography and Paleoenvironment

Whereas *Pholidocercus* is only known from Messel to date, *Macrocranion* is widespread, known from at least eight additional species from the early and middle Eocene of North America and Europe. Recent studies (Smith et al. 2002; Maître et al. 2008) suggest that *Macrocranion* could be of European origin and immigrated to North America, via Greenland, in the early Eocene (Chapter 2). Amphilemuridae decline in diversity during the Eocene and are replaced by Erinaceidae from Asia in the wake of the "Grande Coupure" (Oligocene).

Extant lipotyphlans are primarily omnivorous and ground dwellers. Except for moles, they show no special adaptations to their environment (Storch 1996). Therefore, it is surprising that the three Messel species show three different ecological specializations. *M. tupaiodon* displays features reminiscent of those found in moonrats (Galericinae), with a predilection for fish; *M. tenerum* looks more like real hedgehogs (Erinaceinae) and might have avoided the lake shore; and *Pholidocercus* is unique. With their extinction, the Erinaceomorpha experienced a loss of ecological diversity.

> **Box 12.3.1: Like Superman's X-ray vision**
> In Messel, most vertebrates are preserved with clenched jaws, so that the occlusal surface of their teeth is hidden. Mechanical extraction would result in the partial destruction the fossil, but µCT methodology (Chapter 5) enables digital preparation of the teeth and comparison with other amphilemurids. In *Macrocranion tupaiodon*, for instance, the back part of the lower molars (green and purple) is broader than the front part. The three bulbous cusps on the back part of the molars are also connected by distinct crests, unlike in *Pholidocercus*.
>
>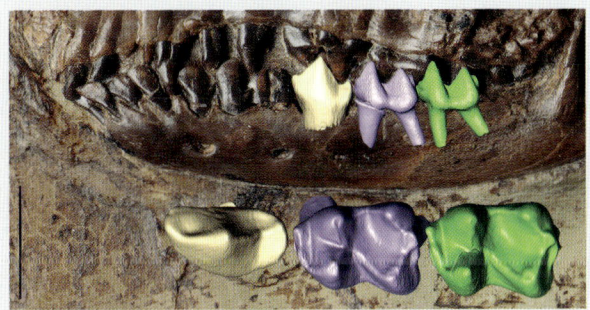

zations than that of *Macrocranion*. Neither the olecranon (elbow) nor the heel is particularly long, the tibia is fused distally with the fibula. The terminal phalanges of *Pholidocercus* are deeply fissured, perhaps to accommodate strong claws (Storch 1996). The animal is therefore reconstructed as a quadrupedal ground dweller that could easily paw through the forest floor in search for food (Koenigswald et al. 1992)

Pholidocercus has small orbits, a large brain, and a short snout. As in *M. tenerum*, the antemolar teeth show a low degree of differentiation (Storch 1996). The molars and last premolars are inflated and bear bulbous cusps. *Pholidocercus* can be distinguished from *Macrocranion* on the basis of certain crests on the molars (Koenigswald & Storch 1987) (Box 12.3.1). Accordingly, *Pholidocercus* was omnivorous (Storch 1996), but the species must have been opportunistic in its food choice, as two of the three specimens with gut contents ate predominantly insects such as beetles, whereas the third fed mostly on plant matter such as fruit and leaves (Storch & Richter 1994).

Fig. 12.3.8: *Pholidocercus hassiacus*, showing coarse, bristly fur on its back. Scale: 1 cm.

Chapter 12.4
Primates – Rarities in Messel

Jens Lorenz Franzen, Philip D. Gingerich

The title sounds paradoxical, since primates today typically inhabit tropical rain forest, where they are abundant in regard to species and individuals. However, at the primeval forest fossil site of Messel the situation was evidently different. During more than 40 years of intensive excavation, only eight primate specimens have been found. These belong to three species: *Europolemur koenigswaldi*, *Europolemur kelleri*, and *Darwinius masillae*. Each of the Messel primates had a cat-like body, and a skull that is relatively short compared to that of lemurs. The limbs are also short in comparison to the body length, and the tail is long. The Messel species differ little in size. Body length from the tip of the nose to the end of the tail ranges from about 46 cm in *E. koenigswaldi* to an estimated 66 cm in *E. kelleri*. *Darwinius* is intermediate, with a body length of about 58 cm, but it should be considered that *Darwinius* as well as *E. koenigswaldi* are represented by juvenile individuals that would still grow a little.

All three primate species from Messel – *Europolemur koenigswaldi*, *E. kelleri*, and *Darwinius* – belong to the Cercamoniinae, a subfamily of Adapiformes. Adapiform primates first appeared at the beginning of the Eocene, about 56 million years ago. They became extinct in Europe about 20 million years later at the Eocene-Oligocene boundary.

It is a matter of debate whether Adapiformes are more closely related to Strepsirrhini ("wet-nosed" primates) or to Haplorhini ("dry nosed" primates). This differentiation refers to living primates, which possess either a wet rhinarium such as lemurs and lorises, or a dry rhinarium such as tarsiers and anthropoids (monkeys and great apes). The question is, what dental or bony characteristics correspond to these features when dealing with fossils. Some scientists focus on primitive characteristics and the lemur-like appearance of adapiforms and consider them to be Strepsirrhini closely related to lemurs (Fig. 12.4.2, right). Others emphasize the advanced or 'derived' characters of adapiforms, such as the short rostrum, spatulate incisors, projecting dimorphic canines, and quadrate molars. These characteristics are more monkey-like and point to a relationship to Haplorhini and specifically anthropoids (Fig. 12.4.2, left).

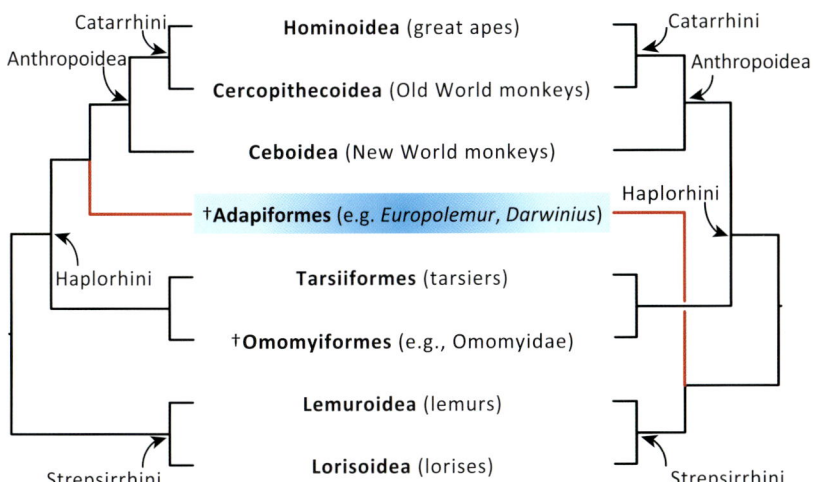

Fig. 12.4.2: Phylogeny of primates, showing alternative positions for Adapiformes, the only taxa known from Messel. Top: the haplorhine hypothesis. Bottom: the strepsirrhine hypothesis.

Fig. 12.4.1: *Darwinius masillae*, consisting of two plates. The almost complete skeleton on plate A is surrounded by the almost complete body silhouette. Scale: 10 cm.

The first discoveries

The first fossil primate at Messel was found in 1975, almost 100 years after the discovery of the first fossils there. The specimen was excavated by a team from the Hessisches Landesmuseum in Darmstadt (HLMD) (Koenigswald 1979). It includes both hind limbs, grasping feet, the pelvis, and a rather large penis bone (Fig. 12.4.3). The penis bone (or baculum) is found in males of most living primates and in rodents (Chapter 12.6) and carnivores. Consequently, the primate is undoubtedly a male. The size is that of a domestic cat, but the genus and species remain indeterminate because there are no remains of the dentition.

Subsequent discoveries of primates at Messel were also fragmentary. In 1982 a student participating in a Senckenberg excavation found the anterior half of a skeleton within hours of being trained (Franzen 1987). This specimen includes the skull, which displays a complete dentition (Fig. 12.4.4, Fig. 12.4.5). Now for the first time, it was possible to identify a primate to genus and species. This was named *Europolemur koenigswaldi* in honor of the famous Senckenberg paleoanthropologist Gustav Heinrich Ralph von Koenigswald. *Europolemur koenigswaldi* is clearly smaller than the HLMD find, being only half the size of a domestic cat, and it preserves no trace of a baculum. Could this be a female of the same species? This is possible, albeit improbable considering the great difference in size.

A third primate was excavated by the Staatliches Museum für Naturkunde Karlsruhe (SMNK) in 1984 (Fig. 12.4.6). It resembles the large Darmstadt specimen, but is about 10% smaller. A baculum is present, which is even smaller yet (by about 15%). The SMNK specimen is clearly male, but its systematic position is uncertain because the skull and dentition are missing (Koenigswald 1985).

Fig. 12.4.3: The first primate from the Messel quarry, discovered in 1975 by the Hessisches Landesmuseum of Darmstadt (arrow points to the preserved baculum). Scale: 1 cm.

Fig. 12.4.4: Type specimen of *Europolemur koenigswaldi* from 1982, plate A. Scale: 1 cm.

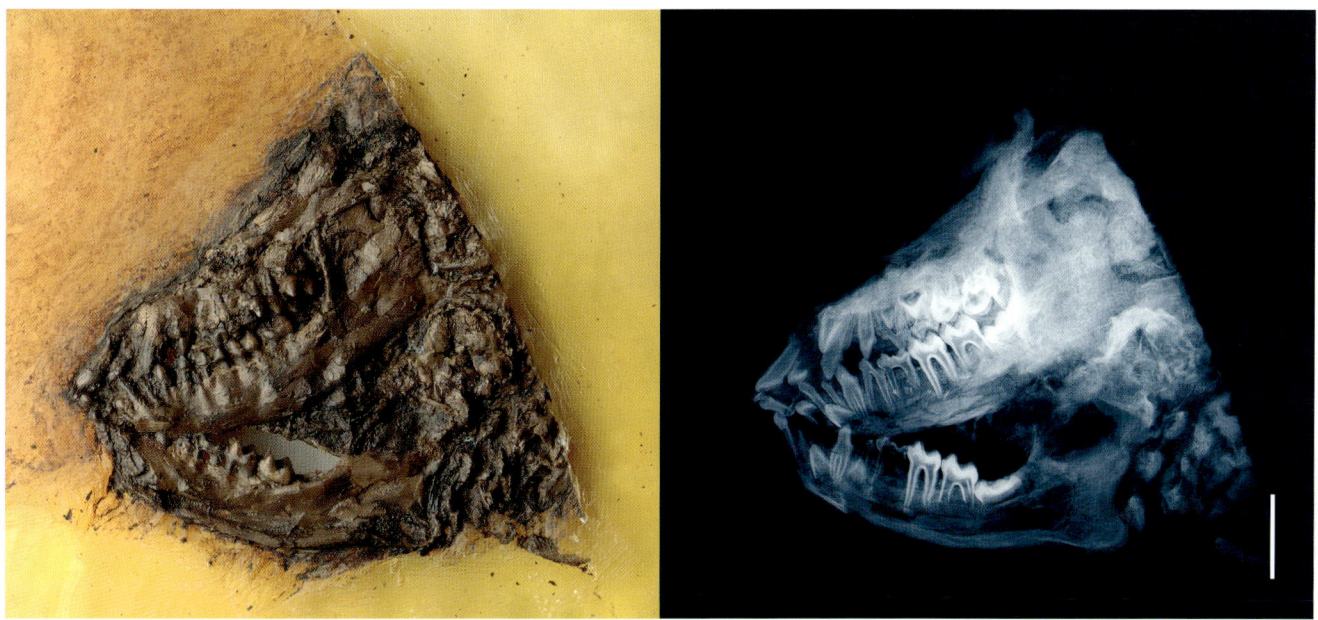

Fig. 12.4.5: The skull of *Europolemur koenigswaldi* (left) and X-ray image (right), plate B of the same individual as shown in Fig. 12.4.4. Scale: 1 cm.

Fig. 12.4.7: A right forearm with grasping hand was discovered in 1987. Left: coated with ammonium chloride, right: drawing. Scales: 1 cm.

Fig. 12.4.6: The third primate unearthed in 1984 was tentatively assigned to *Europolemur kelleri*. Scale: 5 cm.

  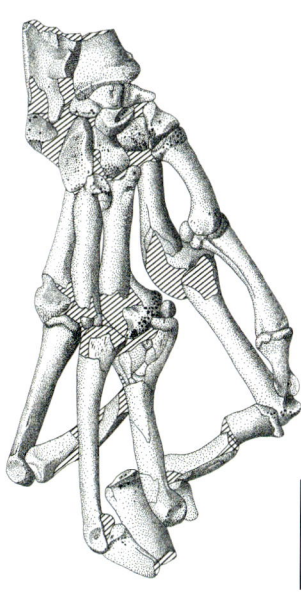

The fourth primate from Messel was discovered by Senckenberg in 1987 (Franzen 1988). It is the right forearm with a grasping hand displaying flat nails, which is one of the oldest primate hands ever found (Fig. 12.4.7). Again, no teeth were found and the specimen can therefore not be identified with certainty. It is clearly larger than *E. koenigswaldi*. The articulated, three-dimensional state of preservation surpasses the hands of other primates known from the Eocene. Also of interest are the bite marks of some carnivore, and the radius is split along its entire length. It is possible that the carnivore was a crocodile, but this is not known for certain.

The fifth primate from Messel is another discovery by the SMNK (Franzen & Frey 1993). Again, it is a partial skeleton, consisting of the posterior half beginning with a fragmentary pelvis. Contrary to the first and third primates, the whole vertebral column of the tail is preserved, as are both hind limbs (Fig. 12.4.8). Considering its dimensions and proportions, the specimen corresponds perfectly to *E. koenigswaldi*, whose skeleton is thus completed. Whether the specimen is male or female remains uncertain, since neither a baculum nor the pubic area is preserved. Conspicuous are many small dents that occur all over the skeleton. SEM analyses show that these are real impressions in the bone surface, and hence bite marks. One of the bite marks holds the tip of a broken tooth (Fig. 12.4.9). The perpetrator left behind his 'fingerprint.' It was a crocodile (Franzen & Frey 1993).

Ida, the little diva of Messel

The sixth primate from Messel includes the anterior and posterior parts of a skeleton, with missing parts replaced by a replica as revealed in a radiograph (Fig. 12.4.10). This specimen made history. A small round object in the abdomen looked like the scale of a fish at first glance, but closer examination through a microscope revealed the hardened endocarp (seed) of a fruit (Fig. 12.4.11, right). Additionally, tiny yellow fragments were visible under a fluorescence microscope. SEM analyses proved that these are fragments of leaves (Figs. 12.4.11, left and center). Here, for the first time, are direct observations of the content of the digestive tract of a fossil primate (Franzen & Wilde 2003). This primate clearly fed on leaves and fruit!

The whole significance of the sixth primate from Messel became clear in 2006 when its counterpart came to light at a fossil fair in Hamburg. Together, part and counterpart represent the most complete fossil primate ever discovered. The left foot and a short part of the left lower leg are missing, but the skeleton is otherwise complete (Fig. 12.4.1, Fig. 12.4.10, left). Extensive studies revealed that it represents a new genus and species (Franzen et al. 2009). The genus name *Darwinius* was given in honor of Charles Darwin (1809–1882) on his 200th birthday. The species name *masillae* recognizes the site yielding the fossil (Masilla is an old Latin name for the village of Messel, taken from the codex of the Lorsch Monastery). The

Fig. 12.4.8: The articulated posterior part of *Europolemur koenigswaldi* was found in 1990. Scale: 1 cm.

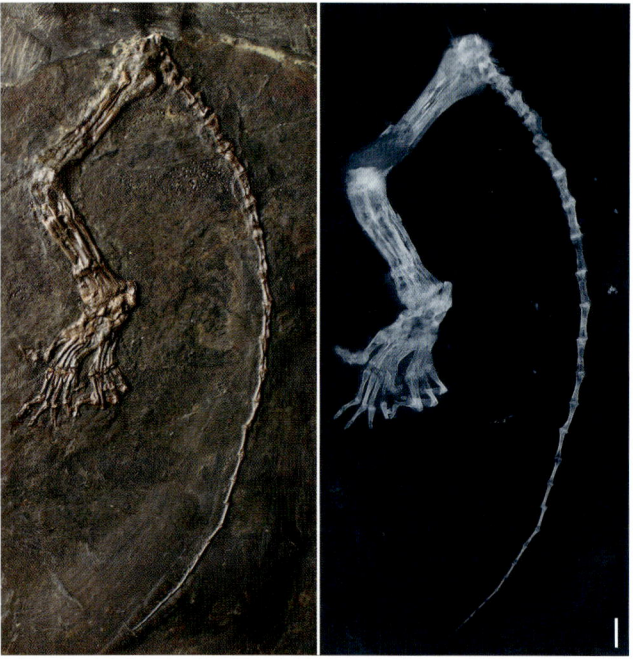

Fig. 12.4.9: The broken tip of a crocodile tooth was embedded in a bite mark on *Europolemur koenigswaldi*, the fifth primate from Messel. Scale: 1 mm.

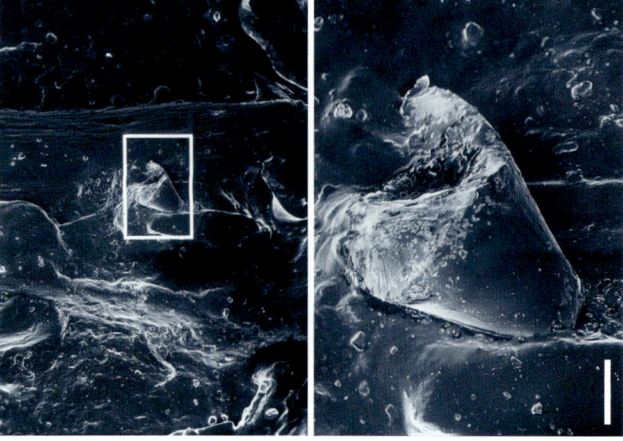

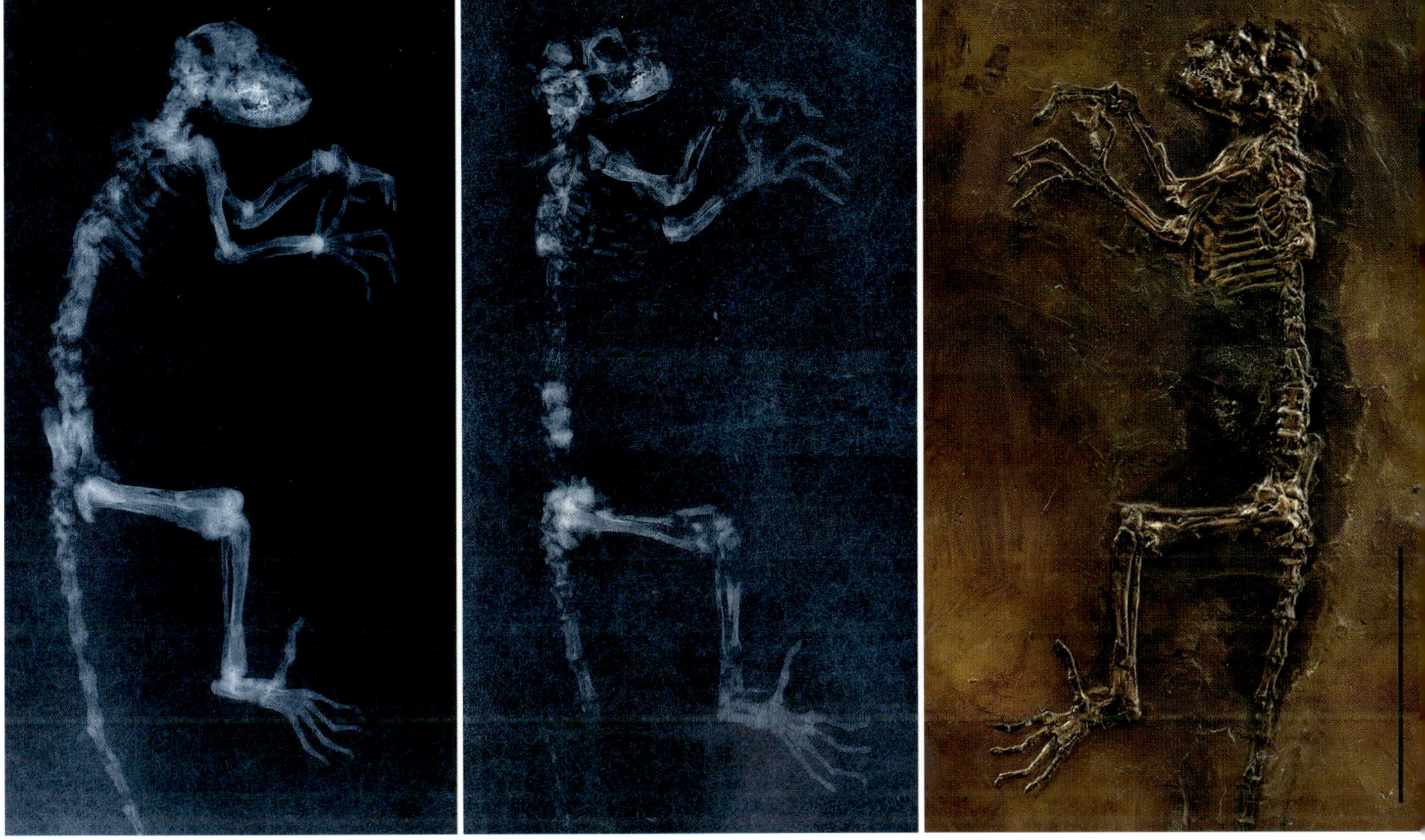

Fig. 12.4.10: *Darwinius masillae*, plate A (left: X-ray image) and plate B (center: X-ray image, backside view; right: photo). Scale: 10 cm.

Fig. 12.4.11: Contents of the intestine of *Darwinius masillae*: leaf particles (left, scale: 30 µm; center, scale: 10 µm) and the hardened endocarp (seed) of a fruit (right, scale: 0.5 mm).

combined name is thus *Darwinius masillae* (and its popular name is 'Ida'). The specimen was purchased by the University of Oslo, where it is now on exhibition in the Museum of Natural History.

Darwinius has the body proportions of a lemur or a South American ceboid monkey. Like these, it was a quadrupedal climber lacking specializations for slow climbing and for extreme leaping. *Darwinius*, with a head-and-body length of about 28 cm, and an estimated body weight of 660 g, resembled small lemurs and small ceboids. The body mass of *Darwinius* was clearly above Kay's threshold of 500 g that separates smaller insectivorous primates from larger folivorous primates today (Kay 1975). This is consistent with the intestinal contents known for *Darwinius*.

Unique for a fossil primate is not only the nearly complete skeleton, but also the fact that this is surrounded by the almost complete silhouette of soft body tissue and fur up to the tips of the hairs (Fig. 12.4.1). For this reason, we know that the external ears were rather small and round, and for the most part hidden in the fur. Only the hairs of the tail are not complete. The gender is determinable, because the pubic area is well preserved and does not display any trace of a baculum. The age of the individual can be determined by the stage of tooth development, showing that Ida was an immature female. She died at an estimated age of 9-10 months. Long bones still have some open end plates or epiphyses, as is also typical for a juvenile (Fig. 12.4.12).

Why are the skeleton of Ida and even her soft body tissues so perfectly preserved? Why did Ida seemingly suffer a fate different than that of other primates at Messel? There is no evidence that a crocodile or any other amphibious carnivore killed her or participated in transporting her carcass into Lake Messel. There are no bite marks. However, a swelling near the wrist of the right hand turned out to be significant (Franzen et al. 2012). A μCT scan revealed that the lower ends of the ulna and radius are smashed and shifted to overlie the carpals, with which they fused by callus tissue (Fig. 12.4.12, Fig. 12.4.13). The following scenario may explain what occurred: the young female primate may have broken her wrist during a fall from a tree. Following the accident, she could no longer climb in the trees as she had before, but spent more time moving on the ground. On the lakeshore, Ida may have been overcome by poisonous gases of post-volcanic or biogenic origin, particularly carbon dioxide (CO_2), lost consciousness, and drowned. Then her carcass, buoyed by digestive gases, floated into the lake until decomposition allowed the digestive gases to escape. The carcass sank down to the bottom of the lake where it was finally buried and fossilized.

Further discoveries

The seventh primate from Messel came to light during the Senckenberg excavations of 1989 (Franzen 1997). It is the fragment of a mandible with a few cheek teeth, which enabled its identification as *Europolemur koenigswaldi*. The mandible is unusual in being embedded in the middle of a petrified fecal pellet or coprolite (Fig. 12.4.14). It is unlikely that a crocodile or carnivorous bird produced the coprolite, because crocodiles and birds normally demineralize tooth enamel during digestion (Fisher 1981). Thus, it is more likely that the primate was consumed by an amphibious mammalian carnivore, such as the otter-like mammal *Buxolestes piscator* known from Messel.

The eighth primate from Messel was found by a private collector. It is an isolated but complete skull, split during excavation into upper and lower parts, which were then prepared separately (Fig. 12.4.15). Dental characteristics indicate that the skull belongs to a juvenile *Europolemur* that was much larger than *Europolemur koenigswaldi*. The combination of differing dental characteristics and size suggests that the skull belongs to a different species, which was

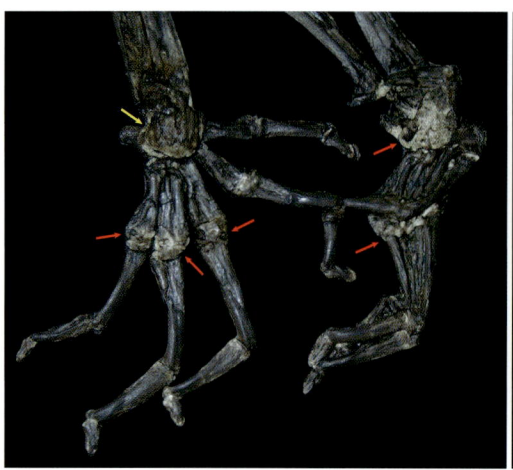
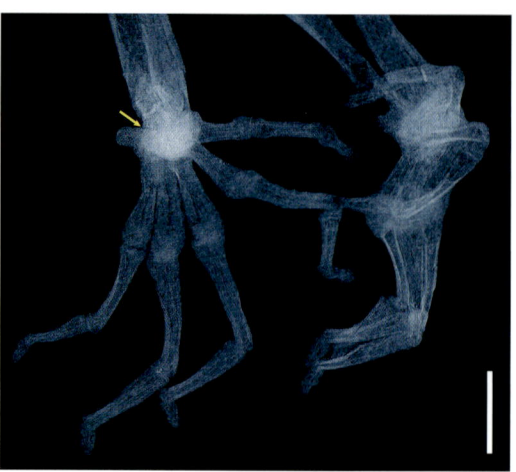

Fig. 12.4.12: Left and right hand of *Darwinius masillae*. Siderite (iron carbonate) fills the growth fissures (red arrows), whereas callus tissue indicates a healed fracture between the forearm and the hand (yellow arrow). Scale: 1 cm.

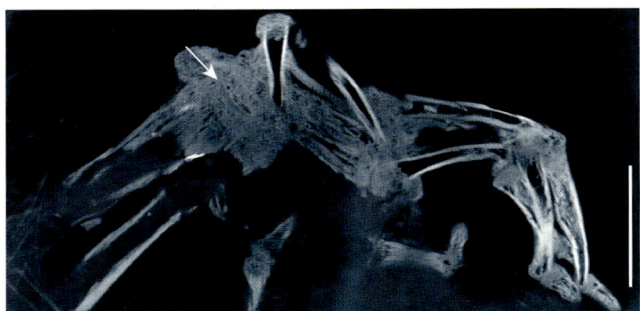

Fig. 12.4.13: µCT scan (single slice) of the right wrist of *Darwinius masillae,* seen from the side.

Fig. 12.4.14: Partial left mandible with cheek teeth of *Europolemur koenigswaldi* (top: plate A, bottom: plate B). Scale: 1 cm.

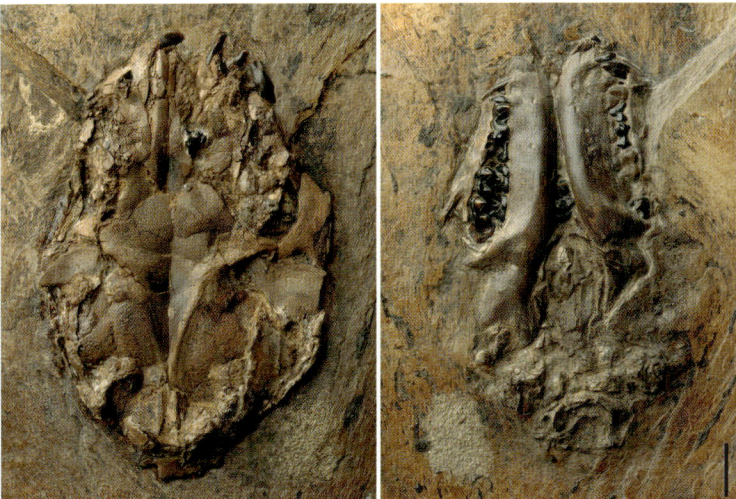

Fig. 12.4.15: Type specimen of *Europolemur kelleri*, plate A with the top of the cranium (left) and plate B with the underside covered by both mandibles (right). Scale: 1 cm.

named *E. kelleri,* in honor of the fossil's collector, who donated it to Senckenberg (Franzen 2000). Dimensions and proportions make it highly probable that the HLMD specimen of 1975 and the forearm with grasping hand of the Senckenberg excavation of 1987 belong to the same species. All in all, *E. kelleri* and *E. koenigswaldi* differ in size but otherwise resemble each other closely in terms of morphology.

The primate assemblage of Messel is not large, with only eight specimens representing three known species, but tarsier-like omomyiforms are surprisingly absent. In the Eocene, at the time of Messel, omomyiforms were relatively common in Europe and elsewhere on northern continents. Why do they not appear at Messel? One explanation could be their small body size. They may have lived high up in trees, where they were relatively safe from the dangers lurking lower down in the Messel forest. These dangers may have included floods, poisonous gases, and carnivorous animals. Death high up in the trees might also mean a decreased chance that their carcasses would reach Lake Messel. The fact that omomyiforms are still missing in the Messel fossil record is probably a taphonomic effect (Chapter 4) rather than an ecological or paleobiogeographic problem.

Chapter 12.5
Bats – Highly Specialized Nocturnal Hunters with Echolocation

Jörg Habersetzer, Renate Rabenstein, Gregg F. Gunnell †

Bats are the only flying mammals. Their extraordinary evolutionary success, with nearly 1,400 extant species and over 400 extinct species, is a testament to their adaptive capabilities. Their adaptive strategies can be best documented by the development of diverse wing forms and associated hunting and flight behaviors as well as by the development of multiple levels of echolocation capabilities across bat species. These latter abilities range from ultrasonic sounds, which enable them to avoid obstacles during their nocturnal hunting flights, to much more sophisticated systems that allow for detection and even acoustic identification of insect prey in the open air some hundreds of meters above the ground as well as in dense vegetation. The oil shales of Messel have provided hundreds of exceptionally well-preserved bat skeletons, including many with wing membranes and complete wing skeletons, three dimensionally preserved inner ears and gut contents. Bats are the most frequently discovered mammals at Messel.

The bats at the Messel Lake

Fig. 12.5.1 shows the Messel bat *Tachypteron franzeni*, which was assigned to the sheath-tailed bats (Emballonuridae), a living family (Storch et al. 2002). In contrast, all other bats found in Messel belong to extinct families. The phylogenetic relationships of bats are shown in simplified form in Fig. 12.5.2 (modified after Simmons et al. 2008). *T. franzeni* is beautifully preserved, with the dark brown parts representing the folded wing membranes (Fig. 12.5.1) and large parts of the tail membrane, which was spread between the hind legs in life. The black area below the ribs is preserved gastrointestinal contents. The wing shape can be reconstructed by using X-ray images and virtually unfolding the completely preserved wing skeleton. On the left, on top of the head of *T. franzeni* the streamlined contour of the outer ear is visible (a contour that, by comparison with modern bats, suggests that this bat was a fast flyer).

It is a stroke of luck that in the only two known specimens of *T. franzeni*, the bony capsule of the inner ear (cochlea) is completely and three-dimensionally preserved. Ultrasonic sounds are processed in the cochlea, and the size and shape of the cochlea therefore reveal details of acoustic specializations. On the left side, enlarged details of the 2-mm-sized inner ear are shown, which were obtained by micro-tomography (Chapter 5). The virtual endocasts are shown (from top to bottom) in front, back, left side and right side views. Volume and cross-sectional areas of this spiral channel –filled with lymph fluid in life – correspond well to those of extant relatives from the same family. This is also true for the length of the basilar membrane (carrying the hearing receptor cells), which can be measured after uncoiling the spiral channel. From all of this we can conclude that *Tachypteron franzeni* likely captured insects at high altitudes of ten to a hundred meters or more, utilizing its high-speed flight capabilities – like its extant relatives – and an echolocation system well adapted to hunting over long distances (Habersetzer et al. 2012).

All other bat species from Messel can be compared to extant bats in a similar way, and also to fossils from other localities with the same, older or younger age. Step-by-step conclusions can be drawn regarding their flight biology and echolocation abilities, and finally, the general ecology of the Messel bat community may be reconstructed.

Similar to *Tachypteron*, *Hassianycteris messelensis* (Fig. 12.5.3) was a fast and high-flying species. Together, *H. messelensis* and the considerably larger *H. magna* compose the extinct family Hassianycteridae (Fig. 12.5.2). These two species are more advanced than the other extinct Messel bats in the construction of their skeletons and teeth. Their forearms are long

---

Fig. 12.5.1: *Tachypteron franzeni* (scale: 2 cm) with 3-D views of the cochlea, left row (scale: 2 mm). This bat species is a member of an extant family and had a wing span of 35 cm.

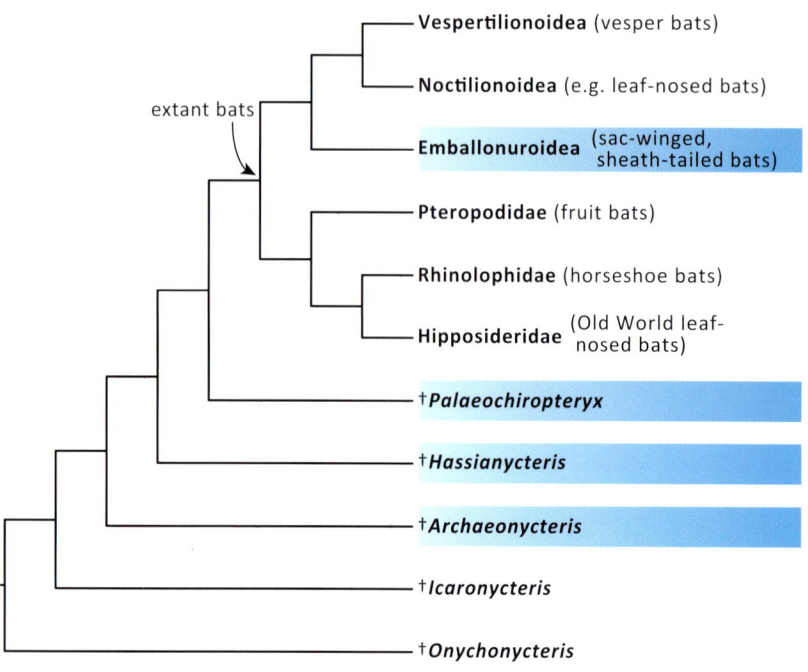

Fig. 12.5.2: Relationships of fossil and recent bats (highly simplified). Taxa known from Messel are highlighted in blue.

The most primitive known bat to date is *Onychonycteris finneyi* (Fig. 12.5.7) from the early Eocene of North America (Green River Formation in Wyoming, 52.5 million years). This fossil is several million years older than those from Messel. A unique character of *O. finneyi* is that all of the finger-tips bear small claws, a condition unknown in any other extant or extinct bat species. On the whole, the wings are very short and the legs are exceptionally long (Simmons et al. 2008). Also found in the Green River Formation in Wyoming is the slightly more advanced bat *Icaronycteris index* (Fig. 12.5.8). It has – like the Messel bat *Archaeonycteris* – an unspecialized wing shape and, in addition to a thumb claw, it also has a claw on the index finger, which led to the species name. These Green River bats allow useful comparisons with the somewhat younger bats from Messel.

and curved, and the second finger consists of one phalanx only. By comparison, *Archaeonycteris trigonodon* (Fig. 12.5.4) is a primitive species with a short and straight radius (and so, forearm). This extinct family (Archaeonycteridae) includes another, larger species, *A. pollex*, with body proportions very similar to *A. trigonodon* (Storch & Habersetzer 1988).

Two species are included in the third extinct Messel bat family (Palaeochiropterygidae), which are – in comparison to other Messel bats – small and specialized. Their forearm is long compared to the delicate body and curved in the proximal part. The slightly more robust *Palaeochiropteryx spiegeli* (Fig. 12.5.5) has the same body proportions as the smaller *P. tupaiodon* (Fig. 12.5.6). With several hundred specimens, *P. tupaiodon* is the most abundant mammal species found in Messel (Habersetzer & Storch 1987). On slab A (above) the fossil is seen from the underside, while slab B (bottom) displays the animal from the backside. The wing membranes and soft tissue remains of the outer ears are excellently preserved.

Wing shapes and hunting modes

As mentioned in the beginning, Messel bat families have quite different wing proportions. One of the parameters that is used for comparative investigation of wing morphology is the "aspect ratio," which is the ratio of the square of the wingspan to the wing area: broad wings have a low aspect ratio, whereas narrow wings have a higher aspect ratio. This is quite similar to the "aspect ratio index," which is also used to quantify wing shape, especially in fossil bats. In Fig. 12.5.9, it becomes apparent that Messel bats cover nearly the entire range of variation seen in extant bats. That means that the majority of differentiation of wing forms shown by almost 1,400 living species was already documented by the early middle Eocene, even given the rather small number of species. Interestingly, *Onychonycteris* does not fit into this spectrum of extant and fossil bats, due to the extremely short wings in combination with a very large tail membrane. Nonetheless, based on other morphological adaptations, *Onychonycteris* was capable of active and sustained flight, although it likely

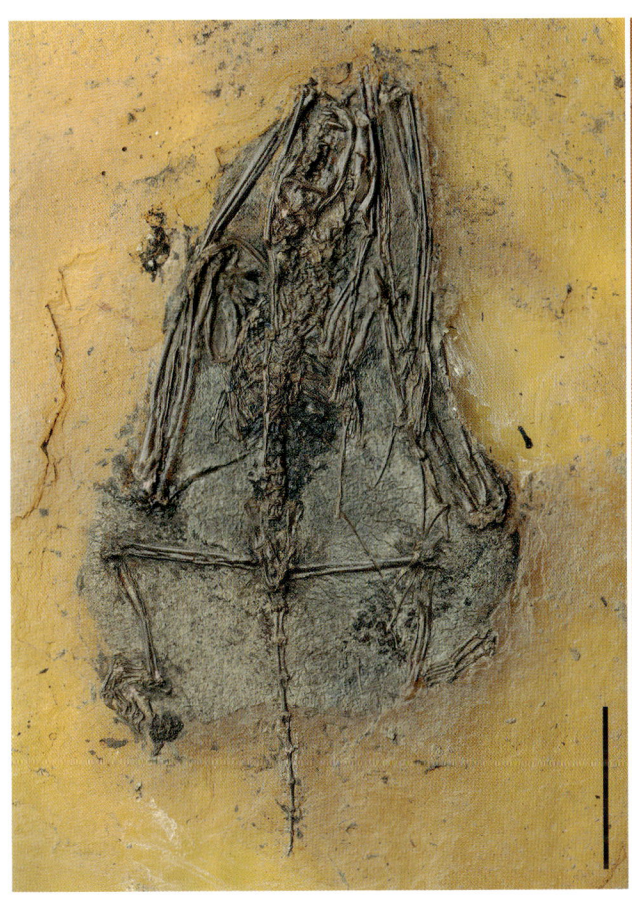

Fig. 12.5.3: *Hassianycteris messelensis* is a large bat species from Messel with a wing span of 39–48 cm. It is characterized by a bowed forearm. Scale: 2 cm.

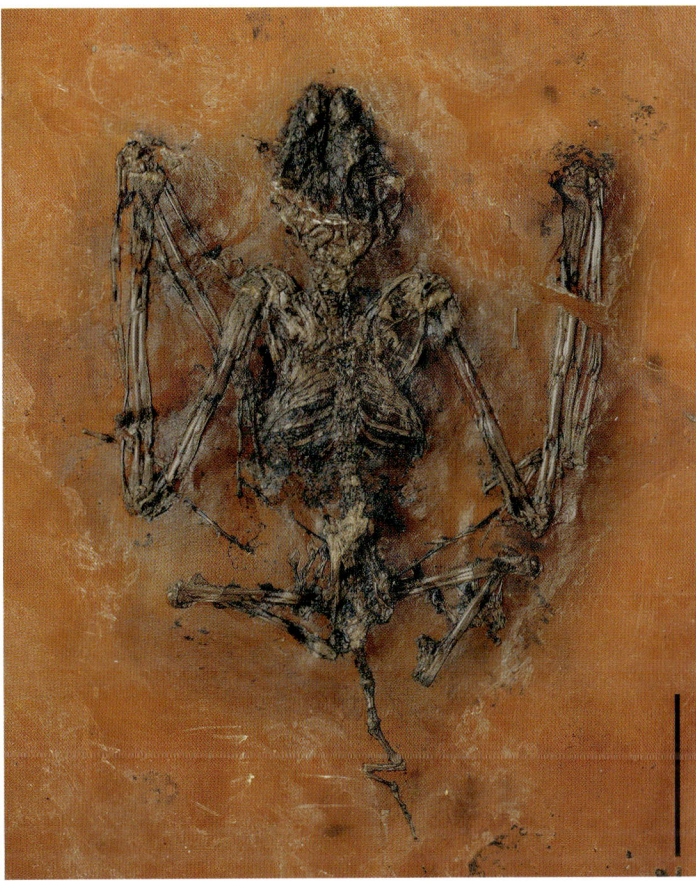

Fig. 12.5.4: *Archaeonycteris trigonodon* had a wing span of 37–39 cm. Scale: 2 cm.

practiced a fluttering flight style that resulted in high energy loss. In contrast, *Icaronycteris*, which is of the same geological age, had a more "modern" wing, similar to most extant and other fossil bats.

Fig. 12.5.10 shows the relation of wing shape and hunting style based on three extant bat species. *Taphozous melanopogon* is directly comparable with *Tachypteron*, which belongs to the same family (top row, from left to right), suggesting that its preferred hunting habitat was an open landscape, similar to the extant species, as it had the characteristic narrow wing shape of fast and high-flying bats. *Pipistrellus dormeri*, a typical extant vesper bat, is depicted in the middle row. This bat mainly hunts at intermediate heights in the open spaces between trees. Its wing is relatively unspecialized. The bottom row shows the delicate and broad-winged Old World leaf-nosed bat *Hipposideros speoris*, which hunts for prey in dense vegetation, utilizing an agile and highly maneuverable flight style at very low speed of less than 1 m/s. These differences in flight style among various species are also reflected in the size and species composition of prey items utilized by extant bats (Eckrich 1988).

Stomach contents

Some bat specimens from Messel provide direct evidence for their dietary preferences through study of the composition of preserved gut contents. From the

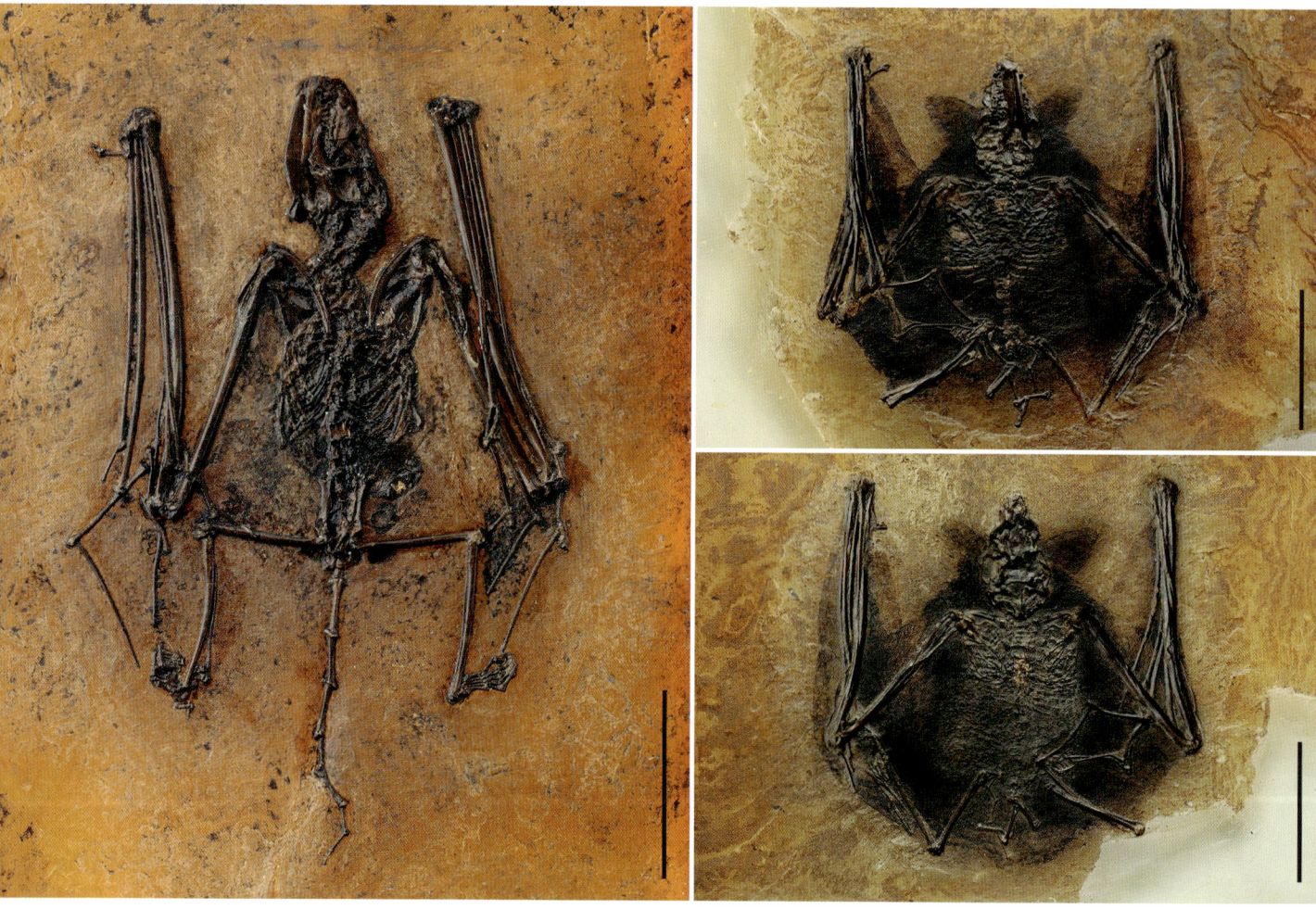

Fig. 12.5.5: *Palaeochiropteryx spiegeli* is the larger of two closely related species. Scale: 2 cm.

Fig. 12.5.6: *Palaeochiropteryx tupaiodon* had a wing span between 26–29 cm and together with *P. spiegeli* it represents the smallest genus among Messel bats. Scale: 2 cm.

blackish discolorations of the gastrointestinal area (see Figs. 12.5.1, 12.5.3, 12.5.5, 12.5.6), minute samples can be taken and investigated with a scanning electron microscope. For details, see SEM (Chapter 5).

Some of the most interesting SEM results are shown in Fig. 12.5.11. One individual of *P. tupaiodon* (left column: top, center, bottom) shows evidence of its last meal as well as fur cleaning. In the stomach content, a cuticle fragment with sensory organs was found (top). Presumably, it derives from the wing base of a lepidopteran. Very often, insect hairs (center) are found, but also included are hairs from the bat itself (bottom). These can be unmistakably identified as mammal hairs by their horsetail-like structure. Stomach contents of the extant African bat species *Lavia frons* revealed that hairs are often swallowed while grooming the fur.

In many individuals of *P. tupaiodon*, long and narrow scales from microlepidopteran moths are found (center column, top and center). Both SEM images show the distal part of a simple lepidopteran scale, each with four partially destroyed tips. Such scales consist of upper and lower membranes (lamellae), which are supported by ribs. In the scale's distal part, the majority of both lamellae are preserved, i.e., the tips are fully three-dimensionally preserved. The scale's outermost part shows the minor structures of the inner surface of the lower lamella.

Food remains also include insect hairs (center column, center) along with the scales. Hairy fragments of cuticulae (center column, bottom) with distinctive pores are also often found in *Palaeochiropteryx*. Comparisons with extant material show that, for instance, mosquitos (*Aedes* sp.) have a similar hairy cuticula with pores, in which scales are inserted. This suggests that *Palaeochiropteryx* was feeding on insects such as mosquitos, i.e., species that are still common around lakes today.

The right column shows gut contents of the large bat *Hassianycteris messelensis*. Stomach and gut contents of these rare bats often consist of massive insect fragments, which may display a strongly ornamented surface (top right). The fossil from which this sample was taken is shown in Fig. 12.5.3 (top left). Another *H. messelensis* specimen fed on a massive, armored insect. Longitudinal striations of the surface and the clearly rounded shape of the fragment indi-

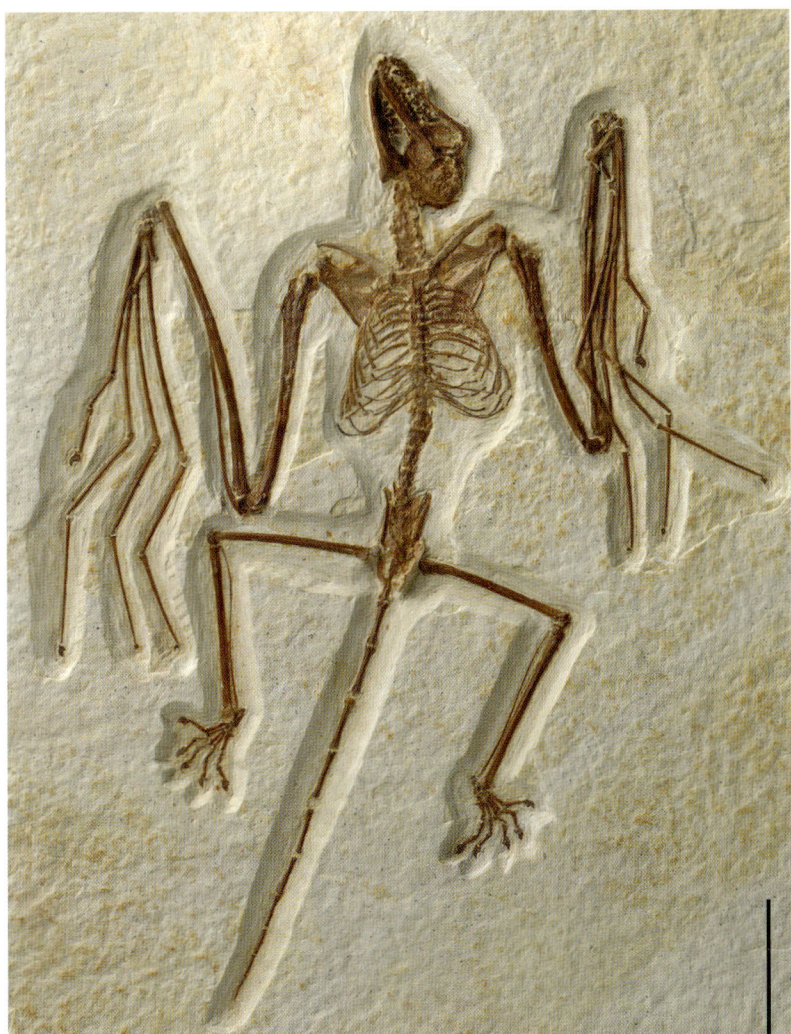

Fig. 12.5.7: The paratype of *Onychonycteris finneyi* is embedded in light-colored sediment and one body side is completely prepared. Scale: 2 cm.

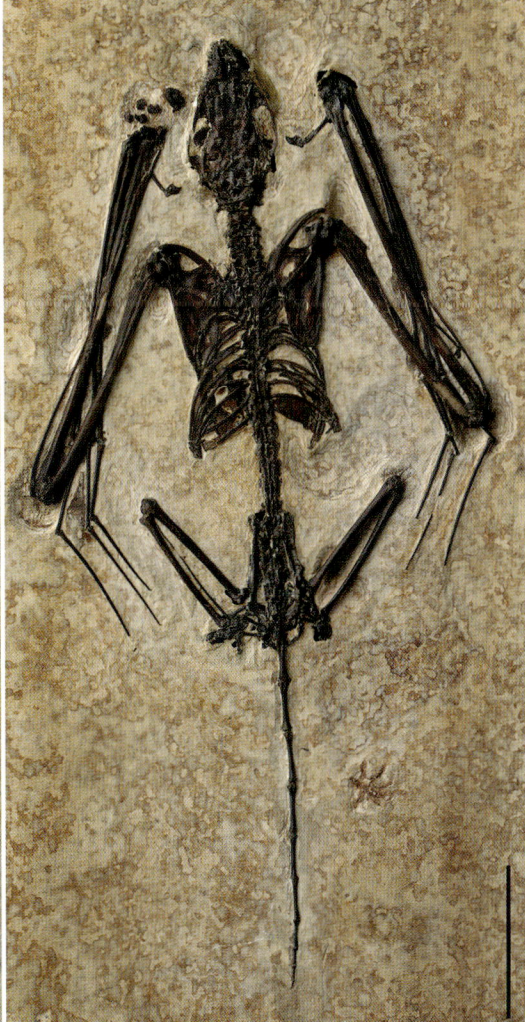

Fig. 12.5.8: *Icaronycteris index* originates - like *Onychonycteris finneyi* - from the Green River Formation. Scale: 2 cm.

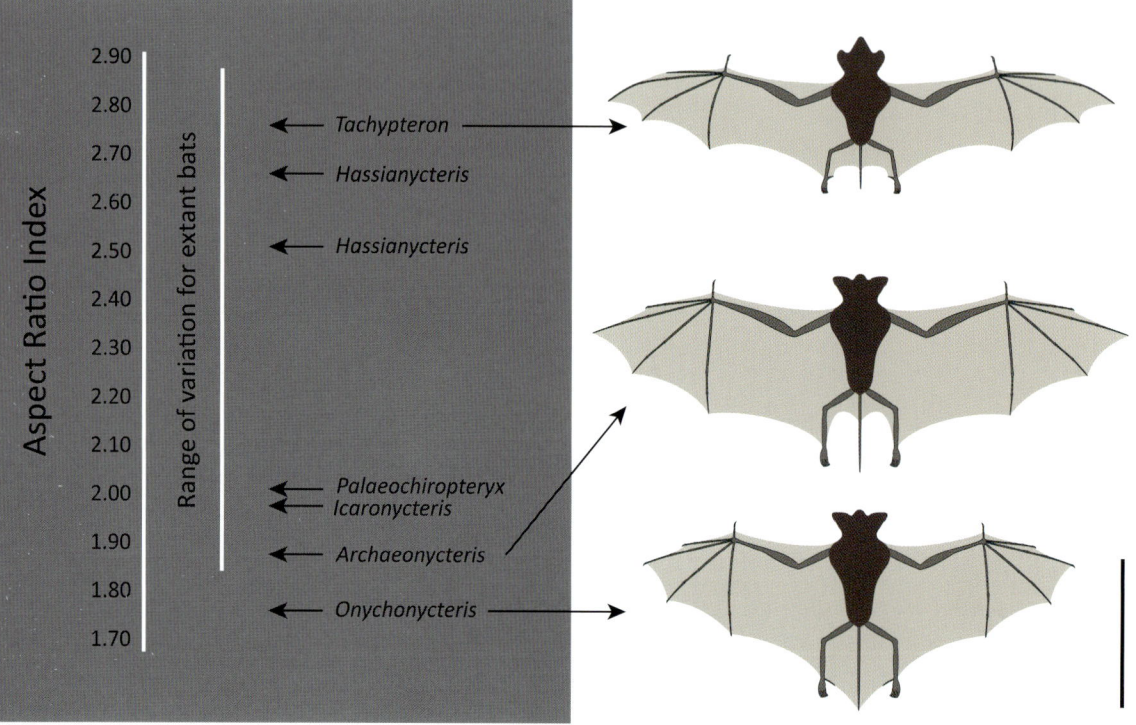

Fig. 12.5.9: Wing shape of *Onychonycteris* in comparison to other bats shows that *Onychonycteris* is characterized by a very short wing and a large tail membrane. Scale: 10 cm.

cate an elytron of a beetle (right, center). A curiosity is shown in the last picture (bottom right): this pollen cluster from *P. tupaiodon* was swallowed most likely with a nocturnal moth that it preyed on.

As can be expected on the basis of tooth morphology, all Messel bats fed predominantly on insects, as do most extant echolocating bat species. The analyses of the gut contents of *Palaeochiropteryx* documents that both species fed on slow and low-flying insects, as predicted based on their wing morphology. The scarce food remains found for *Archaeonycteris* species consist of pieces of the cuticles of beetles only, whereas *Hassianycteris* species apparently preyed on beetles and moths as well. The preserved gut contents suggest that after successful hunting, these particularly slow-flying bats accidentally fell into the Messel Lake and subsequently died (Habersetzer et al. 1994). The identification of nocturnal microlepidopterans (Micropterigydae) among fossilized prey is very revealing, as these noiselessly flying insects give indirect evidence for nocturnal predation via echolocation. In general, 70% of all extant lepidopterans and beetles are nocturnal (Hölker et al. 2010).

What the cochlea reveals

A more direct indicator of echolocation in fossil bats is provided by the morphology of the inner ear region, as already demonstrated for *Tachypteron* above (Fig. 12.5.1). Echolocation performance and acoustic adaptations of Messel bat species can be evaluated and compared with other fossil bats. However, 3-D preservation of inner ears is very rare in Messel, and this is generally true for most fossil bat specimens because of the delicate nature of these structures. Nonetheless, even when compressed to a nearly 2-D state of preservation, bat cochleae still may contribute valuable information. Many individuals can provide partial information that can be collected and used to produce a composite 3-D inner ear region.

In addition to cochlear dimensions, the size and shape of the semicircular canals of bats can also be instructive in determining hearing and flight capabilities. However, none of the Messel bats have preserved semicircular canals that show the equilibrium (=vestibular) organ sufficiently enough to reconstruct it. Therefore, an extraordinarily well-preserved Eocene bat, *Stehlinia minor* (from Quercy in southern

France), is used here to show the complete acoustic and vestibular organs. This bat can serve as a model for *Palaeochiropteryx*, especially because the skull size as well as the cochlear size is almost the same compared to the Messel bat.

The skull of *Stehlinia minor* (Fig. 12.5.12) is depicted in micro-CT models from below (top center) and in both side views (top right and left). Both isolated acoustic and vestibular areas were virtually prepared from the micro-CT data set (lower) and then were rendered semitransparent (center) for better comparison. Cochlear width is defined as the oblique diameter (green line) along the second half-turn of the coiled cochlear duct (Habersetzer & Storch 1992). It is in this area of the cochlea that high-frequency sounds are received by the basilar membrane (carrying the hearing receptor cells) – for typical living bats these (extremely) high frequencies represent echolocation calls in the range of 20–200 kHz. Morphological changes in this area, for example an increase in size, indicate special adaptations of the ultrasonic system of the animals. The best-suited reference for skull size in relation to cochlear diameter is the basicranial width, which is indicated by the horizontal green line (Fig. 12.5.12).

In contrast to Messel bats, both vestibular organs of *Stehlinia* are complete and three-dimensionally preserved, which allows for comparison of these structures with extant and other fossil bats. Both cochlear sizes of *Stehlinia* and *Palaeochiropteryx* are in the lower range of typical extant aerial hawking insectivorous bats. Aerial hawking (catching insects in flight) is by far the most common feeding behavior among echolocating bats today. Interestingly, the majority of Messel bat species show a similar relation of cochlear width to basicranial width, with the exception of *Archaeonycteris* (for details, see Fig. 12.5.15).

Tanzanycteris mannardi (Fig. 12.5.13) is the only Eocene bat known from sub-Saharan Africa. Although it is preserved as a partial skeleton only, it is of great importance for understanding the evolution of echolocation in bats. Bones are mainly dissolved, and inner ears are therefore preserved as impressions in the sediment, with a few remaining bone fragments (Gunnell et al. 2003). However, using 2-D as well as 3-D X-ray methods, the cochlear size (as defined above, e.g., for *Stehlinia*) can be measured with high accuracy. The inner ear has a enormous relative volume and surpasses all other known early and middle Eocene bats. Based on further morphological characters such as thickness and shape of the inner ear

Fig. 12.5.10: Different hunting habitats and wing shapes in extant bats, for details see text. Scales: 10 cm.

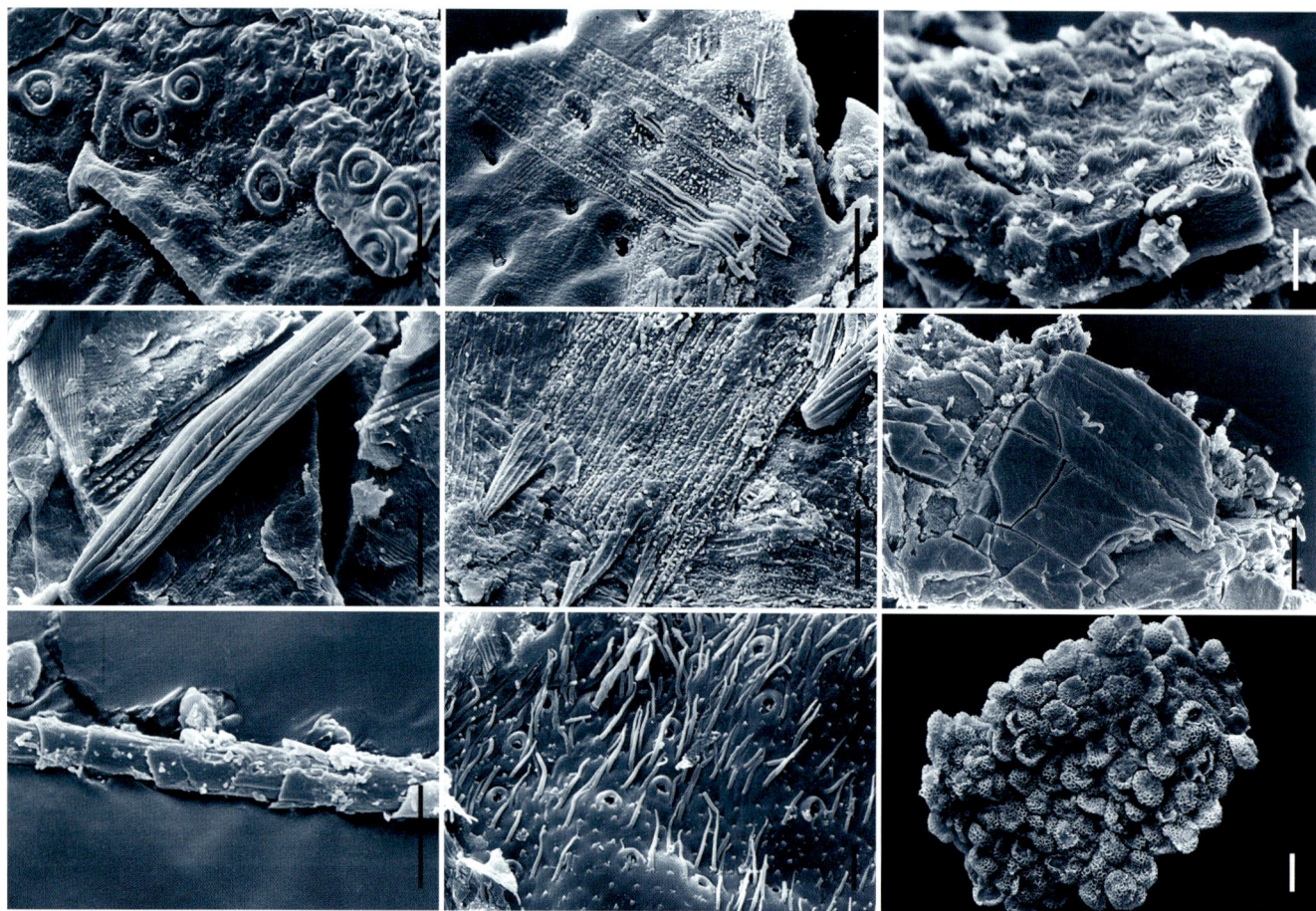

Fig. 12.5.11: Gut contents of Messel bats (SEM) prove feeding on insects. The pollen lump is an extremely rare find. Scales: 10 µm, exceptions: left center and right bottom 20 µm, right center 100 µm.

capsule, this species can be assigned to Old World leaf-nosed bats (Hipposideridae). This family and the horseshoe bats (Rhinolophidae) are part of the superfamily Rhinolophoidea (Fig. 12.5.2), representing the most highly specialized echolocators among extant bats worldwide. Although preservation is fragmentary, this fossil clearly demonstrates that the evolution of echolocation occurred rapidly during the early radiation of bats, and that only 1-2 million years after Messel, bats had achieved a range of echolocating abilities comparable to that of living bats.

In contrast, *Onychonycteris* (Fig. 12.5.7) represents a bat that was most probably not able to actively echolocate at a time only a few million years before Messel. Its cochlea was small, similar to those of non-echolocating, extant flying foxes (i.e., fruit bats or Pteropodidae) or those species of New World leaf-nosed bats (Phyllostomidae) that occupy a similar ecological niche to the Old World fruit bats (e.g., those feeding on fruit, pollen and nectar). Interestingly, the relative cochlear size of *Icaronycteris* and *Archaeonycteris* is intermediate between *Onychonycteris* and all other Messel bat species (for details, see also Fig. 12.5.15). This indicates that they represent an intermediate state of acoustic specialization with simply-structured echolocation sounds.

The evolution of echolocation

The three most important examples of the evolution of echolocation during the Eocene, namely *Onychonycteris* (Fig. 12.5.7, without echolocation), *Tachypteron* (Fig. 12.5.1, member of modern family, acoustic features like living relatives) and *Tanzanycteris* (Fig. 12.5.13, acoustically extremely specialized), can be compared to three extant bats, which were selected from a large sample because of their similar skull and inner ear sizes. Fig. 12.5.14 shows 2-D micro-radiographs of three skulls (top) and details of the hearing region (bottom) of extant bats. The flying fox *Micropteropus pusillus* (left), which has a normal-sized cochlea similar to that of other mammals, does not echolocate. Only very few species of flying foxes

Fig. 12.5.12: µCT models of the extremely rare complete 3-D preservation of the skull of *Stehlinia minor* (top) and its ear region (bottom). Scales: 1 mm.

Fig. 12.5.13: *Tanzanycteris mannardi*, partial skeleton (top, scale: 1 cm), and µCT-model of ear region (bottom, scale: 1 mm).

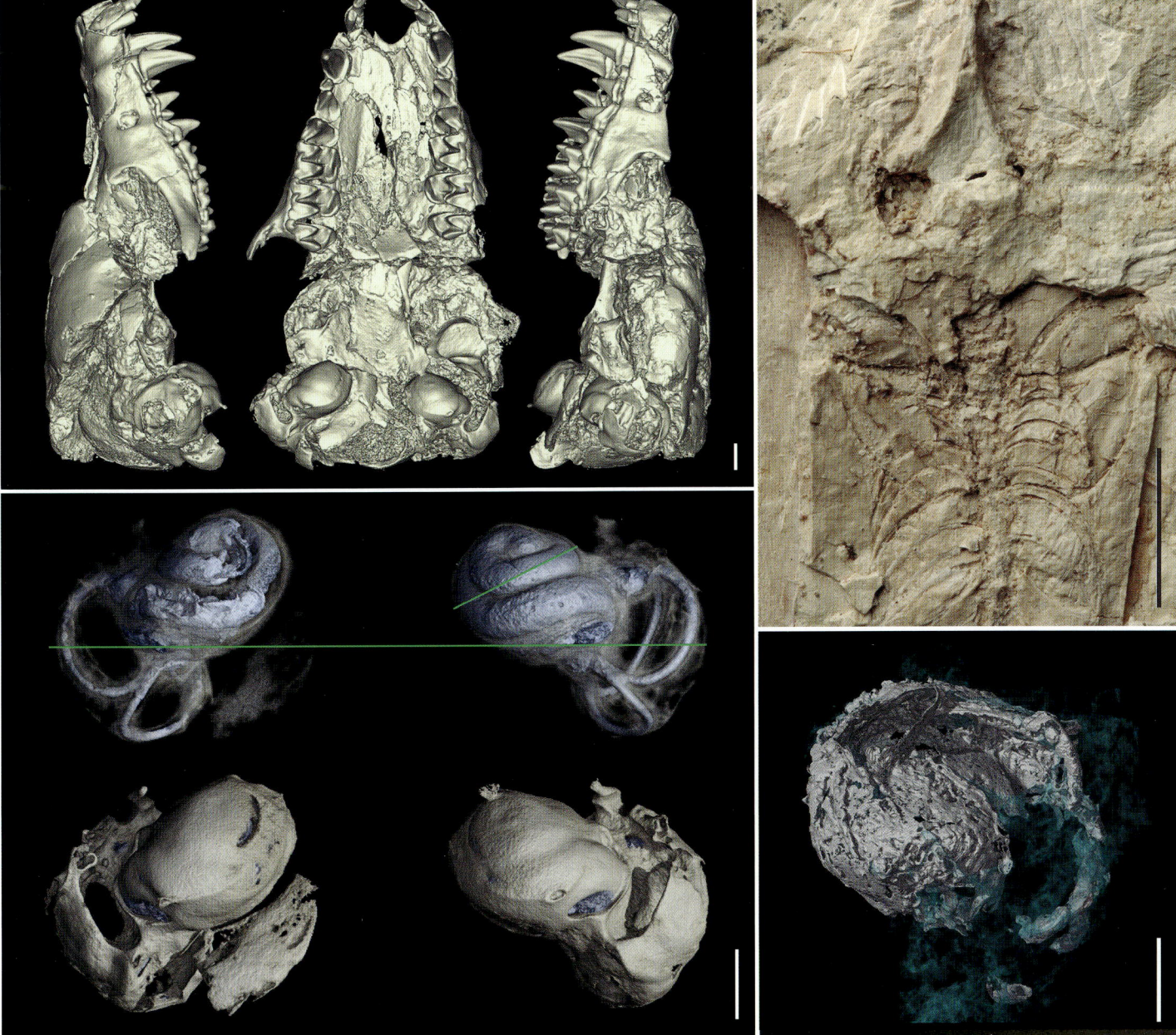

practice a rudimentary form of echolocation by producing simple clicking sounds with their tongues, which serve as a guide for primitive acoustic orientation. The high and fast-flying *Taphozous melanopogon* (center), already mentioned for comparison above, hunts prey with relatively low-frequency and shallow frequency-modulated ultrasonic sounds. It has an enlarged cochlea. A very large inner ear and a relatively small vestibular organ characterizes the Old World leaf-nosed bat *Hipposideros jonesi* (right). This bat uses long, pure tone signals, the so called CF sounds (CF = constant frequency), which are emitted through the nose after appropriate filtering. In contrast to normal echolocation, these long-duration sounds cover an extended time interval during one breath of air (high duty cycle). These sounds enable the very precise acoustic discrimination of flying prey by processing Doppler-shift effects in the echoes of prey against static background echoes. Thus, novel ecological niches, for example hunting prey in dense vegetation, can be made acoustically accessible, niches that cannot be utilized by other echo-

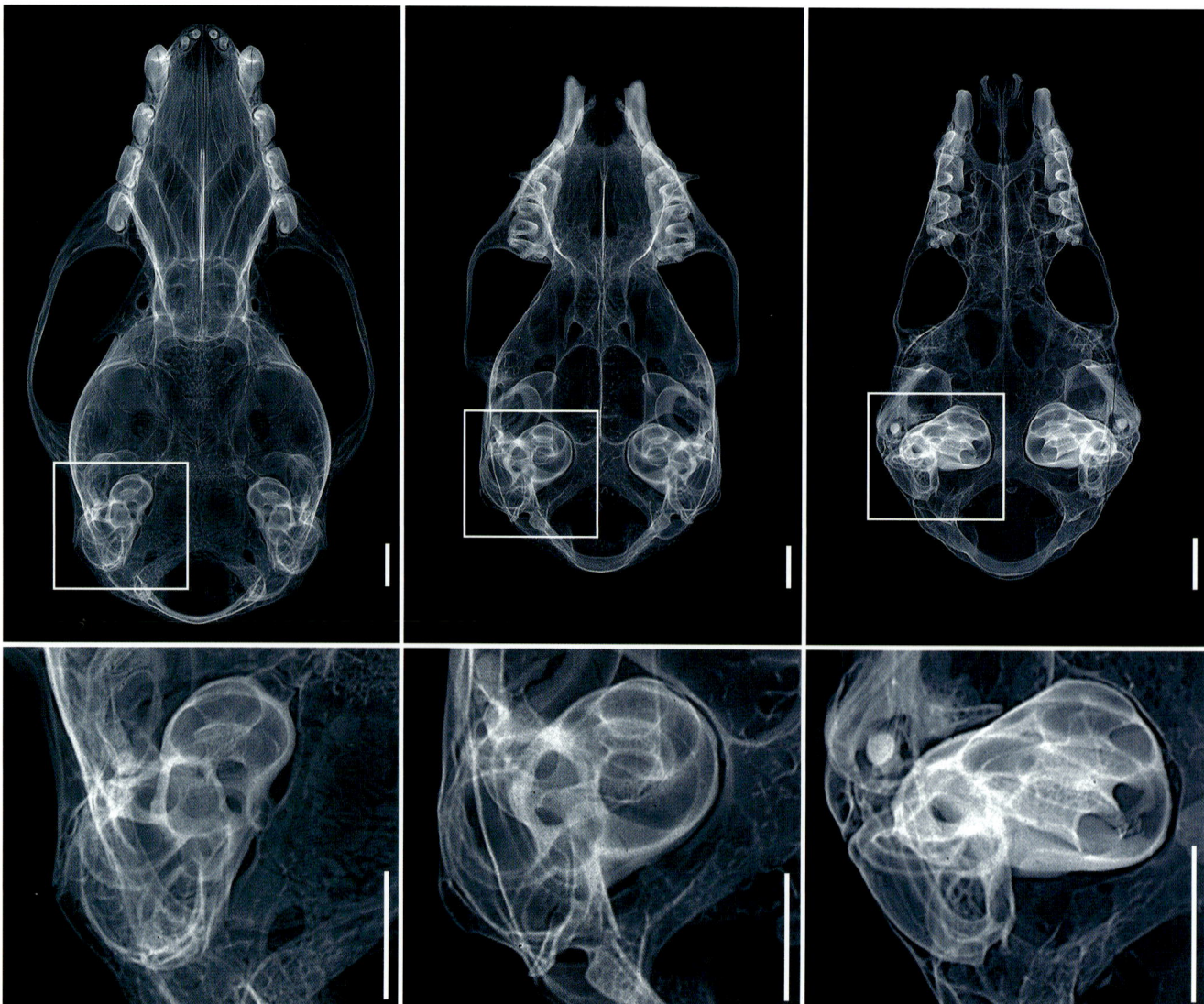

Fig. 12.5.14: Microradiographs of skull (top) and details of the ear region (bottom) of extant bats with different acoustic specializations. From left to right, *Micropteropus pusillus*, *Taphozous melanopogon*, *Hipposideros jonesi*. Scale: 2 cm.

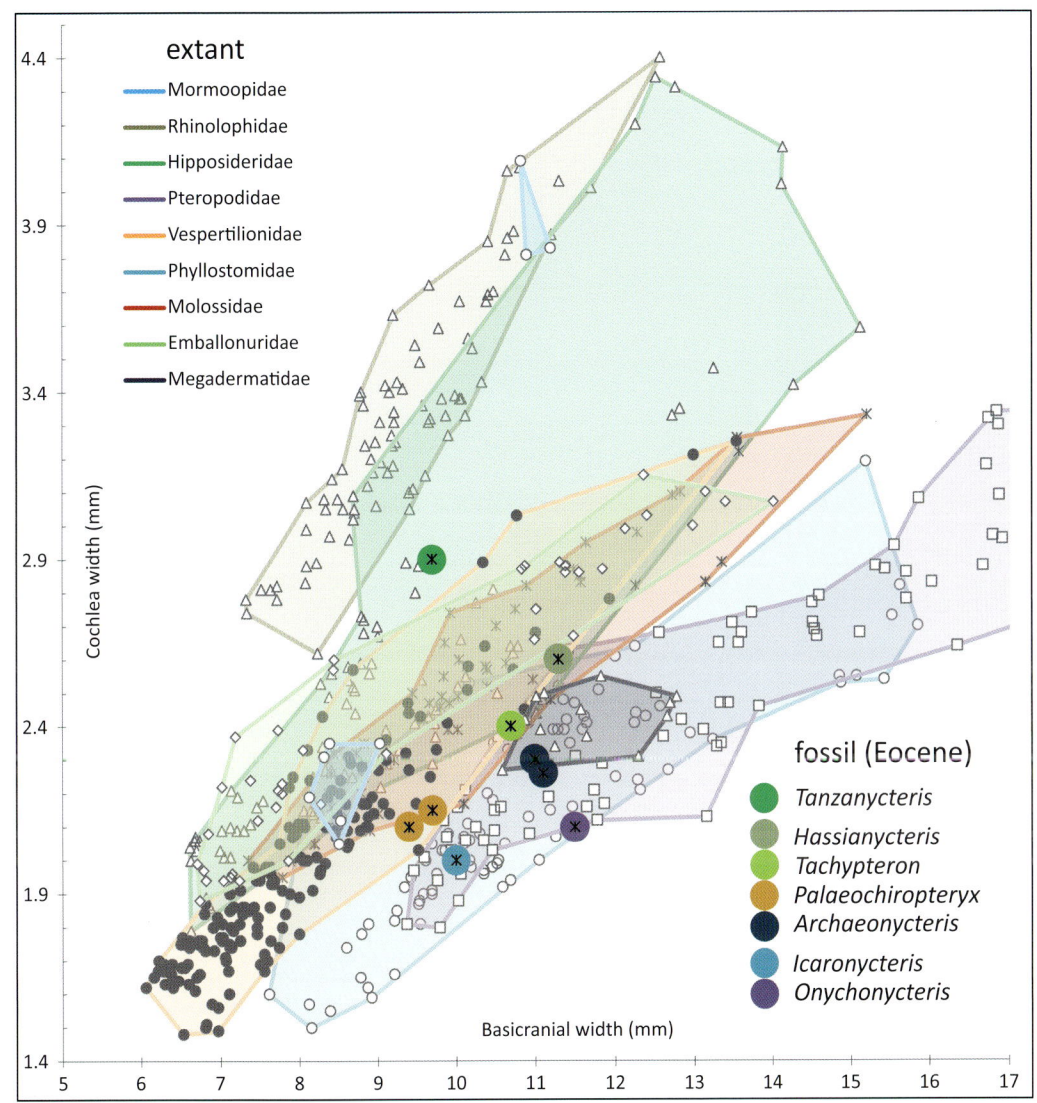

Fig. 12.5.15: Inner ear width in relation to basicranial width of extant and Eocene bats, explanations see text.

locating species (Habersetzer et al. 1984). These functional relationships between cochlear size and echolocation systems in living bats are a key tool in interpreting the bioacoustic capabilities of bats in the fossil record.

To further demonstrate the usefulness of cochlear size in defining acoustic abilities, inner ear width and basicranial width were measured for approximately half of the morphologically distinguishable extant bat species. Fig. 12.5.15 displays a fan-shaped distribution of these two parameters for the most species-rich extant bat families. Family names are given in the upper left image caption, and different colors clarify the polygons (envelopes) for the particular families. The illustration reflects different stages of acoustic specialization, ranging from non-echolocators (small relative cochlear width) to extreme specialists (extremely high values of relative cochlear width). The measurements for fossil bats are indicated by star symbols on large colored circles and the caption is given at the lower right (Fig. 12.5.15). This shows, in ascending order from *Onychonycteris* to *Tanzanycteris*, the increase of relative cochlear width. The small gaps separating the symbols of

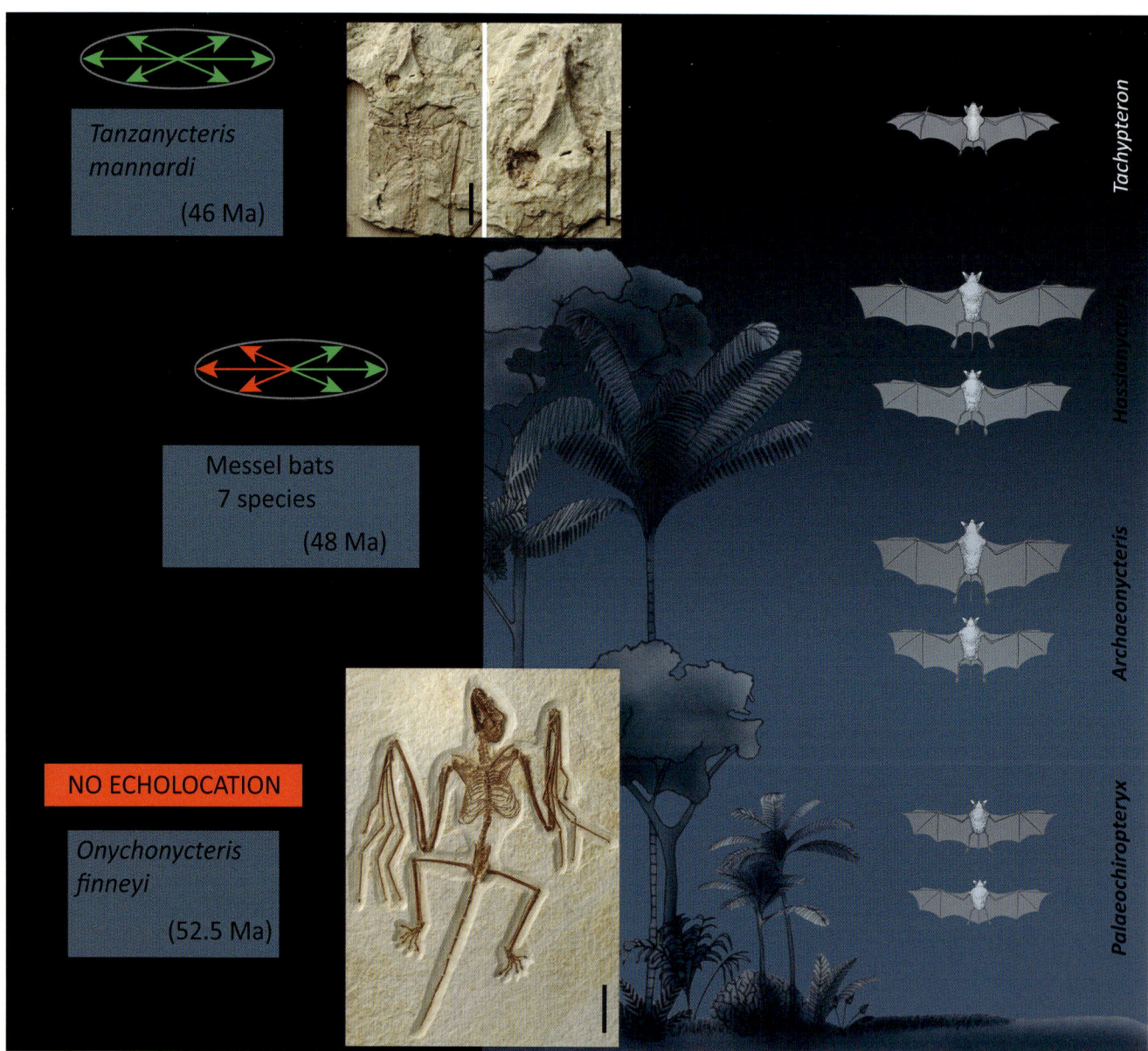

Fig. 12.5.16: Differentiated hunting habitats of Messel bats according to their wing shapes and functional properties (right) and echolocation abilities of Eocene bats (left).

fossil bats in the legend correspond to the different ages of excavation sites (52.5, 48, 46 million years). Compared to living bats, the Eocene species cover a large range of variation, indicating that a similar degree of sophisticated acoustic specialization already existed very early in bat diversification. Only the family of the horseshoe bats (Rhinolophidae) and some of the South American mustached bats (Mormoopidae) show an even more advanced enlargement of the inner ear than *Tanzanycteris*.

All Messel bats were aerial hawking echolocators. Recent reconstructions of body mass indicate that bats had achieved a body mass similar to the average extant echolocating bat (~14 g) by the early middle Eocene, and that this average was maintained throughout their phylogenetic history, probably due to physical constraints on size imposed by the combination of flight and echolocation (Giannini et al. 2012). Besides the previously introduced measurement for wing shape, the aspect ratio index (Fig. 12.5.9), additional parameters can be obtained for the reconstruction of the flight style and the flight speed of bats. These include wing loading relative to body weight and the tip index describing the ratio of hand wing area compared to the whole wing membrane. When sorting the reconstructed wing shapes of the Messel bats according to all of these parameters (Fig. 12.5.16, right), a clear spatial separation suggests differentiated characteristic hunting styles for the different species. This indicates that an ecologically diversified bat fauna similar to that found in extant bat communities already existed in the Messel bat assemblage, with one notable exception.

Messel bats, as typical aerial hawking echolocators, were able to exploit open air and vegetation-edge habitats but lacked echolocation abilities sufficiently sophisticated to allow them to hunt in dense vegetation. This is shown schematically in Fig. 12.5.16 (center, left) by the green arrows indicating one acoustically undisturbed spatial hemisphere. A "complete" acoustic world with both hemispheres available for active hunting for insects by ultrasonic echolocation was available in the Eocene for *Tanzanycteris*, Fig. 12.5.16 (top left), shortly after the time of Messel, but none of the Messel bats were capable of exploiting the densely vegetated portion of this acoustical world. Only a couple of million years earlier, "real" insect-feeding bats such as *Onychonycteris* existed, which were not capable of echolocation. Thus, the time interval for the evolution of echolocation, from its beginnings with simply-structured sounds (see above *Icaronycteris* and *Archaeonycteris*) to extremely sophisticated CF-FM echolocation is represented by the fossil examples compared here.

Summary of Eocene bats worldwide

Archaic bats similar to those found at Messel are also known from around the world. The very earliest known bats come from South America (Tejedor et al. 2005), Australia (Hand et al. 1994), North America (Jepsen 1966; Beard et al. 1992; Simmons et al. 2008), North Africa (Ravel et al. 2011), India (Smith et al. 2007), and Europe (Russell et al. 1973; Hooker 1996), and all appear in the fossil record at about the same time in the early Eocene.

Early-appearing species of the family Onychonycteridae (Simmons et al. 2008) are among the most primitive bats known and include taxa from North America (*Onychonycteris* and *Honrovits*) and Europe (*Eppsinycteris* and *Ageina*). Also appearing early in the record are two other primitive families, Icaronycteridae and Archaeonycteridae. Icaronycteridae includes a single genus, *Icaronycteris*, and is known from three geographically widespread species – *I. index* from North America (Jepsen 1966), *I. menui* from Europe (Russell et al. 1973) and *I. sigei* from India (Smith et al. 2007). Archaeonycteridae are represented by the two Messel species of *Archaeonycteris*, along with two other *Archaeonycteris* species from elsewhere in Europe (Russell et al. 1973; Harrison & Hooker 2010) and potentially by two other species, one from Europe (Tabuce et al. 2009) and one from India (Smith et al. 2007). In addition, the primitive genus *Protonycteris* from India (Smith et al. 2007) is also included within Archaeonycteridae.

Palaeochiropterygidae are by far the best-represented bats from Messel, with several hundred specimens known. However, palaeochiropterygids are also known from elsewhere in the Old World, with an early Eocene taxon, *Microchiropteryx*, appearing in India (Smith et al.2007) and a late middle Eocene form, *Lapichiropteryx*, known from China (Tong 1997). In addition to the well-known species from Messel, two other taxa from the middle Eocene are represented from Geiseltal in Germany, *Cecilionycteris* and *Matthesia* (Heller 1935; Sigé & Russell 1980). *Stehlinia*, discussed in detail above, could also be a member of Palaeochiropterygidae (Smith et al. 2012).

Finally, Hassianycteridae is represented by the two (perhaps three) Messel species, but in addition is also known by two taxa from India, *Hassianycteris kumari* and *Cambaya complexus* (Smith et al. 2007; Smith et al. 2012).

Despite a global distribution in the Eocene, detailed information concerning ancient bat biodiversity, morphology and behavior – based on complete skeletons, preserved gastrointestinal contents, accurate reconstructions of wing membranes and complete inner ears – is only provided by the fossil assemblage from Messel.

Chapter 12.6
Rodents – Gnawing Their Way to Success

Irina Ruf, Thomas Lehmann

Today, rodents (Rodentia) constitute by far the most species-rich order among mammals, comprising about 42 % of all living mammal species (Carlton & Musser 2005). Their first representatives in the upper Paleocene had already developed the dentition features characteristic for rodents, which, among others things, are responsible for the order's evolutionary success: one chisel-shaped, ever-growing incisor in each jaw quadrant, covered with enamel only on the front side and separated from the cheek teeth (premolars and molars) by a large gap (diastema) (Fig. 12.6.1, Fig. 12.6.5).

While three of the four rodent species found in Messel are also known from other localities, the record there is restricted to teeth, jaw fragments, and isolated skeletal elements. Thanks to their outstanding preservation as fully articulated skeletons, partly even with soft-tissue remains and gut contents, the rodent specimens from Messel provide a unique opportunity for gaining a deeper insight into the body build and lifestyle of these animals.

Systematics

The traditional classification of the main groups within Rodentia is based on characters in the jaw closure apparatus (e.g., Brandt 1855). In this regard, the pathway of the masseter muscle is of particular importance, as it can be found in four character states among rodents (Box 12.6.1) and was formerly used as the main basis for classification. However, it has since become apparent that the three derived character states are evolved repeatedly in different rodent lineages (convergence) and therefore cannot be considered relevant for distinguishing the major systematic groups.

Genetic analyses in recent years revealed a very different picture of the major systematic divisions within the rodents (e.g., Huchon et al. 2007; Blanga-Kanfi et al. 2009). The recent taxa are now divided

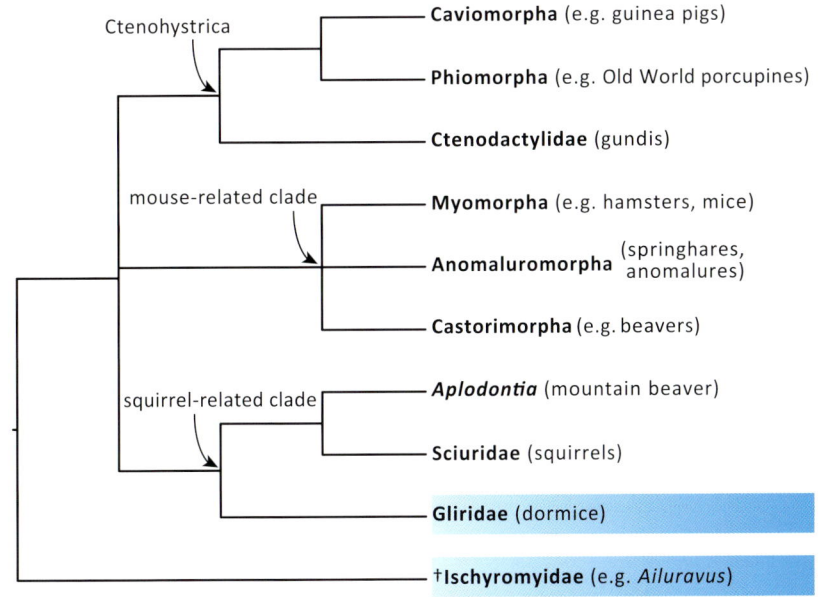

Fig. 12.6.2: Simplified phylogenetic tree of the rodents. Taxa known from Messel are highlighted in blue.

Fig. 12.6.1: *Masillamys beegeri* with soft-tissue remains. The counterpart bears the right half of the skull (inset). Scale: 1 cm.

Box 12.6.1: The unique chewing musculature of the rodents

Rodents possess highly specialized chewing muscles with a dominant masseter muscle. In Rodentia, this muscle consists of two parts and enables particularly efficient movement of the lower jaw when gnawing and chewing. In the plesiomorphic state (protrogomorph condition, top left), the masseter muscle extends from the zygomatic arch to the lower jaw (red arrows); the infraorbital foramen (marked in white) on the anterior edge of the zygomatic arch is small. In the sciuromorph type (bottom left), a part of the muscle runs in front of the zygomatic arch, increasing its leverage. In the hystricomorph type (bottom right), one part of the muscle runs through the greatly enlarged infraorbital foramen. The myomorph type (top right) shows a similar pattern, but the infraorbital foramen is constricted and takes a key-hole shape here. In this regard, the rodents from Messel show different character states.

Contrary to recent Gliridae, which show a myomorph or hystricomorph pattern, *Eogliravus wildi* is protrogomorph (Storch & Seiffert 2007). Thus, the dormouse from Messel confirms that the extension of the masseter muscle in modern Gliridae developed convergently with other myo- and hystricomorph rodent taxa (Vianey-Liaud 1989; Maier et al. 2002; Hautier et al. 2008).

Ailuravus macrurus and *Hartenbergeromys parvus* are considered protrogomorph, whereas *Masillamys beegeri* is considered sciuromorph (Hartenberger 1968; Koenigswald et al. 1988; Paus 2002). However, a closer look reveals that *Masillamys* possesses a very large infraorbital foramen (center), comparable to hystricomorph rodents; therefore, a revision of this structural complex appears to be necessary.

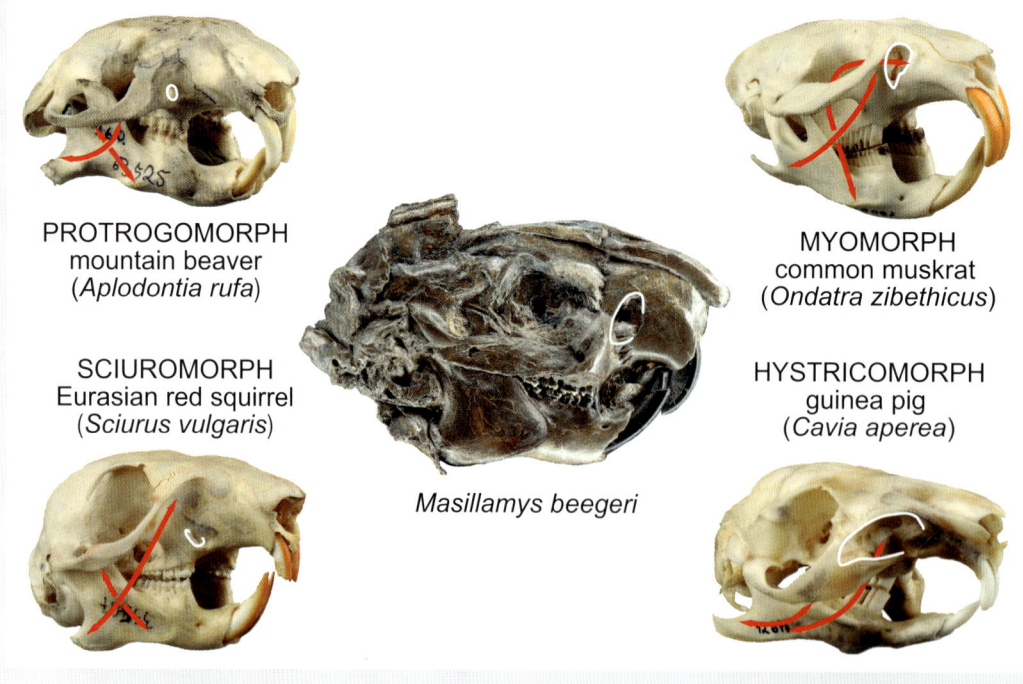

PROTROGOMORPH
mountain beaver
(*Aplodontia rufa*)

MYOMORPH
common muskrat
(*Ondatra zibethicus*)

SCIUROMORPH
Eurasian red squirrel
(*Sciurus vulgaris*)

HYSTRICOMORPH
guinea pig
(*Cavia aperea*)

Masillamys beegeri

into three main groups: the mouse-related clade, the squirrel-related clade, and the Ctenohystrica (Fig. 12.6.2), whose sister-group relationships, however, have not yet been definitively clarified. In part, the composition of these groups shows significant differences with the traditional phylogenetic trees; for example, the Gliridae (dormice) are now considered more closely related to the squirrels and the Mountain Beaver than to the Myomorpha (mouse-like rodents).

The rodent fauna that inhabited the area around Lake Messel 48 million years ago is known by four species so far. *Ailuravus macrurus*, *Masillamys beegeri* and *Hartenbergeromys parvus* belong to the family Ischyromyidae, which originated from the first large radiation of rodents during the Paleogene, thereby constituting representatives of the stem group of modern Rodentia (Rodrigues et al. 2010). *Eogliravus wildi* is the oldest, most basal species of the Gliridae, which also include the native European Hazel Dormouse (*Muscardinus avellanarius*) and the Edible Dormouse (*Glis glis*).

The large leaf-eater *Ailuravus*

To date, more than 20 specimens of *Ailuravus macrurus* have been found in the Messel Pit, several with soft-tissue remains and gut contents (Fig. 12.6.3, Fig. 12.6.4). *Ailuravus* is the largest rodent from the Messel Pit, with a head-body length of about 40 cm and an almost 60-cm-long tail (Weitzel 1949; Gwosdek 1996; Paus 2002). At first glance, the species resembles a robust squirrel (Sciuridae), although it is not closely related to this group. *Ailuravus* shows adaptations both to a slow but powerful mode of locomotion on the ground as well as to an arboreal lifestyle. For the latter, however, climbing was slow and performed with the aid of claws hooked into the substrate, rather than by agile jumping through the branches. The tail is covered with long, bushy hair and served as a balancing organ. Although the tip of the tail is curled in most individuals, the existence of a prehensile tail could not be confirmed to date. The forelimbs are about two thirds the length of the powerful hind limbs. The thumb is greatly reduced, and the fingers and toes were equipped with sharp claws. The fingers could be spread, giving *Ailuravus* a secure grip when climbing among the branches. The East Asian giant squirrel (*Ratufa*) could be a living analog occupying a comparable ecological niche, although this species moves through the treetops in a much more agile manner, performing long jumps.

The skull of *Ailuravus* is about 9 cm long and shows a domed nasal ridge, a massive zygomatic arch, a strong sagittal crest and a robust lower jaw with a prominent coronoid process. The temporalis muscle, which runs from the top of the skull (including sagittal crest) to this process and controls the vertical movement of the lower jaw, was particularly well-developed in *Ailuravus*. With regard to the number of cheek teeth, the dentition corresponds to the plesiomorphic state in Rodentia. The upper jaw shows two premolars and three molars, whereas the lower jaw has one premolar and three molars. They are brachydont (low-crowned), with minor wrinkling of the enamel on the chewing surface, and show a specific pattern of cusps with sharp ridges (Weitzel 1949; Tobien 1954; Paus 2002) (Fig. 12.6.5).

Fig. 12.6.3: *Ailuravus macrurus* is the largest rodent species known from Messel. Gut contents are preserved in the abdominal region. Scale: 5 cm.

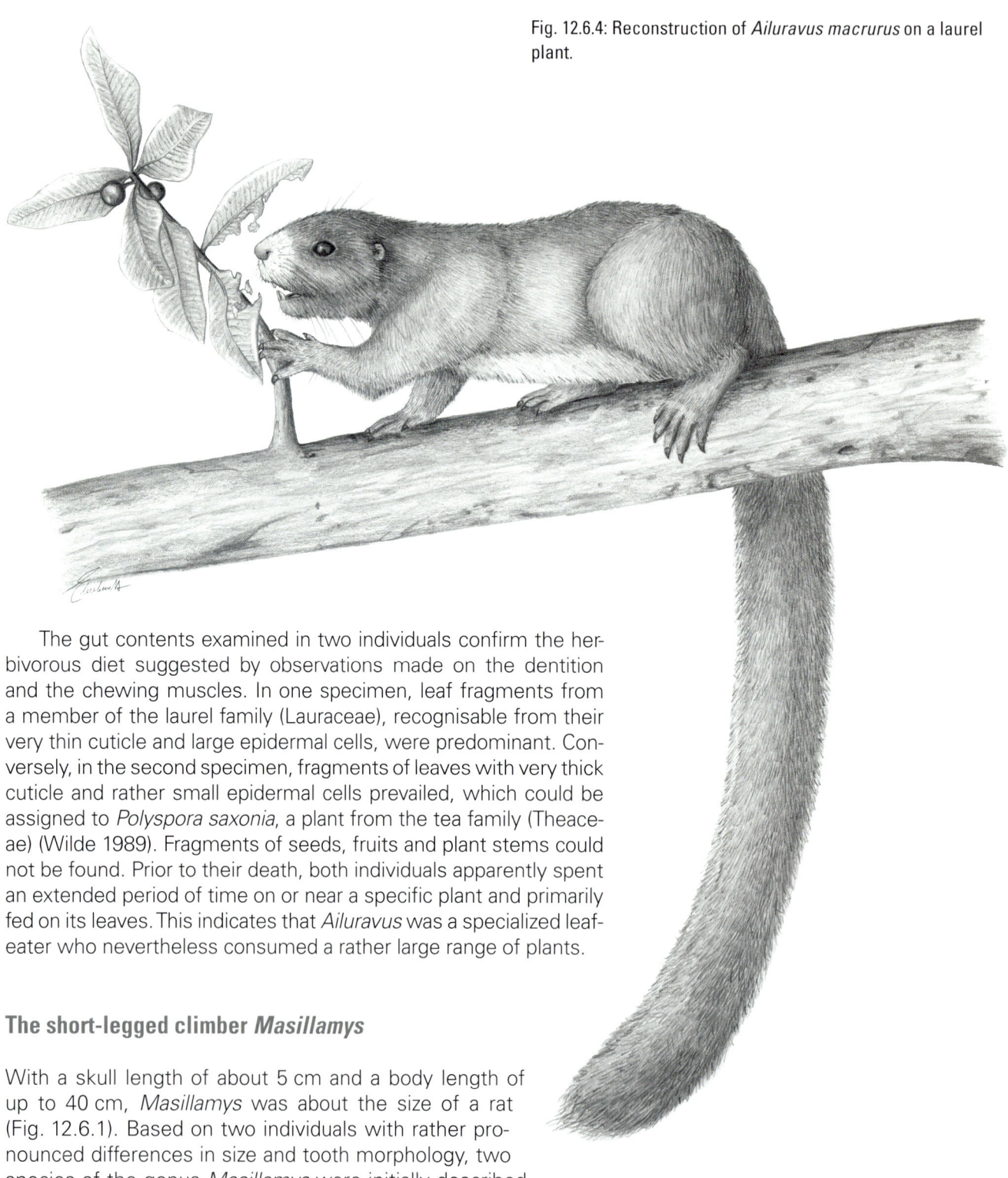

Fig. 12.6.4: Reconstruction of *Ailuravus macrurus* on a laurel plant.

The gut contents examined in two individuals confirm the herbivorous diet suggested by observations made on the dentition and the chewing muscles. In one specimen, leaf fragments from a member of the laurel family (Lauraceae), recognisable from their very thin cuticle and large epidermal cells, were predominant. Conversely, in the second specimen, fragments of leaves with very thick cuticle and rather small epidermal cells prevailed, which could be assigned to *Polyspora saxonia*, a plant from the tea family (Theaceae) (Wilde 1989). Fragments of seeds, fruits and plant stems could not be found. Prior to their death, both individuals apparently spent an extended period of time on or near a specific plant and primarily fed on its leaves. This indicates that *Ailuravus* was a specialized leaf-eater who nevertheless consumed a rather large range of plants.

The short-legged climber *Masillamys*

With a skull length of about 5 cm and a body length of up to 40 cm, *Masillamys* was about the size of a rat (Fig. 12.6.1). Based on two individuals with rather pronounced differences in size and tooth morphology, two species of the genus *Masillamys* were initially described (Tobien 1954). The five cheek teeth in the upper and four cheek teeth in the lower jaw arc brachydont, with cusps connected by transverse ridges. The enamel on the chewing surfaces shows pronounced grooves and wrinkling. In the meantime, a certain degree of variability in the dentition and body size has been shown in this genus. Therefore, only a single species, *Masillamys beegeri*, is considered valid today, of which almost 20 skeletons have been found to date (Escarguel 1999).

Masillamys probably shared the branches of the tropical forest in Messel with *Ailuravus*, although the two species do not show identical adaptations to an arboreal locomotion. *Masillamys* had a rather plump body and short limbs with broad hands and feet. These proportions most closely correspond to those found among recent burrowing and primarily subterranean species of voles. On the other hand, the slender, high claws and the tail length contradict this assumption (Koenigswald et al. 1988). The tail of *Masillamys* makes up about 50 % of the body length and was only sparsely covered with hair, as revealed by soft tissue remains. Thus, it was better-suited to serve as a balance organ than to steer the animal when performing long jumps. This habitus can be compared to that of the modern climbing rats (*Tylomys*), which climb through the tree tops in the tropical forests of Central America instead of jumping from branch to branch.

Hartenbergeromys: a still enigmatic rodent

Initially assigned to the genus *Masillamys*, then to *Microparamys*, the second-smallest rodent known from Messel, *Hartenbergeromys parvus*, is placed in its own genus today, since it shows clear differences in its tooth morphology (Tobien 1954; Hartenberger 1968; Escarguel 1999). *Hartenbergeromys*, known by at least two skeletons, had a skull length of 3.5 cm and an overall length of about 26 cm (Fig. 12.6.6), making it somewhat larger than a modern House Mouse (*Mus musculus*). At first glance, *Hartenbergeromys*

Fig. 12.6.5: Detailed view of the dentition of *Ailuravus macrurus*, showing the occlusal surface of the low-crowned molars. Scale: 1 cm.

resembles *Masillamys* in its build, which prompted the suggestion of a similar lifestyle. However, *Hartenbergeromys* has a somewhat shorter tail, and the hind limbs are clearly longer than the forelimbs. Although a detailed examination of the skeleton is still pending, an exclusively arboreal lifestyle can already be ruled out.

Eogliravus: The oldest dormouse

Eogliravus wildi was first described on the basis of a few isolated teeth from the lower Eocene of France (Hartenberger 1971). Though the *Eogliravus wildi* specimen from Messel (Fig. 12.6.7) is slightly younger, it constitutes the oldest completely articulated glirid skeleton (Storch & Seiffert 2007). *Eogliravus* reached an overall length of about 11 cm, even smaller than the Hazel Dormouse. Both the skull and the skeleton of *Eogliravus* already show many anatomical similarities with recent Gliridae. However, the craniofacial region and the diastema between the incisors and premolars are rather short, causing the strikingly large eye sockets to be placed near the front of the skull. This suggests a potentially crepuscular or nocturnal lifestyle for *Eogliravus*. The upper jaw still holds two premolars, while modern forms only have one.

Moreover, *Eogliravus* was an agile climber, living in the shrubs and trees around Lake Messel, as inferred from various skeletal traits: e.g., the short olecranon (elbow) is noticeably curved upwards, the phalanges of the fingers and toes (except the thumb) are rather long and show prominent tubercles as attachment points for the flexor tendons, and the toes were likely armed with long claws. The proportions of the limbs also show clear adaptations to a scansorial (climbing) lifestyle. For example, the length ratio of each limb and that of the fifth toe are comparable to those of extant climbing rodents. Some of these values for *Eogliravus* fall between the Forest Dormouse (*Dryomys nitedula*) and the Hazel Dormouse within the recent Gliridae. The distal ends of the tibia and fibula are not fused with each other, which constitutes a plesiomorphic state. In *Gliravus micio* from the lower Oligocene, these bones are fused over a short distance, while in extant members of the Gliridae the tibia and fibula are fused along one fourth to one third of the entire tibia length (Fejfar & Storch 1994).

The soft-tissue remains clearly show that *Eogliravus* had a bushy tail, similar to the Edible Dormouse (*Glis glis*). Compared to extant Gliridae, the tail was relatively short, with only 16 vertebrae, but it served as a steering organ for climbing and jumping among the branches. The present specimen is obviously a male, as shown by the presence of a baculum (penis bone), whose slightly curved tip is similar to that of most recent Gliridae. The gut contents of *Eogliravus* reveal that the animal apparently favored a diet of fruits, seeds and buds, which is similar to the dietary range of the Edible Dormouse and the Hazel Dormouse.

Paleobiogeography and paleoenvironment

The Gliridae are one of the oldest extant rodent families; they probably originated in Europe, since the oldest representatives are found in the Eocene of France, Spain and Germany. They were endemic to Europe during the Eocene, whereas they are found across Eurasia and Africa today (Storch & Seiffert 2007). Ischyromyidae, on the other hand, apparently had a holarctic distribution, although they are primarily known from North America (Escarguel 1999; Rana et al. 2008). For example, the subfamily Ailuravinae, which includes *Ailuravus macrurus*, is mainly known from North America and Europe since the early Eocene. The taxon that includes *Masillamys* and *Hartenbergeromys* already occurs in the upper Paleocene in North America (Escarguel 1999). This paleogeographic distribution can be explained by a faunal exchange via the northern land bridges (Chapter 2). Conversely, a faunal exchange between Europe and Asia during the Paleocene and Eocene was long deemed improbable, since the Turgai Strait was considered a natural

Fig. 12.6.6: One of the rare specimens of *Hartenbergeromys parvus*. Scale: 3 cm.

Fig. 12.6.7: *Eogliravus wildi* (holotype) with soft-tissue remains. The bushy tail is clearly visible. Scale: 1 cm.

barrier. Meanwhile, however, a growing number of discoveries suggests that faunal exchange must have occurred between Europe and Asia as well (Rana et al. 2008; Solé et al. 2016). In particular, newly discovered primitive Ailuravinae from the lower Eocene in India show a closer relationship to European forms than to North American ones. Together with some bats (Chapter 12.5), the Ailuravinae could add support to the hypothesis of such a faunal exchange in the Paleocene and Eocene, documenting the multi-regional origin of the rodents from Messel.

The arboreal adaptations and the gut contents of *Ailuravus macrurus* and *Eogliravus wildi* fit the reconstruction of a paratropical environment in the vicinity of Lake Messel, densely covered with vegetation (Paus 2002; Storch & Seiffert 2007). Hence, well-adapted for climbing, the first representatives of the Gliridae, which originated in Europe, already occupied a habitat similar to that of many extant species of dormice.

Chapter 12.7
Ferae – Animals that Eat Animals

Gregg F. Gunnell †, Thomas Lehmann, Irina Ruf, Jörg Habersetzer, Michael Morlo, Kenneth D. Rose

Each of the species included in this chapter preyed on other animals: four carnivorous species specialized on vertebrates and invertebrates, and the other three species were insect-eating specialists. Some may have also included vegetable matter in their diets. At least five of the species may represent stem members of the crown groups Carnivora and Pholidota while the other two species are members of a group that may be related to these two orders. All are relatively small to medium sized mammals, three of them being terrestrial, fossorial (digging) forms, two being generalized terrestrial quadrupeds, and two being likely arboreal climbers.

Systematics of Carnivoraformes and Pholidotamorpha

Most of the animals in this chapter (Fig. 12.7.2) can be placed in the higher level clade Laurasiatheria (Springer et al. 2003).

Paroodectes and *Messelogale* belong to Carnivoraformes with the latter perhaps being closer to crown Carnivora than the former (Morlo et al. 2004; Solé 2014; Solé et al. 2016a). These taxa are members of a broadly defined, paraphyletic family "Miacidae" (Smith & Smith 2010; Solé 2014; Solé et al. 2014, 2016a). Springhorn (1982, 2000) interpreted *Messelogale* as having only two upper and lower molars, a similarity shared with *Quercygale*. The latter may have been a stem member of crown Carnivora (Wesley-Hunt & Werdelin 2005), a position that *Messelogale* potentially shares as well (Solé et al. 2014, 2016a).

Lesmesodon belongs to the extinct mammalian order Hyaenodonta. The relationships of Hyaenodonta are problematic, but recent work has suggested that hyaenodontans may have originated in Africa in the Paleocene Solé et al. 2016b). If this turns out to be true, it suggests that the group could be included as a clade within the endemic African radiation known as Afrotheria (Springer et al. 2003; Solé et al. 2016b). Nonetheless, according to O'Leary et al. (2013) all Messel Ferae (including *Lesmesodon*) belong to Laurasiatheria.

Hyaenodonta are characterized by modifications to two or three pairs of upper and lower molars to form carnassials (opposing upper and lower teeth with elongated cutting blades that function together as a slicing pair). Hyaenodontans may be related to true Carnivora (typified by a single pair of carnassial teeth), but this relationship remains tentative (Solé et al. 2014; Spaulding & Flynn 2012; O'Leary et al. 2013). *Lesmesodon* falls within the hyaenodontan subfamily Proviverrinae (Morlo & Habersetzer 1999).

The other three taxa from Messel included in Ferae are all potential members of Pholidota (pangolins). *Eomanis* was originally described as a pangolin (Storch 1978) based on the species *E. waldi*. *Eomanis krebsi* was added by Storch & Martin (1994) but later moved to a new genus, *Euromanis*, by Gaudin et al. (2009). *Eurotamandua joresi* was proposed as a myrmecophagid (anteater) xenarthran by Storch (1981). Many of the features that link these three taxa also characterize Palaeanodonta, an extinct Eocene group mostly known from North America that resembles extant xenarthrans in many ways but is best supported as a sister taxon of Pholidota (Gaudin et al. 2009). However, Gaudin et al. (2009) note that the relationships among all of the Messel taxa and those making up palaeanodonts are unclear, partly because it is difficult to score characters on the Messel skeletons as preserved.

Ferae as recognized here may well be a monophyletic group, especially if all of the Messel taxa can be recognized as members of Carnivoramorpha (potentially including Hyaenodonta) and Pholidotamorpha (Fig. 12.7.2).

Fig. 12.7.1 – Holotype of *Eurotamandua joresi*, a probable pangolin relative from Messel. Scale: 5 cm.

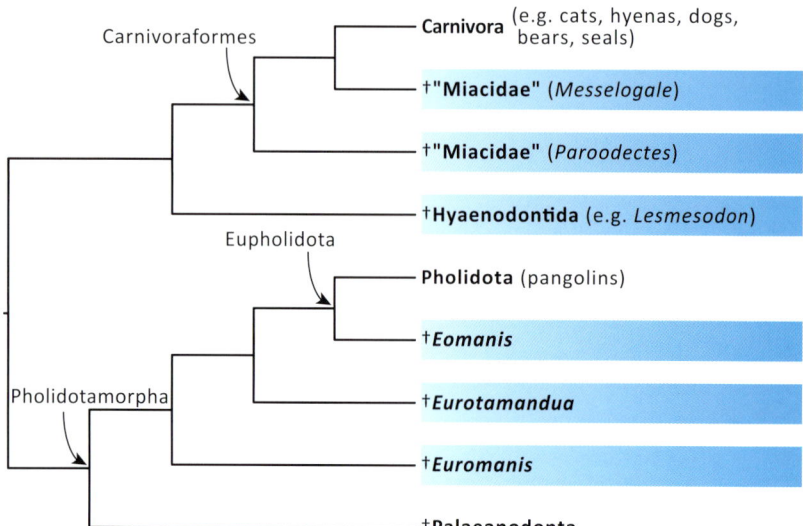

Fig. 12.7.2: Simplified phylogenetic tree depicting potential relationships of Messel Ferae. The paraphyletic "Miacidae" appears twice here. Taxa known from Messel are highlighted in blue.

Lesmesodon: the Messel hyaenodontan

The hyaenodontan *Lesmesodon* is represented by two species, *L. edingeri* (known from four specimens) and *L. behnkeae* (known only from the holotype) (Fig. 12.7.3 & 12.7.4). The five specimens are all juveniles, and the genus is only known from Messel. *Lesmesodon* is a relatively small hyaenodontan, *L. edingeri* having molar tooth proportions nearly as small as *Proviverra typica*, the smallest European proviverrine. *Lesmesodon behnkeae* is relatively larger, with the upper first molar being approximately 40% longer than in *L. edingeri* (Morlo & Habersetzer 1999). *Lesmesodon behnkeae* (estimated body length about 45 cm, including tail) is comparable in size to the medium-sized European proviverrines *Quercitherium tenebrosum* and *Paracynohyaenodon*, but smaller than the largest members of the subfamily. *Lesmesodon edingeri* is smaller, with an estimated body length of 30 cm. In terms of body mass, *L. edingeri* is estimated to have weighed approximately 300 g, based on lower molar size (about the size of the living Dwarf Mongoose, *Helogale parvula*), while *L. behnkeae* was larger, with an estimated mass of 1.5 kg (approximately the size of a modern Malagasy civet, *Fossa fossana*) (Morlo & Habersetzer 1999).

Perhaps the most unusual attribute of the skeleton of *Lesmesodon* is the lack of fissured distal phalanges (claws), a feature otherwise shared by other known hyaenodontans. Beyond this, the hind limbs were slightly longer than the forelimbs, the clavicle was relatively small, the humeral deltopectoral crest (for muscle attachment) was relatively weak, the ulna slightly curved, with a prolonged olecranon (elbow), the radius slightly curved, with an elevated capitular eminence, radius and ulna closely approximated, the femur with a distinct third trochanter (attachment for superficial gluteus), the manus slightly shorter in overall proportions compared to the pes, and the tail relatively short and bushy, as revealed by soft tissue preservation (Fig. 12.7.3).

Locomotor patterns for *Lesmesodon* can best be thought of as similar to those of small living viverrids such as mongooses (*Herpestes*) or genets (*Genetta*) and therefore similar to generalized terrestrial quadrupeds, although some capacity for climbing was likely also present. Combining body size with what is known of the dentition, and keeping in mind that all specimens represent juveniles, it can be inferred that *Lesmesodon* probably was a generalized, opportunistic carnivore that concentrated on vertebrate food but almost certainly took non-vertebrate and non-animal foods as well (Box 12.7.1).

Figure 12.7.3: Holotype skeleton of *Lesmesodon behnkeae*, showing body proportions and presence of a relatively short and bushy tail. Scale: 5 cm.

Fig. 12.7.4: Skeleton of *Lesmesodon edingeri*. Note that parts of the hind limb and the tail are missing and have been painted on the slab. Scale: 5 cm.

Paroodectes feisti: an agile climber

The carnivoraform *Paroodectes feisti* is known from a single specimen (Fig. 12.7.5) discovered in 1974 (Springhorn 1980). As a miacid it represents the most basal carnivoraform from Messel. *Paroodectes* is a relatively small mammal with an estimated body length (including tail) of 55 cm. With an estimated body weight of approximately 600 grams, *Paroodectes* would have been about the size of an extant Indonesian weasel (*Mustela lutreolina*).

Paroodectes can be distinguished from Carnivora by the presence of a complete clavicle and the lack of fusion of two carpal (wrist) bones (scaphoid and lunate) that articulate with the radius. However, the specimen is a juvenile (based on the presence of a deciduous dentition), so subsequent ontogenetic changes, including potential carpal fusion, cannot be excluded.

The skull of *Paroodectes* is short and compact, with a proportionally large and high brain case. Based on X-ray images of the inner ear, the cochlea shows 1.75 turns, quite a low number compared to crown Carnivora, and therefore probably a plesiomorphic character of *Paroodectes*. The basal turn has a relatively large diameter of 1.4 mm, which is comparable to some extant carnivorans such as the Red Fox (*Vulpes vulpes*), even though *Paroodectes* was a much smaller animal (probably about 10% of the body weight of the Red Fox). This suggests that *Paroodectes* possessed quite acute hearing capabilities.

Box 12.7.1: Enjoy your meal!

The reconstruction of fossil food webs is an important contribution to the understanding of ancient ecosystems (Chapter 13), but often it is only possible to do so by examining proxies instead of direct evidence. One of these cases is represented by a juvenile *Lesmesodon edingeri* specimen from Messel (scale: 2 cm). This specimen has been interpreted to have been swallowed and regurgitated by a snake, probably *Palaeopython*, as inferred from the body length of the mammal at approximately 30 cm. The skeleton of *Lesmesodon* is mostly intact, indicating that little digestion had occurred prior to its regurgitation, though it is deformed to a slender carcass with forelimbs pressed against the body. The specimen shows gut contents that have been investigated by μCT. Before the unfortunate *Lesmesodon* met with the snake, it had eaten its fill. Indeed, its gut contents includes numerous teeth and tiny pieces of crushed bones of several vertebrates. One jaw with a row of pointed teeth belongs to a small amphibian or reptile (left inset). The lower jaw of a small insectivorous mammal showing its molar dentition is also present, as are other isolated teeth (right inset).

Fig. 12.7.5: Holotype of *Paroodectes feisti,* with detailed view of skull and dentition (inset). Scale: 5 cm.

The lower jaw is slender but robust, with a deep masseteric fossa (for attachment of chewing muscles). The tooth count is typical for primitive placentals and includes three incisors, one canine, four premolars and three molars in each jaw quadrant. Some of the erupted premolars are clearly deciduous (milk teeth). X-ray images show the premolars of the adult dentition still hidden in the maxilla. The upper third and lower fourth deciduous premolars temporarily form the carnassial complex until it is replaced by the adult carnassial series, comprising the upper fourth premolar and the lower first molar.

Paroodectes was an agile climber. Its postcranial anatomy shows a mix of arboreal and terrestrial characters. Among the potential arboreal adaptations are a long tail, strong and slightly curved clavicle, unfused radius and ulna, high and rostrally inclined ulnar olecranon process, and long and slender metacarpals (middle hand bones). Terrestrial indicators include a hip joint and foot morphology similar to small extant carnivorans that are plantigrade (walking on palms) but practice some arboreal locomotion as well, a pattern viewed as plesiomorphic for Carnivoraformes (Wang 1993).

Messelogale kessleri: a small predator

Four specimens represent the Messel carnivoraform *Messelogale kessleri* (Springhorn 1982, 1985, 2000). It is a small animal that did not exceed 21 cm in total body length (including tail). The estimated body mass of *Messelogale*, based on the lower first molar length, is about 200 g, approximately one-third the mass calculated for *Paroodectes*. This would make *Messelogale* about the size of a modern Least Weasel (*Mustela nivalis*).

Messelogale is characterized by a short braincase, ventrally bowed mandible, short forearms and

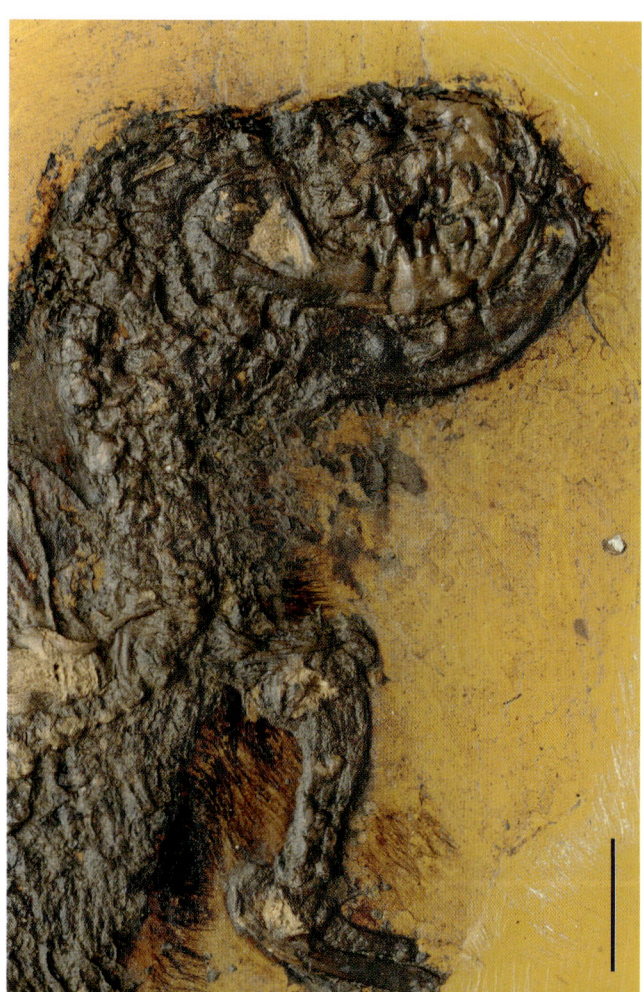

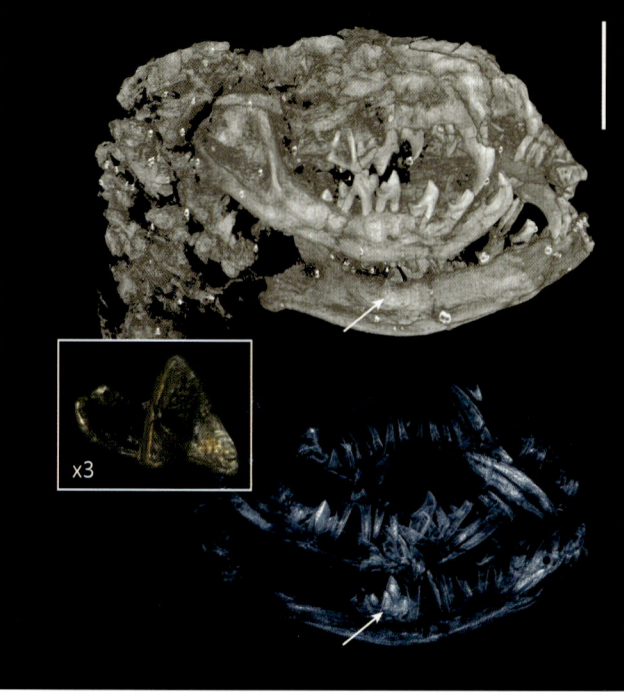

Fig. 12.7.6: *Messelogale kessleri*. Top: Head and torso of a specimen (note the remains of hair in the neck and chest region); center: 3-D model of the skull obtained from recent μCT scans; bottom: same with special filter to reduce the contrast of bony elements. Arrows point to the unerupted left lower first molar. Inset: 3-D model of the unerupted left lower first molar in occlusal view. Scale: 10 mm.

lower legs, strong claws, as well as fluffy fur and a bushy tail (Fig. 12.7.6). Like many other mammals from Messel, all four specimens of *M. kessleri* are juveniles. In the absence of permanent teeth, *Messelogale* has, as in modern juvenile carnivores and as previously described for *Paroodectes*, a specialized pair of deciduous teeth that assume carnassial functions later taken over by the permanent teeth. No second molar is yet known for *Messelogale,* but the first lower molar can be seen on X-ray images and has been digitally extracted after recent μCT scans (Fig. 12.7.6, inset).

The skull of *Messelogale* is rather compact and had relatively short parietal bones that are sub-equal in size to the frontal bone, a feature found in crown carnivorans (Wesley-Hunt & Werdelin 2005). The parietal also bears a weak sagittal crest for the temporal muscles involved in mastication. The other major chewing muscle, the masseter, was certainly well developed, as suggested by the deep attachment fossa on the lateral side of the mandible.

In one specimen, Springhorn (2000) was able to observe structures of the middle and inner ear. The cochlea, for instance, has three turns, more than in *Paroodectes*, and is relatively broad, with a basal turn diameter of 0.63 mm, similar in size to some extant species that are twice as large in body size. This indicates that *Messelogale* had an acute sense of hearing.

The postcranial skeleton of *Messelogale* shows features that reveal its scansorial (scrambling through low branches and undergrowth) locomotor pattern. The tail is relatively short for a carnivoraform, being shorter than the rest of the body and head combined and, like in the other three Messel carnivorous taxa, was not prehensile. Such a configuration is seen in extant carnivores that are partially arboreal. Like all of the Messel carnivorans, no known skeleton of *Messelogale* preserves evidence of a baculum (penis bone), which tends to be prominent in living Carnivora, indicating either that these taxa lacked bacula or that all specimens are females.

The forelimb of *Messelogale* shows similarities to modern carnivores adapted to climbing (e.g., African Palm Civet, *Nandinia binotata*), as well as to arboreal and terrestrial quadrupedal walking as in certain genet species (Taylor 1974). Like these mammals, its hands were plantigrade and the whole forelimb displays adaptations for increased mobility such as the ability to pronate and supinate (rotate) the forearm, and enlarged attachment surfaces on the humerus and carpal bones to accommodate strong flexor and extensor muscles of the hand. Moreover, enlarged processes on the scapula (shoulder blade) for attachment of flexor muscles of the arm as well as the elongated and angled olecranon on the ulna enabled powerful shoulder and arm movements.

The hind limb of *Messelogale* shows similarities to modern carnivores adapted to climbing, but not as much as the forelimb, and are rather comparable to genets that spend as much time on the ground as they spend in trees (Taylor 1976). The thumb and big toe of *Messelogale* show enlarged articulation surfaces, suggesting some sort of manipulation abilities, but they are too short to be able to grasp large branches. Finally, claws could be retracted relatively far on the dorsal aspect of the fingers, a feature typical of modern felids (cats) that enhances their ability to climb and to defend themselves and at the same time protects the claws when not in use (when retracted).

Eomanis waldi: the oldest pangolin

The holotype of *Eomanis waldi,* even though crushed and distorted, was the first articulated and substantially complete skeleton of a fossil pangolin to be described (Storch 1978). Several better-preserved specimens of *Eomanis* have been found since the holotype (Fig. 12.7.7, 12.7.8), and all exhibit anatomical specializations characteristic of pangolins, including a tubular, edentulous (lacking teeth) skull with very slender mandibles; a robust, fossorially adapted skeleton with relatively short limbs and large claws on the terminal phalanges (larger on forelimb and middle digits); and the presence of small, imbricating keratinous (horny) epidermal scales like those that cover extant pangolins (Storch 1978; Koenigswald et al. 1981; Storch & Richter 1992a). It has also been suggested that a small tooth-like projection near the front of the lower jaw could be an incipient mandibular prong, a unique hallmark of pholidotans (Storch 2003). Some of these features also characterize palaeanodonts, the sister taxon of Pholidota (Gaudin et al. 2009). In addition, *Eomanis* shares with palaeanodonts several other unusual features suggest-

Fig. 12.7.7: *Eomanis waldi* skeleton (cast) in dorsal view, showing the tubular skull and edentulous jaws, the relatively robust shoulders and limbs, and the large claw-bearing terminal phalanges on the forefeet. Scale: 5 cm.

ing a close relationship, including a thickening ("medial buttress") on the inner surface of the lower jaw (Fig. 12.7.8, inset).

Eomanis was about half the size of most living pangolins, with a total length of about 50 cm (including tail). The tail was shorter and less developed than in most living pangolins, and had only about half as many vertebrae (24) as in the living Long-tailed Pangolin, *Uromanis tetradactyla*. The shorter and weaker tail of *Eomanis* suggests it was mainly terrestrial, like some extant pangolins. The claw-bearing terminal phalanges of the forefeet are markedly larger than those of hind feet and are elongate, shallow, and only slightly curved, closely resembling those of living armadillos. Such claws are well suited for digging and may have been used to tear open insect mounds similar to pangolins today that often prey on termites found in mounds (see Chapter 7 for discussion of Messel insects).

The anatomy of *Eomanis* – in particular the edentulous jaws and fossorially adapted skeleton – indicates that, like present-day pangolins and anteaters, it was myrmecophagous, adapted to consume only flightless social insects such as ants and termites (see Chapter 7). Debris found in the gut contents of

Fig. 12.7.8: Isolated skull of *Eomanis waldi*. The mandible shows the presence of a medial buttress (arrow) as in palaeanodonts. Scale: 1 cm.

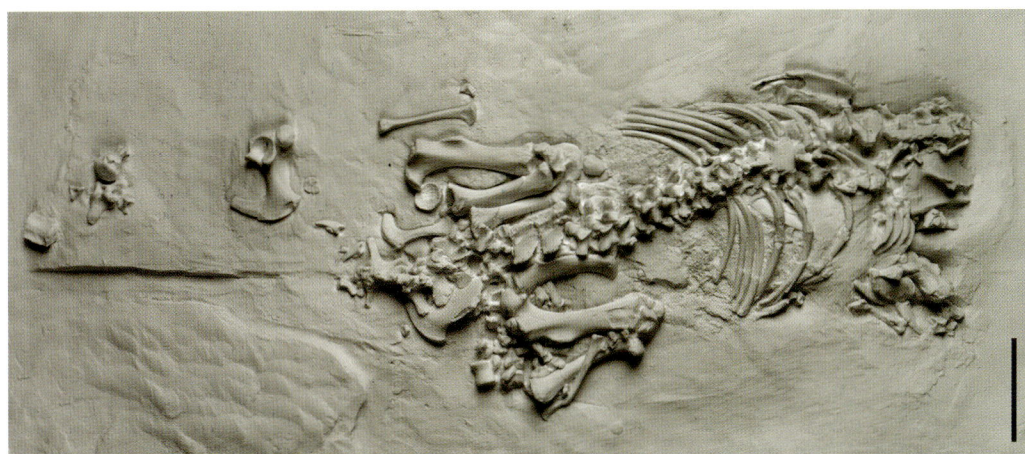

Fig. 12.7.9: Holotype of *Euromanis krebsi* (cast). Scale: 5 cm.

several skeletons of *Eomanis* includes sand, which helps to break down the chitin of ingested insects in living pangolins, and in one specimen, insect chitin (Koenigswald et al. 1981; Storch & Richter 1992a). Oddly, most specimens also contain leaf fragments and other plant debris but no insect remains. Since *Eomanis*, being edentulous, lacked the ability to comminute leaves, one possible explanation for this paradox is that *Eomanis* may have maintained a partly herbivorous diet by intentionally raiding lines of leafcutter ants (Storch & Richter 1992a). However, these ants are not present in the Messel insect assemblage, so the presence of this plant debris remains mysterious (Chapter 7).

Euromanis krebsi: the headless anteater

Known from a single skeleton (Fig. 12.7.9) lacking the skull, parts of the forelimbs, and tail (Storch & Martin 1994), *Euromanis krebsi* is somewhat larger than *Eomanis waldi* and smaller than *Eurotamandua joresi*. The missing parts, together with the immature state of the skeleton, complicate efforts to establish its identity.

Euromanis has
    pangolins

Like *Eomanis* and *Eurotamandua*, *Euromanis* is primitive in having unfissured terminal phalanges and a femoral third trochanter located proximal to midshaft. In other fossil pholidotans, the third trochanter is situated distal to midshaft. Present-day manids differ in having deeply fissured terminal phalanges, and they lack a third trochanter because the superficial gluteal muscle inserts near the distal end of the femur.

The scapula of *Euromanis* possesses a very prominent spine, which terminates in an elongate acromion process that overhangs the shoulder joint, and the posteroventral border has a sharp, elevated margin. These features (also seen in *Eomanis* and *Eurotamandua*) contrast with later pangolins but resemble palaeanodonts. Minor differences in foot bones and some other elements separate *Euromanis* from the other two genera (Gaudin et al. 2009).

Based on the essential similarity of the incomplete skeleton of *Euromanis* to that of *Eomanis* and *Eurotamandua*, it can be assumed that its lifestyle and behavior were similar. Its powerfully built skeleton is evidence that *Euromanis* was a capable digger, and it was presumably myrmecophagous, though discovery of a skull is required to confirm this.

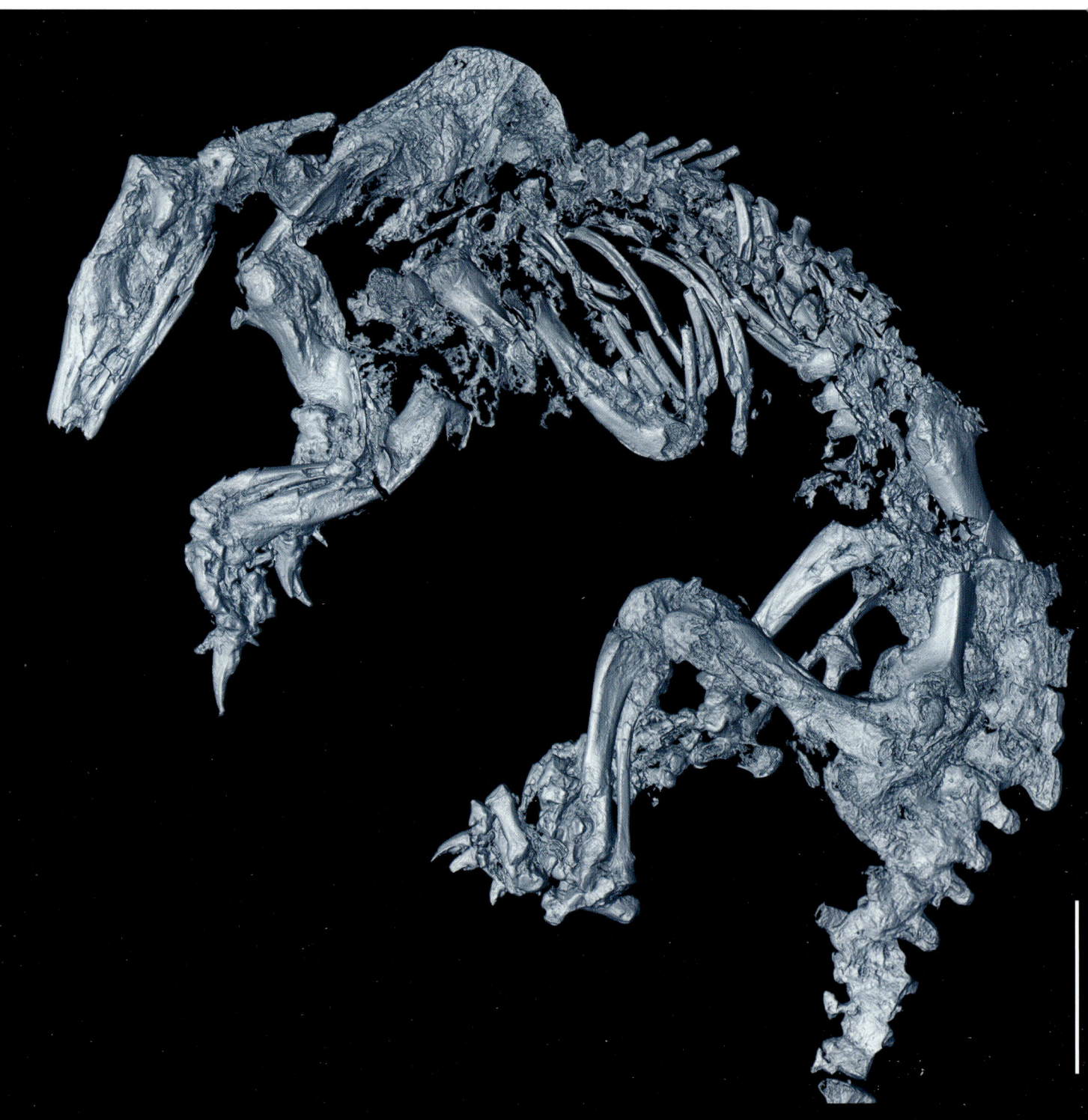

Fig. 12.7.10: 3-D model of the holotype skeleton of *Eurotamandua joresi,* showing the animal from its hidden left side. Scale: 10 cm.

Eurotamandua joresi: a doubtful South American

Eurotamandua (Storch 1981) was first described as an anteater (Myrmecophagidae), based on similarities to living anteaters that include a tubular, edentulous skull, a powerful, fossorially adapted skeleton and details of forelimb anatomy as well as the presence of accessory (xenarthrous) intervertebral joints in the lumbar region, the hallmark of the order Xenarthra, to which anteaters belong. As we have seen, these features, except for the last one, also characterize pangolins.

Eurotamandua is still known only from the holotype skeleton from Messel (Fig. 12.7.1, 12.7.10) and two isolated forelimb bones (humerus and ulna) from Geiseltal (Storch & Haubold 1989). Because Xenarthra are otherwise confined to the New World, the potential relationship of *Eurotamandua* to anteaters has been carefully scrutinized, and challenged, by several authors (Shoshani et al. 1997; Szalay & Schrenk 1998; Rose 1999; Gaudin et al. 2009). Specific details of the forelimb – humeral shaft, crests, processes, and distal articulation, as well as shape of the radius, and metacarpal and phalangeal anatomy – are much closer to those of palaeanodonts and pangolins than to those of xenarthrans (Rose & Emry 1993; Rose 1999). Most significantly, upon reassessment, the evidence for xenarthrous joints appears to be spurious (Fig. 12.7.11) and was likely based on preservational artifacts (Szalay & Schrenk 1998; Storch 2003). In addition, evidence is now available that suggests that *Eurotamandua* possessed a medial buttress on the dentary, as in palaeanodonts (and *Eomanis* as well), and might have had an incipient mandibular prong, a feature found in living pangolins and also in the North American fossil pangolin *Patriomanis* (Fig. 12.7.12). The weight of evidence at present favors the interpretation that *Eurotamandua*, like *Eomanis* and *Euromanis*, is a basal pholidotan (Gaudin et al. 2009). With the reinterpretation of *Eurotamandua* as a pholidotan, the dilemma of explaining how a xenarthran reached Europe from South America in the Eocene is avoided.

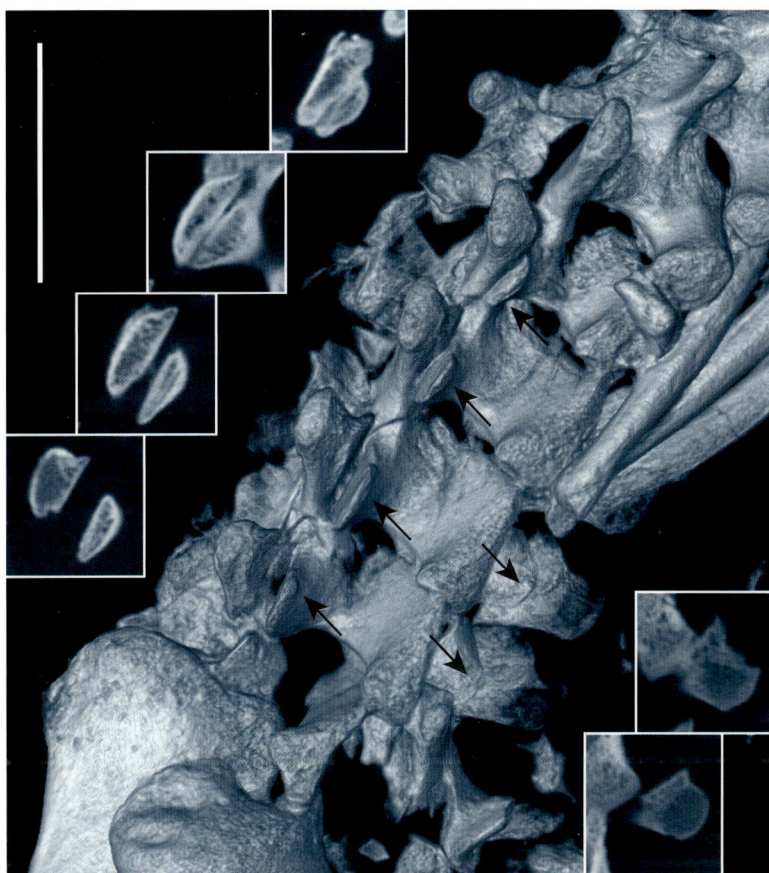

Fig. 12.7.11: Details of 3-D model of *Eurotamandua joresi*, showing the lower thoracic/lumbar region. Insets: slices through the normal articular surfaces corresponding to the position of the arrows. Scale: 2 cm.

Overall body proportions of the three Messel pholidotans are very similar, although *Eurotamandua* is a larger and more robust animal than the other two. It is possible that the three Messel pholidotans are more closely related than has been recognized, and that apparent differences are exaggerated by size and ontogenetic differences as well as preservation (note that virtually all of the Messel pholidotan skeletons known have at least some unfused epiphyses indicating that none were fully grown). In any case, their similar skeletal anatomy indicates that they had similar lifestyles. Like *Eomanis*, *Eurotamandua* was certainly a myrmecophagous animal, which probably used its enlarged central foreclaws (closely resem-

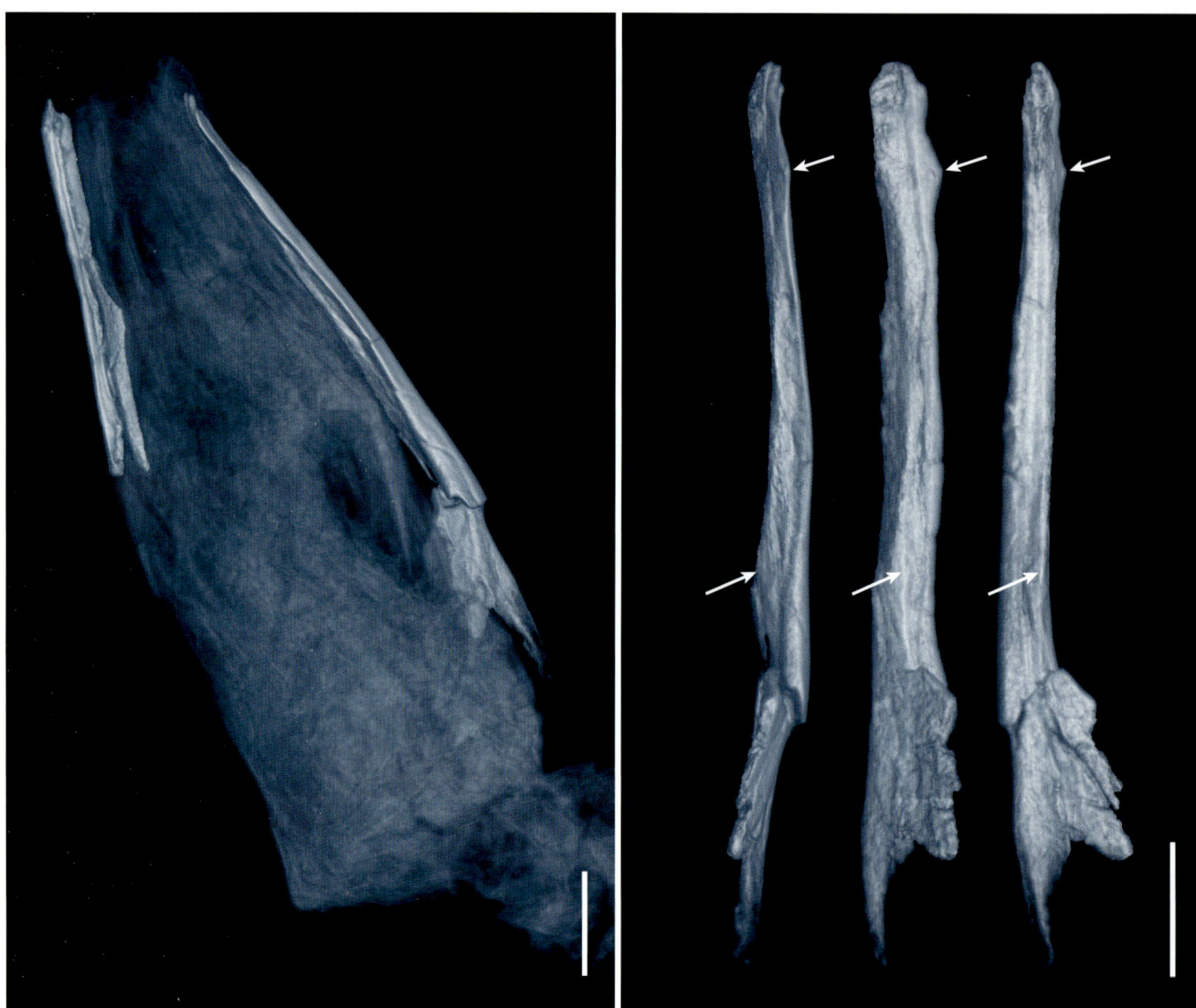

Fig. 12.7.12: Translucent μCT of the head of *Eurotamandua joresi* (left), with both mandibles. Right mandible (right) in three different medial views (from left to right: +45°, 0° = medial, -45° in clockwise rotation). Upper arrows show the prong, and lower arrows show the position of a medial buttress, similar to palaeanodonts. Scales: 1 cm.

bling those of *Eomanis*) to tear open insect nests and mounds. Fossilized stomach contents reveal the presence of sand grains and leaf fragments, as in *Eomanis*, but also a considerable amount of insect cuticle (Storch & Richter 1992b). The tail of *Eurotamandua*, albeit robust, is shorter than head and body length, more consistent with terrestrial than arboreal habits, although occasional climbing cannot be ruled out.

Paleogeography

As skeletal remains from European "miacid" Carnivoraformes are scarce, the single specimen of *Paroodectes feisti* is very important for elucidating carnivoraform evolution and paleobiogeography. According to Springhorn (1980) *Paroodectes* shows significant resemblance to middle Eocene North American basal carnivoraforms such as *Oodectes*, *Vulpavus* and *Miacis* in terms of dental and postcranial characters as well as limb proportions. These observations support a relationship of the middle Eocene carnivoraform fauna of Europe and North America. Thus, the common ancestor of *Paroodectes* and *Oodectes* may have migrated from North America to Europe before the connecting land bridges disappeared in the late early Eocene (McKenna 1983; Brikiatis 2014). However, according to Solé et al. (2016a), a faunal dispersal from Asia to Europe – and only later to North America – during the Paleocene and Eocene was also possible via a land connection across the Turgai Strait (Chapter 2).

European hyaenodontans with a similar age as *Lesmesodon* are best known from Geiseltal (*Matthodon*, *Oxyaenoides*, *Eurotherium*). However, these animals are much larger than *Lesmesodon* and are more distantly related within Proviverrinae (Solé 2012). *Lesmesodon* seems to be most closely related to *Proviverra* and *Leonhardtina* among European proviverrines (Solé et al. 2014). North American taxa contemporary to *Lesmesodon* include sinopines (*Sinopa* and *Tritemnodon*) and limnocyonines (*Limnocyon* and *Thinocyon*). Limnocyonines can easily be distinguished from *Lesmesodon* by the lack of upper and lower third molars in the North American taxa. *Lesmesodon* also differs from all of these North American taxa in retaining a double-rooted first lower premolar. It appears likely that *Lesmesodon* was an endemic taxon that derived from a small, as yet unknown proviverrine at some point in the early Eocene.

The two German fossiliferous sites of Geiseltal and Messel have no carnivorous species in common (Lange-Badré & Haubold 1990). However, *Quercygale* from Geiseltal (and other European sites) is probably morphologically the closest carnivoraform genus to *Messelogale*, despite their size differences (Solé 2014). *Quercygale* is characterized by the absence of an upper third molar, which is probably also the case in *Messelogale*. An Asian origin has been suggested for *Quercygale* (Solé 2014). *Quercygale* is also very important phylogenetically, because it is the closest carnivoraform to the crown-group Carnivora (Spaulding & Flynn 2012). If the affinities with *Quercygale* are confirmed, *Messelogale* would represent one of the most derived stem Carnivora, and its ancestry could likely be traced to Asia.

Living pangolins occur in tropical and subtropical environments in Africa and Asia, and fossil pangolins are very rare. *Eomanis* pushed the record of pangolins back to the middle Eocene and, together with late Eocene *Patriomanis* from western North America, helped to show that pangolins once had a much wider geographic distribution.

The Messel skeletons are still the oldest known pholidotans in the world, and exactly when and where the order evolved is not known. Some insight can be gained from the probability that Paleogene palaeanodonts are the sister taxon of Pholidota. Palaeanodonts are mainly North American, though two significant, recent discoveries highlight the possibility of dispersal around the beginning of the Eocene – a palaeanodont jaw from lower Eocene beds of Ellesmere Island, Arctic Canada (Rose et al. 2004), near the Thule land bridge (McKenna 1983; Brikiatis 2014), and a jaw and a few foot bones of *Palaeanodon* from the earliest Eocene of France (Gheerbrant et al. 2005). Current evidence favors an origin of Pholidota from a palaeanodont similar to *Palaeanodon*, although a more complete fossil record is needed to confirm or reject this hypothesis.

Chapter 12.8
The Advent of Even-toed Hoofed Mammals

Thomas Lehmann, Irina Ruf

Even-toed hoofed mammals (Artiodactyla) are among the most diverse and abundant ungulates, especially their ruminant representatives (bovids, deer, giraffes, goats, etc.). Their success was acquired gradually, to the detriment of the perissodactyls (horses, rhinos, tapirs), which had dominated the large herbivore ecological niche well into the Eocene (Foss & Prothero 2007). Today, artiodactyls dominate this niche on several continents. Artiodactyls are called "even-toed" because the axis of symmetry of the foot runs between the strengthened third and fourth digits (Fig. 12.8.2), whereas the other digits are reduced. However, the chief synapomorphy of artiodactyls and their most iconic innovation is certainly the "double-pulley" astragalus (ankle bone) (Fig. 12.8.1, inset), which is already present not only in the oldest known representative (*Diacodexis*), but also in the earliest cetaceans (whales), making them – as also supported by genetic evidence – members of this clade (Nikaido et al. 1999; Gingerich et al. 2001; Thewissen et al. 2001).

The four species of Artiodactyla found at Messel belong to the families Dichobunidae and Choeropotamidae (Fig. 12.8.3). Along with the oldest-known artiodactyls, the Diacodexeidae, they are often grouped together as Dichobunoidea, which is, however, regarded as a paraphyletic group of stem artiodactyls (Theodor et al. 2007). Dichobunoids are nonetheless crucial for understanding the origin and early evolution of Artiodactyla. Within Dichobunidae, *Aumelasia* and *Messelobunodon* are members of the subfamily Dichobuninae, whereas *Eurodexis* belongs to the Eurodexeinae (Theodor et al. 2007). Finally, *Masillabune* belongs to the Choeropotamidae (Erfurt & Métais 2007).

Messelobunodon: a primitive even-toed ungulate

The holotype and only known specimen of *Messelobunodon schaeferi* is more than 60 cm long (including tail), which corresponds to a medium-sized dichobunine (Franzen 1981) (Fig. 12.8.4). It was a slender animal with elongated hind limbs and a very long tail (Fig. 12.8.5). Compared to modern artiodactyls, its skeleton shows many primitive features, e.g., its forearm, leg bones, and two ankle bones (cuboid and navicular) are not fused (Theodor et al. 2007). The feet bear four toes, whereas the hands have

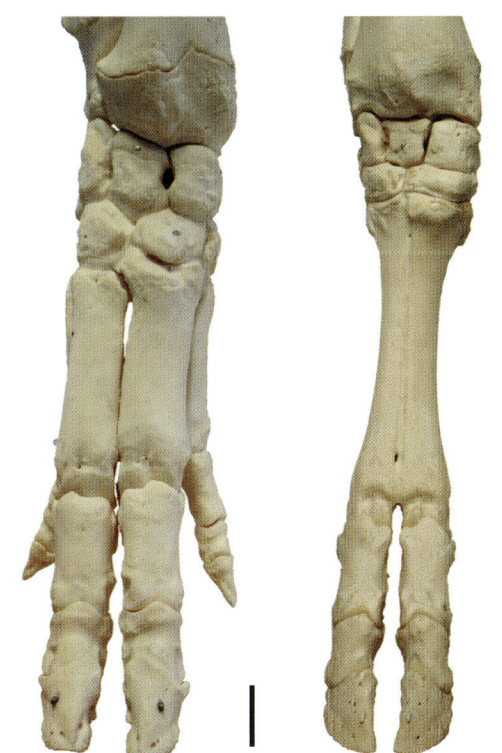

Fig. 12.8.2: Forefoot of a pig (*Sus scrofa*) (left) with strengthened but not fused fingers III and IV. Forefoot of an elk (*Cervus canadensis*) (right), showing the derived fusion of the enlarged fingers III and IV in what is called a "cannon bone." Scale: 2 cm.

Fig. 12.8.1: *Eurodexis* sp. The astragalus (insert; arrow indicates the viewing direction; scale: 1 cm) shows the double-pulley joints typical for Artiodactyla. Scale: 10 cm.

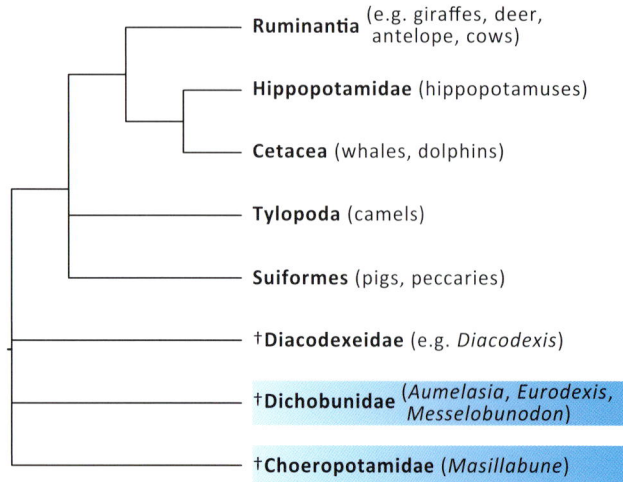

Fig. 12.8.3: Simplified phylogeny of Artiodactyla and selected stem taxa. Taxa known from Messel are highlighted in blue.

a reduced thumb. The hind limb is longer than the forelimb, which is mostly due to the lengthening of the tibia (even longer than the femur) and metatarsal bones of the foot. These proportions are closest to those seen in modern mouse-deer (*Tragulus*). The calcaneus (heel bone) shows a very long heel with deep muscular insertion grooves on the sides. Such a configuration helps to stabilize the hock and boosts the action of the extensor muscles of the foot. Thus, *Messelobunodon* was a fleet-footed forest-dweller that was able to sprint and jump quite well.

The skull of *Messelobunodon* is about 11 cm long and has a long snout. Franzen (1981) described the acoustic canal as a simple opening, unlike in modern artiodactyls, which possess an external ossified ear canal and an ossified auditory bulla. The dentition shows three incisors (lost in ruminants), a canine, four premolars and three molars. The diastema (gap) between the second and third upper premolars is particularly diagnostic for *Messelobunodon* (Franzen 1983). The low-crowned molars with bulbous cusps

Fig. 12.8.4: Holotype of *Messelobunodon schaeferi*. Note the long limbs and tail. Scale: 10 cm.

Fig. 12.8.5: Life reconstruction of an adult *Messelobunodon schaeferi* with calf foraging for fungi on the Messel forest floor.

(i.e., bunodont) are consistent with the diet shown by gut contents: fungi, leaves, and fruits (Box 12.8.1). The upper molars of *Messelobunodon* have only three major cusps (plus two accessory ones) and are missing a fourth one (the hypocone) in the lingual posterior corner, an innovation that is widespread among placental mammals.

Aumelasia: a cousin from France

Aumelasia was first described from France on the basis of several jaws and teeth (Sudre 1980; Sudre et al. 1983). Three sub-adult skeletons from Messel are referred to *Aumelasia gabineaudi* (Franzen 1988) (Fig. 12.8.6). *Aumelasia* is a primitive dichobunine of medium size (ca. 60 cm long, including tail), with long limbs and a very long tail. As in *Messelobunodon*, the postcranial skeleton is largely primitive. The radius and tibia of *Aumelasia* are proportionally shorter than in *Messelobunodon*, more similar to pigs and to *Diacodexis* (Franzen 1988; Thewissen & Hussain 1990). Since the length of the radius increases with running ability (Howell 1944), *Aumelasia* was probably not as fast as *Messelobunodon*. Overall, the skeleton of *Aumelasia* is suggestive of an agile and versatile animal, which was perhaps living in the undergrowth (Franzen 1988). The preserved gut contents suggest a browser (Franzen & Richter 1992).

The skull of *Aumelasia* is massive, shorter than in *Messelobunodon* but with a similarly long snout. It differs however, in having pronounced crests on the top of the skull, as well as a broad zygomatic arch (Franzen 1988). The mandible is correspondingly massive, with deep flanges for muscle attachment but no hook-like posterior projection. These features suggest strongly developed chewing muscles. *Aumelasia* retains the same number of teeth as *Messelobunodon*, but with fewer gaps. The upper molars lack a hypocone, but display a large central accessory cusp on the labial side (mesostyle), diagnostic of the genus.

Fig. 12.8.6: Completely extracted specimen of *Aumelasia* cf. *gabineaudi*. The skull and neck have been virtually repositioned in the correct anatomical position. Scale: 5 cm.

Eurodexis: the smallest artiodactyl from Messel

The second specimen of *Messelobunodon* from Messel (Fig. 12.8.1) differed significantly from *Me. schaeferi* (Franzen 1983), e.g., in its shorter body size (estimated length of 50 cm with tail), and was later moved to the genus *Eurodexis*, after the revision of the species *E. ceciliensis* from Geiseltal, but a species attribution is still pending (Franzen & Krumbiegel 1980; Erfurt & Sudre 1996).

The limbs of *Eurodexis* are slender and elongated, the hind limb is longer than the forelimb, the tibia is longer than the femur, and the tail is very long. Remarkably, radius and tibia are proportionally very long in comparison to other Messel artiodactyls. This pattern is comparable to the modern Steinbock (*Capra ibex*) and Roe Deer (*Capreolus capreolus*) and suggests an animal highly adapted to running and jumping (Erfurt & Métais 2007). Moreover, rotation of the forearm was reduced, though radius and ulna are not fused. The femur shows several plesiomorphic features, e.g., a greater trochanter (attachment of hip muscles) that does not project over the femoral head as in modern artiodactyls. More surprising is the presence of a small third trochanter on the femoral shaft (Fig. 12.8.7), a primitive character that sets *Eurodexis* apart from all other artiodactyls from Messel.

Skull proportions as well as the shape of the mandible, with a hook-like posterior projection, are comparable in *Eurodexis* and *Messelobunodon*. The upper molars of *Eurodexis* are more crescentiform (i.e., bear distinct crests on their central and external cusps) than in *Messelobunodon*. The gut contents show that Messel *Eurodexis* consumed fruit (Box 12.8.1), which is in accordance with its tooth morphology (Theodor et al. 2007). The first and second upper molars of *Eurodexis* also have an hypocone, unlike any other Messel artiodactyl (Erfurt & Haubold 1989) (Fig. 12.8.7).

Fig. 12.8.7: Top: Cranium of *Eurodexis* sp. revealing the small hypocone on the second upper molar (arrow). Bottom: Femur showing the third trochanter (arrow). Scale: 1 cm.

Masillabune: a robust browser

Masillabune martini is a medium-sized choeropotamid, about 55 cm with tail (Tobien 1980; Gentry & Hooker 1988) (Fig. 12.8.7). As in other early artiodactyls, its forelimb is shorter than its hind limb, and its postcranial skeleton is fairly primitive (see above). The femur of *Masillabune* is proportionally the longest among early artiodactyls. It is unknown whether a third trochanter was present on the shattered femurs of the holotype (Tobien 1985), but it is clearly absent in the closely related form from Geiseltal (Erfurt & Haubold 1989). The loss of the third trochanter appears to be a convergent trait among artiodactyl lineages. In contrast to every known early artiodactyl, the tibia of *Masilabune* is reported to be shorter than the femur (Tobien 1985), with proportions similar to modern pigs. Finally, the tail is shorter than in other Messel artiodactyls. Overall, the skeleton shows adaptations for terrestrial locomotion but probably not for running and jumping (Tobien 1985).

The skull of *Masilabune* is almost as long as that of *Messelobunodon*, but its snout is much shorter. *Masillabune* most likely has the complete placental dental count (see above). The tooth rows are very compact, with molars longer than in *Messelobunodon* and lacking spaces between the premolars. The lower first premolar and upper canine are strong and caniniform, whereas the lower canine is incisiform and appressed to the incisors (Fig. 12.8.8). The upper molars of *Masillabune* show bulbous cusps, some of which are further connected by crests and form a specific pattern (bunoselenodont) that evokes the more derived state (selenodont) seen in modern ruminants. Dental wear and gut contents suggest that *Masillabune* was a browser (Box 12.8.1).

Box 12.8.1: Fungi, leaves, and fruits on the menu

Artiodactyls are rare elements of the Messel fauna. Could this be explained by the absence of adequate or sufficient food resources around the Messel Lake? The gut contents of *Messelobunodon* consisted primarily of fungi (or lichen) fragments (top; scale: 100 μm), along with some rotten leaves (bottom; scale: 100 μm), and seeds. This rotten foodstuff was probably taken in as by-product while ingesting fungi from the forest floor (Richter 1987). *Eurodexis* and Aumelasia had a fruit-based diet (as revealed by the recovered seeds and fruit pulp), with the addition of fungi. Like *Messelobunodon*, they probably foraged on the forest floor, because the pulp of some of these fruits had started to decay (presence of saprophile fungi; white tubular structures, right) (Franzen & Richter 1992). Laurel (Lauraceae) leaf fragments, a plant family abundant at Messel (Chapter 6), were found in the gut of *Masillabune* (Tobien 1980). The absence of any signs of decay on these leaves suggests that, contrary to the other Messel artiodactyls, *Masillabune* picked fresh leaves directly from the plant. A sufficient food supply was thus available to Messel artiodactyls. The cause of their rarity must be looked for elsewhere.

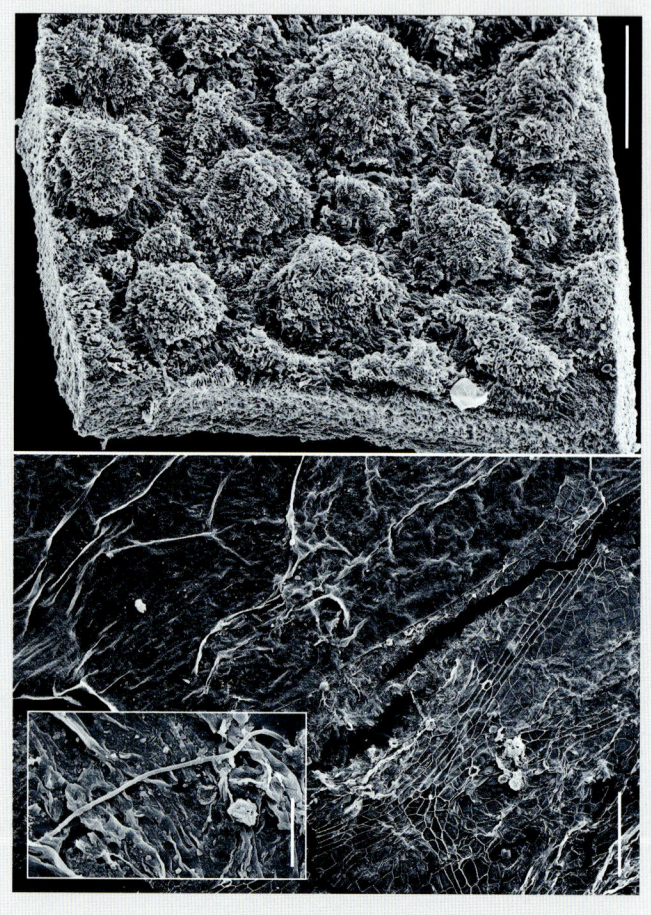

Paleobiogeography and Paleoenvironment

By the middle Eocene, when Europe was separated from the other continents (Chapter 2), Dichobunidae were isolated and underwent a rapid radiation that resulted in unique dental and postcranial adaptations (Theodor et al. 2007). This apparent endemism lasted until the Eocene-Oligocene boundary and the "Grande Coupure," when climate change and/or competition with more derived taxa arriving from Asia led to the extinction of several Dichobunidae. To date, *Messelobunodon* is only known from Messel and by fragmentary material from Geiseltal. Conversely, *Aumelasia* is known from several species in France and Germany. *Eurodexis* is known from Germany and France, but Eurodexeinae also have representatives in Spain (Theodor et al. 2007). From the early Eocene until their extinction near the Eocene-Oligocene boundary, Choeropotamidae remained geographically restricted to Europe (Erfurt & Métais 2007). Though only one choeropotamid species lived in Messel, several were recorded from Geiseltal, thus documenting the radiation of the family, which reached its highest diversity around 40 million years ago (Erfurt & Métais 2007).

The artiodactyls from Messel document the first stages toward hegemony of artiodactyls over the large herbivore niche. If all early artiodactyls were

Fig. 12.8.8: Holotype of *Masillabune martini*. The canine-shaped front teeth are actually the upper canine and the lower first premolar. Scale: 10 cm.

more or less adapted to running, at least two morphologically different groups emerged from the European Eocene radiation: small taxa (such as *Eurodexis*) with crescentiform molars (probably mainly insectivorous, frugivorous); and stockier taxa (such as *Aumelasia*, *Masillabune*, *Messelobunodon*) with bunodont teeth (probably mainly omnivorous, herbivorous) (Theodor et al. 2007). Messel vividly documents this ecologically diverse artiodactyl community, although the low number of recovered specimens (fewer than 10) compared to the numerous horses (Chapter 12.9) illustrates the dominance of the latter over the herbivorous ecological niche. This has been interpreted as evidence for a solitary lifestyle in early artiodactyls in contrast to an already gregarious life in herds, as found in horses (Franzen & Richter 1992). On the other hand, a niche partitioning argument could be made, since Messel artiodactyls are much smaller than their perissodactyl counterparts and might have occupied different ecological niches (Tobien 1985).

Chapter 12.9
Odd-toed Ungulates –
Early Horses and Tapiromorphs

Jens Lorenz Franzen

Among the ungulates, the odd-toed ungulates (Perissodactyla) are characterized by a particularly well-developed central (3rd) digit on their fore- and hind limbs (Fig. 12.9.2). This differentiates them from the even-toed ungulates, in which two central digits, the 3rd and 4th, are both strongly developed (Chapter 12.8). Modern representatives of the Perissodactyla include the horses, tapirs, and rhinoceroses. They are divided into two subgroups, the Equoidea (horse-like animals) and the Tapiromorpha (tapirs and rhinoceroses).

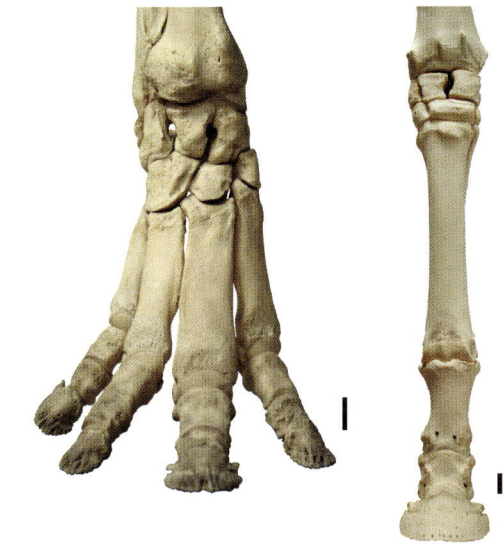

Fig. 12.9.2: In odd-toed ungulates, the central or third digit of the originally five-toed extremities is bigger than the others (example: forefoot of a tapir, left, and a horse, right). Scale: 2 cm.

The early horses (Equoidea)

According to certain phylogenetic analyses, in the early Eocene the Equoidea split into the North American Equidae and the European Palaeotheriidae (Hooker 1994; Froehlich 2002). However, these analyses were mainly based on characteristics of the skull and dentition. The inclusion of additional characters of the body skeleton (Franzen 1989) leads to the phylogenetic tree advocated here. This places several purported Palaeotheriidae with the Equidae, while the true Palaeotheriidae only immigrated into Europe in the middle Eocene after the Messel era. Their area of origin is unknown. Their physical design differs markedly from that of the early Equidae. Some of the differentiating traits, e.g., the fact that in the "true" Palaeotheriidae the metacarpals are always longer than the metatarsals, while this ratio is reversed in all Equidae, make it impossible to trace both the Palaeotheriidae and the Equidae back to the same origin. Undoubtedly, the fossil horses from Messel do

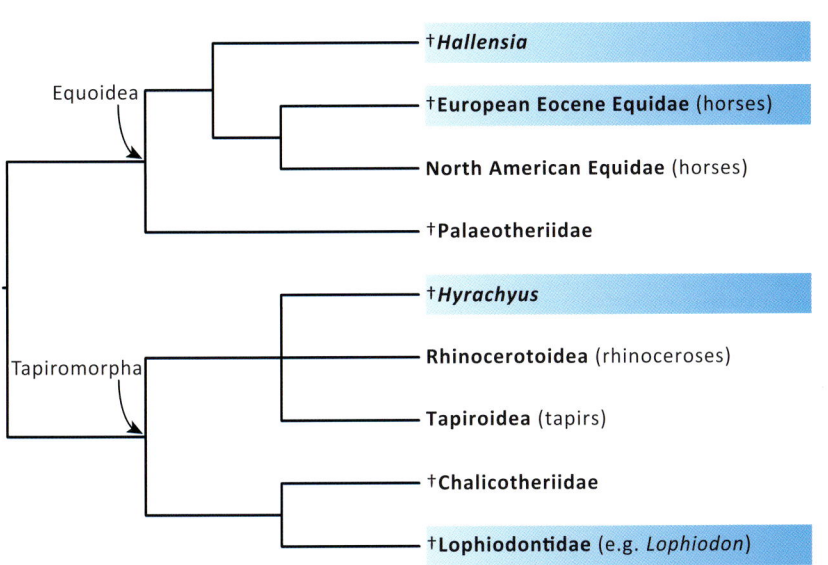

Fig. 12.9.3: Simplified phylogeny of the odd-toed ungulates (Perissodactyla). Groups documented in Messel are marked in blue.

Fig. 12.9.1: Pregnant mare of *Eurohippus messelensis*. Scale: 10 cm.

Chapter 12.9 Odd-toed Ungulates – Early Horses and Tapiromorphs

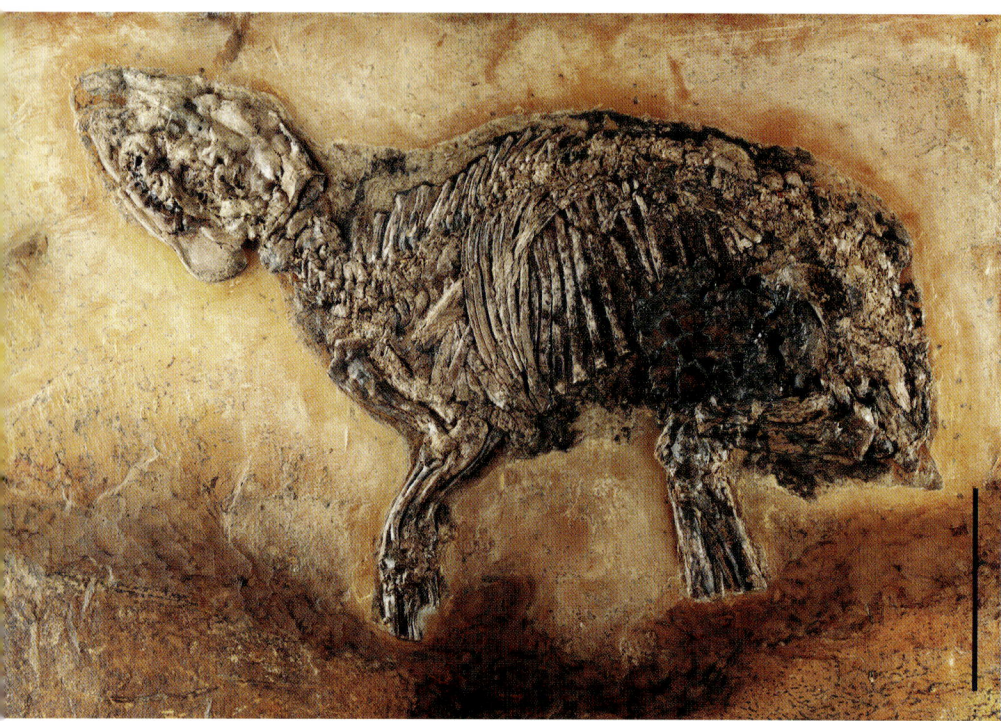

Fig. 12.9.4: The first discovery of an early horse by Senckenberg from the year 1975 (*Eurohippus messelensis*). Scale: 10 cm.

Fig. 12.9.5: Skeleton of the "Hessian" early horse *Propalaeotherium hassiacum*. Scale: 10 cm.

not represent Palaeotheriidae. This is not only evidenced by the contrasting development of the metacarpals and metatarsals, but also by the pelvic region. The question remains, however, whether all species from Messel can be assigned to the Equidae. The genus *Hallensia* represents an independent branch among the Equoidea, which cannot be traced back to the oldest known Equidae (Franzen 1990).

The early horses are the "icons" of the Messel fossil site. Jaw bones with teeth as well as individual teeth of fossil horses count among the earliest mammalian discoveries from this location (Haupt 1911). They offered the first indication as to the age of the Messel deposits, since these early horses are only known from the Eocene epoch. Already during the former mining days, four skeletons of fossil horses were unearthed, although they are rather incomplete due to the poor retrieval conditions. Since the beginning of scientific excavations, a total of 63 skeletons has been found to date. Not all are complete, but their number is so high that the Messel Pit must now be considered the most important discovery site of early fossil horses in the world.

Meanwhile, four species of early horses have been documented from Messel. With their strongly arched back, relatively short legs and a short neck, they are more reminiscent of a duiker antelope than a horse. Moreover, their front legs still show four, and the hind legs three, hooves. That means that these

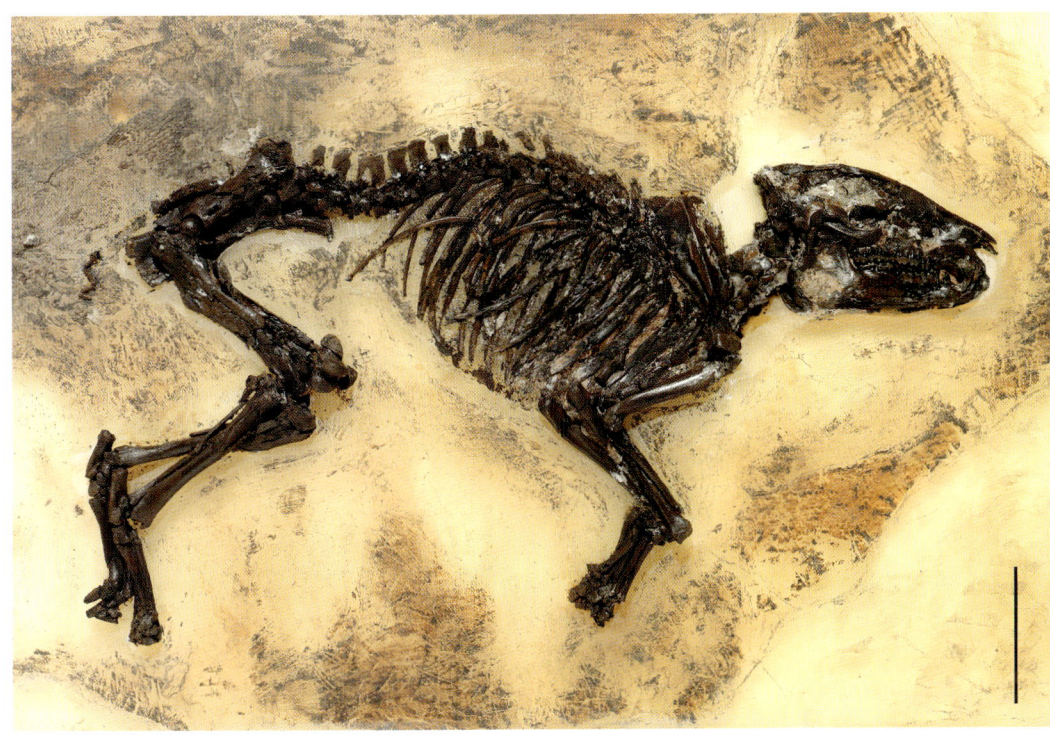

Fig. 12.9.6: First skeleton of *Propalaeotherium voigti*. Scale: 10 cm.

early horses each bore a total of 14 hooves, compared to 4 in the modern horses! In addition, unlike in modern horses, the ulna and radius of the forelegs and the tibia and fibula of the hind legs were not yet fused in the early horses from the Messel Pit.

The most common fossil horse from the Messel Pit, with a total of 45 skeletons to date, is the smallest species, *Eurohippus messelensis* (Fig. 12.9.1, Fig. 12.9.4, Fig. 12.9.12). With a shoulder height of 30-35 cm, it was about the size of a fox terrier. The "Hessian" early horse, *Propalaeotherium hassiacum* (Fig. 12.9.5), which is represented by 14 skeletons, was noticeably larger, reaching the size of a German shepherd. It was rather massively built. *Propalaeotherium voigti* was only slightly smaller, but clearly more slender than *P. hassiacum*. *Propalaeotherium voigti* is now represented from Messel by a complete skeleton (Fig. 12.9.6), which significantly increases our knowledge of this species, discoveries of which had previously been restricted to a skull and a skull fragment from the former mining days (Franzen 2007b). Among the early horses from Messel, *Hallensia matthesi* shows the most primitive dentition structure (Franzen 1990). This species is known from two skeletons (Fig. 12.9.7). They show that *Hallensia* also possessed such primitive skeletal traits that the genus cannot be derived from any of the oldest known early horses. Instead, *Hallensia* reveals similarities to specimens from the lower Eocene of India, thereby offering an indication of the horses' initial place of origin (Rose et al. 2014).

The life of the early horses

The unique preservation, both in regard to quantity as well as quality, of the 48-million-year-old early horses from the Messel Pit allows far-reaching conclusions regarding the animals' way of life. Many of these insights would have been deemed impossible only a short while ago. Examples of the particularly high level of preservation in Messel horses include the outline of an external ear (Fig. 12.9.8) and the hair on the tail of *Eurohippus messelensis* (Fig. 12.9.9). In the specimen of *Hallensia matthesi* from Brussels, the outline of the cecum could be identified (Fig. 12.9.10). It is even possible to draw conclusions about the ani-

Chapter 12.9 Odd-toed Ungulates – Early Horses and Tapiromorphs

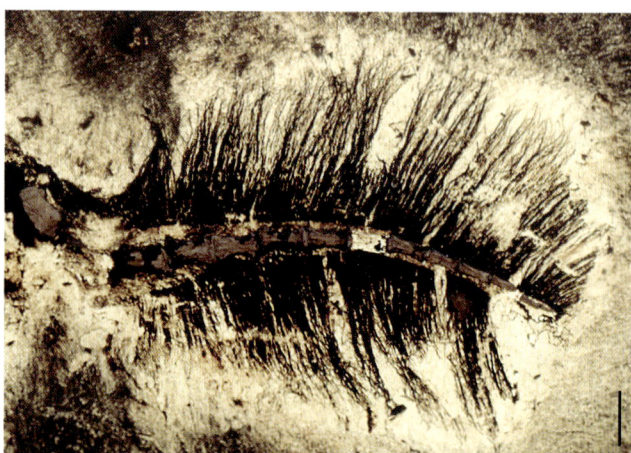

Fig. 12.9.9: Tail hairs of a mare of *Eurohippus messelensis*. Scale: 1 cm.

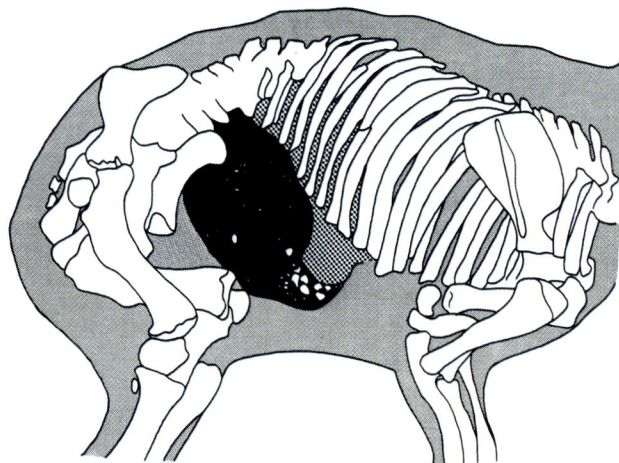

Fig. 12.9.7: Fully grown specimen of *Hallensia matthesi*. Scale: 10 cm.

Fig. 12.9.8: External ear of a *Eurohippus messelensis*. Scale: 1 cm.

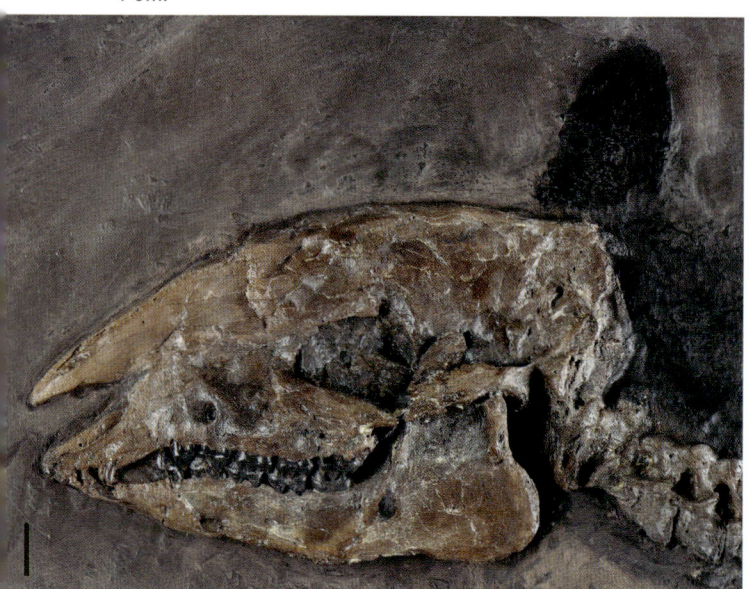

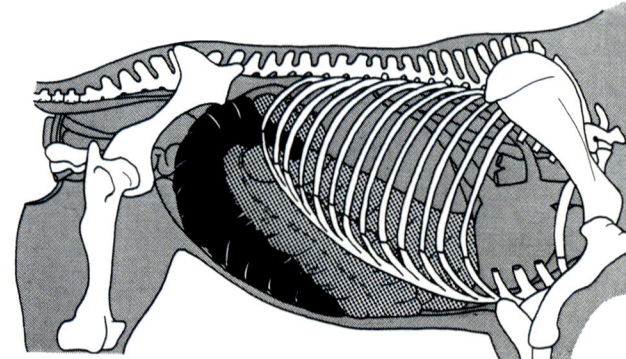

Fig. 12.9.10: Cecum (black) of *Hallensia matthesi* (top), compared to a modern horse (bottom).

mals' social behavior. For example, the large number of skeletons of *Eurohippus messelensis,* compared to other dawn horse species from Messel, indicates that the species lived at least in groups, even if not in herds. Based on its relative abundance, it is also possible that *Propalaeotherium hassiacum* lived in social groups as well (Franzen 2007b). On the other hand, *Propalaeotherium voigti* and *Hallensia matthesi* presumably had a solitary lifestyle, or they lived further from the Eocene Lake Messel, which may account for these species' rarity in the fossil record.

Besides statements regarding the phylogeny, skeletons of different age classes allow insights into the individual development (ontogeny), at least regarding the more common species. Moreover, the completeness of several skeletons made it possible to determine that the sexes already showed the same differences regarding the design of their pelvic region, in particular the width of the pelvic inlet, that are present among modern horses (Fig. 12.9.11). Proof of this came with the discovery of skeletons of pregnant mares (Franzen 2007a, b). The determination of the gender, in turn, allowed conclusions regarding the social behavior, as expressed by the structure of animals that apparently lived in social communities with each other. The equal numeric ratio of the sexes and the approximately equal size of the canine teeth indicate that the early horses from Messel (Franzen 2007a), contrary to those from North America (Gingerich 1981), had not yet developed the dominance of stallions ("harem structure") so commonly found in modern equine herds. This may be due to environmental conditions, since herds that gather around a stallion, as shown by the development of environmental conditions in the early Eocene of North America, apparently only developed in conjunction with wider open landscapes that required improved protection against predators (Gingerich 1981). In the surroundings of the Messel freshwater lake, on the other hand, primeval forests predominated.

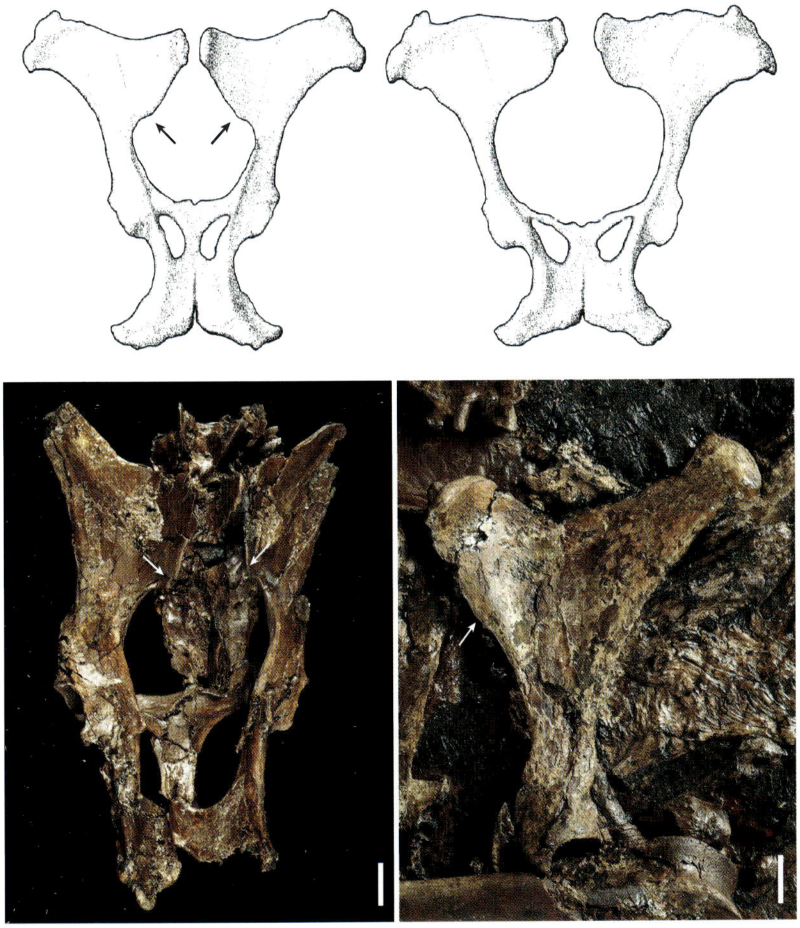

A particularly high level of soft-tissue preservation can be seen in the uteri that were found in several pregnant mares (Franzen et al. 2015). In one case, the uteroplacenta is almost completely preserved, allowing the reconstruction of the fetus' appearance and position (Fig. 12.9.12). In this context, even the broad ligament (ligamentum latum) is visible, which attaches the uterus to the lumbar spinal cord, just as it does in modern-day mares. In another case, the uterus has only been preserved in parts – which, however, even allow the differentiation of separate layers of the uterine wall (Fig. 12.9.13). The fact that all of the eight pregnant mares of *E. messelensis* only bear a single fetus each (Franzen 2007b)

Fig. 12.9.11: Pelvic inlet and sex in early horses from Messel (bottom), compared to modern horses (top). The arrows indicate protrusions of the ilium that restrict the pelvic inlet in stallions (left), as opposed to mares (right). Scale: 1 cm.

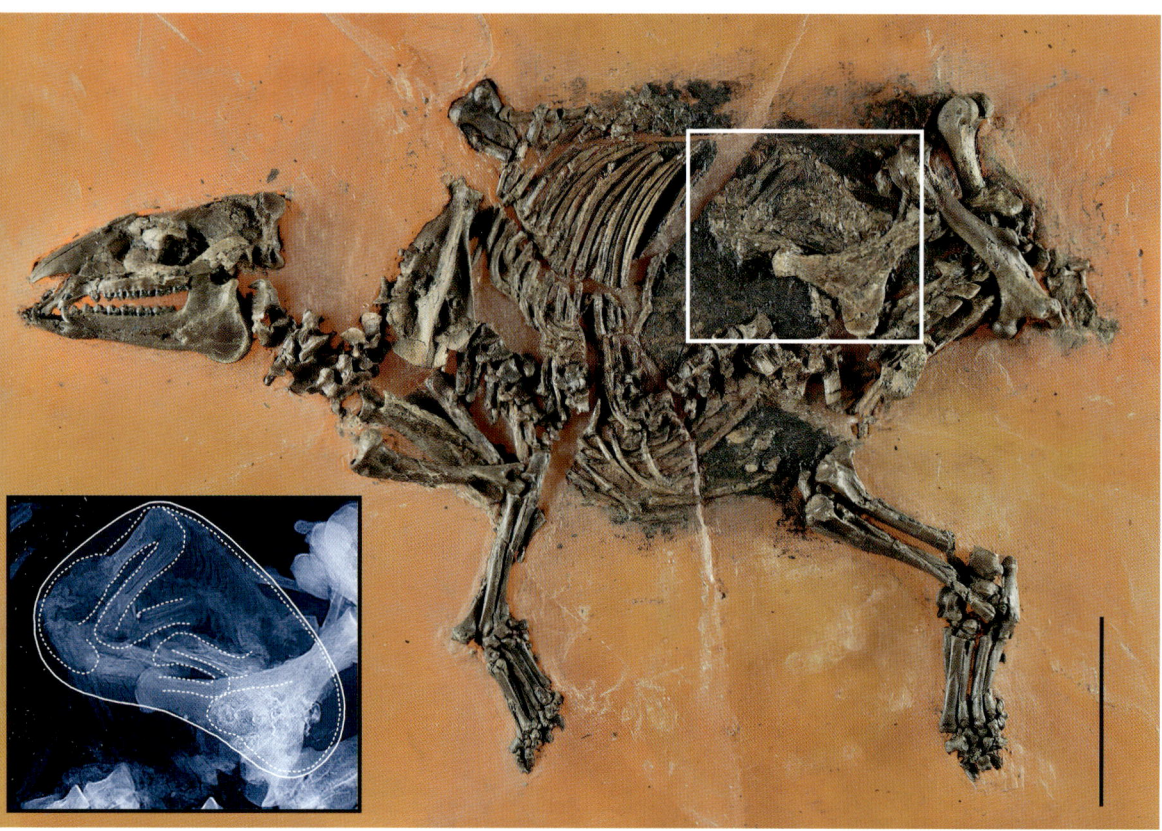

Fig. 12.9.12: Pregnant mare of *Eurohippus messelensis*. The fetus and its position within the uterus have been reconstructed in the X-ray image (bottom left). Scale: 10 cm.

Fig. 12.9.13: Pregnant mare of *Eurohippus messelensis*. Composition of the uterine wall: dark red = external wall of the uterus; light red = interior wall of the placenta. Scale: 2 cm.

is connected to the reproductive strategy already followed by these early horses. Here, the entire energy is directed toward the raising and survival of a single offspring, whereas other organisms produce as many offspring as possible to increase the chances of survival of at least a few of them to the reproductive age when parental care is lacking (Koenigswald 1987).

From leaf browser to grass eater

A Messel specimen settled the scientific dispute regarding the diet of early horses (Box 12.9.1), but it did not decide the ultimate question about the reason for the horses' phylogenetic development. Subsequent discoveries showed that the transition from browsing horses to grazers occurred during a geologically brief span of time, at the beginning of the middle Miocene around 18 million years ago (Franzen 2007a, 2010). Yet, the analysis of plant phytoliths (abrasive, microscopic inclusions of silica) revealed that grasslands in Nebraska already started to develop in the

Box 12.9.1: Hypotheses and reality
Eurohippus messelensis has the distinction of being the first fossil mammal in which contents of the intestine were discovered. This settled a long-term scientific dispute. As early as 1876, the Russian mammalian paleontologist Vladimir Kovalevsky had proposed the hypothesis that the evolution of horses had to be understood against the background of the development of the flora, as a specialization from initial omnivores to the modern-day grazers. On the other hand, fifty years later, William Diller Matthew of the American Museum of Natural History in New York believed he could prove that the earliest horses were not omnivorous but browsed on leaves, based on comparisons with the dentition of South American mazama deer. Therefore, the evolution of horses could be explained by changes in the feeding habits in conjunction with a habitat transition from originally predominant primeval forests to open savannas and prairies. The first discovery of an early horse by the Senckenberg Research Institute in 1975 decided this dispute in favor of Matthew, since the content of the intestine consisted of a dense packet of deciduous leaves, as a view through the microscope at the cell structure shows.

late Eocene, i.e., more than 17 million years prior to the development of the high-crowned molars (Strömberg 2004), even though these were considered an adaptation to a grass diet. Moreover, the evolution of horses as documented by fossil discoveries extends over a span of no less than 56 million years and is therefore by no means restricted to the time of dietary transition! On the other hand, across the entire time span their evolution is characterized by an increasing specialization of the body's design toward a more efficient mode of locomotion. In particular, the lower extremities elongated, while the number of toes was ultimately reduced to a single toe per foot. These toes only initially rested on a sole pad. In the course of evolution, they were increasingly carried by an elastic tendon that released the energy stored during the setting down of the hoof upon the foot's subsequent relifting (Camp & Smith 1942). Such developments were necessary to ensure that despite the phylogenetically increasing body size, the animals were still able to reach the speed required to survive when fleeing from predators and to maintain this speed until the pursuers had been left in the dust.

Without such specializations towards a rapid and energy-efficient locomotion, the increasingly larger and heavier horses would have become slower and slower, since the body weight is proportional to the volume (the third power of length), while the muscle strength is only proportional to the cross-section (the second power of length) (Franzen 2007a, 2010).

The tapir-like animals (Tapiromorpha)

Much more sparsely represented in the fossil record than the Equoidea are the two remaining groups of odd-toed ungulates, the tapir-like Lophiodontidae and the genus *Hyrachyus*, which are both close to the phylogenetic root of the tapirs and rhinoceroses. Both groups are represented in Messel by only a few skeletons, although other fossil sites indicate that these animals were certainly not rare in Europe at that time. Apparently, they were simply too large for Messel, which seems to have had a limit in regard to the body size of terrestrial vertebrates, extending to a shoulder height of about 65 cm and a body length of

Fig. 12.9.14: The only specimen of *Lophiodon* from Messel, a young animal. Scale: 10 cm.

100 cm. This limit is likely connected to the shallow depth and/or the flow rates of creeks that transported the carcasses of terrestrial vertebrates into Lake Messel. Larger bodies obviously did not manage to reach the lake. It is significant that the only skeleton of a *Lophiodon* found in Messel to date falls just under this limit, since it is a juvenile (Fig. 12.9.14). Besides a skeleton, the only other fossil of a *Lophiodon* is a single lower premolar. This may have been part of a crocodile's prey, since, due to its low buoyancy, it could not have been transported by the rather weak currents.

Lophiodontidae are only known from Europe, where they appeared for the first time around the end of the early Eocene. Their area of origin is unknown. During the middle Eocene, the group underwent a rapid diversification. However, this ended only a few million years later during the late Eocene with a species that reached the size of a modern horse. Regarding their build, the Lophiodontidae showed no similarities to horses. Instead, they were rather stocky and carried a remarkably massive skull. The skull lacked the deep nasal cleft characteristic of modern-day tapirs, where it is connected to the development of a trunk. The similarity of lophiodonts to tapirs is mainly in the low-crowned teeth, which as in tapirs are characterized by transverse ridges. It is still unknown how the lophiodonts used their massive

Fig. 12.9.15: Fully grown specimen of *Hyrachyus minimus*. Scale: 10 cm.

skull and body when feeding. It is possible that they dug around in muddy ground like modern pigs. This corresponds to the fact that lophiodonts are particularly frequently found in sites whose faunal composition and type of embedding sediments reflect the presence of wet habitats.

Members of the genus *Hyrachyus* also showed a fondness for water. This genus is represented in Messel by two very well-preserved, complete skeletons, which at first glance only show minor differences to early fossil horses (Fig. 12.9.15). It comes as no surprise that the first discovery of this genus in Messel made the headlines as an early horse. The secure differentiation from early horses is mainly based on dentition details (Franzen 1981). The same details also indicate a close phylogenetic relationship to early tapirs and rhinoceroses. According to recent studies, the relatively high-crowned, triangular last molar in the upper jaw and the detailed structure of the tooth enamel are the primary indicators of a close relationship between *Hyrachyus* and the rhinoceroses (Koenigswald et al. 2011).

Chapter 13
The Messel Ecosystem

Krister T. Smith, Thomas Lehmann, Gerald Mayr, Norbert Micklich, Renate Rabenstein, Sonja Wedmann

An ecosystem is the totality of the living organisms in an area together with the non-living components such as air, water, and soil matter, all of which are linked by flows of energy and nutrients. Although the boundaries of any ecosystem may be somewhat arbitrary, this cannot mask the large differences observed between different ecosystems. The tundra of the vast North Slope of Alaska (nearly 250,000 km^2) is characterized by some 20 mammal species, a mean (or average) annual temperature below freezing, and tremendous yearly fluctuations in light levels. In contrast, a tiny fragment of tropical rainforest such as La Selva, Costa Rica (15 km^2), hosts 113 mammal species (Wilson 1990), with a mean annual temperature near 25 °C and little change throughout the year. One of the great aims of ecology is to explain why ecosystems differ so drastically in species richness (Chapter 5).

Messel offers one of the best opportunities to study a terrestrial ecosystem in a "greenhouse" climate. That is because the oil shale layers were deposited some 48 million years ago during the Eocene, the most recent time when the Earth was characterized by high levels of CO_2 in the atmosphere, high sea levels, warm and equable temperatures, and little ice at the poles (Zachos et al. 2001; Chapter 3). Specifically, it is believed that at Messel the mean annual temperature was around 20°C, winter temperatures rarely dropped below 10°C, and frost almost never occurred (Grein et al. 2011). Yet Messel also represents conditions that no longer exist anywhere on Earth: despite the fairly equable temperatures, daylight levels fluctuated throughout the year in correspondence to the geographic position of Messel at middle latitudes. The Eocene is therefore of keen interest to earth scientists, who want to understand how the Earth System functions under conditions similar to those projected for the next 200 years (Zachos et al. 2008; Blois et al. 2013).

Topography and lake chemistry

At a local level, the ecosystems at Messel were strongly influenced by the lay of the land – and especially by the topography of the crater created by the volcanic eruptions (Chapter 2). The margins of the crater were initially steep (Fig. 13.2), so that sediment not fixed in place by vegetation would collapse under its own weight. The lake that formed in the crater was surrounded by a ring wall. However, by the time the layers were deposited from which macroscopic fossils are excavated, the ring wall was considerably eroded and possibly absent in some places (Micklich 2012; Tütken 2014).

Detailed studies of the lake sediments show that in the so-called Early Initial Lake Phase, shortly after the crater walls had been stabilized, the lake water was mixed from top to bottom. Such a lake is called "holomictic." However, the deep water of the lake was slightly denser than the surface water, because it was cooler and saturated with dissolved minerals

Fig. 13.1: Reconstruction of Lake Messel.

Fig. 13.2: Modern-day maar lake, Ranu Lading (Java).

(Goth 1990). Within a few hundred years at most, it was devoid of oxygen, and it was impossible for animal life to exist there. Thus, in the Late Initial Lake Phase and continuing throughout most of its history, Lake Messel was a so-called "meromictic" lake, one in which the water was nearly continually stratified (Goth 1990). The uppermost several meters of the lake water remained thoroughly mixed by the wind and absorbed oxygen from the atmosphere, forming the mixolimnion ("mixed lake"). This transitioned rather abruptly to the vast monimolimnion ("stable lake"), which extended down to the lake bottom.

The mixolimnion and monimolimnion make up two of the compartments of the physical environment of Messel (Fig. 13.3). In addition, there are the land, its interface with water (the shore), the air, and the subsurface. In a meromictic lake such as Messel the monimolimnion is largely a one-way street. When nutrients or food particles enter the monimolimnion, they are isolated there (Goth 1990). Nitrogen and phosphorous, for instance, are typically not present in sufficient quantities in surface waters (so-called "limiting nutrients"). All algal cells that did not decay in the mixolimnion itself ended up at the lake bottom, and even after those cells decayed, the released nutrients, including nitrogen and phosphorous, did not return to the mixolimnion. Therefore, the nitrogen and phosphorous in the mixolimnion had to be replenished year after year. The abundant green alga *Tetraedron minimum*, from which most of the organic matter in the Messel Formation derives, responds with particularly strong growth to the addition of dissolved phosphorous in modern lakes such as the Sea of Galilee (Berman et al. 1991), so perhaps the phosphorous introduced from the land was especially abundant. Nitrogen gas makes up nearly 80% of the air we breathe, but it only becomes useful for growth when it is transformed ("fixed") into a biologically available form, a function performed especially by bacteria associated with plants in the family Leguminosae, which were diverse around Lake Messel (Chapter 6).

Although no multicellular life existed in the monimolimnion, it is the only place around Messel where animals and plants are preserved. Thus, it is our only source of information about the aquatic and terrestrial ecosystems. In order to reconstruct them, we must carefully read the fossils and sediments from the monimolimnion. We must be sensitive to potential biases induced by taphonomic processes (Chapter 4), which influenced the type and abundance of different organisms that came to rest there. Then, on the basis of diversity, abundance, the adaptations of individual species, the chemistry of the bones, and the contents of the digestive tract, we can begin to reconstruct the ecosystem in all its splendor: the diversity of life at Messel, the habitat and behavior of the individual species, and the energy flows – acts of predation and of consumption – that linked the living community (Box 13.1).

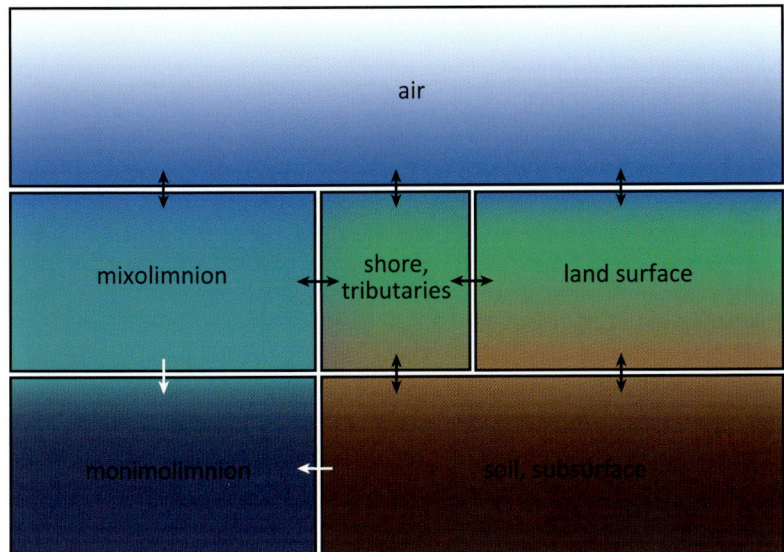

Fig. 13.3: Compartments of the Messel ecosystem.

The aquatic ecosystem

The basis of the aquatic food web (Fig. 13.4, top) was primarily the green algae that captured the sun's energy, extracted CO_2 and other nutrients from the water and thus formed living matter. In the Early Initial Lake Phase, single-celled dinoflagellates such as *Messelodinium* played an important role in primary production (Lenz et al. 2007a, b; Richter et al. 2017). Probably beginning in the Late Initial Lake Phase and continuing throughout the Main Lake Phase (Lenz et al. 2007a), the primary production was dominated by *Tetraedron minimum* (Goth 1990; Richter & Wedmann 2005; Lenz et al. 2007b). Diatoms, single-celled algae whose shells are composed of silicon dioxide (SiO_2), were also present, but the shells were dissolved after being buried in the sediment, leaving behind at best a hollow form, so their relative abundance is uncertain (Goth 1990). Living *Tetraedron minimum* grow best in the presence of ample light (Wall & Briand 1979), so we can assume that they did not live deep in the mixolimnion. As light levels rose in the spring, they began to multiply, taking advantage of the nutrients washed into the lake during the winter. They were most productive in the summer, and then died off as light levels decreased and nutrients were used up (Goth 1990). This seasonal "boom-and-bust" cycle repeated itself, year after year, in a more or less pronounced way for over a million years.

The aquatic food web is best documented by the fossil feces (coprolites) of fish. However, these are absent from the first 50,000 years of Lake Messel's history, suggesting that stable fish populations had not yet become established. In the absence of fish, a large species of water-flea was present in the lake, as shown by egg casings; this species disappeared when fish populations became established (Richter & Wedmann 2005). The primary consumers were all invertebrates (Chapter 7), especially water-fleas. Occasionally, caddisfly larvae and other invertebrates may have attached to floating mats of green algae and lived for a time in the open lake (Rietschel 1988; Richter & Baszio 2001a).

With the exception of very small fish larvae, the secondary consumers were mainly the predatory larval stages of phantom midges (Richter & Baszio 2001a; Richter & Wedmann 2005). Small fish, such as the teleosts *Thaumaturus* (Richter & Baszio 2001b) and *Rhenanoperca*, ate the secondary consumers and occasionally primary consumers and were in turn eaten by large predatory fish, soft-shelled and pig-nosed turtles (Chapter 10.2), and small crocodylians. Larger crocodylians such as *Diplocynodon darwini* were presumably the apex predators in the lake (Chapter 10.3), although their gut contents have not yet been described.

The shore and possible tributaries

The water of the shore area was shallow (Fig. 13.1). The producers and primary consumers there probably included many of the same species mentioned above for the open water of the lake. However, the shallow water was a habitat for various other animals and plants that required a substrate in the oxygenated zone. Geologically speaking, the shore had both pelagic and benthic elements (those living in the water column and those living on or in a substrate, respectively). Thus, it is treated as a separate zone (Fig. 13.4, center).

Box 13.1: Producers and consumers

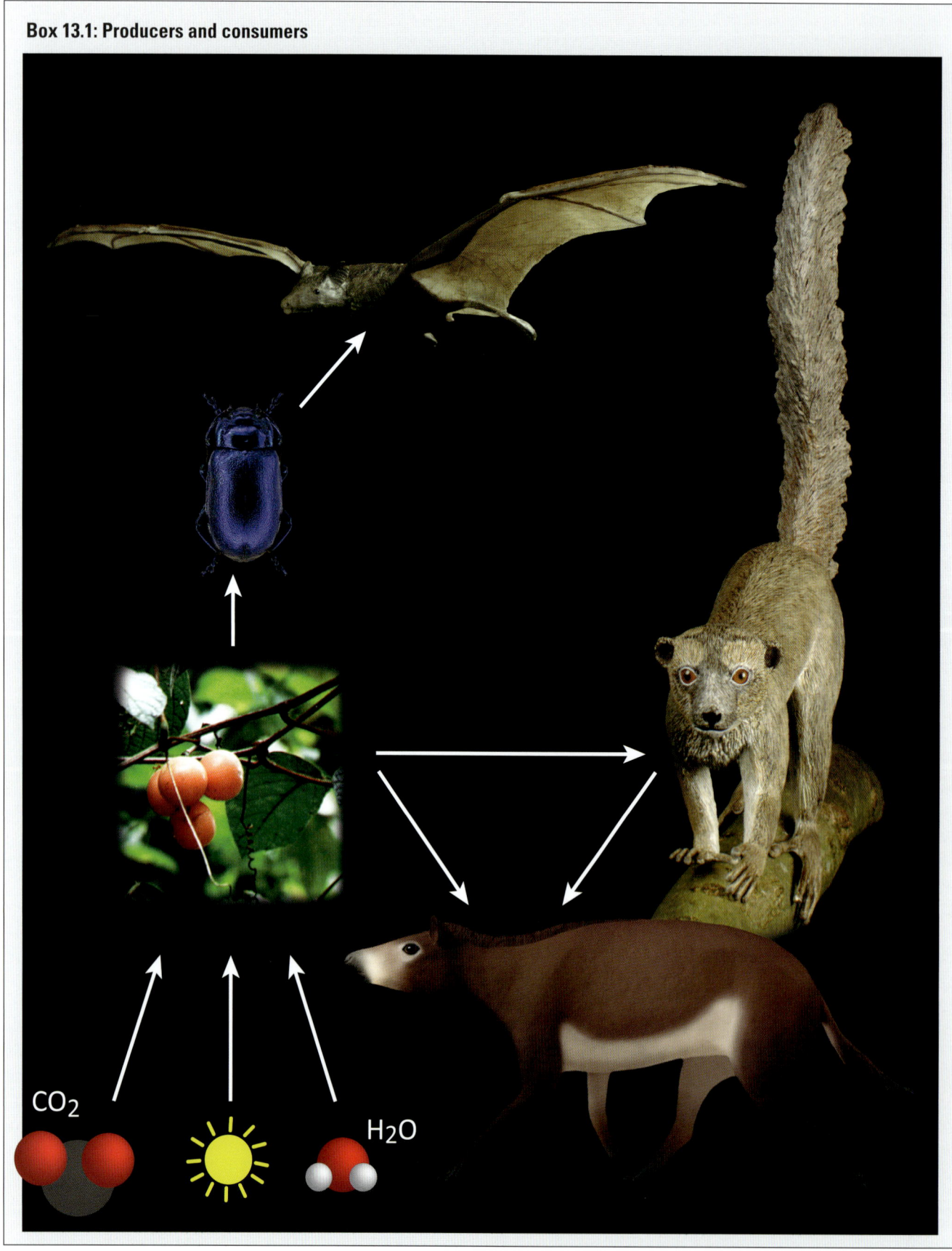

> **Box 13.1: Producers and consumers**
> The flow of energy in an ecosystem is commonly visualized as a food chain, and a set of terms is used to describe the different levels of that chain. At the base of the chain in most ecosystems are the *primary producers* (such as plants) that use water, carbon dioxide, soil nutrients, and the energy from sunlight to create living tissue, producing oxygen gas as a by-product. *Primary consumers*, or herbivores, consume plants, breaking them down into their elementary building blocks, and forging new living tissue. The process is not always very efficient, and a lot of energy is lost, but if sufficient plant matter is available, it can sustain a large number of herbivores. In turn, the *secondary consumers*, or first-level carnivores, consume the herbivores, going through the same process of breaking down prey tissue and forging new tissue. Again, energy is lost, one reason that carnivores are generally less abundant than herbivores. Higher-level carnivores take up places near the top of the food chain.
>
> This representation is, of course, highly simplified, because nature is messy. Many animals do not fit cleanly into one of these levels of the food chain. Instead, omnivores of various types are very common. A primate, for instance, might consume fruit, but also plant-eating insects. Furthermore, any one species will probably consume, and in turn be consumed, by a variety of other species. Thus, it is more common to speak of a food web, rather than a food chain. Messel illustrates these principles well.
>
> An example of multispecies interaction was proposed by Rabenstein et al. (2007). The grape family (Vitaceae) grew densely around Lake Messel, absorbing water and carbon dioxide. They commonly flowered and produced fruit, high in the canopy where they received the highest amount of light. Insects such as leaf beetles (Chrysomelidae) foraged on the grape leaves (cf. Rabenstein & Schöller 1996) and, while moving from plant to plant, were consumed by bats (*Tachypteron*) that flew in and above the canopy. The fruit of Vitaceae, wild grapes, were consumed by primates such as *Europolemur* and birds such as *Primozygodactylus*. Grapes dropped by the primates from the canopy were in turn consumed by primeval horses (*Eurohippus*), which conceivably tracked primates in the trees, picking off the grapes as they fell (cf. Heymann & Hsia 2015; Chapter 12.9).

Particularly noteworthy about the shore of Lake Messel were stands of water lilies that were anchored in the sediment at <2 m water depth and covered the water surface with their leaves and flowers (Wilde 1989; Collinson et al. 2012; Chapter 6). The water lilies and other aquatic plants such as Hydrocharaceae created a dense, three-dimensional habitat in which the shore-dwelling animals lived and sought shelter. In addition to the suspended green algae, benthic algae presumably lived on substrates such as plants, rocks and sediment along the shore (Richter & Baszio 2001a).

Primary consumers include caddisfly larvae, with one type of larva associated with the water lilies (Lutz 1990); clam-like conchostracans (Richter & Baszio 2001a); and molluscs such as ramshorn snails and mystery snails. Another group of primary consumers near the shore are freshwater sponges (Chapter 7). Their spicules occur rarely in much of the Messel oil shale, but in two several-meter-thick horizons they are much more abundant (Richter & Baszio 2009).

Secondary consumers likely included the same ones listed above as well as small frogs and very small fishes. These, in turn, were probably consumed by other small fishes. Higher-level consumers included gars, bowfins, percoids (*Amphiperca*; Gruber & Micklich 2007) and possibly birds (*Messelornis*; Morlo 2004), small crocodylians, amphibious turtles such as *Palaeoemys* and *Neochelys*, and mammals such as *Buxolestes* and *Macrocranion*. As in the open water, large crocodiles probably constituted apex predators, but large boas (*Palaeopython*) also consumed small crocodiles (Greene 1983; Chapter 10.1).

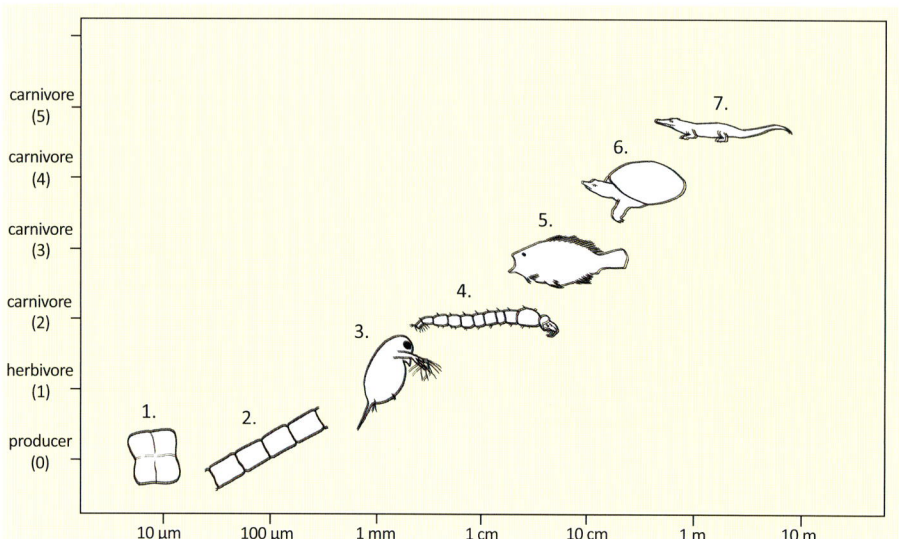

Fig. 13.4: Ecospace occupation at Messel: top, the aquatic ecosystem; center, the shore zone (simplified); bottom, the terrestrial ecosystem (highly simplified).

1. green alga *Tetraedron minimum* 2. diatom *Melosira* 3. water-flea (Cladocera) 4. phantom midge larva (Chaoboridae) 5. fish h *Amphiperca* (Perciformes 6. softshell turtle (Trionychidae) 7. crocodylian *Diplocynodon* (Alligatoroidea

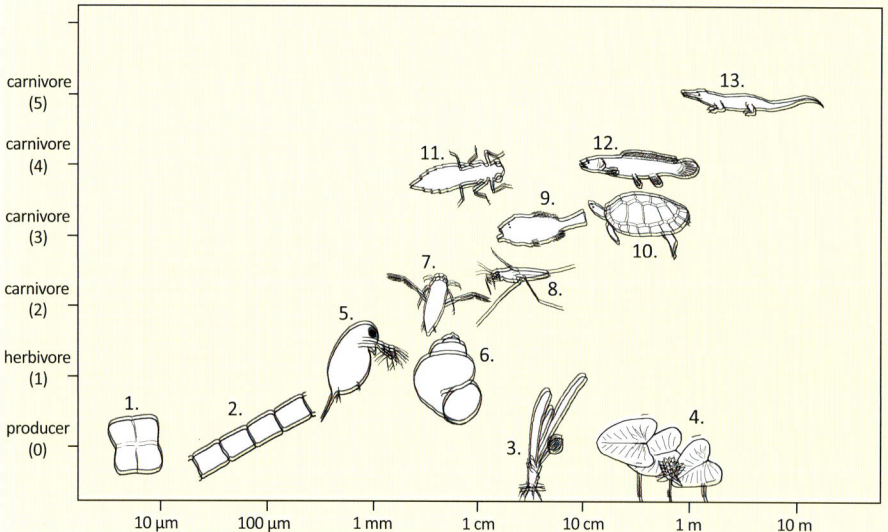

1. green alga *Tetraedron minimum* 2. diatom *Melosira* 3. water-plants (Hydro-charitaceae) 4. water lily (Nymphaeales) 5. water-flea (Cladocera) 6. mystery snail (Viviparidae) 7. backswimmer (Notonect-idae) 8. water strider (Gerridae) 9. fish *Rhenanoperca* (Centrarchiformes) 10. turtle *Neochelys* (Podocnemididae) 11. dragonfly larva (Odonata) 12. bowfin *Cyclurus* (Amiiformes) 13. crocodylian *Asiatosuchus* (Crocodyloidea)

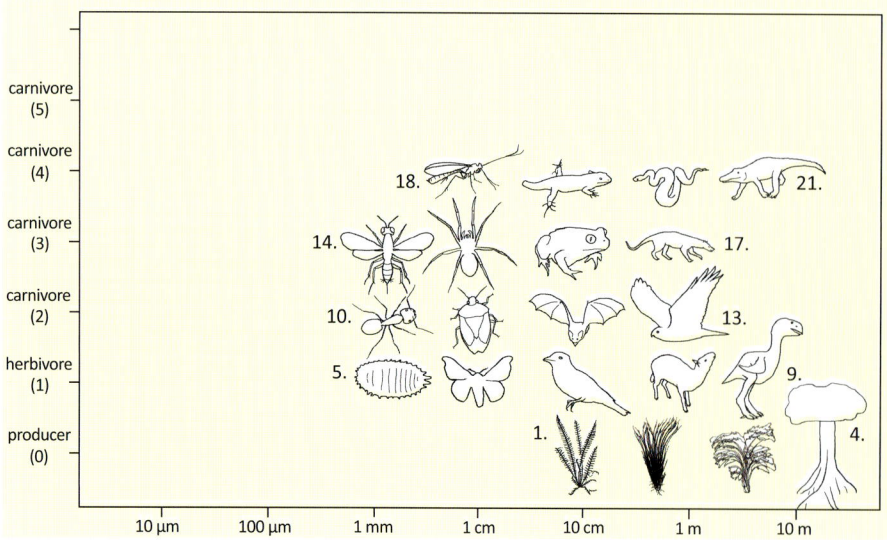

1. fern (Osmundaceae) 2. sedge (Cyperaceae) 3. palm (Arecaceae) 4. tree *Sloanea* (Elaeocarpaceae) 5. scale insect (Coccoidea) 6. moth (Lepidoptera) 7. roller (Coraciiformes) 8. dawn horse (Equioidea) 9. flightless bird (*Gastornis*) 10. ant (Ponerinae) 11. bug (Pentatomidae) 12. bat *Archaeonycteris* (Chiroptera) 13. seriama *Dynamopterus* (Idiornithidae) 14. chalcid wasp (Chalcidoidea) 15. true spider (Araneae) 16. frog *Eopelobates* (Eopelobatidae) 17. carnivorous mammal *Lesmesodon* (Hyaenodontida) 18. ichneumon wasp (Ichneumonidae) 19. carnivorous lizard *Necrosaurus* (Anguimorpha) 20. snake *Palaeopython* (Boidae) 21. terrestrial crocodyliform *Boverisuchus* (Planocraniidae)

The terrestrial ecosystem

Shortly after the cessation of the volcanic eruptions, the crater had filled with water, but the surroundings were devastated. During the Early Initial Lake Phase, a pioneer vegetation first settled in this ruined landscape that consisted of small herbaceous plants, especially ferns, but also palms and the rush-like monocots Restionaceae (Lenz et al. 2007a). After stabilization of the topography, a swampy vegetation – especially monocots – grew densely in parts of the lake margins (Fig. 13.5): Restionaceae, sedges, the conifer *Doliostrobus*, palm-like Pandanaceae, true palms, including *Nypa*, shrubby waterwillows (*Decodon*), and deciduous *Nyssa* trees. Other inhabitants of moister biotopes included cypress and a large variety of herbaceous plants such as ferns and horsetails (Collinson et al. 2012; Wilde 1989; Smith et al. 2009).

Farther away from the lake margins, the primeval forest grew tall and dense (Fig. 13.6; Chapter 6). The light-suffused margins of the forest formed a dense curtain, with a large varieties of lianas (woody vines), which otherwise preferred clearings caused by treefalls. The most abundant groups of plants here were laurels, walnuts, tea relatives as well as leguminous plants. Trees of the genus *Sloanea*, like their extant relatives, probably formed buttress roots near the forest floor. The diverse forms of these plants – herbaceous, shrubs, trees – imply a multistratal forest (Wilde 1989).

Farther away, there must have been stands of pines and Fagaceae (beech, oak, etc.), which are known only from their pollen at Messel. The dry season was not dry enough to promote wildfires, for charcoal is unknown in the Middle Messel Formation (Wilde & Riegel 2010).

In this forest lived a flourishing community of primary consumers (Fig. 13.4, bottom). The arthropods included major herbivorous groups such as weevils, jewel beetles, scarab beetles, several bug groups and plant-sucking insects such as scale insects (Chapter 7). Vertebrate herbivores inhabited the forest floor, like even-toed and odd-toed ungulates (Chapters 12.8, 12.9). In addition, there was a variety of herbivorous ground-dwelling birds, the most impressive of which was the giant, flightless *Gastornis* (Chapter 11).

Various omnivores consumed both plant matter and herbivores, including lizards (Chapter 10.1), birds (Chapter 11), and mammals (Chapter 12.7). These inhabited both the forest floor (like *Pholidocercus*) and the trees (like *Ornatocephalus* and primates). Secondary consumers (first-level carnivores) were plentiful: particularly ants, which are represented by a large number of species, but also various stinkbugs and click beetles (Chapter 7). Many insectivorous birds may be placed here, including those that caught insects on the wing (the swift *Scaniacypselus* and the potoo *Paraprefica*) and picked them from leaves (the stem passeriform *Primozygodactylus*), as well as some archaic (Chapter 12.2) and lipotyphlan (Chapter 12.3) mammals. Other arthropods consumed the secondary consumers and thus count as higher-level consumers: particularly spiders and ichneumon wasps. Any tetrapod insectivore (amphibian, squamate, mammal or bird) that consumed these would also belong to high-level consumers.

Reasons for the great species diversity in Messel

The aquatic food web (Fig. 13.4, top) differed fundamentally from the terrestrial food web (Fig. 13.4, bottom) in that the basis – the primary producers – were exclusively very small (much less than 1 mm) and unicellular. The herbivores, while larger, were also very small (1 cm or less), and the same is true of the carnivores above them. The concept of a food chain came closest to reality there. In contrast, the largest primary producers on land (various trees) were tens of meters in size. Both large and small herbivores can feed on the same large plant. All of the vertebrate primary and secondary consumers were significantly larger than their counterparts in the lake. That is, the ecological space is considerably populated to the right of the diagonal. This pattern is consistently observed in modern ecosystems as well (Lindeman 1942; Shurin et al. 2006). The shore ecosystem (Fig. 13.4, center) was intermediate between the two.

Although this pattern may be familiar, the extraordinary species richness of the Messel terrestrial ecosystem is unknown in similar latitudes at the present time. In all of Germany today, there are about 260 species of bird that regularly breed in the country, and more than one-quarter of that number is already known from Messel. In all of Germany today, there are not even 100 species of mammal, and nearly half that number is already known from Messel. In all of

Chapter 13 The Messel Ecosystem

Fig. 13.5: Reconstruction of a swampy area around Lake Messel.

Germany today, there are only 11 species of reptile, and nearly thrice that number is already known from Messel. This diversity is more astonishing in view of the fact that the Messel ecosystems existed merely 18 million years after the extinction of the dinosaurs at the end of the Cretaceous, when mammals and birds were rapidly diversifying. And this is more astonishing still considering the small geographic area that Messel represents!

Why are there such great differences in species richness among ecosystems today? Geographers and ecologists have grappled with this contentious question for over 200 years, and numerous explanations have been explored to account for known patterns in the present. One leading theory holds that higher temperatures and/or water availability allow for higher species richness at a local scale (e.g., Hawkins et al. 2003). The greater species richness of the tropics may be promoted by different mechanisms, such as greater speciation rates (Jablonski et al. 2006; Allen et al. 2006; Mayr 2017), and magnified by historical effects, such as the post-Eocene global cooling (Smith 2009; Jetz & Fine 2012). The study of ancient ecosystems sheds new light on this question.

Regardless of the cause, the high species richness of the terrestrial ecosystem calls into question how all the species could have occupied such a small area. Consider, for instance, the crocodylians (Chapter 10.3). In no ecosystem on Earth today, even in the tropics, do so many species of crocodylian co-occur. One explanation is *time-averaging*. This fundamental concept of taphonomy (Chapter 4) holds that most sediment beds in which fossils occur are not instantaneous records of ancient habitats but rather were accumulated over a sometimes considerable period of time (years or more). This factor is very important at Messel. The tetrapod fossils we know were excavated from a sediment profile around 45 m thick, which, at an average sedimentation rate of 0.14 mm/year (Chapter 2), represents a time span of approximately 320,000 years. The climate during the deposition of the Messel Formation was variable (Chapter 3). As noted above, the composition of the flora and fauna was also variable (Chapters 3, 6, 7, 8). Thus, it is not necessarily the case that all seven species of crocodylians (for example) were actively roaming the shores and land around Lake Messel at the same instant. Time averaging can lead to an overestimate of standing (instantaneous) species richness in a community (Kidwell & Bosence 1991).

The role of niches

Yet time-averaging does not solve the problem completely. Such a high number of crocodylian species, to come back to that example, does not occur anywhere today in such close proximity that they would ever be fossilized together. Another factor must have played a role: *niche differentiation*, the tendency of similar species to use the environment in different ways. Ecologists have long studied how similar species can co-exist (e.g., Pianka 1973). It is generally assumed that two species that play the exact same role in an ecosystem (i.e., occupy the same niche) cannot co-exist, because one would (locally) out-compete the other.

Niche differentiation facilitates co-existence and can happen in any number of ways: being active at different times, eating different food, or occupying different microhabitats. For instance, flying tetrapods that take insects on the wing can be active in daylight, at twilight or at night. Many aerial insectivores among birds are active during the day, whereas bats are active at night. Their flying prey presumably also showed different activity patterns. Finally, the bats themselves foraged at different heights above the forest floor (Fig. 13.7). This conclusion is clearly demonstrated by various indices of the wing proportions and body size, which among extant bats are closely correlated with flying ability (Chapter 12.5). The exceptional preservation of Messel thus gives us an early example of niche differentiation among closely related terrestrial vertebrates. Similarly, lizards of similar size and dentition related to Lacertidae also show a spectrum of adaptations to different habitats in the forest: cryptozoic, open terrestrial and arboreal (Chapter 10.1). And two of the seven crocodyliforms were primarily land-dwellers (Chapter 10.3). We can expect that many other groups of animals showed similar adaptations that enabled the co-existence of so many species, although this topic remains to be explored.

Other noteworthy aspects of the Messel terrestrial ecosystem may have been historically contingent (Gould 1989). For instance, no tetrapod known from Messel has gliding adaptations, whereas at least three lineages (the marsupial *Peradectes*, the lizard

Fig. 13.7: Niche differentiation: the bats of Messel hunted at different heights in and above the forest.

Ornatocephalus and a lacertid relative, and potentially the rodent *Ailuravus*; Chapters 10.1, 12.1, and 12.6) independently evolved a prehensile tail. Borneo today has a high percentage of tetrapods with gliding adaptations (frogs, lizards, snakes, rodents, flying lemurs), most of which acquired them independently. These facts have been explained in part by the low abundance of lianas (Emmons & Gentry 1983). Since the trees therefore have few lateral connections, gliding becomes a very energy-efficient means of moving about the forest canopy, leading to the independent evolution of gliding membranes in different groups. In contrast, western Amazonia has a moderate abundance of lianas and many palms (which are more difficult to climb), factors that Emmons & Gentry (1983) used to explain the high number of tetrapod species with prehensile tails. Perhaps the abundance of lianas and palms (Chapter 6) explains the number of tetrapods with prehensile tails at Messel.

Future prospects

Messel represents one of the most exquisitely preserved ecosystems in the fossil record, and one of the most important from the last greenhouse phase of the Earth's climate. Before fossil greenhouse ecosystems can be understood, it is first necessary to describe the various extinct species of plants and animals that occurred in them. This work proceeds apace at Messel as new discoveries are made and new specialists are drawn to it. New discoveries — including specimens of previously known species — aid in this process and tell us more about the way of life of the extinct species. But scientific attention is increasingly turning to the structure and function of the Messel ecosystems (e.g., Schweizer et al. 2007; Dunne et al. 2014). New techniques (Chapters 4, 5) help to elucidate what colors the animals were, how they moved about in their environment, how they partitioned their habitat, and how they were linked in the great food web that united all species through flows of energy. Messel ecosystems were diverse, they were dynamic, and they have much more to teach us about life in greenhouse climates.

Fig. 13.6: Reconstruction of the forest floor near Lake Messel.

References

Chapter 1

Behnke, C., Eikamp, H. & Zollweg, M. (1986) Die Grube Messel. Paläontologische Schatzkammer und unersetzliches Archiv für die Geschichte des Lebens. Geologie Bergbaugeschichte Fossilien, 168; Goldschneck-Verlag, Leinen

Hessisches Landesmuseum Darmstadt (Eds.) (2007) Messel Schätze der Urzeit, 159; Wissenschaftliche Buchgesellschaft, Darmstadt

Kuster-Wendenburg, E. (1969) Fossil-Grabungen in den mitteleozänen Süßwasserpeliten der „Grube Messel" bei Darmstadt (Hessen). *Notizblatt des Hessischen Landesamtes für Bodenforschung zu Wiesbaden*, 97: 65–75

Kühne, W.G. (1961) Präparation von flachen Wirbeltierfossilien auf künstlicher Matrix. *Paläontologische Zeitschrift*, 35 (3/4): 251–252

Lenz, O.K., Wilde, V., Mertz, D.F. & Riegel, W. (2015) New palynology-based astronomical and revised 40Ar/39Ar ages for the Eocene maar lake of Messel (Germany). *International Journal of Earth Sciences*, 104: 873–889

Ludwig, R. (1877) Fossile Crocodiliden aus der Tertiärformation des Mainzer Beckens. *Palaeontographica Supplement*, 3/4 (5): 1–52

Mertz, D.F. & Renne, P.R. (2005) A numerical age for the Messel fossil deposit (UNESCO World Heritage Site) derived from $^{40}Ar/^{39}Ar$ dating on a basaltic rock fragment. *Courier Forschungsinstitut Senckenberg*, 255: 67–75

Mößle, W. & Pagnia, H. (2000) Der Kampf um die Grube Messel. – In: Gemeindevorstand der Gemeinde Messel (Eds.) 1200 Jahre Messel, 404–409. Die Drucker, Reinheim

Schaal, S. & Schneider, U. (Eds.) (1995) Chronik der Grube Messel. Kempkes, Gladenbach

Schaal, S. & Ziegler, W. (Eds.) (1988) Messel – Ein Schaufenster in die Geschichte der Erde und des Lebens, 315; Kramer, Frankfurt a. M.

Schaal, S. & Ziegler, W. (Eds.) (1992) Messel – An insight into the history of life and of the Earth, 322; Clarendon Press, Oxford

Wittich, E. (1898) Beiträge zur Kenntnis der Messeler Braunkohle und ihrer Fauna. Geologie der Messeler Braunkohle. *Abhandlungen der Grossherzoglich hessischen geologischen Landesanstalt zu Darmstadt*, 3: 79–102

Chapter 2

Brikiatis, L. (2014) The De Geer, Thulean and Beringia routes: key concepts for understanding early Cenozoic biogeography. *Journal of Biogeography*, 41: 1036–1054

Büchel, G. (1988) Geophysik der Eifel-Maare 2: Geomagnetische Erkundung von Trockenmaaren im Vulkanfeld der Westeifel. *Mainzer geowissenschaftliche Mitteilungen*, 17: 357–376

Buness, H., Gabriel, G., Pucher, R., Rolf, C., Schulz, R. & Wonik, T. (2004) Grube Messel: Die Geophysik blickt unter die Abbausohle. *Natur und Museum*, 134: 65–76

Felder, M. & Harms, F.-J. (2004) Lithologie und genetische Interpretation der vulkano-sedimentären Ablagerungen aus der Grube Messel anhand der Forschungsbohrung Messel 2001 und weitere Bohrungen. *Courier Forschungsinstitut Senckenberg*, 252: 151–203

Lenz, O.K., Wilde, V., Mertz, D.F. & Riegel, W. (2015) New palynology-based astronomical and revised $^{40}Ar/^{39}Ar$ ages for the Eocene maar lake of Messel (Germany). *International Journal of Earth Sciences*, 104: 873–889

Lippolt, H.J., Baranyi, I. & Todt, W. (1975) Die Kalium-Argon-Alter der postpermischen Vulkanite des nord-östlichen Oberrheingrabens. *Der Aufschluß*, 27 (Sonderband): 205–212

Lorenz, V. (2000) Formation of maar-diatreme-volcanoes. *Terra Nostra*, 6: 284–291

Marell, D. (1989) Das Rotliegende zwischen Odenwald und Taunus. *Geologische Abhandlungen Hessen*, 89: 1–128

Matthess, G. (1966) Zur Geologie des Ölschiefervorkommens von Messel bei Darmstadt. *Abhandlungen des hessischen Landesamtes für Bodenforschung*, 51: 1–87

Mckenna, M.C. (1983) Cenozoic Paleogeography of North Atlantic Land Bridges. – In: Bott, M.H.P., Saxov, S., Talwani, N. & Thiede, J. (1983) Structure and Development of the Greenland-Scotland Ridge, *Nato Conference Series*, 8: 351–399; Springer, New York

Mertz, D.F. & Renne, P.R. (2005) A numerical age for the Messel fossil deposit (UNESCO World Heritage Site) derived from $^{40}Ar/^{39}Ar$ dating on a basaltic rock fragment. *Courier Forschungsinstitut Senckenberg*, 255: 67–75

Mezger, J.E., Felder, M. & Harms, F.-J. (2013) Crystalline rocks in the maar deposits of Messel: key to understand the geometries of the Messel Fault Zone and diatreme and the post-eruptional development of the basin fill. *Zeitschrift der deutschen Gesellschaft für Geowissenschaften*, 164: 639–662

Negendank, J.F.W. (1975) Permische und tertiäre Vulkanite im Bereich des nördlichen Odenwaldes. *Aufschluß*, 27 (Sonderband): 197–204

Ogg, J.G., Ogg, G. & Gradstein, F.M. (2016) A Concise Geologic Time Scale: 2016, 1–213; Elsevier, Amsterdam

Pirrung, M. (1998) Zur Entstehung isolierter alttertiärer Seesedimente in zentraleuropäischen Vulkanfeldern. *Mainzer Naturwissenschaftliche Abhandlungen, Beiheft*, 20: 1–117

Pirrung, M., Büchel, G., Merten, D., Assing, H., Schulte-Vieting, U., Heublein, S., Theune-Hobbs, M. & Boehrer, B. (2004) Morphometry, limnology, hydrology and sedimentology of maar lakes in east Java, Indonesia. *Studia Quaternaria*, 21: 139–152

Rullkötter, J., Littke, R., Hagedorn-Götz, B. & Janowski, B. (1988) Vorläufige Ergebnisse der organisch-geochemischen und organisch-petrologischen Untersuchungen an Kernproben des

Messeler Ölschiefers. *Courier Forschungsinstitut Senckenberg*, **107**: 37–51

Schaal, S., Schmitz-Münker, M. & Wolf, H.G. (1987): Neue Korrelationsmöglichkeiten von Grabungsstellen in der eozänen Fossillagerstätte Grube Messel. *Courier Forschungsinstitut Senckenberg*, **91**: 203–211

Schulz, R., Harms, F.-J. & Felder, M. (2002) Die Forschungsbohrung Messel 2001: Ein Beitrag zur Entschlüsselung der Genese einer Ölschieferlagerstätte. *Zeitschrift für angewandte Geologie*, **4**: 9–17

Scotese, C.R. (2013) PALEOMAP PaleoAtlas for ArcGIS, volumes 1–6, Cenozoic, Cretaceous, Jurassic & Triassic, Late Paleozoic, Early Paleozoic, Precambrian. PALEOMAP Project, Arlington, Texas https://www.researchgate.net/profile/Christopher_Scotese3

Solé, F., Smith, T., De Bast, E., Codrea, V. & Gheerbrandt, E. (2016) New carnivoraforms from the latest Paleocene of Europe and their bearing on the origin and radiation of Carnivoraformes (Carnivoramorpha, Mammalia). *Journal of Vertebrate Paleontology*, **36** (2): e1082480

Stein, E. (2001) Die magmatischen Gesteine des Bergsträßer Odenwaldes und ihre Platznahme-Geschichte. *Jahresberichte und Mitteilungen des Oberrheinischen Geologischen Vereins*, **83**: 267–283

Suhr, P., Goth, K., Lorenz, V. & Suhr, S. (2006) Long lasting subsidence and deformation in and above maar-diatreme volcanoes – a never ending story. *Zeitschrift der deutschen Geowissenschaften*, **157**: 491–511

Chapter 3

Collinson, M.E., Manchester, S.R. & Wilde, V. (2012) Fossil fruits and seeds of the Middle Eocene Messel biota, Germany. *Abhandlungen der Senckenberg Gesellschaft für Naturforschung*, **570**: 1–251

Goth, K. (1990) Der Messeler Ölschiefer – ein Algenlaminit. *Courier Forschungsinstitut Senckenberg*, **131**: 1–143

Grein, M., Utescher, T., Wilde, V. & Roth-Nebelsick, A. (2011) Reconstruction of the middle Eocene climate of Messel using palaeobotanical data. *Neues Jahrbuch für Geologie und Paläontologie Abhandlungen*, **260**: 305–318

IPCC (2014) Climate Change 2014: Synthesis Report. Contribution of Working Groups I, II and III to the Fifth Assessment Report of the Intergovernmental Panel on Climate Change. – In: Pachauri, R.K. & Meyer, L.A. (Eds.). 151; Genf, Schweiz.

Irion, G. (1977) Der eozäne See von Messel. *Natur und Museum*, **107**: 213–218

Lenz, O.K., Wilde, V., Riegel, W. & Schaarschmidt, F. (2005) Klima- und Vegetationsdynamik im Mitteleozän von Messel – eine Neubewertung der palynologischen Daten aus der Kernbohrung 4 von 1980. *Courier Forschungsinstitut Senckenberg*, **255**: 81–101

Lenz, O.K., Wilde, V., Riegel, W. & Harms, F.-J. (2010) A 600 k.y. record of El Niño – Southern Oscillation (ENSO): Evidence for persisting teleconnections during the Middle Eocene greenhouse climate of Central Europe. *Geology*, **38**: 627–630

Lenz, O.K., Wilde, V. & Riegel, W. (2011) Short-term fluctuations in vegetation and phytoplankton during the Middle Eocene greenhouse climate: a 640–kyr record from the Messel oil shale (Germany). *International Journal of Earth Sciences*, **100**: 1851–1874

Lenz, O.K., Wilde, V., Mertz, D.F. & Riegel, W. (2015) New palynology-basedastronomical and revised $^{40}Ar/^{39}Ar$ ages for the Eocene maar lake of Messel (Germany). *International Journal of Earth Sciences*, **104**: 873–889

Lenz, O.K., Wilde, V. & Riegel, W. (2017) ENSO- and solar-driven sub-Milankovitch cyclicity in the Palaeogene greenhouse world; high-resolution pollen records from Eocene Lake Messel, Germany. *Journal of the Geological Society London*, **174**: 110–128

Markwick, P.J. (1998) Fossil crocodilians as indicators of Late Cretaceous and Cenozoic climates: implications for using palaeontological data in reconstructing palaeoclimate. *Palaeogeography, Palaeoclimatology, Palaeoecology*, **137**: 205–271

McInerney, F.A. & Wing, S.L. (2011) The Paleocene-Eocene Thermal Maximum: a perturbation of carbon cycle, climate, and biosphere with implications for the future. *Annual Review of Earth and Planetary Sciences*, **39**: 489–516.

Parrish, J.T. & Soreghan, G.S. (2013) Sedimentary geology and the future of paleoclimate studies. *The Sedimentary Record*, **11**: 4–10

Röhl, U., Westerhold, T., Bralower, T.J. & Zachos, J.C. (2007) On the duration of the Paleocene-Eocene thermal maximum (PETM). *Geochemistry, Geophysics, Geosystems*, **8**: Q12002

Thiele-Pfeiffer, H. (1988) Die Mikroflora aus dem mitteleozänen Ölschiefer von Messel bei Darmstadt. *Palaeontographica B*, **211**: 1–86

Traverse, A. (2007) Paleopalynology (2. Ed). Springer, Dordrecht, Niederlande

Tütken, T. (2014) Isotope compositions (C, O, Sr, Nd) of vertebrate fossils from the Middle Eocene oil shale of Messel, Germany: Implications for their taphonomy and palaeoenvironment. *Palaeogeography, Palaeoclimatology, Palaeoecology*, **416**: 92–109

Wilde, V. (2004) Aktuelle Übersicht zur Flora aus dem mitteleozänen „Ölschiefer" der Grube Messel bei Darmstadt (Hessen, Deutschland). *Courier Forschungsinstitut Senckenberg*, **252**: 109–114

Wolfe, J. (1979) Temperature parameters of humid to mesic forests of eastern Asia and relation to forests of other regions of the Northern Hemisphere and Australasia. *Geological Survey Professional Paper*, **1106**: 1–37

Zachos, J. C., Pagani, M., Sloan, L., Thomas, E. & Billups, K. (2001) Trends, rhythms, and aberrations in global climate 65 Ma to present. *Science*, **292**: 686–693

Zachos, J.C., Dickens, G.R. & Zeebe, R.E. (2008) An early Cenozoic perspective on greenhouse warming and carbon-cycle dynamics. *Nature*, **451**: 279–283

Chapter 4

Cadena, E. (2016) Microscopical and elemental FESEM and Phenom ProX-SEM-EDS analysis of osteocyte- and blood vessel-like microstructures obtained from fossil vertebrates of the Eocene Messel Pit, Germany. PeerJ 4: e1618; doi.org/10.7717/peerj.1618

Colleary, C., Dolocanc, A., Gardnerd, J. Singha, S., Wuttke, M., Rabenstein, R., Habersetzer, J., Schaal, S., Fesehag, M., Clemens, M., Jacobs,B.F., Curranoi, E.D., Jacobs, L.L., Sylvestersen, R.L., Gabbott, S.E. & Vinther, J. (2015) Chemical, experimental, and morphological evidence for diagenetically altered melanin in exceptionally preserved fossils. *Proceedings of the National Academy of Sciences USA*, **104** (2): 565–569

Franzen, J.L. & Frey, E. (1993) *Europolemur* completed. *Kaupia*, **3**: 113–130

Franzen, J.L., Smith, K.T. & Habersetzer, J. (2017) Krokodil gegen Urpferd. *Natur, Forschung, Museum*, **147** (1/2): 34–42

Glass, K., Ito, S., Wilby, P.R., Sota, T., Nakamura, A., Bowers, C.R., Vinther, J., Dutta, S., Summons, R., Briggs, D.E. & Wakamatsu, K. (2012) Direct chemical evidence for eumelanin pigment from the Jurassic period. *Proceedings of the National Academy of Sciences USA*, **109**: 10218–10223

Habersetzer, J., Richter, G. & Storch, G. (1992) Bats: already highly specialized insect predators. Messel – An insight into the history of life and of the Earth. – In: Schaal, S. & Ziegler, W. (eds). Clarendon Press, Oxford: 179–190

Habersetzer, J., Richter, G. & Storch, G. (1994) Paleoecology of early middle Eocene bats from Messel, FRG. Aspects of flight, feeding and echolocation. *Historical Biology*, **8** (1–4): 235–260

Habersetzer, J. & Rabenstein, R. (2011) Studies in taphonomy of extant and Messel bats. *Journal of Vertebrate Paleontology*, **31** (Abstract Book): 120

Joyce, W.G., Micklich, N., Schaal, S.F.K. & Scheyer, T. (2012) Caught in the act: the first record of copulating fossil vertebrates. *Biology Letters*, **8** (5): 846–848

Kaiser, T., Ansorge, J., Arratia, G., Bullwinkel, V., Gunell, G.F., Herendeen, P.S., Jacobs, B., Mingram, J., Msuya, C. Musolff, A., Naumann, R., Schulz, E. & Wilde, V. (2006) The maar lake of Mahenge (Tanzania) – unique evidence of Eocene terrestrial environments in sub-Sahara Africa. *Zeitschrift der deutschen Gesellschaft für Geowissenschaften*, **157** (3): 411–431

Koenigswald, W. V., Braun, A. & Pfeiffer, T. (2004) Cyanobacteria and seasonal death: A new taphonomic model for the Eocene Messel lake. *Paläontologische Zeitschrift*, **78** (2): 417–424

Lutz, H. & Kaulfuss, U. (2006) A dynamic model for the meromictic lake Eckfeld Maar (Middle Eocene, Germany). *Zeitschrift der deutschen Gesellschaft für Geowissenschaften*, **157** (3): 433–450

Mayr, G. (2016) Avian feet, crocodilian food and the diversity of larger birds in the early Eocene of Messel. *Palaeobiodiversity and Palaeoenvironments*, **96** (4): 601–609

Pfanz, H. (2008): Mofetten – Kalter Atem schlafender Vulkane. Rheinischer Verein für Denkmalpflege und Landschaftsschutz, Köln

Pybus, M.J., Hobson, D.P. & Oncerka, D.K. (1986) Mass mortality of bats due to probable blue-green algal toxicity. *Journal of Wildlife and Diseases*, **22**: 449–450

Rabenstein, R., R. Usman, R. & Schaal, S. (2004) Suche nach rezenten Seen als Modelle für den eozänen Lebensraum von Messel. *Courier Forschungsinstitut Senckenberg*, **252**: 115–138

Reisdorf, A.G. & Wuttke, M. (2012) Re-evaluating Moodie's Opisthotonic-Posture Hypothesis in Fossil Vertebrates Part I: Reptiles – the taphonomy of the bipedal dinosaurs *Compsognathus longipes* and *Juravenator starki* from the Solnhofen Archipelago (Jurassic, Germany). *Palaeobiodiversity and Palaeoenvironments*, **92** (1): 119–168

Ruttner, F. (1931) Hydrographische und hydrochemische Beobachtungen auf Java, Sumatra und Bali. – In: Thienemann, A. (Ed.): *Archiv für Hydrobiologie Supplement*, **8**: 197–449

Schindler, T. & Wuttke, M. (2015) A revised sedimentological model for the late Oligocene crater lake Enspel (Enspel Formation, Westerwald Mountains, Germany). *Palaeobiodiversity and Palaeoenvironments*, **95** (1): 5–16

Smith, K.T. & Scanferla, A. (2016) Fossil snake preserving three trophic levels and evidence for an ontogenetic dietary shift. *Palaeobiodiversity and Palaeoenvironments*, **96** (4): 589–599

Smith, K.T. & Wuttke, M. (2012) From tree to shining sea: Taphonomy of the arboreal lizard *Geiseltaliellus maarius* from Messel, Germany. *Palaeobiodiversity and Palaeoenvironments*, **92**: 45-65

Smith, T., Rana, R., Missiaen, P., Rose, K., Sahni, A., Singh, H. & Singh, L. (2007) High bat (Chiroptera) diversity in the Early Eocene of India. *Naturwissenschaften*, **94** (12): 1003–1009

Vinther, J. Briggs, D.E.G., Prum, R. O. & Saranathan, V. (2008) The color of fossil feathers. *Biology Letters*, **4**: 522–525

Vitek, N.S., Vinther, J., Schiffbauer, J.D., Briggs, D.E.G. & Prum, R.O. (2013) Exceptional three-dimensional preservation and coloration of an originally iridescent fossil feather from the Middle Eocene Messel Oil Shale. *Paläontologische Zeitschrift*, **87**: 493–503

Wuttke, M. (1983) ,Weichteil-Erhaltung' durch lithifizierte Mikroorganismen bei mittel-eozänen Vertebraten aus den Ölschiefern der „Grube Messel" bei Darmstadt. *Senckenbergiana Lethaea*, **64**: 509–527

Wuttke, M. & Reisdorf, A.G. (2012) Taphonomic processes in terrestrial and marine environments. *Palaeobiodiversity and Palaeoenvironments*, **92**: 1-3

Chapter 5

Collinson, M.E., Manchester, S.R. & Wilde, V. (2012) Fossil fruits and seeds of the Middle Eocene Messel biota, Germany, *Abhandlungen der Senckenberg Gesellschaft für Naturforschung*, **570**: 1–251

Colwell, R.K., Chao, A., Gotelli, N.J., Lin, S.Y., Mao, C.X., Chazdon, R.L., & Longino, J.T. (2012) Models and estimators linking individual-based and sample-based rarefaction, extrapolation and comparison of assemblages, *Journal of Plant Ecology*, **5** (1): 3–21

Foote, M. (1997) The evolution of morphological diversity. *Annual Review of Ecology and Systematics*, **28**: 129–152

Gotelli, N.J., & Colwell, R.K. (2001) Quantifying biodiversity: procedures and pitfalls in the measurement and comparison of species richness. *Ecology Letters*, **4** (4): 379–391

Goth, K. (1990) Der Messeler Ölschiefer ein Algenlaminit. *Courier Forschungsinstitut Senckenberg*, **131**: 143 S.

Habersetzer, J. (1995) Paläontologie.– In: Heuck, F.H.W. & Macherauch, E. (Eds) Forschung mit Röntgenstrahlen Bilanz eines Jahrhunderts 1895–1992. 633–641; Springer, Heidelberg

Habersetzer, J. (2004) Röntgenverfahren zur Untersuchung Messeler Fossilien. *Courier Forschungsinstitut Senckenberg*, **252**: 211–218

Habersetzer, J., Schlosser-Sturm, E., Storch, G. & Sigé, B. (2012) Shoulder joint and inner ear of *Tachypteron franzeni*, an embalonurid bat from the Middle Eocene of Messel. – In: Gunnell, G.F. & Simmons, N.B. (Eds) Evolutionary History of Bats: Fossils, Molecules and Morphology. 67–104; Cambridge University Press, Cambridge

Hennig, W. (1966) Phylogenetic Systematics. University of Illinois Press, Urbana

Keller, T., Frey, E., Heil, R., Rietschel, S., Schaal, S.F.K. & Schmitz, M. (1991) Ein Regelwerk für paläontologische Grabungen in der Grube Messel. *Paläontologische Zeitschrift*, **65** (1–2): 221–224

Kühne, W.G. (1961) Präparation von flachen Wirbeltierfossilien auf künstlicher Matrix. *Paläontologische Zeitschrift*, **35** (3): 251–252

Linnaeus, C. (1758) Systema naturae per regna tria naturae, secundum classes, ordines, genera, species, cum characteribus, differentiis, synonymis, Locis. 1 (10. Ed.). Salviae, Stockholm

Richter, G. & Storch, G. (1980) Beiträge zur Ernährungsbiologie eozäner Fledermäuse aus der Grube Messel. *Natur und Museum*, 110 (12): 353–367

Rowe, T., Carlson, W., & Bottorff, W. (1995) *Thrinaxodon*: digital atlas of the skull. CD-ROM (2. Ed.). University of Texas Press, Austin, Texas

Ruf, I., Volpato, V., Rose, K.D., Billet, G., De Muizon, C. & Lehmann, T. (2016) Digital reconstruction of the inner ear of *Leptictidium auderiense* (Leptictida, Mammalia) and North American leptictids reveals new insight into leptictidan locomotor agility. *Paläontologische Zeitschrift*, 90 (1): 153–171

Schaal, S. & Habersetzer, J. (1991) Ein neues Verfahren zum Schutz und zur dauerhaften Erhaltung geborgener Fossilien aus der Fundstätte Messel. *Courier Forschungsinstitut Senckenberg*, 139: 165–169

Wuttke, M. (1983) Weichteil-Erhaltung durch lithifizierte Mikroorganismen bei mittel-eozänen Vertebraten aus den Ölschiefern der Grube Messel bei Darmstadt. *Senckenbergiana Lethaea*, 64 (5/6): 509–527

Chapter 6

Adam, P., Schaeffer, P. & Albrecht, P. (2006) C40 monoaromatic lycopane derivatives as indicators of the contribution of the alga *Botryococcus braunii* race L to the organic matter of Messel oil shale (Eocene, Germany). *Organic Geochemistry*, 37: 584–596

APG (Angiosperm Phylogeny Group) IV (2016) An update of the Angiosperm Phylogeny Group classification for the orders and families of flowering plants: APG IV. *Botanical Journal of the Linnean Society*, 181: 1–20

Chelius, C. (1886) Blatt Messel. Erläuterungen zur Geologischen Karte des Grossherzogthums Hessen 1:25000, 1. Lieferung, 1–67, Großherzogliche geologische Landesanstalt, Darmstadt

Chen, I. & Manchester, S.R. (2007) Seed morphology of modern and fossil *Ampelocissus* (Vitaceae) and implications for phytogeography. *American Journal of Botany*, 94: 1534–1553

Chen, I. & Manchester, S.R. (2011) Seed morphology of Vitaceae. *International Journal of Plant Sciences*, 172: 1–35

Christenhusz, M. J. M. & Byng, J. W. (2016) The number of known plants [sic] species in the world and its annual increase. *Phytotaxa*, 261, 201–217

Collinson, M.E. (1982) A preliminary report on the Senkenberg-Museum collection of fruits and seeds from Messel bei Darmstadt. *Courier Forschungsinstitut Senckenberg*, 56: 49–57

Collinson, M.E. (1986) Früchte und Samen aus dem Messeler Ölschiefer. *Courier Forschungsinstitut Senckenberg*, 85: 217–220

Collinson, M.E. (1988) The special significance of the Middle Eocene fruit and seed flora from Messel, West Germany. *Courier Forschungsinstitut Senckenberg*, 107: 187–197

Collinson, M.E. & Gregor, H.J. (1988) Rutaceae from the Eocene of Messel, West Germany. *Tertiary Research*, 9: 67–80

Collinson, M.E., Manchester, S.R. & Wilde, V. (2012) Fossil fruits and seeds of the Middle Eocene Messel biota, Germany. *Abhandlungen der Senckenberg Gesellschaft für Naturforschung*, 570: 1–251.

Engelhardt, H. (1922) Die alttertiäre Flora von Messel bei Darmstadt. *Abhandlungen der Hessischen Geologischen Landesanstalt zu Darmstadt*, 7: 20–128

Ferguson, D.K. (1993) Plant taphonomic studies with special reference to Messel. *Kaupia*, 2: 117–126

Frankenhäuser, H. & Wilde, V. (1993) Farne aus der mitteleozänen Maarfüllung von Eckfeld bei Manderscheid in der Eifel. *Mainzer naturwissenschaftliches Archiv*, 31: 149–167

Frankenhäuser, H. & Wilde, V. (1994) Zweiflügelige Juglandaceen-Früchte aus dem Mitteleozän von Eckfeld bei Manderscheid (Eifel).- Vorläufige Mitteilung. *Mainzer naturwissenschaftliches Archiv Beihefte*, 16:143–150

Franzen, J.L. (1977) Urpferdchen und Krokodile, Messel vor 50 Millionen Jahren. *Kleine Senckenberg-Reihe*, 7: 1–36

Glinka, U. & Walther, H. (2003) *Rhodomyrtophyllum reticulosum* (Rossm.) Knobloch & Z. Kvaček – ein bedeutendes eozänes Florenelement im Tertiär Mitteleuropas. *Feddes Repertorium*, 114: 30–55

Goth, K. (1990) Der Messeler Ölschiefer – ein Algenlaminit. *Courier Forschungsinstitut Senckenberg*, 131: 1–143

Goth, K., De Leeuw, J.W., Püttmann, W. & Tegelaar, E.W. (1988) Origin of Messel oil shale kerogen. *Nature*, 336: 759–761

Grimsson, F., Zetter, R., Labandeira, C.C., Engel, M.S. & Wappler, T. (2017) Taxonomic description of in situ bee pollen from the middle Eocene of Messel and Eckfeld. *Grana*, 56: 37–70

Gruber, G. & Micklich, N. [Eds.] (2007) Messel Treasures of the Eocene. The book to the exhibition Messel on Tour of the Hessisches Landesmuseum Darmstadt. 1–159; Darmstadt (Hessisches Landesmuseum Darmstadt)

Habermehl, G. & Hundrieser, H.-J. (1983a) 50 Millionen Jahre altes Coniferen-Lignin aus Messel. *Naturwissenschaften*, 70: 249–250

Habermehl, G. & Hundrieser, H.-J. (1983b) Fossile Relikte der „Wasserblüte" im Messeler Ölschiefer. *Naturwissenschaften*, 70: 566–568

Harley, M.M. (1997) Ultrastructure of pollen from some Eocene palm flowers (Messel, Germany). *Mededelingen Nederlands Instituut voor Toegepaste Geowetenschappen TNO*, 58: 193–209

Kenrick, P. & Crane, P.R. (1997) The origin and early evolution of plants on land. *Nature* 389: 33–39

Kvaček, Z. (1988) The Lauraceae of the European Paleogene, based on leaf cuticles. *Courier Forschungsinstitut Senckenberg*, 107: 345–354

Kvaček, Z. & Wilde, V. (2010) Foliage and seeds of malvalean plants from the Eocene of Europe. *Bulletin of Geosciences*, 85: 163–182

Leliaert, F., Smith, D. R., Moreau, H., Herron, M. D., Verbruggen, H., Delwiche, C. F. & De Clerck, O. (2012) Phylogeny and molecular evolution of the green algae. *Critical Reviews in Plant Sciences*, 31: 1–46

Lenz, O.K., Wilde, V. & Riegel, W. (2011) Short-term fluctuations in vegetation and phytoplankton during the Middle Eocene greenhouse climate: A 640–kyr record from the Messel oil shale (Germany). International Journal of Earth Sciences, 100: 1851–1874.

Lenz, O.K., Wilde, V., Riegel, W. & Heinrichs, T. (2007) Distribution and ecologic significance of a new freshwater dinoflagellate cyst (*Messelodinium thielepfeifferae* gen. et sp. nov.) from the Middle Eocene Lake Messel. *Palynology*, 31: 119–134

Manchester, S.R., Collinson, M.E. & Goth, K. (1994) Fruits of the Juglandaceae from the Eocene of Messel, Germany and implications for early Tertiary phytogeographic exchange between Europe and western North America. *International Journal of Plant Sciences*, **155**: 388–394

Manchester, S.R., Wilde, V. & Collinson, M.E.C. (2007) Fossil cashew nuts from the Eocene of Europe: Biogeographic links between Africa and South America. *International Journal of Plant Sciences*, **168**: 1199–1206

Matthess, G. (1966) Zur Geologie des Ölschiefervorkommens von Messel bei Darmstadt. *Abhandlungen des Hessischen Landesamtes für Bodenforschung*, **51**: 1–87

Otto, A., Simoneit, B.R.T., Wilde, V., Kunzmann, L. & Püttmann, W. (2002) Terpenoid composition of three fossil resins from Cretaceous and Tertiary conifers. *Review of Palaeobotany and Palynology*, **120**: 203–215; Amsterdam

Richter, G., Schiller, W. & Baszio, S. (2013) A green alga of the genus *Coelastrum* Naegeli from the sediments of the Tertiary Lake Messel. *Palaeobiodiversity and Palaeoenvironment*, **93**: 285–298

Schaal, S. & Ziegler, W. (Eds.) (1992) Messel: An insight into the history of life and of the Earth. 1–322; Clarendon Press, Oxford

Schaarschmidt, F. (1982) Präparation und Untersuchung der eozänen Pflanzenfossilien von Messel bei Darmstadt. *Courier Forschungsinstitut Senckenberg*, **56**: 59–77

Schaarschmidt, F. (1984) Flowers from the Eocene oil-shale of Messel: A preliminary report. *Annals of the Missouri Botanical Garden*, **71**: 599–606

Schaarschmidt, F. (1986) Blüten von Pflanzen des Messeler Ölschiefers. *Courier Forschungsinstitut Senckenberg*, **85**: 214–216

Schaarschmidt, F. (1988) Der Wald, fossile Pflanzen als Zeugen eines warmen Klimas. – In: Schaal, S. & Ziegler, W. (Eds.): Messel: Ein Schaufenster in die Geschichte der Erde und des Lebens: 27–52; Frankfurt am Main (Verlag Waldemar Kramer)

Schaarschmidt, F. (1992) The vegetation: Fossil plants as witnesses of a warm climate. – In: Schaal, S. & Ziegler, W. (Eds.): Messel: An insight into the History of Life and of the Earth: 27–52; Oxford (Clarendon Press)

Schaarschmidt, F. & Wilde, V. (1986) Palmenblüten und -blätter aus dem Eozän von Messel. *Courier Forschungsinstitut Senckenberg*, **86**: 177–202

Smith, S.Y., Collinson, M.E. & Rudall, P.J. (2008) Fossil *Cyclanthus* (Cyclanthaceae, Pandanales) from the Eocene of Germany and England. *American Journal of Botany*, **95**: 688–699

Smith, S.Y., Collinson, M.E., Simpson, D.A., Rudall, P.J., Marone, F. & Stampanoni, M. (2009b) Elucidating the affinities and habitat of ancient, widespread Cyperaceae: *Volkeria messelensis* gen. et sp. nov., a fossil mapanioid sedge from the Eocene of Europe. *American Journal of Botany*, **96**: 1506–1518

Storch, G. & Schaarschmidt, F. (1988) Fauna Und Flora Von Messel – Ein Biogeographisches Puzzle. – In: Schaal, S. & Ziegler, W. (Eds.): Messel: Ein Schaufenster in die Geschichte der Erde und des Lebens: 291–297; Frankfurt am Main (Verlag Waldemar Kramer)

Storch, G. & Schaarschmidt, F. (1992) Fauna and flora from Messel – a biogeographical puzzle. – In: Schaal, S. & Ziegler, W. (Eds.): Messel: An insight into the history of life and of the Earth: 291–297; Oxford (Clarendon Press)

Sturm, M. (1971) Die eozäne Flora von Messel bei Darmstadt I. Lauraceae. *Palaeontographica B*, **134**: 1–60

Sturm, M. (1978) Maw contents of an Eocene horse (*Propalaeotherium*) out of the oil shale of Messel near Darmstadt. *Courier Forschungsinstitut Senckenberg*, **30**: 120–122

Thiele-Pfeiffer, H. (1988) Die Mikroflora aus dem mitteleozänen Ölschiefer von Messel bei Darmstadt. *Palaeontographica B*, **211**: 1–86.

Thomson, P.W. & Pflug, H. (1953) Pollen und Sporen des mitteleuropäischen Tertiärs. *Palaeontographica B*, **94**: 1–138

Tobien, H. (1969) Die alttertiäre (mitteleozäne) Fossilfundstätte Messel bei Darmstadt (Hessen). *Mainzer Naturwissenschaftliches Archiv*, **8**: 149–180

Wilde, V. (1989) Untersuchungen zur Systematik der Blattreste aus dem Mitteleozän der Grube Messel bei Darmstadt (Hessen, Bundesrepublik Deutschland). *Courier Forschungsinstitut Senckenberg*, **115**: 1–215

Wilde, V. (2004) Aktuelle Übersicht zur Flora aus dem mitteleozänen „Ölschiefer" der Grube Messel bei Darmstadt (Hessen, Deutschland). *Courier Forschungsinstitut Senckenberg*, **252**: 109–114

Wilde, V. & Frankenhäuser, H. (1995) Flügelfrüchte engelhardioider Juglandaceen aus dem Mitteleozän von Eckfeld bei Manderscheid (Eifel). *Mainzer naturwissenschaftliches Archiv*, **33**: 47–52, 2 Abb.; Mainz

Wilde, V. & Frankenhäuser, H. (2000) *Comptonia*-like leaves from the German Middle Eocene. *Acta Palaeobotanica Supplementum*, **2** (for 1999): 447–463

Wilde, V, Kvaček, Z. & Bogner, J. (2005) Fossil leaves of Araceae from the European Eocene and notes on other aroid fossils. *International Journal of Plant Sciences*, **166**: 157–183

Wilde, V. & Manchester, S.R. (2003) *Cedrelospermum* fruits (Ulmaceae) and related leaves from the Middle Eocene of Messel (Hesse, Germany). *Courier Forschungsinstitut Senckenberg*, **241**: 147–153

Wilde, V. & Micklich, N. (2007a) The lake and its immediate surroundings. – In: Gruber, G. & Micklich, N. [Eds.]: Messel. Treasures of the Eocene. The book to the exhibition Messel on tour of the Hessisches Landesmuseum Darmstadt: 52–55; Darmstadt (Hessisches Landesmuseum Darmstadt)

Wilde, V. & Micklich, N. (2007b) The rain forest of Messel. – In: Gruber, G. & Micklich, N. (Eds.): Messel. Treasures of the Eocene. 80–84; Hessisches Landesmuseum, Darmstadt

Wilde, V. & Schaarschmidt, F. (1993a) Neue Möglichkeiten zur Untersuchung von Pollen in situ an Pflanzenresten aus dem „Ölschiefer" von Messel. *Mitteilungen für Wissenschaft und Technik*, **10**: 209–214

Wilde, V. & Schaarschmidt, F. (1993b) High resolution CLSM-studies of in-situ preserved pollen from the Messel oil shale. – *Scientific and Technical Information*, Wetzlar, **10**: 209–214

Wilde,V. & Süss, H. (2001) First wood with anatomically preserved details from the Middle Eocene oilshale of Messel (Hesse, Germany). *Acta Palaeobotanica*, **41**: 133–139

Wolfe, J.A. (1977) Paleogene floras from the Gulf of Alaska Region. *United States Geological Survey Professional Paper*, **997**: 1–107

Chapter 7

Archibald, S.B., Cover, S.P. & Moreau, C.S. (2006) Bulldog ants of the Eocene Okanagan Highlands and history of the subfamily (Hymenoptera: Formicidae: Myrmeciinae). *Annals of the Entomological Society of America*, **99** (3): 487–523

Archibald, B. & Farrel, B.D. (2003) Wheeler's Dilemma. In: Krzeminska, E. & Krzeminski, W. (Eds.) Proceedings of the Second Congress on Palaeoentomology "Fossil Insects", Krakow, Poland, 5–9 September, 2001. *Acta Zoologica Cracoviensia*, **46** (supplement): 17–23

Archibald, S.B., Johnson, K.R., Mathewes, R.W. & Greenwood, D.R. (2011) Intercontinental dispersal of giant thermophilic ants across the Arctic during early Eocene hyperthermals. *Proceedings of the Royal Society B*, **278**: 3679–3686

Bellmann, H., Honomichl, K. & Jacobs, W. (2007) Biologie und Ökologie der Insekten: ein Taschenlexikon. Elsevier, Spektrum Akademischer Verlag

Dathe, H. (Eds.) (2005) Lehrbuch der speziellen Zoologie, Band I, Wirbellose Tiere. 5. Teil: Insecta (Vol. 1). Spektrum Akademischer Verlag

Dlussky, G M. (2012) New fossil ants of the subfamily Myrmeciinae (Hymenoptera, Formicidae) from Germany. *Paleontological Journal*, **46** (3): 288–292

Dlussky, G.M., Wappler, T. & Wedmann, S. (2008) New Middle Eocene formicid species from Germany and the evolution of weaver ants. *Acta Palaeontolonica Polonica*, **53**: 615–626

Dlussky, G. M., Wappler, T. & Wedmann, S. (2009) Fossil ants of the genus *Gesomyrmex* Mayr (Hymenoptera, Formicidae) from the Eocene of Europe and remarks on the evolution of arboreal ant communities. *Zootaxa*, **2031**: 1–20

Dlussky, G.M. & Wedmann, S. (2012) Poneromorph ants (Hymenoptera, Formicidae: Amblyoponinae, Ectatomminae, Ponerinae) of Grube Messel, Germany: Diversification during the Eocene. *Journal of Systematic Palaeontology*, **10**: 725–753

Dunn, C. W., Giribet, G., Edgecombe, G. D. & Hejnol, A. (2014) Animal phylogeny and its evolutionary implications. *Annual Review of Ecology, Evolution, and Systematics*, **45**: 371–395

Eskov, K.Y. (2002) Geographical history of insects. – In: Rasnitsyn AP, Quicke DLJ (eds.) History of Insects, 427–435, Dordrecht: Kluwer

Fikácek, M., Wedmann, S. & Schmied, H. (2010) Diversification of the greater hydrophilines clade of giant water scavenger beetles dated back to the Middle Eocene (Coleoptera: Hydrophilidae: Hydrophilina). *Invertebrate Systematics*, **24**: 9–22

Fittkau, E.J. & Klinge, H. (1973) On biomass and trophic structure of the Central Amazonas rain forest ecosystem. *Biotropica*, **5**: 2–14

Garrouste, R. & Nel, A. (2015) New Eocene damselflies and first Cenozoic damsel-dragonfly of the isophlebiopteran lineage (Insecta: Odonata). *Zootaxa*, ***4028*** (3): 354–366

Gebeshuber, I.C. (2008) Strukturfarben in der Biologie. *PLUS LUCIS*, **1–2/2008**: 44–47

Goth, K. (1990) Der Messeler Ölschiefer- ein Algenlaminit. *Courier Forschungsinstitut Senckenberg*, **131**: 1–143

Grabatin, C. (2015) Systematic and palaeobiology of *Pseudotettigonia* n. sp. from the Eocene of Messel, Germany. Rheinische Friedrich-Wilhelms-Universität Bonn, 38 S., unveröffentlichte Bachelor-Arbeit

Greschbach, J. (2015) Biodiversity and systematics of fossil wasps (Hymenoptera: Vespoidea) from the Messel Pit, Germany. Goethe-Universität Frankfurt, 79 S., Unveröffentlichte Masterarbeit

Grimaldi, D. & Engel, M. S. (2005) Evolution of the Insects. Cambridge University Press

Günther, K., Hannemann, H-J, Hieke, F., Königsmann, E. & Schumann, H. (1994) Urania Tierreich in sechs Bänden. Insekten, Urania Verlagsgesellschaft

Habersetzer, J., Richter, G. & Storch, G. (1994) Paleoecology of early Middle Eocene bats from Messel, FRG. Aspects of flight, feeding and echolocation. *Historical Biology*, **8**: 235–260

Hörnschemeyer, T. (1994) Ein fossiler Tenebrionide *Ceropria? messelense* n. sp. (Coleoptera: Tenebrionidae: Diaperinae) aus dem Mitteleozän von Messel bei Darmstadt. *Courier Forschungsinstitut Senckenberg*, **170**: 75–83

Hörnschemeyer, T., Tröster, G. & Wedmann, S. (1995) Die eozänen Käferfaunen des Geiseltales und der Grube Messel – ein Vergleich unter systematischen und paläoökologischen Gesichtspunkten. *Hallesches Jahrbuch der Geowissenschaften B*, **17**: 107–119

Hörnschemeyer, T. & Wedmann, S. (1994) Fossile Prachtkäfer (Coleoptera: Buprestidae: Buprestinae) aus dem Mitteleozän der Grube Messel bei Darmstadt, Teil 1. *Courier Forschungsinstitut Senckenberg*, **170**: 85–136

Hughes, D.P., Wappler, T., & Labandeira, C.C. (2011) Ancient death-grip leaf scars reveal ant-fungal parasitism. *Biology Letters*, **7** (1): 67–70

Ivany, L.C., Patterson, W.P. & Lohmann, K C. (2000) Cooler winters as a possible cause of mass extinctions at the Eocene/Oligocene boundary. *Nature*, **407**: 887–890

Janis, C.M. (1993) Tertiary mammal evolution in the context of changing climates, vegetation, and tectonic events. *Annual Review of Ecology and Systematics*, **24**: 467–500

Kinzelbach, R.K. (1970a) Wanzen aus dem eozänen Ölschiefer von Messel (Insecta: Heteroptera). *Notizblatt des hessischen Landesamtes für Bodenforschung*, **98**: 9–18

Kinzelbach, R.K. (1970b) Eine Gangmine aus dem eozänen Ölschiefer von Messel (Insecta: ? Lepidoptera). *Paläontologische Zeitschrift*, **44**: 93–96

Kinzelbach, R.K. & Pohl, H. (1994) The fossil Strepsiptera (Insecta: Strepsiptera). *Annals of the Entomological Society of America*, **87**: 59–70

Krell, F.T. (2006) Fossil record and evolution of Scarabaeoidea (Coleoptera: Polyphaga). *The Coleopterists Bulletin*, **60** (sp5): 120–143

Labandeira, C. C., & Currano, E. D. (2013) The fossil record of plant-insect dynamics. *Annual Review of Earth and Planetary Sciences*, **41**: 287–311

Lutz, H. (1986) Eine neue Unterfamilie der Formicidae (Insecta: Hymenoptera) aus dem mittel-eozänen Ölschiefer der „Grube Messel" bei Darmstadt (Deutschland, S-Hessen). *Senckenbergiana lethaea*, **67**: 177–218

Lutz, H. (1988) Riesenameisen und andere Raritäten – die Insektenfauna. – In: Schaal, S. & Ziegler, W. (Eds.) Messel – ein Schaufenster in die Geschichte der Erde und des Lebens, 55–67, W. Kramer, Frankfurt

Lutz, H. (1990) Systematische und palökologische Untersuchungen an Insekten aus dem Mittel-Eozän der Grube Messel bei Darmstadt. *Courier Forschungsinstitut Senckenberg*, **124**: 1–165

Lutz, H. (1991) Autochthone aquatische Arthropoda aus dem Mittel-Eozän der Fundstätte Messel (Insecta: Heteroptera; Coleoptera; cf. Diptera-Nematocera; Crustacea: Cladocera). *Courier Forschungsinstitut Senckenberg*, **139**: 119–125

Martini, E. & Richter, G. (1996) Gehäuse von Köcherfliegenlarven aus den Ablagerungen von Messel und Sieblos/Rhön. *Natur und Museum*, **126** (8): 262–266

McNamara, M.E., Briggs, D.E. G., Orr, P.J., Wedmann, S., Noh, H., Cao, H. (2011) Fossilised biophotonic nanostructures reveal the original colors of 47 million-year-old moths. *PloS Biology*, **9** (11): e1001200

Meunier, F. (1921) Die Insektenreste aus dem Lutetien von Messel bei Darmstadt. *Abhandlungen der hessischen geologischen Landesanstalt*, **7** (3): 1–15

Mosbrugger, V. Utescher, T. & Dilcher, D.L. (2005) Cenozoic continental climatic evolution of Central Europe. *Proceedings of the Academy of Sciences USA*, **102**: 14964–14969

Neubert, E. (1999) The mollusca of the Eocene lake of Messel. *Courier Forschungsinstitut Senckenberg*, **216**: 167–181

Parker, A.R. & McKenzie, D.R. (2003) The cause of 50 million-year-old colour. *Proceedings of the Royal Society of London B*, **270** (Suppl 2): 151–153

Petrulevicius, J. F., Wappler, T., Wedmann, S., Rust, J. & Nel, A. (2008) New megapodagrionid damselflies (Odonata: Zygoptera) from the Paleogene of Europe. *Journal of Paleontology*, **82**: 1173–1181

Prum, R. O., Quinn, T. & Torres, R.H. (2006) Anatomically diverse butterfly scales all produce structural colours by coherent scattering. *Journal of Experimental Biology*, **209** (4): 748–765

Rheinheimer, J. (2007) Neue fossile Rüsselkäfer (Coleoptera: Curculionidae) aus dem Eozän des Baltischen Bernsteins und der Grube Messel bei Darmstadt.*Stuttgarter Beiträge zur Naturkunde B*, **365**: 1–24

Richter, G. (1993) Proof of feeding specialism in Messel bats? *Kaupia*, **3**: 107–112

Richter, G. & Baszio, S. (2001) Traces of a limnic food web in the Eocene lake Messel- a preliminary report based on fish coprolite analyses. *Palaeogeography, Palaeoclimatology, Palaeoecology*, **166**: 345–368

Richter, G. & Baszio, S. (2002) Beiträge zur Ökologie des tertiären Messelsees. *Natur und Museum*, **132** (4): 137–149

Richter, G. & Baszio, S. (2009) Geographic and stratigraphic distribution of spongillids (Porifera) and the leit value of spiculites in the Messel Pit Fossil Site. *Palaeobiodiversity and Palaeoenvironments*, **89** (1–2): 53–66

Richter, G. & Krebs, G. (1999) Larvenstadien von Eintagsfliegen (Insecta: Ephemeroptera) aus Sedimenten des eozänen Messelsees. *Natur und Museum*, **129** (1): 21–28

Richter, G. & Wedmann, S. (2005) Ecology of the Eocene Lake Messel revealed by analysis of small fish coprolithes and sediments from a drilling core. *Palaeogeography, Palaeoclimatology, Palaeoecology*, **223**: 147–161

Richter, G. & Wuttke, M. (1995) Der Messeler Süßwasser-Schwamm Spongilla gutenbergiana, eine Ephydatia. *Natur und Museum*, **125**: 134–135

Richter, G. & Wuttke, M. (1999) *Lutetiospongilla heili* n. gen. n. sp. und die eozäne Spongillidenfauna von Messel. *Courier Forschungsinstitut Senckenberg*, **216**: 183–195

Satoh, N., Rokhsar, D. & Nishikawa, T. (2014) Chordate evolution and the three-phylum system. *Proceedings of the Royal Society B*, **281**: 20141729

Schaarschmidt, F. & Wilde, V. (1986) Palmenblüten und-blätter aus dem Eozän von Messel. *Courier Forschungsinstitut Senckenberg*, **86**: 177–202

Schmied, H. (2009) Cockroaches (Blattodea) from the middle Eocene of Messel (Germany). Rheinische Friedrich-Wilhelms-Universität Bonn, 78 S., unveröffentlichte Diplom-Arbeit

Skartveit, J. & Wedmann, S. (2015) Fossil Bibionidae (Insecta: Diptera) from the Eocene of Grube Messel, Germany. *Studia Dipterologica*, **22**: 77–83

Szwedo, J. & Wappler, T. (2006) New planthoppers (Insecta: Hemiptera: Fulgoromorpha) from the Middle Eocene Messel Maar. *Annales Zoologici*, **56**: 555–566

Tröster, G. (1991) Eine neue Gattung der Elateridae (Insecta: Coleoptera) *Macropunctum* gen. n. aus der Messel-Formation des unteren Mittel-Eozän der Fundstätte Messel. *Courier Forschungsinstitut Senckenberg*, **139**: 99–117

Tröster, G. (1993) Zwei neue mitteleuropäische Arten der Gattung Tenomerga Neboiss 1984 aus dem Mitteleozän der Grube Messel und des Eckfelder Maares (Coleoptera: Archostemmata: Cupedidae). *Mainzer naturwissenschaftliches Archiv*, **31**: 169–176

Tröster, G. (1994) Fossile Elateridae (Insecta: Coleoptera) aus dem Unteren Mitteleozän (Lutetium) der Grube Messel bei Darmstadt. *Courier Forschungsinstitut Senckenberg*, **170**: 11–64

Tröster, G. (1999) An unusual new fossil click-beetle (Coleoptera: Elateridae) from the Middle Eocene of the Grube Messel (Germany). *Neues Jahrbuch für Geologie und Paläontologie Monatshefte*, **1999** (1): 11–20

Wanninger, A. (2016) Twenty years into the "new animal phylogeny": Changes and challenges. Introductory chapter to the Special Issue The new animal phylogeny: The first 20 years. *Organisms, Diversity and Evolution*, **16** (2): 315

Wappler, T. (2003a) New fossil lace bugs (Heteroptera: Tingidae) from the Middle Eocene of the Grube Messel (Germany), with a catalog of fossil lace bugs. *Zootaxa*, **374**: 1–26

Wappler, T. (2003b) Die Insekten aus dem Mittel-Eozän des Eckfelder Maares, Vulkaneifel. *Mainzer Naturwissenschaftliches Archiv, Beiheft*, **27**: 1–234

Wappler, T. & Andersen, N.M. (2004) Fossil water striders from the Middle Eocene fossil sites of Eckfeld and Messel, Germany (Hemiptera, Gerromorpha). *Paläontologische Zeitschrift*, **78**: 41–52

Wappler, T. (2006) Lutetiacader, a puzzling new genus of cantacaderid lace bugs (Heteroptera: Tingidae) from the Middle Eocene Messel Maar, Germany. *Palaeontology*, **49** (2): 435–444

Wappler, T. & Ben-Dov, Y. (2008) Preservation of armoured scale insects on angiosperm leaves from the Eocene of Germany. *Acta Palaeontologica Polonica*, **53** (4): 627–634

Wappler, T. & Engel, M.S. (2003) The middle Eocene bee faunas of Eckfeld and Messel, Germany (Hymenoptera: Apoidea). *Journal of Paleontology*, **77** (5): 908–921

Wappler, T. & Heiss, E. (2006) Flatbugs from Paleocene limnic sediments. I. Messel maar (Heteroptera: Aradidae). *Polish Journal of Entomology*, **75**: 207–217

Wappler, T., Heiss, E. & Wedmann, S. (2015a) New flatbug (Hemiptera: Heteroptera: Aradidae) records from the Middle Eocene Messel Maar, Germany. *Paläontologische Zeitschrift*, **89**: 653–660

Wappler, T., Labandeira, C. C., Engel, M.S., Zetter, R. & Grímsson, F. (2015b) Specialized and Generalized Pollen-Collection Strategies in an Ancient Bee Lineage. *Current Biology*, **25** (23): 3092–3098

Wappler, T., Labandeira, C. C., Rust, J., Frankenhäuser, H. & Wilde, V. (2012) Testing for the effects and consequences of mid Paleogene climate change on insect herbivory. *PLoS One*, **7** (7): e40744

Wappler, T., Tokuda, M. & Yukawa, J. (2010) Insect herbivores on *Laurophyllum lanigeroides* (Engelhardt 1992) Wilde: a role of a distinct plant-insect associational suite in host taxonomic assignment. *Palaeontographica B*, **283**: 137–155

Wedmann, S. (2005) Annotated taxon-list of the invertebrate animals from the Eocene fossil site Grube Messel near Darmstadt, Germany. *Courier Forschungsinstitut Senckenberg*, 255: 103–110

Wedmann, S. (2007) A nemestrinid fly (Insecta: Diptera: Nemestrinidae: cf. *Hirmoneura*) from the Eocene Messel Pit (Germany). *Journal of Paleontology*, 81: 1114–1117

Wedmann, S., Bradler, S. & Rust, J. (2007) The first fossil leaf insect: 47 million years of specialized cryptic morphology and behavior. *Proceedings of the National Academy of Sciences of the United States of America*, 104 (2): 565–569

Wedmann, S. & Hörnschemeyer, T. (1994) Fossile Prachtkäfer (Coleoptera: Buprestidae: Buprestinae und Agrilinae) aus dem Mitteleozän der Grube Messel bei Darmstadt, Teil 2. *Courier Forschungsinstitut Senckenberg*, 170: 137–187

Wedmann, S. & Makarkin, V. (2007) A new genus of Mantispidae (Insecta: Neuroptera) from the Eocene of Germany, with a review of the fossil record and palaeobiogeography of the family. *Zoological Journal of the Linnean Society*, 149: 701–716

Wedmann, S. & Richter, G. (2007) The ecological role of immature phantom midges (Diptera: Chaoboridae) in the Eocene Lake Messel, Germany. *African Invertebrates*, 48: 59–70

Wedmann, S. & Yeates, D. (2008) Eocene records of bee flies (Insecta, Diptera, Bombyliidae, *Comptosia*): Their paleobiogeographic implications and remarks on the evolutionary history of bombyliids. *Palaeontology*, 51: 231–240

Wedmann, S., Hörnschemeyer, T. & Schmied, H. (2011) Fossil water-penny beetles (Coleoptera: Psephenidae: Eubrianacinae) from the Eocene of Europe, with remarks on their phylogenetic position and biogeography. *Palaeontology*, 54: 965–980

Wedmann, S., Pouillon J.-M. & Nel, A. (2014) New Palaeogene horntail wasps (Hymenoptera, Siricidae) and a discussion of their fossil record. *Zootaxa*, 3869 (1): 33–43

Wedmann, S., Wappler, T. & Engel, M.S. (2009) Direct and indirect fossil records of megachilid bees from the Paleogene of Central Europe (Hymenoptera: Megachilidae). *Naturwissenschaften*, 96: 703–712

Westheide, W. & Rieger, G. (Hrsg.) (2013) Spezielle Zoologie. Teil 1. Einzeller und Wirbellose Tiere. 3. Auflage. 1–892; Springer-Spektrum, Berlin-Heidelberg.

Wunderlich, J. (1986) Die ersten Spinnen aus dem Mittel-Eozän der Grube Messel. *Senckenbergiana Lethaea*, 67 (1/4): 171–176

Wuttke, M. (1988) Erhaltung – Lösung – Umbau. Zum Verhalten biogener Stoffe bei der Fossilisation. – In: Schaal, S. & Ziegler, W. (Eds.) Messel – ein Schaufenster in die Geschichte der Erde und des Lebens. 263–276; W. Kramer, Frankfurt

Zachos, J., Pagani, M., Sloan, L. Thomas, E. & Billups, K. (2001) Trends, Rhythms, and Aberrations in Global climate 65 Ma to present. *Science*, 292: 686–692

Chapter 8

Andreae, A. (1893) Vorläufige Mittheilung über die Ganoiden (Lepidosteus und Amia) des Mainzer Beckens. Verhandlungen des naturhistorisch-medizinischen Vereins Heidelberg, 2: 7–15

Gaudant, J. (1993) The freshwater fish-fauna of Europe: from palaeobiogeography to palaeoclimatology. *Kaupia*, 3: 231–244

Gaudant, J. (2000) Nouvelles observations sur quelques Percoidei (Poissons téléostéens) des eaux douces et saumâtres du Cénozoïque européen. *Neues Jahrbuch für Geologie und Paläontologie, Abhandlungen,* 217 (2): 199–244

Grande, L. (1980) Palaeontology of the Green River Formation, with a review of the fish fauna. *Bulletin of the Geological Survey of Wyoming*, 63: 1–333

Grande, L. (2010) An empirical synthetic pattern study of gars (Lepisosteiformes) and closely related species, based mostly on skeletal anatomy: the resurrection of Holostei. *Copeia* 10 (2A Suppl.): 1–871

Grande, L. (2013) The Lost World of Fossil Lake. University of Chicago Press, Chicago

Grande, L. & Bemis, W.E. (1998) A comprehensive phylogenetic study of amiid fishes (Amiidae) based on comparative skeletal anatomy. An empirical search for interconnected patterns of natural history. *Journal of Vertebrate Paleontology*, 18 (1, Suppl.): 1–690

Koenigswald, W. v. & Rust, J. (2007) Globale Aspekte zur Paläogeographie und Evolution in der Fauna von Messel. – In: Gruber, G. & Micklich, N. (Eds.) Messel – Schätze der Urzeit, 29–34; Hessisches Landesmuseum, Darmstadt

Leidy, J. (1873) Notice of remains of fishes in the Bridger Tertiary Formation of Wyoming. Proceedings of the Academy of Natural Sciences, Philadelphia, 1873: 97–99

Martens, K., Goddeeries, B. & Coulter, G. (Eds.) (1994) Speciation in Ancient Lakes. E. Schweizerbart, Stuttgart

Micklich, N. (1978) Palaeoperca proxima, ein neuer Knochenfisch aus dem Mittel-Eozän von Messel bei Darmstadt. *Senckenbergiana lethaea*, 59: 483–501

Micklich, N. (1983) Ein Aal aus der Grube Messel: Gedanken und Probleme bei Aussagen zu Fossilienfunden. *Natur und Museum*, 113: 211–221

Micklich, N. (1985) Biologisch-paläontologische Untersuchungen zur Fischfauna der Messeler Ölschiefer (Mittel-Eozän, Lutetium). *Andrias*, 4: 1–171

Micklich, N. (1988): Urtümliche Panzerträger und moderne Kannibalen. – In: Schaal, S. & Ziegler, W. (Eds.) Messel – ein Schaufenster in die Geschichte der Erde und des Lebens, 71–92; Waldemar Kramer, Frankfurt a. M.

Micklich, N. (1996) Percoids (Teleostei, Perciformes) from the oilshale of the Messel-Formation (Middle Eocene, Lower Geiseltalian): An ancient speciation? *Publicaciones especiales de la Instituto Español de Oceanografía*, 21: 113–127

Micklich, N. (2002) The fish fauna of Messel Pit: A nursery school? *Courier Forschungsinstitut Senckenberg*, 237: 97–127

Micklich, N. (2005) Umwelt-Informanten im Messel-See. Die Fische. *Vernissage*, 21/05: 26–31

Micklich, N. (2007a) Ursprüngliche Knochenfische. – In: Gruber, G. & Micklich, N. (Eds.) Messel – Schätze der Urzeit, 56–58; Hessisches Landesmuseum, Darmstadt

Micklich, N. (2007b) Moderne Knochenfische. – In: Gruber, G. & Micklich, N. (Eds.) Messel – Schätze der Urzeit, 59–61; Hessisches Landesmuseum, Darmstadt

Micklich, N. (2007c) Zufälle oder Notwendigkeiten? – In: Gruber, G. & Micklich, N. (Eds.) Messel – Schätze der Urzeit, 38–48; Hessisches Landesmuseum, Darmstadt

Micklich, N. (2012a) Peculiarities of the Messel fish fauna and their palaeoecological implications: a case study. *Palaeodiversity and Palaeoenvironments*, 92 (4): 585–629

Micklich, N. (2012b) Der Messel-See: eine Artenwiege? – In: Martin, T., Koenigswald W. von, Radtke, G. & Rust, J. (Eds.) Paläontologie. 100 Jahre Paläontologische Gesellschaft, 64–65; Friedrich Pfeil, München

Micklich, N. & Klappert, G. (2001) *Masillosteus kelleri*, a new gar (Actinopterygii, Lepisosteidae) from the Middle Eocene of Grube Messel (Hessen, Germany). *Kaupia*, **11**: 73–81

Micklich, N. & Klappert, G. (2004) Character variation in some Messel fishes. – In: Arratia, G. & Tintori, A. (Eds.) Mesozoic Fishes 3 – Systematics, Palaeoenvironment and Biodiversity, 137–163; Freidrich Pfeil, München

Micklich, N. & Mentges, S. (2012) Fin ray fractures in Messel fishes. *Kaupia*, **18**: 19–27

Near, T.J., Eytan, R.I., Dornburg, A, Kuhn, K.L, Moore J.A., Davis, M.P., Wainwright, P., Friedman, M. & Smith, W.L. (2012) Resolution of ray-finned fish phylogeny and timing of diversification. *Proceedings of the National Academy of Sciences USA*, **109** (34): 13698–13703

Romer, A.S. & Fryxell, F.M. (1928) *Paramiatus gurleyi*, a deep-bodied amiid fish from the Eocene of Wyoming. American Journal of Science, **16**: 519–527

Sauvage, M.H.E. (1880) Notes sur les poissons fossiles. XVI. Sur les perches tertiaires. Bulletin de la Société géologique de France (3), **8**: 451–452

Schliewen, U.K., Rassmann, K., Markmann, M., Markert, J., Kocher, T. & Tautz, D. (2001) Genetic and ecological divergence of a monophyletic cichlid species pair under fully sympatric conditions in Lake Ejagham, Cameroon. *Molecular Ecology*, **10**: 14/1–1488

Weitzel, K. (1933) *Amphiperca multiformis* n. g. n. sp. und *Thaumaturus intermedius* n. sp., Knochenfische aus dem Mitteleozän von Messel. Notizblatt des Vereins für Erdkunde und der hessischen geologischen Landesanstalt, **5**: 89–97

Whitlock, J.A. (2010) Phylogenetic relationships of the Eocene percomorph fishes †*Priscacara* and †*Mioplosus*. *Journal of Vertebrate Palaeontology*, **30**: 1037–1048

Chapter 9

Davic, R.D. & Welsh, H.H. (2004) On the ecological roles of Salamanders. *Annual Review of Ecology, Evolution, and Systematics*, **35**: 405–34

Keller T. & Wuttke M. (1997) Ein Messeler Frosch mit Beutetier (Grube Messel, Mittel-Eozän, Hessen, BRD). *Courier Forschungsinstitut Senckenberg*, **201**: 237–242

Maus M. & Wuttke, M. (2004) The ontogenetic development of *Pelobates* cf. *decheni* tadpoles from the Upper Oligocene of Enspel (Westerwald/Germany). *Neues Jahrbuch für Geologie und Paläontologie, Abhandlungen*, **232**: 215–230

Ovaska, K. & Gregory, P.T. (1989) Population structure, growth, and reproduction in a Vancouver Island population of the salamander *Plethodon vehiculum*. *Herpetologica*, **45**: 133–143

Pomel, M. (1853) Catalogue méthodique et descriptif des vertébrés fossiles decouverts dans le basin hydrographique superieur de la Loire, et surtout dans la Valèe de son affluent principal, l'allier. Bailliers, Paris

Roček, Z. & Wuttke, M. (2010) Amphibia of Enspel (Late Oligocene, Germany). *Palaeobiodiversity and Palaeoenvironments*, **90**: 321–340

Roček, Z., Wang, Y. & Dong, L. (2012) Post-metamorphic development of Early Cretaceous frogs as a tool for taxonomic comparisons. *Journal of Vertebrate Paleontology*, **32**: 1285–1292

Roček, Z., Wuttke, M., Gardner, J.D. & Bhullar, B.-A.S. (2014) The Euro-American genus *Eopelobates*, and a re-definition of the family Pelobatidae (Amphibia, Anura). *Palaeobiodiversity and Palaeoenvironments*, **94** (4): 529–567

Roček, Z., Boistel, R., Lenoir, N., Mazurier, A., Pierce, S.E., Rage, J.-C., Smirnov, S.V., Schwermann, A.H., Valentin, X., Venczel, M., Wuttke, M. & Zikmund, T. (2015) Frontoparietal bone in extinct Palaeobatrachidae (Anura): its variation and taxonomic value. *The Anatomical Record*, **298**: 1848–1863

Schoch. R.R., Poschmann, M. & Kupfer, A. (2015) The salamandrid *Chelotriton paradoxus* from Enspel and Randeck Maars (Oligocene – Miocene, Germany). *Palaeobiodiversity and Palaeoenvironments*, **95**: 77–86

Smith, K.T. & Wuttke, M. (2015) Avian pellets from the late Oligocene of Enspel, Germany – ecological interactions in deep time. *Palaeobiodiversity and Palaeoenvironments*, **95** (1): 103–113

Villa, A., Roček, Z., Tschopp, E., Van Den Hoek, L.W. Ostende & Delfino, M. (2016) *Palaeobatrachus eurydices*, sp. nov. (Amphibia, Anura), the last Western European palaeobatrachid. *Journal of Vertebrate Paleontology*, **36** (6): e1211664

Westphal, F. (1988) *Chelotriton robustus* n. sp., ein Salamandride aus dem Eozän der Grube Messel bei Darmstadt. *Senckenbergiana Lethaea*, **60** (4/5): 475–487

Wuttke, M. (2012a) Redescription of the Middle Eocene frog *Lutetiobatrachus gracilis* Wuttke in Sanchiz, 1998 (Lower Geiseltalian, "Grube Messel", near Darmstadt, southern Hesse, Germany). *Kaupia*, **18**: 29–41

Wuttke, M. (2012b) The genus *Eopelobates* (Anura, Pelobatidae) from Messel, Geiseltal, and Eckfeld (Middle Eocene, Germany). Part I: Redescription of *Eopelobates wagneri* (Weitzel, 1938) from Messel (Lower Geiseltalium, Germany). *Kaupia*, **18**: 43–71

Wuttke, M. & Poschmann, M. (2010) First finding of fish in the diet of a water-dwelling extinct frog *Palaeobatrachus* from the Upper Oligocene Fossil-Lagerstätte Enspel (Westerwald Mountains, Western Germany). *Palaeobiodiversity and Palaeoenvironments*, **90** (1): 59–64

Wuttke, M., Prikyl, T., Ratnikov, V.Yu., Dvorak, Z. & Roček, Z. (2012) Generic diversity and distributional dynamics of the Palaeobatrachidae (Amphibia: Anura). *Palaeobiodiversity and Palaeoenvironments*, **92**: 367–395

Chapter 10

Benson, R.B., Campione, N.E., Carrano, M.T., Mannion, P.D., Sullivan, C., Upchurch, P. & Evans. D.C. (2014) Rates of dinosaur body mass evolution indicate 170 million years of sustained ecological innovation on the avian stem lineage. *PloS Biology*, **12**: e1001853

Benton, M.J. & Donoghue, P.C.J. (2007) Paleontological evidence to date the tree of life. *Molecular Biology and Evolution*, **24**: 25–53

Bever, G.L., Lyson, T.R., Field, D.J. & Bhullar, B.-A.S. (2015) Evolutionary origin of the turtle skull. *Nature*, **525**: 239–242

Brusatte, S.L., Nesbitt, S.J., Irmis, R.B., Butler, R.J., Benton, M.J. & Norell, M.A. (2010) The origin and early radiation of dinosaurs. *Earth-Science Reviews*, **101**: 68–100

Brusatte, S.L., Butler, R.J., Barrett, P.M., Carrano, M.T., Evans, D.C., Lloyd, G.T., Mannion, P.D., Norell, M.A., Peppe, D.J., Upchurch, P. & Williamson, T.E. (2015) The extinction of the dinosaurs. *Biological Reviews*, **90**: 628–642

Carroll, R.L. (1997) Patterns and Processes of Vertebrate Evolution. Cambridge University Press, Cambridge, U.K.

Clack, J.A. (2002) Gaining Ground: The Origin and Evolution of Tetrapods. Indiana University Press, Bloomington, Indiana.

Fröbisch, J. (2013) Vertebrate diversity across the end-Permian mass extinction – Separating biological and geological signals. *Palaeogeography, Palaeoclimatology, Palaeoecology*, **372**: 50–61

Gauthier, J.A., Kluge, A.G. & Rowe, T. (1988) Amniote phylogeny and the importance of fossils. *Cladistics*, **4**: 105–209

Norell, M.A. & Xu, X. (2005) Feathered dinosaurs. *Annual Review of Earth and Planetary Sciences*, **33**: 277–299

Roos, J., Aggarwal, R.K. & Janke, A. (2007) Extended mitogenomic phylogenetic analyses yield new insight into crocodylian evolution and their survival of the Cretaceous – Tertiary boundary. Molecular *Phylogenetics and Evolution*, **45**: 663–673

Sumida, S.S. & Martin, K.L.M. (Eds.) (1997) Amniote Origins: Completing the Transition to Land. Academic Press, New York.

Chapter 10.1

Augé, M. (1990) La faune de lézards et d'Amphisbenes (Reptilia, Squamata) du gisement de Dormaal (Belgique, Eocene inférieur). *Bulletin de l'Institut Royal des Sciences Naturelles de Belgique, Sciences de la Terre*, **60**: 161–173

Augé, M. (2005) Évolution des lézards du Paléogène en Europe. *Mémoires du Muséum National d'Histoire Naturelle*, **192**: 1–369

Baszio, S. (2004) *Messelophis variatus* n. gen. n. sp. from the Eocene of Messel: a tropidopheine snake with affinities to Erycinae (Boidae). *Courier Forschungsinstitut Senckenberg*, **252**: 47–66

Beck, D.D. (2005) Biology of Gila Monsters and Beaded Lizards. University of California Press, Berkeley, California

Beck, D.D. & Lowe, C.H. (1991) Ecology of the Beaded Lizard, *Heloderma horridum*, in a tropical dry forest in Jalisco, México. *Journal of Herpetology*, **25**: 395–406

Blackburn, D.G. & Stewart, J.R. (2011) Viviparity and placentation in snakes. – In: Aldrich, R. D. & Sever, D.M. (Eds.) Reproductive Biology and Phylogeny of Snakes. 119–181; Science Publishers, Enfield, New Hampshire

Bolet, A. & Evans, S.E. (2013) Lizards and amphibians (Reptilia, Squamata) from the late Eocene of Sossís (Catalonia, Spain). *Palaeontologia Electronica*, **16** (1): 1–23

Borsuk-Białynicka, M., Lubka, M. & Böhme, W. (1999) A lizard from Baltic amber (Eocene) and the ancestry of the crown group lacertids. *Acta Palaeontologica Polonica*, **44**: 349–382

Caldwell, M.W. & Lee, M.S.Y. (2001) Live birth in Cretaceous marine lizards (mosasauroids). *Proceedings of the Royal Society of London, Series B*, **268**: 2397–2401

Camp, C.L. (1923) Classification of the lizards. *Bulletin of the American Museum of Natural History*, **48**: 289–481

Čerňanský, A. & Bauer, A.M. (2010) *Euleptes gallica* Müller (Squamata: Gekkota: Sphaerodactylidae) from the lower Miocene of north-west Bohemia, Czech Republic. *Folia Zoologica*, **59**: 323–328

Čerňanský, A. & Augé, M. (2013) New species of the genus *Plesiolacerta* (Squamata: Lacertidae) from the upper Oligocene (MP28) of southern Germany and a revision of the type species *Plesiolacerta lydekkeri*. *Palaeontology*, **56**: 79–94

Čerňanský, A. & Smith, K.T. (2017) Eolacertidae: a new extinct clade of lizards from the Palaeogene; with comments on the origin of the dominant European reptile group- Lacertidae. Historical Biology, DOI: 10.1080/08912963.2017.1327530

Close, M. & Cundall, D. (2012) Mammals as prey: estimating ingestible size. *Journal of Morphology*, **273**: 1042–1049

Conrad, J.L. (2008) Phylogeny and systematics of Squamata (Reptilia) based on morphology. *Bulletin of the American Museum of Natural History*, **310**: 1–182

Conrad, J.L. & Norell, M.A. (2006) High-resolution X-ray computed tomography of an Early Cretaceous gekkonomorph (Squamata) from Öösh (Övökhangai; Mongolia). *Historical Biology*, **18**: 405–431

Cundall, D. & Greene, H.W. (2000) Feeding in snakes. – In: Schwenk, K. (Eds.) Feeding: Form, Function and Evolution in Tetrapod Vertebrates. 293–333; Academic Press, San Diego

Estes, R. (1983) Sauria Terrestria, Amphisbaenia (Handbuch der Paläoherpetologie, v. 10A). Gustav Fischer Verlag, Stuttgart

Evans, S.E. (2003) At the feet of the dinosaurs: the early history and radiation of lizards. *Biological Reviews*, **78**: 513–551

Gamble, T., Greenbaum, E., Jackman, T.R., Russell, A.P. & Bauer, A.M. (2012) Repeated origin and loss of adhesive toepads in geckos. *PLoS One*, **7** (6): e39429

Gauthier, J., Kearney, M., Maisano, J.A., Rieppel, O. & Behlke, A. (2012) Assembling the squamate tree of life: Perspectives from the phenotype and the fossil record. *Bulletin of the Peabody Museum of Natural History*, **53**: 3–308

Gauthier, J.A., Estes, R. & De Queiroz, K. (1988) A phylogenetic analysis of Lepidosauromorpha. – In: Estes, R. & Pregill, G. K. (Eds.) Phylogenetic Relationships of the Lizard Families. 15–98; Stanford University Press, Stanford

Greene, H.W. (1983) Dietary correlates of the origin and radiation of snakes. *American Zoologist*, **23**: 431–441

Greer, A.E. (1991) Limb reduction in squamates: identification of the lineages and discussion of the trends. *Journal of Herpetology*, **25**: 166–173

Head, J.J., Bloch, J.I., Hastings, A.K., Bourque, J.R., Cadena, E.A., Herrera, F.A., Polly, P.D. & Jaramillo, C.A. (2009) Giant boid snake from the Palaeocene neotropics reveals hotter past equatorial temperatures. *Nature*, **457**: 715–717

Ivanov, M., Rage, J.-C., Szyndlar, Z. & Venczel, M. (2000) Histoire et origine géographique des faunes de serpents en Europe. *Bulletin de la Société herpétologique de France*, **96**: 15–24

Jones, M.E.H., Anderson, C.L., Hipsley, C.A., Müller, J., Evans, S.E. & Schoch, R.M. (2013) Integration of molecules and new fossils supports a Triassic origin for Lepidosauria (lizards, snakes, and tuatara). *BMC Evolutionary Biology*, **13**: 208

Keller, T. (2009) Beiträge zur Kenntnis von *Placosauriops abderhaldeni* KUHN, 1940 (Anguidae, Glyptosaurinae MARSH, 1872) aus dem Mitteleozän der Grube Messel - Skelettanatomie, Taphonomie und Biomechanik. *Kaupia*, **16**: 3–145

Kuhn, O. (1940) Die Placosauriden und Anguiden aus dem mittleren Eozän des Geiseltales. *Nova Acta Leopoldina*, **8**: 461–486

Lang, M. (1989) Phylogenetic and biogeographic patterns of basiliscine iguanians (Reptilia: Squamata: „Iguanidae"). Bonner Zoologische Monographien, **28**: 1–172

Longrich, N.R., Vinther, J., Pyron, R.A., Pisani, D. & Gauthier, J.A. (2015) Biogeography of worm lizards (Amphisbaenia) driven by end-Cretaceous mass extinction. *Proceedings of the Royal Society of London B*, **282**: 20143034

Mayr, G. & Schaal, S.F.K. (2016) Gastric pellets with bird remains from the early Eocene of Messel. *Palaios*, **31** (9): 447–451

Müller, J. (2001) Osteology and relationships of *Eolacerta robusta*, a lizard from the middle Eocene of Germany (Reptilia, Squamata). *Journal of Vertebrate Paleontology*, **21**: 261–278

Müller, J. (2002) *Eolacerta* from the Eocene of Prémontré, France (Reptilia, Squamata). *Neues Jahrbuch für Geologie und Paläontologie, Monatshefte*, **2002**: 490–500

Müller, J., Hipsley, C.A., Head, J.J., Kardjilov, N., Hilger, A., Wuttke, M. & Reisz, R.R. (2011) Eocene lizard from Germany revelas amphisbaenian origins. *Nature*, **473**: 364–367

Nöth, L. (1940) *Eolacerta robusta* n. g. n. sp., ein Lacertilier aus dem mittleren Eozän des Geiseltales. *Nova Acta Leopoldina*, **8**: 439–460

Nydam, R.L. (2000) A new taxon of helodermatid-like lizard from the Albian – Cenomanian of Utah. *Journal of Vertebrate Paleontology*, **20**: 285–294

Pough, F.H. (1973) Lizard energetics and diet. *Ecology*, **54**: 837–844

Reynolds, R.G., Niemiller, M.L. & Revell, L.J. (2014) Toward a Tree-of-Life for the boas and pythons: Multilocus species-level phylogeny with unprecedented taxon sampling. *Molecular Phylogenetics and Evolution*, **71**: 201–213

Rieppel, O. (1980) Ein Lacertilier aus dem Eozän von Messel bei Darmstadt. *Beiträge zur naturkundlichen Forschung Südwest-Deutschlands*, **39**: 57–69

Scanferla, A. (2016) Postnatal ontogeny and the evolution of macrostomy in snakes. *Royal Society Open Science*, **3**: 160612

Scanferla, C.A., Smith, K.T. & Schaal, S.F.K. (2016) Revision of the cranial anatomy and phylogenetic relationships of the Eocene minute boas *Messelophis variatus* and *Messelophis ermannorum* (Serpentes, Booidea). *Zoological Journal of the Linnean Society*, **176**: 182–206

Schaal, S. (2004) *Palaeopython fischeri* n. sp. (Serpentes: Boidae), eine Riesenschlange aus dem Eozän (MP 11) von Messel. *Courier Forschungsinstitut Senckenberg*, **252**: 35–45

Schaal, S. & Baszio, S. (2004) *Messelophis ermannorum* n. sp., eine neue Zwergboa (Serpentes: Boidae: Tropidopheinae) aus dem Mittel-Eozän von Messel. *Courier Forschungsinstitut Senckenberg*, **252**: 67–77

Schulte, J.A., Ii, Valladares, J.P. & Larson, A. (2003) Phylogenetic relationships within Iguanidae inferred using molecular and morphological data and a phylogenetic taxonomy of iguanian lizards. *Herpetologica*, **59**: 399–419

Smith, K.T. (2009a) Eocene lizards of the clade *Geiseltaliellus* from Messel and Geiseltal, Germany, and the early radiation of Iguanidae (Squamata: Iguania). *Bulletin of the Peabody Museum of Natural History*, **50**: 219–306

Smith, K.T. (2009b) A new lizard assemblage from the earliest Eocene (zone Wa0) of the Bighorn Basin, Wyoming, USA: Biogeography during the warmest interval of the Cenozoic. *Journal of Systematic Palaeontology*, **7**: 299–358

Smith, K.T. (2011) The evolution of mid-latitude faunas during the Eocene: late Eocene lizards of the Medicine Pole Hills reconsidered. *Bulletin of the Peabody Museum of Natural History*, **52**: 3–105

Smith, K.T. (2017a) The squamation of the stem-basilisk *Geiseltaliellus maarius* (Squamata: Iguanidae) from the Eocene of Messel, Germany. *Salamandra*

Smith, K.T. (2017b) First crocodile-tailed lizard (*Squamata: Pan-Shinisaurus*) from the Paleogene of Europe. *Journal of Vertebrate Paleontology*, DOI:10.1080/02724634.2017.1313743

Smith, K.T. & Wuttke, M. (2012) From tree to shining sea: Taphonomy of the arboreal lizard *Geiseltaliellus maarius* from Messel, Germany. *Palaeobiodiversity and Palaeoenvironments*, **92**: 45–65

Smith, K.T. & Gauthier, J.A. (2013) Early Eocene lizards of the Wasatch Formation near Bitter Creek, Wyoming: diversity and paleoenvironment during an interval of global warming. *Bulletin of the Peabody Museum of Natural History*, **54**: 135–230

Smith, K.T. & Scanferla, A. (2016) Fossil snake preserving three trophic levels and evidence for an ontogenetic dietary shift. *Palaeobiodiversity and Palaeoenvironments*, **96**: 589–599

Stritzke, R. (1983) *Saniwa feisti* n. sp., ein Varanide (Lacertilia, Reptilia) aus dem Mittel-Eozän von Messel bei Darmstadt. *Senckenbergiana Lethaea*, **64**: 497–508

Sullivan, R.M., Keller, T. & Habersetzer, J. (1999) Middle Eocene (Geiseltalian) anguid lizards from Geiseltal and Messel, Germany. I. *Ophisauriscus quadrupes* Kuhn 1940. *Courier Forschungsinstitut Senckenberg*, **216**: 97–129

Szyndlar, Z. & Rage, J.-C. (2003) Non-erycine Booidea from the Oligocene and Miocene of Europe. Polish Academy of Sciences, Krakow

Wang, Y. & Evans, S.E. (2011) A gravid lizard from the Cretaceous of China and the early history of squamate viviparity. *Naturwissenschaften*, **98**: 739–743

Weber, S. (2004) *Ornatocephalus metzleri* gen. et spec. nov. (Lacertilia, Scincoidea) – taxonomy and paleobiology of a basal scincoid lizard from the Messel Formation (middle Eocene: basal Lutetian, Geiseltalium), Germany. *Abhandlungen der senckenbergischen naturforschenden Gesellschaft*, **561**: 1–159

Wiens, J.J. & Slingluff, J.L. (2001) How lizards turn into snakes: A phylogenetic analysis of body-form evolution in anguid lizards. *Evolution*, **55**: 2303–2318

Williams, E.E. (1959) Gadow's arcualia and the development of tetrapod vertebrae. *Quarterly Review of Biology*, **1959**: 1–32

Chapter 10.2

Burke, A.C. (1989) Development of the turtle carapace: implications for the evolution of a novel bauplan. *Journal of Morphology*, **199**: 363–378

Cadena, E. (2015) A global phylogeny of *Pelomedusoides* turtles with new material of *Neochelys franzeni* Schleich, 1993 (Testudines, Podocnemididae) from the middle Eocene, Messel Pit, of Germany. *PeerJ*, **3**: e1221; DOI 10.7717/peerj.1221

Cadena, E. (2016) *Palaeoamyda messeliana* nov. comb. (*Testudines, Pan-Trionychidae*) from the Eocene Messel Pit and Geiseltal localities, Germany, taxonomic and phylogenetic insights. *PeerJ*, **4**: e2647

Cadena, E. & Joyce, W.G. (2015) A review of the fossil record of turtles of the clades Platychelyidae and Dortokidae. *Bulletin of the Peabody Museum of Natural History*, **56**: 3–20

Claude, J. & Tong, H. (2004) Early Eocene testudinoid turtles from Saint-Papoul, France with comments on the early evolution of modern Testudinoidea. *Oryctos*, **5**: 3–45

Ernst, C.H. & Barbour, R.W. (1989) Turtles of the World. 290; Smithsonian Institution Press, Washington, DC

Gaffney, E.S. (1975) A phylogeny and classification of the higher categories of turtles. *Bulletin of the American Museum of Natural History*, **155**: 389–436

Gaffney, E.S., Meylan, P.A., Wood, R., Simons, E. & De Almeida Campos, D. (2011) Evolution of the side-necked turtles: the family Podocnemididae. *Bulletin of the American Museum of Natural History*, **350**: 1–237

Georgalis, G. & Joyce, W.G. (2017) A review of the fossil record of Old World turtles of the clade *Pan-Trionychidae*. *Bulletin of the Peabody Museum of Natural History*, **56** (2): 185–244

Gramann, F. (1956) Schildkröten aus dem Melanienton von Borken (Niederhessische Senke) (*Trionyx, Anosteira*). *Notiz-blatt des hessischen Landes-Amts für Bodenforschung*, **84**: 16–20

Harrassowitz, H. (1922) Die Schildkrötengattung *Anosteira* von Messel bei Darmstadt und ihre stammesgeschichtliche Bedeutung. *Abhandlungen der Hessischen Geologischen Landesanstalt*, **6**: 132–239.

Hay, O.P. (1908) The Fossil Turtles of North America. *Carnegie Institute of Washington Publications*, **75**: 1–568

Hervet, S. (2004) Systématique du groupe «*Palaeochelys* sensu lato – *Mauremys*» (Chelonii, Testudinoidea) du Tertiaire d'Europe occidentale : principaux résultats. *Annales de Paléontologie*, **90**: 13–78

Holroyd, P.A., J.H. Hutchison & S.G. Strait (2001) Turtle diversity and abundance through the lower Eocene Willwood Formation of the southern Bighorn Basin. *University of Michigan Papers on Paleontology*, **33**: 97–107

Hummel, K. (1935) Schildkröten aus der mitteleozänen Braunkohle des Geiseltales. *Nova Acta Leopoldina*, **2**: 457–483

Hutchison, J.H. (1998) Turtles across the Paleocene/Eocene Epoch boundary in west-central North America. – In: Aubry, M.-P., Lucas, S. & Berggren, W.A. (Eds.) Late Paleocene – Early Eocene Climate and Biotic Events in the Marine and Terrestrial Records. 401–408; Columbia University Press, New York

Joyce, W.G. (2014) A review of the fossil record of turtles of the clade *Pan-Carettochelys*. *Bulletin of the Peabody Museum of Natural History*, **55**: 3–33

Joyce, W.G. & Gauthier, J.A. (2004) Paleoecology of Triassic stem turtles sheds new light on turtle origins. Proceedings of the Royal Society of London, Series B, *Biological Sciences*, **271**: 1–5

Joyce, W.G., Micklich, N., Schaal, S.F.K. & Scheyer, T.M. (2012) Caught in the act: the first record of copulating fossil vertebrates. *Biology Letters*, **8**: 846–848

Joyce, W.G., Rabi, M., Clark, J.M. & Xu, X. (2016) A toothed turtle from the Late Jurassic of China and the global biogeographic history of turtles. *BMC Evolutionary Biology*, **16**: 236

Lapparent De Broin, F. De. (2001) The European turtle fauna from the Triassic to the present. *Dumerilia*, **4**: 155–217

Li, C., Wu, X.-C., Rieppel, O., Wang, L.-T. & Zhao, L.-J. (2008) An ancestral turtle from the Late Triassic of southwestern China. *Nature*, **456**: 497–501

Lee, M.S.Y. (1997) Pareiasaur phylogeny and the origin of turtles. *Zoological Journal of the Linnean Society*, **120**: 197–280

Lyson T.R., Bever, G.S., Bhullar, B.-A.S., Joyce, W.G. & Gauthier, J.A. (2010) Transitional fossils and the origin of turtles. *Biology Letters*, **6**: 830–833

Lyson, T.R, Bhullar, B.-A.S., Bever, G.S, Joyce, W.G., De Queiroz, K., Abzhanov, A. & Gauthier, J.A. (2013a) Homology of the enigmatic nuchal bone reveals novel reorganzation of the shoulder girdle in the evolution of the turtle shell. *Evolution and Development*, **15**: 317–325

Lyson, T.R., Bever, G.S., Scheyer, T.M., Hsiang, A.Y. & Gauthier, J.A. (2013b) Evolutionary origin of the turtle shell. *Current Biology*, **23**: 1113–1119

Reinach, A. (1900) Schildkrötenreste im Mainzer Tertiärbecken und in benachbarten, ungefähr gleichalterigen Ablagerungen. *Abhandlungen der Senckenbergischen Naturforschenden Gesellschaft*, **28**: 1–135

Scheyer, T.M., Mörs, T., Einarsson, E. (2012) First record of soft-shelled turtles (Cryptodira, Trionychidae) from the Late Cretaceous of Europe, *Journal of Vertebrate Paleontology*, **32** (5): 1027–1032

Schleich, H.H. (1993) *Neochelys franzeni* n. sp., the first pleurodiran turtle from Messel. *Kaupia*, **3**: 15–21

Schoch, R.R. & Sues, H.-D. (2015) A Middle Triassic stem-turtle and the evolution of the turtle body plan. *Nature*, **523**: 584–587

Staesche, K. (1928) Sumpfschildkröten aus hessischen Tertiärablagerungen. *Abhandlungen der Hessischen Geologischen Landesanstalt*, **8**: 1–72

Werneburg, I., Hinz, J., Gumpenberger, M., Volpato, V., Natchev, N. & Joyce, W.G. (2015) Modeling neck mobility in fossil turtles. *Journal of Experimental Zoology Part B*, **324**: 230–243

Chapter 10.3

Abel, O. (1928) Das biologische Trägheitsgesetz. E. Haim, Vienna

Aráez, J.L.D., Delfino, M., Luján, À.H., Fortuny, J., Bernardini, F. & Alba, D.M. (2017) New remains of *Diplocynodon* (Crocodylia: Diplocynodontidae) from the Early Miocene of the Iberian Peninsula. *Comptes Rendus Palevol* **16** (1): 12–26

Berg, D.E. (1966) Die Krokodile, insbesondere *Asiatosuchus* und aff. *Sebecus*?, aus dem Eozän von Messel bei Darmstadt/Hessen. *Abhandlungen des hessischen Landesamtes für Bodenforschung*, **52**: 1–105

Brochu, C.A. (1997a) A review of "*Leidyosuchus*" (Crocodyliformes, Eusuchia) from the Cretaceous through Eocene of North America. *Journal of Vertebrate Paleontology*, **17** (4): 679–697

Brochu, C.A. (1997b) Morphology, fossils, divergence timing, and the phylogenetic relationships of *Gavialis*. *Systematic Biology*, **46** (3): 479–522

Brochu, C.A. (1999) Phylogenetics, taxonomy, and historical biogeography of Alligatoroidea. *Journal of Vertebrate Paleontology*, **19** (Suppl. 2): 9–100

Brochu, C.A. (2000) Phylogenetic relationships and divergence timing of *Crocodylus* based on morphology and the fossil record. *Copeia*, **2000**: 657–73

Brochu, C.A. (2003) Phylogenetic approaches toward crocodylian history. *Annual Review of Earth and Planetary Sciences*, **31**: 357–397

Brochu, C.A. (2004) Alligatorine phylogeny and the status of *Allognathosuchus* Mook, 1921. *Journal of Vertebrate Paleontology*, **24** (4): 857–873

Brochu, C.A. (2007) Systematics and taxonomy of Eocene tomistomine crocodylians from Britain and Northern Europe. *Palaeontology*, **50** (4): 917–928

Brochu, C.A. (2013) Phylogenetic relationships of Palaeogene ziphodont eusuchians and the status of *Pristichampsus* Gervais, 1853. *Earth and Environmental Science Transactions of the Royal Society of Edinburgh*, **103** (3–4): 521–550

Brochu, C.A., Parris, D.C., Grandstaff, B.S., Denton Jr, R.K. & Gallagher, W.B. (2012) A new species of Borealosuchus (Crocodyliformes, Eusuchia) from the Late Cretaceous – early Paleogene of New Jersey. *Journal of Vertebrate Paleontology*, **32** (1): 105–116

Buffetaut, E. (1989) A new ziphodont mesosuchian crocodile from the Eocene of Algeria. *Palaeontographica Abteilung A*, **208**: 1–10

Carpenter, K. & Lindsey, D. (1980) The dentary of *Brachychampsa montana* Gilmore (Alligatorinae; Crocodylidae), a Late Cretaceous turtle-eating alligator. *Journal of Paleontology*, **54** (6): 1213–1217

Cope, E. D. (1896) The Primary Factors of Organic Evolution. Open Court Publishing, Chicago

Dalla Vecchia, F.M. & Cau, A. (2011) The first record of a notosuchian crocodyliform from Italy. *Rivista Italiana di Paleontologia e Stratigrafia (Research In Paleontology and Stratigraphy)*, **117** (2): 309–321

Delfino, M. & Smith, T. (2009) A reassessment of the morphology and taxonomic status of '*Crocodylus*' *depressifrons* Blainville, 1855 (Crocodylia, Crocodyloidea) based on the Early Eocene remains from Belgium. *Zoological Journal of the Linnean Society*, **156** (1): 140–167

Delfino, M. & Smith, T. (2012) Reappraisal of the morphology and phylogenetic relationships of the Middle Eocene alligatoroid *Diplocynodon deponiae* (Frey, Laemmert & Riess, 1987) based on a three-dimensional specimen. *Journal of Vertebrate Paleontology*, **32** (6): 1358–1369

Erickson, G.M., Gignac, P.M., Steppan, S.J., Lappin, A.K., Vliet, K.A., Brueggen, J.D. & Webb, G.J. (2012) Insights into the ecology and evolutionary success of crocodilians revealed through bite-force and tooth-pressure experimentation. *PLoS One*, **7** (3): e31781

Frey, E., Laemmert, A. & Riess, J. (1987) *Baryphracta deponiae* n. gn. sp. (Reptilia, Crocodylia), ein neues Krokodil aus der Grube Messel bey Darmstadt (Hessen, Bundesrepublik Deutschland). *Neues Jahrbuch für Geologie und Paläontologie, Monatshefte*, **1987**: 15–26

Galdikas, B.M. (1985) Crocodile predation on a proboscis monkey in Borneo. *Primates*, **26** (4): 495–496

Gold, M.E.L., Brochu, C.A. & Norell, M.A. (2014) An expanded combined evidence approach to the *Gavialis* problem using geometric morphometric data from crocodylian braincases and eustachian systems. *PloS One*, **9** (9): e105793

Grigg, G. & Kirshner, D. (2015) Biology and Evolution of Crocodylians. CSIRO Publishing, Clayton South, Australia

Hastings, A.K. & Hellmund, M. (2017) Evidence for prey preference partitioning in the middle Eocene high-diversity crocodylian assemblage of the Geiseltal-Fossillagerstätte, Germany utilizing skull shape analysis. *Geological Magazine*, **154** (1): 119–146

Hutchison, J.H. (1982) Turtle, crocodilian, and champsosaur diversity changes in the Cenozoic of the north-central region of western United States. *Palaeogeography, Palaeoclimatology, Palaeoecology*, **37** (2): 149–164

Jouve, S. (2016) A new basal tomistomine (Crocodylia, Crocodyloidea) from Issel (Middle Eocene; France): palaeobiogeography of basal tomistomines and palaeogeographic consequences. *Zoological Journal of the Linnean Society*, **177** (1): 165–182

Kuhn, O. (1938) Die Crocodilier aus dem mittlerern Eozän des Geiseltales bei Halle. *Nova Acta Leopoldina* **6**: 313–329

Langston, W. (1975). Ziphodont crocodiles, *Pristichampsus vorax* (Troxell), new combination, from the Eocene of North America. *Fieldiana, Geology*, **1222**: 1–24

Li, J. (1984) A new species of *Planocrania* from Hengdong, Hunan. *Vertebrata PalAsiatica*, **22**: 123–133

Ludwig, R. (1877) Fossile Crocodiliden aus der Tertiärformation des mainzer Beckens. *Palaeontographica, Supplement*, **3** (4–7): 1–52

Mannion, P.D., Benson, R.B., Carrano, M.T., Tennant, J.P., Judd, J. & Butler, R.J. (2015) Climate constrains the evolutionary history and biodiversity of crocodylians. *Nature Communications*, **6**: 8438

Markwick, P.J. (1998) Fossil crocodilians as indicators of Late Cretaceous and Cenozoic climates: implications for using palaeontological data in reconstructing palaeoclimate. *Palaeogeography, Palaeoclimatology, Palaeoecology*, **137** (3): 205–271

Martin, J.E., Smith, T., Lapparent De Broin, F., Escuillié, F. & Delfino, M. (2014) Late Palaeocene eusuchian remains from Mont de Berru, France, and the origin of the alligatoroid *Diplocynodon*. *Zoological Journal of the Linnean Society*, **172** (4): 867–891

Meredith, R.W., Hekkala, E.R., Amato, G. & Gatesy, J. (2011) A phylogenetic hypothesis for *Crocodylus* (Crocodylia) based on mitochondrial DNA: evidence for a trans-Atlantic voyage from Africa to the New World. *Molecular Phylogenetics and Evolution*, **60** (1): 183–191

Micklich, N. (2007) Krokodile. p. 72–76 In: Gruber, G. & Micklich, N. (eds) Messel: Schätze der Urzeit. Hessisches Landesmuseum, Darmstadt

Mook, C.C. (1955) Two new genera of Eocene crocodilians. *American Museum Novitates*, **1727**: 1–4

Norell, M.A. (1989) The higher level relationships of the extant Crocodylia. *Journal of Herpetology*, **23** (4): 325–335

Oaks, J.R. (2011) A time-calibrated species tree of Crocodylia reveals a recent radiation of the true crocodiles. *Evolution*, **65** (11): 3285–3297

Ősi, A. (2014) The evolution of jaw mechanism and dental function in heterodont crocodyliforms. *Historical Biology*, **26** (3): 279–414

Pierce, S.E., Angielczyk, K.D. & Rayfield, E.J. (2008) Patterns of morphospace occupation and mechanical performance in extant crocodilian skulls: a combined geometric morphometric and finite element modeling approach. *Journal of Morphology*, **269** (7): 840–864

Pol, D. & Leardi, J.M. (2015) Diversity patterns of Notosuchia (Crocodyliformes, Mesoeucrocodylia) during the Cretaceous of Gondwana. *Publicación Electrónica de la Asociación Paleontológica Argentina*, **15** (1): 172–186

Puértolas-Pascual, E., Blanco, A., Brochu, C.A. & Canudo, J.I. (2016) Review of the Late Cretaceous-early Paleogene crocodylomorphs of Europe: Extinction patterns across the K-PG boundary. *Cretaceous Research*, **57**: 565–590

Rabi, M. & Sebők, N. (2015) A revised Eurogondwana model: Late Cretaceous notosuchian crocodyliforms and other vertebrate taxa suggest the retention of episodic faunal links between Europe and Gondwana during most of the Cretaceous. *Gondwana Research*, **28** (3): 1197–1211

Rachmawan, D. & Brend, S. (2009) Human-*Tomistoma* interactions in Central Kalimantan, Indonesian Borneo. *Crocodile Specialist Group Newsletter*, **28** (1): 9–11

Rauhe, M. (1990) Habit-Habitus-Wechselbeziehung von *Allognathosuchus gaudryi* Stefano 1905 (= *Allognathosuchus haupti* Weitzel 1935). *Geologisches Jahrbuch Hessen*, **118**: 53–61

Rauhe, M. & Rossmann, T. (1995) News about fossil crocodiles from the middle Eocene of Messel and Geiseltal, Germany. *Hallesches Jahrbuch für Geowissenschaften*, **17**: 81–92

Rebêlo, G. H. & Lugli, L. (2001) Distribution and abundance of four caiman species (Crocodylia: Alligatoridae) in Jaú National Park, Amazonas, Brazil. *Revista de Biología Tropical*, **49** (3–4): 1096–1109

Roos, J., Aggarwal, R.K. & Janke, A. (2007) Extended mitogenomic phylogenetic analyses yield new insight into crocodylian evolution and their survival of the Cretaceous – Tertiary boundary. *Molecular Phylogenetics and Evolution*, **45** (2): 663–673

Rossmann, T. (1998) Studies on Cenozoic crocodiles: 2. Taxonomical revision of the family Pristichampsidae Efimov (Crocodilia: Eusuchia). *Neues Jahrbuch für Geologie und Paläontologie, Abhandlungen*, **210**: 85–128

Rossmann, T. (1999) Studien an kanozoischen Krokodilen: 1. Die paläoökologische Bedeutung des Eusuchen Krokodils *Pristichampsus rollinatii* (Gray) für die Fossillagerstätte Grube Messel. *Courier Forschungsinstitut Senckenberg*, **216**: 85–96

Rossmann, T. (2000a) Skelettanatomische Beschreibung von *Pristichampus rollinatii* (Gray) (Crocodilia, Eusuchia) aus dem Paläogen von Europa, Nordamerika und Ostasien. *Courier Forschungsinstitut Senckenberg*, **221**: 1–100

Rossmann, T. (2000b) Studies on Cenozoic crocodiles: 5. Biomechanical investigation on the postcranial skeleton of the Palaeogene crocodile *Pristichampsus rollinatii* (Eusuchia: Pristichampsidae). *Neues Jahrbuch für Geologie und Paläontologie, Abhandlungen*, **217** (3): 289–330

Rossmann, T. (2002) Studies on cenozoic crocodiles: 10. First evidence of a tomistomid crocodilian (Eusuchia: Tomistomidae) from the Middle Eocene (Geiseltalian, MP 11) of Grube Messel, Germany. *Neues Jahrbuch fur Geologie und Palaontologie, Monatshefte*, (3): 129–146

Rossmann, T., Rauhe, M. & Ortega, F. (2000) Studies on Cenozoic crocodiles: 8. *Bergisuchus dietrichbergi* Kuhn (Sebecosuchia: Bergisuchidae n. fam.) from the Middle Eocene of Germany, some new systematic and biological conclusions. *Paläontologische Zeitschrift*, **74** (3): 379–392

Salisbury, S.W. & Willis, P.M.A. (1996) A new crocodylian from the early Eocene of south-eastern Queensland and a preliminary investigation of the phylogenetic relationships of crocodyloids. *Alcheringa*, **20** (3): 179–226

Villamarín, F., Marioni, B., Thorbjarnarson, J.B., Nelson, B.W., Botero-Arias, R. & Magnusson, W.E. (2011) Conservation and management implications of nest-site selection of the sympatric crocodilians *Melanosuchus niger* and *Caiman crocodilus* in Central Amazonia, Brazil. *Biological Conservation*, **144** (2): 913–919

Wang, Y.Y., Sullivan, C. & Liu, J. (2016) Taxonomic revision of *Eoalligator* (Crocodylia, Brevirostres) and the paleogeographic origins of the Chinese alligatoroids. *PeerJ*, **4**: e2356

Wassersug, R.J. & Hecht, M.K. (1967) The status of the crocodylid genera *Procaimanoidea* and *Hassiacosuchus* in the New World. *Herpetologica*, **23** (1): 30–34

Weitzel, K. (1935) *Hassiacosuchus haupti* n. gn. sp., ein durophages Krokodil aus dem Mitteleozän von Messel. *Notizblatt des Vereins für Erdkunde und der hessischen geologischen Landesanstalt*, **5**: 1–16

Williamson, T.E. (1996) ?*Brachychampsa sealeyi*, sp. nov. (Crocodylia, Alligatoroidea), from the Upper Cretaceous (lower Campanian) Menefee Formation, northwestern New Mexico. *Journal of Vertebrate Paleontology*, **16** (3): 421–431

Wu, X.-C., Russell, A.P. & Brinkman, D.B. (2001) A review of *Leidyosuchus canadensis* Lambe, 1907 (Archosauria: Crocodylia), and an assesment of cranial variation based upon new material. *Canadian Journal of Earth Sciences*, **38** (12): 1665–1687

Zvonok, E.A. & Skutschas, P.P. (2011) On a tomistomine crocodile (Crocodylidae, Tomistominae) from the Middle Eocene of Ukraine. *Paleontological Journal*, **45** (6): 661–664

Chapter 11

Andors, A. (1992) Reappraisal of the Eocene groundbird *Diatryma* (Aves: Anserimorphae). Natural History Museum of Los Angeles County, *Science Series*, **36**: 109–125

Angst, D., Lécuyer, C., Amiot, R., Buffetaut, E., Fourel, F., Martineau, F., Legendre, S., Abourachid, A. & Herrel, A. (2014) Isotopic and anatomical evidence of an herbivorous diet in the Early Tertiary giant bird *Gastornis*. Implications for the structure of Paleocene terrestrial ecosystems. *Naturwissenschaften*, **101**: 313–322

Chiappe, L.M. & Meng, Q. (2016) Birds of Stone: Chinese Avian Fossils from the Age of Dinosaurs. Johns Hopkins University Press, Baltimore

Corlett, R.T. & Primack, R.B. (2011) Tropical Rain Forests: An Ecological and Biogeographical Comparison (2. Ed.). Wiley-Blackwell, Chichester

Ericson, P.G.P., Anderson, C.L., Britton, T., Elzanowski, A., Johansson, U.S., Källersjö, M., Ohlson, J.I., Parsons, T.J., Zuccon, D. & Mayr, G. (2006) Diversification of Neoaves: integration of molecular sequence data and fossils. *Biology Letters*, **2**: 543–547

Eriksson, O. (2016) Evolution of angiosperm seed disperser mutualisms: the timing of origins and their consequences for coevolutionary interactions between angiosperms and frugivores. *Biological Reviews*, **91**: 168–186

Franzen, J.L. & Köster, A. (1994) Die eozänen Tiere von Messel – ertrunken, erstickt oder vergiftet? *Natur und Museum*, **124**: 91–97

Grande, L. (2013) The Lost World of Fossil Lake. Snapshots from Deep Time. University of Chicago Press, Chicago

Grimaldi, D. & Engel, M.S. (2005) Evolution of the Insects. Cambridge University Press, New York

Hackett, S.J., Kimball, R.T., Reddy, S., Bowie, R.C.K., Braun, E.L., Braun, M.J., Chojnowski, J.L., Cox, W.A., Han, K.-L., Harshman, J., Huddleston, C.J., Marks, B.D., Miglia, K.J., Moore, W.S., Sheldon, F.H., Steadman, D.W., Witt, C.C. & Yuri, T. (2008) A phylogenomic study of birds reveals their evolutionary history. *Science*, **320**: 1763–1767

Harrison, C.J.O. (1982) The earliest parrot: a new species from the British Eocene. *Ibis*, **124**: 203–210

Hesse, A. (1990) Die Beschreibung der Messelornithidae (Aves: Gruiformes: Rhynocheti) aus dem Alttertiär Europas und Nordamerikas. *Courier Forschungsinstitut Senckenberg*, **128**: 1–176

Mayr, G. (1998) "Coraciiforme" und "piciforme" Kleinvögel aus dem Mittel-Eozän der Grube Messel (Hessen, Deutschland). *Courier Forschungsinstitut Senckenberg*, **205**: 1–101

Mayr, G. (1999) Caprimulgiform birds from the Middle Eocene of Messel (Hessen, Germany). *Journal of Vertebrate Paleontology*, **19**: 521–532

Mayr, G. (2000) Charadriiform birds from the early Oligocene of Céreste (France) and the Middle Eocene of Messel (Hessen, Germany). *Geobios*, **33**: 625–636

Mayr, G. (2001) New specimens of the Middle Eocene fossil mousebird *Selmes absurdipes* Peters 1999. *Ibis*, **143**: 427–434

Mayr, G. (2002a) A new specimen of *Salmila robusta* (Aves: Gruiformes: Salmilidae n. fam.) from the Middle Eocene of Messel. *Paläontologische Zeitschrift*, **76**: 305–316

Mayr, G. (2002b) A skull of a new pelecaniform bird from the Middle Eocene of Messel, Germany. *Acta Palaeontologica Polonica*, **47**: 507–512

Mayr, G. (2002c) A new species of *Plesiocathartes* (Aves: ?Leptosomidae) from the Middle Eocene of Messel, Germany. *PaleoBios*, **22**: 10–20

Mayr, G. (2003a) A postcranial skeleton of *Palaeopsittacus* Harrison, 1982 (Aves incertae sedis) from the Middle Eocene of Messel (Germany). *Oryctos*, **4**: 75–82

Mayr, G. (2003b) A new Eocene swift-like bird with a peculiar feathering. *Ibis*, **145**: 382–391

Mayr, G. (2004a) Phylogenetic relationships of the early Tertiary Messel rails (Aves, Messelornithidae). *Senckenbergiana Lethaea*, **84**: 317–322

Mayr, G. (2004b) New specimens of *Hassiavis laticauda* (Aves: Cypselomorphae) and *Quasisyndactylus longibrachis* (Aves: Alcediniformes) from the Middle Eocene of Messel, Germany. *Courier Forschungsinstitut Senckenberg*, **252**: 23–28

Mayr, G. (2005a) "Old World phorusrhacids" (Aves, Phorusrhacidae): a new look at *Strigogyps* ("*Aenigmavis*") *sapea* (Peters 1987). *PaleoBios*, **25**: 11–16

Mayr, G. (2005b) The A new cypselomorph bird from the Middle Eocene of Germany and the early diversification of avian aerial insectivores. *Condor*, **107**: 342–352

Mayr, G. (2005c) The Palaeogene Old World potoo *Paraprefica* Mayr, 1999 (Aves, Nyctibiidae): its osteology and affinities to the New World Preficinae Olson, 1987. *Journal of Systematic Palaeontology*, **3**: 359–370

Mayr, G. (2005d) A new Eocene *Chascacocolius*-like mousebird (Aves: Coliiformes) with a remarkable gaping adaptation. *Organisms, Diversity & Evolution*, **5**: 167–171

Mayr, G. (2006a) New specimens of the early Eocene stem group galliform *Paraortygoides* (Gallinuloididae), with comments on the evolution of a crop in the stem lineage of Galliformes. *Journal of Ornithology*, **147**: 31–37

Mayr, G. (2006b) A new raptorial bird from the Middle Eocene of Messel, Germany. *Historical Biology*, **18**: 95–102

Mayr, G. (2006c) New specimens of the Eocene Messelirrisoridae (Aves: Bucerotes), with comments on the preservation of uropygial gland waxes in fossil birds from Messel and the phylogenetic affinities of Bucerotes. *Paläontologische Zeitschrift*, **80**: 390–405

Mayr, G. (2008a) First substantial Middle Eocene record of the Lithornithidae (Aves): A postcranial skeleton from Messel (Germany). *Annales de Paléontologie*, **94**: 29–37

Mayr, G. (2008b) Phylogenetic affinities of the enigmatic avian taxon *Zygodactylus* based on new material from the early Oligocene of France. *Journal of Systematic Palaeontology*, **6**: 333–344

Mayr, G. (2008c) The Madagascan "cuckoo-roller" (Aves: Leptosomidae) is not a roller – notes on the phylogenetic affinities and evolutionary history of a "living fossil". *Acta Ornithologica*, **43**: 226–230

Mayr, G. (2009a) Paleogene Fossil Birds. Springer, Heidelberg.

Mayr, G. (2009b) Towards the complete bird – the skull of the middle Eocene Messel lithornithid (Aves, Lithornithidae). *Bulletin de l'Institut Royal des Sciences Naturelles de Belgique, Sciences de la Terre*, **79**: 169–173

Mayr, G. (2009c) A well-preserved skull of the "falconiform" bird *Masillaraptor* from the middle Eocene of Messel (Germany). *Palaeodiversity*, **2**: 315–320

Mayr, G. (2009d) A well-preserved second trogon skeleton (Aves, Trogonidae) from the middle Eocene of Messel, Germany. *Palaeobiodiversity and Palaeoenvironments*, **89**: 1–6

Mayr, G. (2010a) Phylogenetic relationships of the paraphyletic "caprimulgiform" birds (nightjars and allies). *Journal of Zoological Systematics and Evolutionary Research*, **48**: 126–137

Mayr, G. (2010b) A new avian species with tubercle-bearing cervical vertebrae from the Middle Eocene of Messel (Germany). *Records of the Australian Museum*, **62**: 21–28

Mayr, G. (2011a) Well-preserved new skeleton of the Middle Eocene *Messelastur* substantiates sister group relationship between Messelasturidae and Halcyornithidae (Aves, ?Pan-Psittaciformes). *Journal of Systematic Palaeontology*, **9**: 159–171

Mayr, G. (2011b) Two-phase extinction of "Southern Hemispheric" birds in the Cenozoic of Europe and the origin of the Neotropic avifauna. *Palaeobiodiversity and Palaeoenvironments*, **91**: 325–333

Mayr, G. (2014) The Eocene *Juncitarsus* – its phylogenetic position and significance for the evolution and higher-level affinities of flamingos and grebes. *Comptes Rendus Palevol*, **13**: 9–18

Mayr, G. (2015a) The middle Eocene European "ratite" *Palaeotis* (Aves, Palaeognathae) restudied once more. *Paläontologische Zeitschrift*, **89**: 503–514

Mayr, G. (2015b) Eocene fossils and the early evolution of frogmouths (Podargiformes): further specimens of *Masillapodargus* and a comparison with *Fluvioviridavis*. *Palaeobiodiversity and Palaeoenvironments*, **95**: 587–596

Mayr, G. (2015c) Skeletal morphology of the middle Eocene swift *Scaniacypselus* and the evolutionary history of true swifts (Apodidae). *Journal of Ornithology*, **156**: 441–450

Mayr, G. (2015d) A new specimen of the Early Eocene *Masillacolius brevidactylus* and its implications for the evolution of feeding specializations in mousebirds (Coliiformes). *Comptes Rendus Palevol*, **14**: 363–370

Mayr, G. (2015e) A reassessment of Eocene parrotlike fossils indicates a previously undetected radiation of zygodactyl stem group representatives of passerines (Passeriformes). *Zoologica Scripta*, **44**: 587–602

Mayr, G. (2015f) Towards completion of the early Eocene aviary: A new bird group from the Messel oil shale (Aves, Eopachypterygidae, fam. nov.). *Zootaxa*, **4013**: 252–264

Mayr, G. (2016a) Avian Evolution: The Fossil Record of Birds and its Paleobiological Significance. Wiley-Blackwell, Chichester

Mayr, G. (2016b) Fragmentary but distinctive: three new avian species from the early Eocene of Messel, with the earliest record of medullary bone in a Cenozoic bird. *Neues Jahrbuch für Geologie und Paläontologie, Abhandlungen*, **279**: 273–286

Mayr, G. (2016c) Avian feet, crocodilian food and the diversity of larger birds in the early Eocene of Messel. *Palaeobiodiversity and Palaeoenvironments*, **96**: 601–609

Mayr, G. (2017a) A small, "wader-like" bird from the early Eocene of Messel (Germany). *Annales de Paléontologie*, **103**: 141–147

Mayr, G. (2017b) The early Eocene birds of the Messel fossil site: a 48 million-year-old bird community adds a temporal perspective to the evolution of tropical avifaunas. *Biological Reviews*, **92**: 1174–1188

Mayr, G. & Mourer-Chauviré, C. (2000) Rollers (Aves: Coraciiformes s.s.) from the Middle Eocene of Messel (Germany) and the Upper Eocene of the Quercy (France). *Journal of Vertebrate Paleontology*, **20**: 533–546

Mayr, G., Mourer-Chauviré, C. & Weidig, I. (2004) Osteology and systematic position of the Eocene Primobucconidae (Aves, Coraciiformes sensu stricto), with first records from Europe. *Journal of Systematic Palaeontology*, **2**: 1–12

Mayr, G. & Peters, D.S. (1998) The mousebirds (Aves: Coliiformes) from the Middle Eocene of Grube Messel (Hessen, Germany). *Senckenbergiana Lethaea*, **78**: 179–197

Mayr, G., Rana, R.S., Rose, K.D., Sahni, A., Kumar, K. & Smith, T. (2013) New specimens of the early Eocene bird *Vastanavis* and the interrelationships of stem group Psittaciformes. *Paleontological Journal*, **47**: 1308–1314

Mayr, G. & Richter, G. (2011) Exceptionally preserved plant parenchyma in the digestive tract indicates a herbivorous diet in the Middle Eocene bird *Strigogyps sapea* (Ameghinornithidae). *Paläontologische Zeitschrift*, **85**: 303–307

Mayr, G. & Schaal, S.K.F. (2016) Gastric pellets with bird remains from the early Eocene of Messel. *Palaios*, **31**: 447–451

Mayr, G. & Wilde, V. (2014) Eocene fossil is earliest evidence of flower-visiting by birds. *Biology Letters*, **10**: 20140223

Mayr, G. & Zelenkov, N. (2009) New specimens of zygodactylid birds from the middle Eocene of Messel, with description of a new species of *Primozygodactylus*. *Acta Palaeontologica Polonica*, **54**: 15–20

Mittelbach, G.G., Schemske, D.W., Cornell, H.V., Allen, A.P., Brown, J.M., Bush, M.B., Harrison, S.P., Hurlbert, A.H., Knowlton, N., Lessios, H.A., Mccain, C.M., Mccune, A.R., Mcdade, L.A., Mcpeek, M.A., Near, T.J., Price, T.D., Ricklefs, R.E., Roy, K., Sax, D.F., Schluter, D., Sobel, J.M. & Turelli, M. (2007) Evolution and the latitudinal diversity gradient: speciation, extinction and biogeography. *Ecology Letters*, **10**: 315–331

Mourer-Chauviré, C., Essid, E.M., Khayati Ammar, H., Marivaux, L., Marzougui, W., Temani, R., Vianey-Liaud, M. & Tabuce, R. (2016) New remains of the very small cuckoo, *Chambicuculus pusillus* (Aves, Cuculiformes, Cuculidae) from the late Early/early Middle Eocene of Djebel Chambi, Tunisia. *Palaeovertebrata*, **40**: (1)-e2. doi: 10.18563/pv.40.1.e2

Olson, S.L. (1992) A new family of primitive landbirds from the Lower Eocene Green River Formation of Wyoming. *Natural History Museum of Los Angeles County, Science Series*, **36**: 137–160

Peters, D.S. (1983) Die „Schnepfenralle" *Rhynchaeites messelensis* Wittich 1898 ist ein Ibis. *Journal für Ornithologie*, **124**: 1–27

Peters, D.S. (1987) Ein „Phorusrhacide" aus dem Mittel-Eozän von Messel (Aves: Gruiformes: Cariamae). *Documents des Laboratoires de Géologie de Lyon*, **99**: 71–87

Peters, D.S. (1988a) Ein vollständiges Exemplar von *Palaeotis weigelti* (Aves, Palaeognathae). *Courier Forschungsinstitut Senckenberg*, **107**: 223–233

Peters, D.S. (1988b) Die Messel-Vögel – eine Landvogelfauna. – In: Schaal, S. & Ziegler, W. (Eds.) Messel – Ein Schaufenster in die Geschichte der Erde und des Lebens. 135–151; Kramer, Frankfurt a.M.

Peters, D.S. (1992) A new species of owl (Aves: Strigiformes) from the Middle Eocene Messel oil shale. *Natural History Museum of Los Angeles County, Science Series*, **36**: 161–169.

Peters, D.S. (1995) *Idiornis tuberculata* n. spec., ein weiterer ungewöhnlicher Vogel aus der Grube Messel (Aves: Gruiformes: Cariamidae: Idiornithinae). *Courier Forschungsinstitut Senckenberg*, **181**: 107–119

Peters, D.S. (2006) Die eozäne Avifauna von Messel in ökologischer Sicht. *Ökologie der Vögel*, **25**: 195–214

Prum, R.O., Berv, J.S., Dornburg, A., Field, D.J., Townsend, J.P., Moriarty Lemmon, E. & Lemmon, A.R. (2015) A comprehensive phylogeny of birds (Aves) using targeted next-generation DNA sequencing. *Nature*, **526**: 569–573

Smith, K.T. & Wuttke, M. (2012) From tree to shining sea: Taphonomy of the arboreal lizard *Geiseltaliellus maarius* from Messel, Germany. *Palaeobiodiversity and Palaeoenvironments*, **92**: 45–65

Stidham, T.A. & Eberle, J.J. (2016) The palaeobiology of high latitude birds from the early Eocene greenhouse of Ellesmere Island, Arctic Canada. *Scientific Reports*, **6**: 20912

Vinther, J., Briggs, D.E.G., Clarke, J., Mayr, G. & Prum, R.O. (2010) Structural coloration in a fossil feather. *Biology Letters*, **6**: 128–131

Chapter 12

Angielczyk, K.D. & Schmitz, L. (2014) Nocturnality in synapsids predates the origin of mammals by over 100 million years. *Proceedings of the Royal Society of London Series B*, **281**: 20141642

Carrier, D.R. (1987) The evolution of locomotor stamina in tetrapods: circumventing a mechanical constraint. *Paleobiology*, **13**: 326–341

Luo, Z.X. (2007) Transformation and diversification in early mammal evolution. *Nature*, **450**: 1011–1019

Manley, G.A. (2010) An evolutionary perspective on middle ears. *Hearing Research*, **263**: 3–8

O'Leary, M.A., Bloch, J.I., Flynn, J.J., Gaudin, T.J., Giallombardo, A., Giannini, N.P., Goldberg, S.L., Kraatz, B.P., Luo, Z.-X., Meng, J., Ni, X., Novacek, M.J., Perini, F.A., Randall, Z.S., Rougier, G.W., Sargis, E.J., Silcox, M.T., Simmons, N.B., Spaulding, M., Velazco, P.M., Weksler, M., Wible, J.R. & Cirranello, A.L. (2013) The placental mammal ancestor and the post-K-Pg radiation of placentals. *Science*, **339**: 662–667

Rowe, T.B., Macrini, T.E. & Luo, Z.-X. (2011) Fossil evidence on origin of the mammalian brain. *Science*, **332**: 955–957

Rowe, T.B. & Shepherd, G.M. (2016) Role of ortho-retronasal olfaction in mammalian cortical evolution. *Journal of Comparative Neurology*, **524**: 471–495

Tarver, J E., Dos Reis, M., Mirarab, S., Moran, R.J., Parker, S., O'Reilly, J.E., King, B.L., O'Connell, M.J., Asher, R.J., Warnow, T., Peterson, K.J., Donoghue, P.C.J. & Pisani, D. (2016) The interrelationships of placental mammals and the limits of phylogenetic inference. *Genome Biology and Evolution*, **8**: 330–344

Welker, F., Collins, M.J., Thomas, J.A., Wadsley, M., Brace, S., Cappellini, E., Turvey, S.T., Reguero, M., Gelfo, J.N., Kramarz, A., Burger, J., Thomas-Oates, J., Ashford, D.A., Ashton, P.D., Rowsell, K., Porter, D.M., Kessler, B., Fischer, R., Baessmann, C., Kaspar, S., Olsen, J.V., Kiley, P., Elliott, J.A., Kelstrup, C.D., Mullin, V., Hofreiter, M., Willerslev, E., Hublin, J.-J., Orlando, L., Barnes, I. & Macphee, R.D.E. (2015) Ancient proteins resolve the evolutionary history of Darwin's South American ungulates. *Nature*, **522**: 81–84

Chapter 12.1

Case, J.A., Goin, F.J. & Woodburne, M.O. (2005) „South American" marsupials from the Late Cretaceous of North America and the origin of marsupial cohorts. *Journal of Mammalian Evolution*, **12** (3/4): 461–494

Cuvier, G. (1804) Mémoire sur le squelette presque entier d'un petit quadrudède du genre des Sarigues, trouvé dans la pierre à platre des environs de Paris. *Annales du Muséum d'histoire naturelle*, **5**: 277–292

Koenigswald, W.v. (1982) Die erste Beutelratte aus dem mitteleozänen Ölschiefer von Messel bei Darmstadt. *Natur und Museum*, **112** (2): 41–48

Kurz, C. (2007) The opossum-like marsupials (Didelphimorphia and Peradectia, Marsupialia, Mammalia) from the Eocene of Messel and Geiseltal. Ecomorphology, diversity and palaeogeography. *Kaupia*, **15**: 3–64

Kurz, C. & Habersetzer, J. (2004) Untersuchungen der Zahnmorphologie von Beutelratten aus Messel mittels Mikroröntgenmethode CORR. *Courier Forschungsinstitut Senckenberg*, **249**: 13–21

Martin, J.E., Case, J.A., Jagt, J.W.M., Schulp, A.S. & Mulder, E.W.A. (2005) A new European marsupial indicates a Late Cretaceous high-latitude transatlantic dispersal route. *Journal of Mammalian Evolution*, **12** (3/4): 495–511

Ziegler, R. (2006) Insectivores (Lipotyphla) and bats (Chiroptera) from the Late Miocene of Austria. *Annalen des Naturhistorischen Museums in Wien*. **107A**: 93–196

Chapter 12.2

Bloch, J.I. & Boyer, D.M. (2001) Taphonomy of small mammals in freshwater limestones from the Paleocene of the Clarks Fork Basin. *University of Michigan Papers on Paleontology*, **33**: 185–198

Bown, T.M. & Simons, E.L. (1987) New Oligocene Ptolemaiidae (Mammalia:? Pantolesta) from the Jebel Qatrani Formation, Fayum Depression, Egypt. *Journal of Vertebrate Paleontology*, **7** (3): 311–324

Christian, A. (1999) Zur Biomechanik der Fortbewegung von *Leptictidium* (Mammalia, Proteutheria), *Courier Forschungsinstitut Senckenberg*, **216**: 1–18

Clemens, W.A & Koenigswald, W.v. (1993) A new skeleton of *Kopidodon macrognathus* from the Middle Eocene of Messel and the relationships of paroxyclaenids and pantolestids based on postcranial evidence. *Kaupia*, **3**: 57-73

Eberle, J.J. & McKenna, M.C. (2002) Early Eocene Leptictida, Pantolesta, Creodonta, Carnivora, and Mesonychidae (Mammalia) from the Eureka Sound Group, Ellesmere Island, Nunavut. *Canadian Journal of Earth Sciences*, **39** (6): 899–910

Frey, E., Herkner, B., Schrenk, F. & Seiffert, C. (1993) Reconstructing organismic constructions and the problem of *Leptictidium*'s locomotion. *Kaupia*, **3**: 89–95

Gunnell, G.F., Bown, T.M., Block, J.I. (2008) Chapter 6: Leptictida. – In: Janis, C.M., Gunnell, G.F. & Uhen, M.D. (Eds.) Evolution of Tertiary Mammals of North America, Vol. 2. 82–88; Cambridge University Press, Cambridge

Hooker, J.J. (2013) Origin and evolution of the Pseudorhyncocyonidae, a European Paleogene family of insectivorous placental mammals. *Palaeontology*, **56** (4): 807–835

Kalthoff, D.C., Koenigswald, W.v. & Kurz, C. (2004) A new specimen of *Heterohyus nanus* (Apatemyidae, Mammalia) from the Eocene of Messel (Germany) with an unusual soft-part preservation. *Courier Forschungsinstitut Senckenberg*, **252**: 1–12

Koenigswald, W.v. (1980) Das Skelett eines Pantolestiden (Proteutheria, Mammalia) aus dem mittleren Eozän von Messel bei Darmstadt. *Paläontologische Zeitschrift*, **54** (3/4): 267–287

Koenigswald, W.v. (1983a) Skelettfunde von *Kopidodon* (Condylarthra, Mammalia) aus dem mitteleozänen Ölschiefer von Messel bei Darmstadt. *Neues Jahrbuch für Geologie und Paläontologie, Abhandlungen*, **167**: 1-39

Koenigswald, W.v. (1983b) Der erste Pantolestide (Proteutheria, Mammalia) aus dem Eozän des Geiseltals bei Halle. *Zeitschrift für Geologische Wissenschaften, Berlin*, **11**: 787-793

Koenigswald, W.v. (1987a) Apatemyiden-Skelette aus dem Mitteleozän von Messel und ihre paläobiologische Aussage. *Carolinea*, **45**: 31-35

Koenigswald, W.v. (1987b) Das zweite Skelett eines Pantolestiden (Pantolestidae, Proteutheria, Mammalia) aus dem Mitteleozän von Messel bei Darmstadt. *Carolinea*, **45**: 36–42

Koenigswald, W.v. (1988) *Kopidodon*, ein Verwandter der Urhuftiere, der auf Bäumen lebte. – In: Schaal, S. & Ziegler, W. (Eds.) Messel – Ein Schaufenster in die Geschichte der Erde und des Lebens. 235-237; Kramer Verlag, Frankfurt a.M.

Koenigswald, W.v. (1990) Die Paläobiologie der Apatemyiden (Insectivora s. l.) und die Ausdeutung der Skelettfunde von *Heterohyus nanus* aus dem Mitteleozän von Messel bei Darmstadt. *Palaeontographica A*, **210**: 41-77

Koenigswald, W.v. & Schierning, H.-P. (1987) The ecological niche of an extinct group of mammals, the early Tertiary apatemyids. *Nature*, **326**: 595-597

Koenigswald, W.v. & Storch, G. (1987) *Leptictidium tobieni* n. sp., ein dritter Pseudorhyncocyonide (Proteutheria, Mammalia) aus dem Eozän von Messel. *Courier Forschungsinstitut Senckenberg*, **91**: 107–116

Koenigswald, W.v., Rose, K.D., Grande, L. & Martin, R.D (2005) First apatemyid skeleton from the lower Eocene Fossil Butte Member, Wyoming, compared to the European apatemyid from Messel, Germany, *Palaeontographica*, **272**: 149–169

Koenigswald, W.v., Ruf, I. & Gingerich P.D. (2009) Cranial morphology of a new apatemyid, *Carcinella sigei* n. gen. n. sp. (Mammalia, Apatotheria) from the late Eocene of southern France, *Palaeontographica*, **288**: 53–91

Koenigswald, W.v. & Rust, J. (2011) Globale Aspekte zur Paläogeographie und Evolution der Fauna von Messel. – In: Gruber, G. & Micklich, N. (Eds.): Messel Schätze der Urzeit. 29–34; Hessisches Landesmuseum, Darmstadt

Maier, W., Richter, G. & Storch, G. (1986) *Leptictidium nasutum* – ein archaisches Säugetier aus Messel mit außergewöhnlichen biologischen Anpassungen. *Natur und Museum,* 116: 1–19

Matthew, W.D. (1909) The Carnivora and Insectivora of the Bridger Basin. Middle Eocene. *Memoirs of the American Museum of Natural History*, 9/4: 291–567

McKenna, M.C. (1980) Eocene paleolatitude, climate, and mammals of Ellesmere Island. *Palaeogeography, Palaeoclimatology, Palaeoecology*, 30: 349–362

O'Leary, M.A., Bloch, J.I., Flynn, J.J., Gaudin, T.J., Giallombardo, A., Giannini, N.P., Goldberg, S.L., Kraatz, B.P., Luo, Z.-X., Meng, J., Ni, X., Novacek, M.J., Perini, F.A., Randall, Z.S., Rougier, R.W., Sargis, E.J., Silcox, M.T., Simmons, N.B., Spaulding, M., Velazco, P.M., Weksler, M., Wible, J.R. & Cirranello, A.L. (2013) The placental mammal ancestor and the post-K-Pg radiation of placentals. *Science*, 339: 662–667

Pfretzschner, H.-U. (1993) Muscle reconstruction and aquatic locomotion in the Middle Eocene *Buxolestes piscator* from Messel near Darmstadt. *Kaupia*, 3: 75–87

Pfretzschner, H.-U. (1999) *Buxolestes minor* n. sp. - ein neuer Pantolestide (Mammalia, Proteutheria) aus der eozänen Messelformation. *Courier Forschungsinstitut Senckenberg*, 216: 19–29

Richter, G. (1987) Untersuchungen zur Ernährung eozäner Säuger aus der Fossilfundstätte Messel bei Darmstadt, *Courier Forschungsinstitut Senckenberg*, 91: 1–33

Rose, K.D. (1999) Postcranial skeleton of Eocene Leptictidae (Mammalia), and its implications for behavior and relationships. *Journal of Vertebrate Paleontology*, 19: 355–372

Rose, K.D. (2006a) The Beginning of the Age of Mammals. The Johns Hopkins University Press, Baltimore

Rose, K.D. (2006b) The postcranial skeleton of Early Oligocene *Leptictis* (Mammalia: Leptictida), with a preliminary comparison to *Leptictidium* from the Middle Eocene of Messel. *Palaeontographica A*, 278: 37–56

Rose, K.D. & Koenigswald, W.v. (2005) An exceptionally complete skeleton of *Palaeosinopa* (Mammalia, Cimolesta, Pantolestidae) from the Green River Formation, and other postcranial elements of the Pantolestidae from the Eocene of Wyoming (USA). *Palaeontographica A*, 273 (3–6): 55–96

Rose, K.D., Dunn, R.H. & Grande, L. (2014) A new skeleton of *Palaeosinopa didelphoides* (Mammalia, Pantolesta) from the early Eocene Fossil Butte Member, Green River Formation (Wyoming), and skeletal ontogeny in Pantolestidae. *Journal of Vertebrate Paleontology*, 34 4: 932–940

Ruf, I., Volpato, V., Rose, K.D., Billet, G., De Muizon, C. & Lehmann, T. (2016) Digital reconstruction of the inner ear of *Leptictidium auderiense* (Leptictida, Mammalia) and North American leptictids reveals new insight into leptictidan locomotor agility. *Paläontologische Zeitschrift*, 90 (1): 153–171

Russell, D.E. (1964) Les mammifères Paléocène d'Europe. *Mémoires du Muséum National d'Histoire Naturelle C*, 13: 1–324

Russell, D.E. & McKenna, M.C. (1961) Étude de *Paroxyclaenus*, mammifère des phosphorites du Quercy. *Bulletin de la Société Géologique de France*, 3 (7): 274–282

Russell, D.E. & Godinot, M. (1988) The paroxyclaenidae (Mammalia) and a new form from the early Eocene of Palette, France. *Paläontologische Zeitschrift*, 62 (3–4): 319–331

Russell, D.E., Godinot, M., Louis, P. & Savage, D.E. (1979) Apatotheria (Mammalia) de l'Eocène inférieur de France et de Belgique. *Bulletin Muséum National d'Histoire Naturelle*, 1: 203–243

Silcox, M.J., Bloch, J.J., Boyer, D.M. & Houde, P. (2010) Cranial anatomy of Paleocene and Eocene *Labidolemur kayi* (Mammalia: Apatotheria), and the relationships of the Apatemyidae to other mammals. *Zoological Journal of the Linnean Society*, 160: 773–825

Smith, T., Rose, K.D. & Gingerich, P.D. (2006) Rapid Asia – Europe – North America geographic dispersal of earliest Eocene primate *Teilhardina* during the Paleocene-Eocene thermal maximum. *Proceedings of the National Academy of Sciences*, 103 (30): 11223–11227

Stefen, C. & Lehmann, T. (2011) On new material of *Kopidodon macrognathus* (Mammalia, Paroxyclaenidae) from Messel. – In: Lehmann, T. & Schaal, S.F.K. (Eds.) The World at the Time of Messel: Puzzles in Palaeobiology, Palaeoenvironment and the History of Early Primates. 157–158; Senckenberg Gesellschaft für Naturforschung, Frankfurt a.M.

Storch, G. & Lister, A.M. (1985) *Leptictidium nasutum*, ein Pseudorhyncocyonide aus dem Eozän der „Grube Messel" bei Darmstadt (Mammalia, Proteutheria). *Senckenberg Lethaea*, 66: 1–37

Strait, S.G. (2001) New Wa-0 mammalian fauna from Castle Gardens in the southeastern Bighorn Basin. *University of Michigan, Papers on Paleontology*, 33: 127–143

Tobien, H. (1962) Insectivoren (Mammalia) aus dem Mitteleozän (Lutetium) von Messel bei Darmstadt. *Notizblatt des hessischen Landesamtes für Bodenforschung zu Wiesbaden*, 90: 7–47

Tobien, H. (1969) *Kopidodon* (Condylarthra, Mammalia) aus dem Mitteleozän (Lutetium) von Messel bei Darmstadt (Hessen). *Notizblätter der hessischen Landesanstalt für Bodenforschung*, 97: 7–37

Weitzel, K. (1933) *Kopidodon macrognathus* Wittich, ein Raubtier aus dem Mitteleozän von Messel. *Notizblätter des Vereins für Erdkunde der hessischen geologischen Landesanstalt Darmstadt*, 14: 81–88.

Wittich, E. (1902) *Cryptopithecus macrognathus* n. spec., ein neuer Primate aus den Braunkohlen von Messel. *Centralblatt für Mineralogie, Geologie und Paläontologie*, 10: 289–294

Chapter 12.3

Brace, S., Thomas, J.A., Dalén, L., Burger, J., MacPhee, R.D., Barnes, I. & Turvey, S.T. (2016) Evolutionary history of the Nesophontidae, the last unplaced recent mammal family. *Molecular Biology and Evolution*, 33 (12): 3095–3103

Gould, G.C. (1995) Hedgehog phylogeny (Mammalia, Erinaceidae) – the reciprocal illumination of the quick and the dead. *Novitates* 3131: 1–45

Hooker, J.J. (2014) New postcranial bones of the extinct mammalian family Nyctitheriidae (Paleogene, UK): primitive euarchontans with scansorial locomotion. *Palaeontologia Electronica*, 17 (3): 1–79

Hooker, J.J. & Russell, D.E. (2012) Early Palaeogene Louisinidae (Macroscelidea, Mammalia), their relationships and north European diversity. *Zoological Journal of the Linnean Society*, **164** (4): 856–936

Koenigswald, W.v. & Storch G. (1983) *Pholidocercus hassiacus*, ein Amphilemuride aus dem Eozän der „Grube Messel" bei Darmstadt (Mammalia, Lipotyphla). *Senckenbergiana Lethaea*, **6**: 447–495

Koenigswald, W.v., Storch, G. & Richter, G. (1992) Primitive insectivores, extraordinary hedgehogs, and long fingers. – In: Schaal, S. & Ziegler, W. (Eds.) Messel – an Insight into the History of Life and of the Earth. 159–177; Clarendon Press, Oxford

Lopatin, A.V. (2006) Early Paleogene insectivore mammals of Asia and establishment of the major groups of Insectivora. *Paleontological Journal*, **40** (3): 205–405

Maier, W. (1979) *Macrocranion tupaiodon*, an adapisoricid? Insectivore from the Eocene of ‚Grube Messel' (Western Germany). *Paläontologische Zeitschrift*, **53** (1/2): 38–62

Maître, E., Escarguel, G. & Sige, B. (2008) Amphilemuridae éocènes d'Europe occidentale – Nouvelles données, Formes affines, Systématique et Phylogénie. *Palaeontographica A*, **283**: 35–82

Manz, C.L. & Bloch, J.I. (2015) Systematics and Phylogeny of Paleocene-Eocene Nyctitheriidae (Mammalia, Eulipotyphla?) with Description of a new Species from the Late Paleocene of the Clarks Fork Basin, Wyoming, USA. *Journal of Mammalian Evolution*, **22** (3): 307–342

Novacek, M.J. (1986) The skull of leptictid insectivorans and the higher-level classification of eutherian mammals. *Bulletin of the American Museum of Natural History*, **183**: 1–112

Roca, A.L., Bar-Gal, G.K., Eizirik, E., Helgen, K.M., Maria, R., Springer, M.S., O'Brien, S.J. & Murphy, W.J. (2004) Mesozoic origin for West Indian insectivores. *Nature*, **429** (6992): 649–651

Smith, T., Bloch, J.I., Strait, S.G. & Gingerich, P.D. (2002) New species of *Macrocranion* (Mammalia, Lipotyphla) from the earliest Eocene of North America and its biogeographic implications. *Contributions from the Museum of Paleontology, University of Michigan*, **30**: 373–384

Stanhope, M.J., Waddell, V.G., Madsen, O., De Jong, W., Hedges, S.B., Cleven, G. C., Kao, D. & Springer, M.S. (1998) Molecular evidence for multiple origins of Insectivora and for a new order of endemic African insectivore mammals. *Proceedings of the National Academy of Sciences USA*, **95** (17): 9967–9972

Storch, G. (1993) Morphologie und Paläobiologie von *Macrocranion tenerum*, einem Erinaceomorphen aus dem Mittel-Eozän von Messel bei Darmstadt (Mammalia, Lipotyphla). *Senckenbergiana Lethaea*, **73**: 61–81

Storch, G. (1996) Paleobiology of Messel erinaceomorphs. *Palaeovertebrata*, **25**: 215–224

Storch, G. & Richter, G. (1994) Zur Paläobiologie der Messeler Igel. *Natur und Museum*, **124** (3): 81–90

Symonds, M.R. (2005) Phylogeny and life histories of the 'Insectivora': controversies and consequences. *Biological Reviews*, **80** (1): 93–128

Weitzel, K. (1949) Neue Wirbeltiere (Rodentia, Insectivora, Testudinata) aus dem Mitteleozän von Messel bei Darmstadt. *Abhandlungen der senckenbergischen naturforschenden Gesellschaft*, **480**: 1–24

Chapter 12.4

Fisher, D.C. (1981) Crocodilian scatology, microvertebrate concentrations, and enamel-less teeth. *Palaeobiology*, **7**: 262–275

Franzen, J.L. (1987) Ein neuer Primate aus dem Mitteleozän der Grube Messel (Deutschland, S-Hessen). *Courier Forschungsinstitut Senckenberg*, **91**: 151–187

Franzen, J.L. (1988) Ein weiterer Primatenfund aus der Grube Messel bei Darmstadt. *Courier Forschungsinstitut Senckenberg*, **107**: 275–298

Franzen, J.L. (1997) Ein Koprolith als Leckerbissen. Der siebte Primatenfund aus Messel. *Natur und Museum*, **127**: 46–53

Franzen, J.L. (2000) *Europolemur kelleri* n. sp. von Messel und ein Nachtrag zu Europolemur koenigswaldi (Mammalia, Primates, Notharctidae, Cercamoniinae). *Senckenbergiana Lethaea*, **80**: 275–287

Franzen, J.L. & Frey, E. (1993) *Europolemur* Completed. *Kaupia*, **3**: 113–130

Franzen, J.L., Gingerich, P.D., Habersetzer, J., Hurum, J.H., Koenigswald, W.v. & Smith, H. (2009) Complete Primate Skeleton from the Middle Eocene of Messel in Germany: Morphology and Paleobiology. *PLOS ONE*, **4**: e5723

Franzen, J.L., Habersetzer, J., Schlosser-Sturm, E. & Franzen, E.L. (2012) Palaeopathology and fate of Ida (*Darwinius masillae*, Primates, Mammalia). *Palaeobiodiversity and Palaeoenvironments*, **92**: 567–572

Franzen, J.L. & Wilde, V. (2003) First gut content of a fossil primate. *Journal of Human Evolution*, **44**: 373–378

Kay, R.F. (1975) The functional adaptations of primate molar teeth. *American Journal of Physical Anthropology*, **43**: 195–216

Koenigswald, W.v. (1979) Ein Lemurenrest aus dem eozänen Ölschiefer der Grube Messel bei Darmstadt. *Paläontologische Zeitschrift*, **53**: 63–76

Koenigswald, W.v. (1985) Der dritte Lemurenrest aus dem mitteleozänen Ölschiefer der Grube Messel bei Darmstadt. *Carolinea*, **42**: 145–147

Chapter 12.5

Beard, K.C., Sigé, B. & Krishtalka, L. (1992) A primitive vespertilionoid bat from the early Eocene of central Wyoming. *Comptes Rendus de l'Academie des Sciences, Serie II*, **314**: 735–741

Eckrich, M. (1988) Untersuchungen zur Räuber-Beute-Beziehung zwischen Fledermäusen und Nachtfaltern in Südindien. 153; Dissertation Universität München

Giannini, N.P., Gunnell, G.F., Habersetzer, J. & Simmons, N.B. (2012) Early evolution of body size in bats. – In: Gunnell, G.F. & Simmons, N.B. (Eds.) Evolutionary History of Bats: Fossils, Molecules and Morphology. 530–555; Cambridge University Press, Cambridge

Gunnell, G.F., Jacobs, B.F., Herendeen, P.S., Head, J.J., Kowalski, E., Msuya, C.P., Mizambwa, F.A., Harrison, T., Habersetzer, J. & Storch, G. (2003) Oldest Placental Mammal from Sub-Saharan Africa: Eocene Microbat from Tanzania – Evidence from Early Evolution of Sophisticated Echolocation. *Palaeontologia Electronica*, **5** (3): 1–10

Habersetzer, J., Schuller, G. & Neuweiler, G. (1984) Foraging behavior and Doppler shift compensation in echolocating hippo-

siderid bats, *Hipposideros bicolor* and *Hipposideros speoris*. *Journal of Comparative Physiology A*, **155** (4): 559–567
Habersetzer, J. & Storch, G. (1987) Klassifikation und funktionelle Flügelmorphologie paläogener Fledermäuse (Mammalia, Chiroptera). Forschungsergebnisse zu den Grabungen in der Grube Messel bei Darmstadt. *Courier Forschungsinstitut Senckenberg*, **91**: 117–150
Habersetzer, J. & Storch, G. (1992) Cochlea Size in Extant Chiroptera and Middle Eocene Microchiropterans from Messel. *Naturwissenschaften*, **79** (10): 462–466
Habersetzer, J., Richter, G. & Storch, G. (1994) Paleoecology of early middle Eocene bats from Messel, FRG. Aspects of flight, feeding and echolocation. *Historical Biology*, **8** (1–4): 235–260
Habersetzer, J., Schlosser-Sturm, E., Storch, G. & Sigé, B. (2012) Shoulder joint and inner ear of *Tachypteron franzeni*, an emballonurid bat from the Middle Eocene of Messel. – In: Gunnell, G.F. & Simmons, N.B. (Eds.) Evolutionary History of Bats: Fossils, Molecules and Morphology. 67–104; Cambridge University Press, Cambridge
Hand, S., Novacek, M., Godthelp, H. & Archer, M. (1994) First Eocene bat from Australia, *Journal of Vertebrate Paleontology*, **14**: 375–381
Harrison, D. L. & Hooker, J.J. (2010) Late middle Eocene bats from the Creechbarrow Limestone Formation, Dorset, South England with description of a new species of *Archaeonycteris* (Chiroptera: Archaeonycteridae). *Acta Chiropterologica*, **12**: 1–18
Heller, F. (1935) Fledermäuse aus der eozänen Braunkohle des Geiseltales bei Halle a. d. S. *Nova Acta Leopoldina*, **2**: 301–314
Hölker, F., Wolter, C., Perkin, E. K. & Tockner, K. (2010) Light pollution as a biodiversity threat. *Trends in Ecology and Evolution*, **25**: 681–682
Hooker, J.J. (1996) A primitive emballonurid bat (Chiroptera, Mammalia) from the earliest Eocene of England. *Palaeovertebrata*, **25**: 287–300
Jepsen, G.L. (1966) Early Eocene bat from Wyoming. *Science*, **154**: 1333–1339
Ravel, A., Marivaux, L., Tabuce, R., Adaci, M., Mahboubi, M., Mebrouk, F. & Bensalah, M. (2011) The oldest African bat from the early Eocene of El Kohol (Algeria). *Naturwissenschaften*, **98**: 397–405
Russell, D.E., Louis, P. & Savage, D.E. (1973) Chiroptera and Dermoptera of the French Early Eocene. *University of California Publications in Geological Sciences*, **95**: 1–57
Sigé, B. & Russell, D.E. (1980) Compléments sur les chiroptères de l'Eocène moyen d'Europe. Les genres *Palaeochiropteryx* et *Cecilionycteris*. *Palaeovertebrata, Mémoire Jubilaire*, **1980**: 81–126
Simmons, N.B., Seymour, K.L., Habersetzer, J. & Gunnell, G.F. (2008) Primitive Early Eocene bat from Wyoming and the evolution of flight and echolocation. *Nature*, **451**: 818–821
Smith, T., Rana, R.S., Missiaen, P., Rose, K.D., Sahni, A., Singh, H. & Singh, L. (2007) High bat (Chiroptera) diversity in the Early Eocene of India. *Naturwissenschaften*, **94**: 1003–1009
Smith, T., Habersetzer, J., Simmons, N.B. & Gunnell, G.F. (2012) Systematics and paleobiogeography of early bats. – In: Gunnell, G.F. & Simmons, N.B. (Eds.) Evolutionary History of Bats: Fossils, Molecules and Morphology. 23–66; Cambridge University Press, Cambridge
Storch, G. & Habersetzer, J. (1988) *Archaeonycteris pollex* (Mammalia, Chiroptera), eine neue Fledermaus aus dem Eozän der Grube Messel bei Darmstadt. *Courier Forschungsinstitut Senckenberg*, **107**: 263–273
Storch, G., Sigé, B. & Habersetzer, J. (2002) *Tachypteron franzeni* n. gen., n. sp., earliest emballonurid bat from the Middle Eocene of Messel (Mammalia, Chiroptera). *Paläontologische Zeitschrift*, **76** (2): 189–199
Tabuce, R., Antunes, M.T. & Sigé, B. (2009) A new primitive bat from the earliest Eocene of Europe. *Journal of Vertebrate Paleontology*, **29**: 627–630
Tejedor, M.F., Czaplewski, N.J., Goin, F.J. & Aragon, E. (2005) The oldest record of South American bats. *Journal of Vertebrate Paleontology*, **25**: 990–993
Tong, Y.-S. (1997) Middle Eocene small mammals from Liguanqiao Basin of Henan Province and Yuanqu Basin of Shanxi Province, Central China. *Palaeontologica Sinica 18, New Series C*, **26**: 1–256

Chapter 12.6

Blanga-Kanfi, S., Miranda, H., Penn, O., Pupko, T., Debry, R.W. & Huchon, D. (2009) Rodent phylogeny revised: analysis of six nuclear genes from all major rodent clades. *BMC Evolutionary Biology*, **9**: 71
Brandt, J.K. (1855) Beiträge zur nähern Kenntniss der Säugethiere Russlands. *Mémoires de l'Académie Impériale des Sciences de St. Petersbourg*, **69**: 1–375
Carlton, M.D. & Musser, G.G. (2005) Order Rodentia. – In: Wilson. D.E. & Reeder, D.M. (Eds.) Mammal Species of the World: a Taxonomic and Geographic Reference (Vol. 2). 745–1600; Johns Hopkins University Press, Baltimore
Escarguel, G. (1999) Les rongeurs de l'Éocène inférieur et moyen d'Europe occidentale. Systématique, phylogénie, biochronologie et paléobiogéographie des niveaux-repères MP 7 à MP 14. *Palaeovertebrata*, **28** (2–4): 89–351
Fejfar, O. & Storch, G. (1994) Das Nagetier von Valec-Waltsch in Böhmen – ein historischer fossiler Säugetierfund (Rodentia: Myoxidae). *Münchner Geowissenschaftliche Abhandlungen*, **A26**: 5–34
Gwosdek, S. (1996) Das postcraniale Skelett von *Ailuravus macrurus* (Rodentia, Mammalia) aus dem unteren Mittel-Eozän (Lutet) von Messel – Eine taphonomische, vergleichend anatomische und biometrische Studie. Dissertation, Universität Tübingen
Hartenberger, J.-L. (1968) Les Pseudosciuridae (Rodentia) de l'Éocène moyen et le genre *Masillamys* Tobien. *Comptes rendus de l'Académie des Sciences Paris, D*, **267**: 1817–1820
Hartenberger, J.-L. (1971) Contribution à l'étude des genres *Gliravus* et *Microparamys* (Rodentia) de l'Éocène d'Europe. *Palaeovertebrata*, **4**: 97–135
Hautier, L., Michaux, J., Marivaux L. & Vianey-Liaud, M. (2008) Evolution of the zygomasseteric construction in Rodentia, as revealed by a geometric morphometric analysis of the mandible of *Graphiurus* (Rodentia, Gliridae). *Zoological Journal of the Linnean Society*, **154**: 807–821
Huchon, D., Chevret, P., Jordan, U., Kilpatrick, C.W., Ranwez, V., Jenkins, P.D., Brosius, J. & Schmitz, J. (2007) Multiple molecular evidences for a living mammalian fossil. *Proceedings of the National Academy of Sciences*, **104** (18): 7495–7499
Koenigswald, W.v., Storch, G. & Richter, G. (1988) Nagetiere – am Beginn einer großen Karriere. – In: Schaal, S. & Ziegler, W.

(Eds.) Messel – Ein Schaufenster in die Geschichte der Erde und des Lebens. 219–222; Waldemar Kramer, Frankfurt a.M.

Maier, W., Klingler, P., Ruf, I. (2002) Ontogeny of the medial masseter muscle, pseudo-myomorphy, and the systematic position of the Gliridae. *Journal of Mammalian Evolution*, 9 (4): 253–269

Paus, S.H. (2002) Reconstruction of the skull of *Ailuravus macrurus* (Rodentia) from the Eocene of Messel, Germany. *Kaupia Darmstädter Beiträge zur Naturgeschichte*, 11: 125–152

Rana, R.S., Kumar, K., Escarguel, G., Sahni, A., Rose, K.D., Smith, T., Singh, H. & Singh, L. (2008) An ailuravine rodent from the Lower Eocene Cambay Formation at Vastan, Western India, and its palaeobiogeographic implications, *Acta Palaeontologica Polonica*, 53 (1): 1–14

Rodrigues, H.G., Marivaux, L. & Vianey-Liaud, M. (2010) Phylogeny and systematic revision of Eocene Cricetidae (Rodentia, Mammalia) from Central and East Asia: on the origin of cricetid rodents. *Journal of Zoological Systematics and Evolutionary Research*, 48 (3): 259–268

Solé, F., Smith, T., De Bast, E., Codrea, V. & Gheerbrandt, E. (2016) New carnivoraforms from the latest Paleocene of Europe and their bearing on the origin and radiation of Carnivoraformes (Carnivoramorpha, Mammalia). *Journal of Vertebrate Paleontology*, 36 (2): e1082480

Storch, G. & Seiffert, C. (2007) Extraordinarily preserved specimen of the oldest known glirid from the Middle Eocene of Messel (Rodentia). *Journal of Vertebrate Paleontology*, 27 (1): 189–194

Tobien, H. (1954) Nagerreste aus dem Mitteleozän von Messel bei Darmstadt. *Notizblatt des Hessischen Landesamtes für Bodenforschung zu Wiesbaden*, 82: 13–29

Vianey-Liaud, M. (1989) Parallelism among Gliridae (Rodentia): the genus *Gliravus* Stehlin and Schaub. *Historical Biology*, 2: 213–226

Weitzel, K. (1949) Neue Wirbeltiere (Rodentia, Insectivora, Testudinata) aus dem Mitteleozän von Messel bei Darmstadt. *Abhandlungen der Senckenbergischen Naturforschenden Gesellschaft*, 480: 1–24

Wilde, V. (1989) Untersuchungen zur Systematik der Blattreste aus dem Mitteleozän der Grube Messel bei Darmstadt (Hessen, Bundesrepublik Deutschland). *Courier Forschungsinstitut Senckenberg*, 115: 1–213

Chapter 12.7

Brikiatis, L. (2014) The De Geer, Thulean and Beringia routes: key concepts for understanding early Cenozoic biogeography. *Journal of Biogeography*, 41: 1036–1054

Gaudin, T.J., Emry, R.J. & Wible, J.R. (2009) The phylogeny of living and extinct pangolins (Mammalia, Pholidota) and associated taxa: A morphology based analysis. *Journal of Mammalian Evolution*, 16: 235–305

Gheerbrant, E., Rose, K.D. & Godinot, M. (2005) First palaeanodont (?pholidotan) mammal from the Eocene of Europe. *Acta Palaeontologica Polonica*, 50: 209–218

Koenigswald, W.v., Richter, G. & Storch, G. (1981) Nachweis von Hornschuppen bei *Eomanis waldi* aus der „Grube Messel" bei Darmstadt (Mammalia, Pholidota). *Senckenbergiana lethaia*, 61: 291–298

Lange-Badré, B. & Haubold, H. (1990) Les créodontes (mammifères) du gisement du Geiseltal (Éocène moyen, RDA). *Géobios*, 23: 607–637

McKenna, M.C. (1983) Cenozoic paleogeography of North Atlantic land bridges. – In: Bott, M.H.P., Saxov, S., Talwani, M. & Thiede, J. (Eds.) Structure and development of the Greenland-Scotland Ridge. 351–399; Plenum Press, New York

Morlo, M., Schaal, S., Mayr, G. & Seiffert, C. (2004) An annotated taxonomic list of the Middle Eocene (MP 11) Vertebrata of Messel. *Courier Forschungsinstitut Senckenberg*, 252: 95–108

Morlo, M. & Habersetzer, J. (1999) The Hyaenodontidae (Creodonta, Mammalia) from the lower Middle Eocene (MP 11) of Messel (Germany) with special remarks on new X-ray methods. *Courier Forschungsinstitut Senckenberg*, 216: 31–73

O'Leary, M.A., Bloch, J.I., Flynn, J.J., Gaudin, T.J., Giallombardo, A., Giannini, N.P., Goldberg, S.L., Kraatz, B.P., Luo, Z.X., Meng, J., Ni, X., Novacek, M.J., Perini, F.A., Randall, Z.S., Rougier, R.W., Sargis, E.J., Silcox, M.T., Simmons, N.B., Spaulding, M., Velazco, P.M., Weksler, M., Wible, J.R. & Cirranello, A.L. (2013) The placental mammal ancestor and the post-K-Pg radiation of placentals. *Science*, 339 (6120): 662–667

Rose, K.D. (1999) *Eurotamandua* and Palaeanodonta: convergent or related. *Paläontologische Zeitschrift*, 73: 395–301

Rose, K.D. & Emry, R.J. (1993) Relationships of Xenarthra, Pholidota, and fossil "edentates": the morphological evidence. – In: Szalay, F.S., Novacek, M.J., & Mckenna, M.C. (Eds.) Mammal phylogeny: Placentals. 81–102; Springer, New York

Rose, K.D., Eberle, J.J. & McKenna, M.C. (2004) *Arcticanodon dawsonae*, a primitive new palaeanodonts from the lower Eocene of Ellesmere Island, Canadian High Arctic. *Canadian Journal of Earth Science*, 41: 757–763

Shoshani, J., McKenna, M.C., Rose, K.D. & Emry, R.J. (1997) *Eurotamandua* is a pholidotan not a xenarthran. *Journal of Vertebrate Paleontology* 17 (Suppl.): 76A

Smith, T. & Smith, R. (2010) A new genus of "miacid" carnivoran from the earliest Eocene of Europe and North America. *Acta Palaeontologica Polonica*, 55 (4): 761–764

Solé, F. (2012) New proviverrine genus from the Early Eocene of Europe and the first phylogeny of late Palaeocene-middle Eocene hyaenodontidans (Mammalia). *Journal of Systematic Palaeontology*, 11 (4): 375–398

Solé, F. (2014) New carnivoraforms from the early Eocene of Europe and their bearings on the evolution of the Carnivoraformes. *Palaeontology*, 57:963–978

Solé, F., Falconnet, J. & Yves, L. (2014) New proviverrines (Hyaenodontida) from the early Eocene of Europe; phylogeny and ecological evolution of the Proviverrinae. *Zoological Journal of the Linnean Society*, 171: 878–917

Solé, F. Smith, T., De Bast, E., Codrea, V. & Gheerbrandt, E. (2016a) New carnivoraforms from the latest Paleocene of Europe and their bearing on the origin and radiation of Carnivoraformes (Carnivoramorpha, Mammalia). *Journal of Vertebrate Paleontology*, 36 (2): e1082480

Solé, F., Essid, E.M., Marzougui, W., Temani, R., Ammar, H.K., Mahboubi, M., Marivaux, L., Vianey-Liaud, M. & Tabuce, R. (2016b) New fossils of Hyaenodonta (Mammalia) from the Eocene localities of Chambi (Tunisia) and Bir el Ater (Algeria), and the evolution of the earliest African hyaenodonts. *Palaeontologia Electronica*, 19 (2): 38A

Spaulding, M. & Flynn, J.J. (2012) Phylogeny of the Carnivoramorpha: the impact of postcranial characters. *Journal of Systematic Palaeontology*, 10:653–677

Springer, M.S., Murphy, W.J., Eizirik, E. & O'Brien, S.J. (2003) Placental mammal diversification and the Cretaceous-Tertiary boundary. *Proceedings of the National Academy of Sciences, USA*, **100**: 1056–1061

Springhorn, R. (1980) *Paroodectes feisti*, der erste Miacide (Carnivora, Mammalia) aus dem Mittel-Eozän von Messel. *Paläontologische Zeitschrift*, **54**: 171–198

Springhorn, R. (1982) Neue Raubtiere (Mammalia: Creodonta et Carnivora) aus dem Lutetium der Grube Messel (Deutschland). *Palaeontographica A*, **179**: 105–141

Springhorn, R. (1985) Zwei neue Skelette von *Miacis? kessleri* (Mammalia, Carnivora) aus den lutetischen Ölschiefern der „Grube Messel". *Senckenbergiana lethaea*, **66** (1/2): 121–141

Springhorn, R. (2000) *Messelogale*, eine neue Raubtiergattung aus dem Mittel-Eozän von Messel (Deutschland). *Paläontologische Zeitschrift*, **74**: 425–439

Storch, G. (1978) *Eomanis waldi*, ein Schuppentier aus dem Mittel-Eozän der „Grube Messel" bei Darmstadt (Mammalia: Pholidota). *Senckenbergiana lethaea*, **59**: 503–529

Storch, G. (1981) *Eurotamandua joresi*, ein Myrmecophagide aus dem Eozän der „Grube Messel" bei Darmstadt (Mammalia: Xenarthra). *Senckenbergiana lethaea*, **61**: 247–289

Storch, G. (2003) Fossil Old World edentates (Mammalia). *Senckenbergiana biologica*, **83**: 51–60

Storch, G. & Haubold, H. (1989) Additions to the Geiseltal mammalian faunas, middle Eocene: Didelphidae, Nyctitheriidae, Myrmecophagidae. *Palaeovertebrata*, **19**: 95–114

Storch, G. & Martin, T. (1994) *Eomanis krebsi*, ein neues Schuppentier aus dem Mittel-Eozän der Grube Messel bei Darmstadt (Mammalia: Pholidota). *Berliner Geowissenschaftliche Abhandlungen*, **E13**: 83–97

Storch, G. & Richter, G. (1992a) Pangolins: almost unchanged for 50 million years. – In: Schaal, S. & Ziegler, W. (Eds.) Messel: an Insight into the History of Life and of the Earth. 201–207; Clarendon Press, Oxford

Storch, G. & Richter, G. (1992b) The ant-eater *Eurotamandua*: a South American in Europe. – In Schaal, S. & Ziegler, W. (Eds.) Messel: an Insight into the History of Life and of the Earth. 209–215; Clarendon Press, Oxford

Szalay, F.S. & Schrenk, F. (1998) The middle Eocene *Eurotamandua* and a Darwinian phylogenetic analysis of "edentates". *Kaupia*, **7**: 97–186

Taylor, M.E. (1974) The functional anatomy of the forelimb of some African Viverridae (Carnivora). *Journal of Morphology*, **143**: 307–336

Taylor, M.E. (1976) The functional anatomy of the hindlimb of some African Viverridae (Carnivora). *Journal of Morphology*, **148**: 227–254

Wang, X. (1993) Transformation from plantigrady to digitigrady: functional morphology of locomotion in *Hesperocyon* (Canidae: Carnivora). *American Museum Novitates*, **3069**: 1–23

Wesley-Hunt, G. & Werdelin, L. (2005) Basicranial morphology and phylogenetic position of the upper Eocene carnivoramorphan *Quercygale*. *Acta Palaeontologica Polonica*, **50**: 837–846

Chapter 12.8

Erfurt, J. & Haubold, H. (1989) Artiodactyla aus den eozänen Braunkholen des Geiseltales bei Halle (DDR). *Paleovertebrata*, **19**: 131–160

Erfurt, J. & Sudre, J. (1996) Eurodexeinae, eine neue Unterfamilie der Artiodactyla (Mammalia) aus dem Unter- und Mitteleozän Europas. Palaeovertebrata, **25**: 371–390

Erfurt, J. & Métais, G. (2007) Endemic European Paleogene Artiodactyls: Cebochoeridae, Choeropotamidae, Mixtotheriidae, Cainotheriidae, Anoplotheriidae, Xiphodontidae, and Amphimerycidae. – In: Prothero, D.R. & Scott, E.F. (Eds.) The Evolution of Artiodactyls. 59–84; John Hopkins University Press, Baltimore

Foss, S.E. & Prothero, D.R. (2007) Introduction. – In: Prothero, D.R. & Scott, E.F. (Eds.) The Evolution of Artiodactyls. 1–3; John Hopkins University Press, Baltimore

Franzen, J.L. (1981) Das erste Skelett eines Dichobuniden (Mammalia, Artiodactyla) geborgen aus mitteleozänen Ölschiefern der ‚Grube Messel' bei Darmstadt (Deutschland, S-Hessen). *Senckenbergiana Lethaea*, **61**: 299–353

Franzen, J.L. (1983) Ein zweites Skelett von *Messelobunodon* (Mammalia, Artiodactyla, Dichobunidae) aus der „Grube Messel" bei Darmstadt (Deutschland, S-Hessen). *Senckenbergiana Lethaea*, **64**: 403–445

Franzen, J.L. (1988) Skeletons of *Aumelasia*. (Mammalia, Artiodactyla, Dichobunidae) from Messel (M. Eocene, W. Germany). *Courier Forschungsinstitut Senckenberg*, **107**: 309–321

Franzen, J.L. & Krumbiegel, G. (1980) *Messelobunodon ceciliensis* n. sp. (Mammalia, Artiodactyla)- ein neuer Dichobunide aus der mitteleozänen Fauna des Geiseltales bei Halle (DDR). *Zeitschrift für geologische Wissenschaften, Berlin*, **8**: 1553–1560

Franzen, J.L. & Richter, G. (1992) Primitive even-toed ungulates: loners in the undergrowth. In: Schaal, S., & Ziegler, W. (Eds.) Messel – an Insight into the History of Life and of the Earth. 249–256; Clarendon Press, Oxford

Gentry, A.W. & Hooker, J.J. (1988) The phylogeny of the Artiodactyla. – In: Benton, M.J. (Ed.) The Phylogeny and Classification of the Tetrapods, Vol. 2. 235–271; Clarendon Press, Oxford

Gingerich, P.D., Haq, M.U., Zalmout, I.S., Khan, I.H. & Malakani, M.S. (2001) Origin of whales from early artiodactyls: hands and feet of Eocene Protocetidae from Pakistan. *Science*, **293**: 2239–2242

Howell, A.B. (1944) Speed in Animals, their Specialisations for Running and Leaping. University of Chicago Press, Chicago

Nikaido, M., Rooney, A.P. & Okada, N. (1999) Phylogenetic relationships among cetartiodactyls based on insertions of short and long interpersed elements: hippopotamuses are the closest extant relatives of whales. *Proceedings of the National Academy of Sciences USA*, **96** (18): 10261–10266

Richter, G. (1987) Untersuchungen zur Ernährung eozäner Säuger aus der Fossilfundstätte Messel bei Darmstadt. *Courier Forschungsinstitut Senckenberg*, **91**: 1–31

Sudre, J. (1980) *Aumelasia gabineaudi* n. g. n. sp. nouveau Dichobunidae (Artiodactyla, Mammalia) du gisement d'Aumelas (Herault) d'age lutetien terminal. *Palaeovertebrata*, **9** (Mémoire jubilaire): 197–211

Sudre, J., Russel, D., Louis, P. & Savage, D.E. (1983) Les Artiodactyles de l'Eocène inférieur d'Europe. *Bulletin du Muséum national d'histoire naturelle C,* **5** (3): 281–333

Theodor, J.M., Erfurt, J. & Métais, G. (2007) The Earliest Artiodactyls: Diacodexeidae, Dichobunidae, Homacodontidae, Leptochoeridae, and Raoellidae. – In: Prothero, D.R. & Scott, E.F. (Eds.) The Evolution of Artiodactyls. 32–58; John Hopkins University Press, Baltimore

Thewissen, J.G.M., Williams, E.M., Roe, L.J. & Hussain, S.T. (2001) Skeletons of terrestrial cetaceans and the relationship of whales to artiodactyls. *Nature,* **413**: 277–281

Thewissen, J.G.M. & Hussain, S.T. (1990) Postcranial osteology of the most primitive artiodactyl: *Diacodexis pakistanensis* (Dichobunidae). *Anatomia, Histologia, Embryologia,* **19** (1): 37–48

Tobien, H. (1980) Ein anthracotherioider Paarhufer (Artiodactyla, Mammalia) aus dem Eozän von Messel bei Darmstadt (Hessen). *Geologisches Jahrbuch Hessen,* **108**: 11–22

Tobien, H. (1985) Zur Osteologie von *Masillabune* (Mammalia, Artiodactyla, Haplobunodontidae) aus dem Mitteleozän der Fossilfundstätte Messel bei Darmstadt (S-Hessen, Bundesrepublik Deutschland). *Geologisches Jahrbuch Hessen,* **113**: 5–58

Chapter 12.9

Camp, C.L. & Smith, N. (1942) Phylogeny and functions of the digital ligaments of the horse. *Memoirs of the University of California,* **13**. 69–122

Franzen, J.L. (1981) *Hyrachyus minimus* (Mammalia, Perissodactyla, Helaletidae) aus den mitteleozänen Ölschiefern der „Grube Messel" bei Darmstadt (Deutschland, S-Hessen). *Senckenbergiana Lethaea,* **61**: 371–376

Franzen, J.L. (1989) Origin and systematic position of the Palaeotheriidae. – In: Prothero, D.R. & Schoch, R.M. (Eds.) The Evolution of Perissodactyls. Clarendon Press, Oxford

Franzen, J.L. (1990) *Hallensia* (Mammalia, Perissodactyla) aus Messel und dem Pariser Becken sowie Nachträge aus dem Geiseltal. *Bulletin de l'Institut Royal des Sciences Naturelles de Belgique,* **60**: 175–201

Franzen, J.L. (2007a) Die Urpferde der Morgenröte. Ursprung und Evolution der Pferde. Spektrum, Heidelberg

Franzen, J.L. (2007b) Eozäne Equoidea (Mammalia, Perissodactyla) aus der Grube Messel bei Darmstadt (Deutschland). Funde der Jahre 1969 – 2000. *Schweizerische Paläontologische Abhandlungen,* **127**: 1–245

Franzen, J.L. (2010) The Rise of Horses. 55 Million Years of Evolution. Johns Hopkins University Press, Baltimore

Franzen, J.L., Aurich, C. & Habersetzer, J. (2015) Description of a well preserved fetus of the European Eocene equoid *Eurohippus messelensis*. *PLoS One,* **10**: e0137985

Franzen, J.L. (2017) Report on the discovery of fossil mares with preserved uteroplacenta from the Eocene of Germany. *Fossil Imprint* 73: No. 1–3, pp. 67–75

Froehlich, D.J. (2002) Quo vadis *Eohippus*? The systematics and taxonomy of the early Eocene equids (Perissodactyla). *Zoological Journal of the Linnean Society,* **134**: 141–256

Gingerich, P.D. (1981) Variation, sexual dimorphism, and social structure in the early Eocene horse *Hyracotherium* (Mammalia, Perissodactyla). *Paleobiology,* **7** (4): 443–455

Haupt, O. (1911) *Propalaeotherium* cf. *Rollinati*, Stehlin aus der Braunkohle von Messel bei Darmstadt. *Notizblatt des Vereins für Erdkunde und der Großherzoglichen Geologischen Landesanstalt zu Darmstadt,* **32** (4): 59–70

Hooker, J.J. (1994) The beginning of the equoid radiation. *Zoological Journal of the Linnean Society,* **112**: 29–63

Koenigswald, W.v. (1987) Die Fauna des Ölschiefers von Messel. – In: Heil, R., Koenigswald, W.v., Lippmann, H.G., Graner, D. & Heunisch, C. (Eds.) Fossilien der Messel-Formation. Hessisches Landesmuseum, Darmstadt

Koenigswald, W.v. (2014) Mastication and wear in *Lophiodon* (Perissodactyla, Mammalia) compared with lophodont dentitions in some other mammals. *Annales Zoologici Fennici,* **51**: 162–176

Koenigswald, W.v., Holbrook, L. & Rose, K.D. (2011) The variation of Hunter-Schreger band configuration in tooth enamel of Perissodactyla (Mammalia) and its evolutionary implications. *Acta Palaeontologica Polonica,* **56**: 1–18

Rose, K.D., Holbrook, L.T., Rana, R.S., Kumar, K., Jones, K.E., Ahrens, H.E., Missiaen, P., Sahni, A. & Smith, T. (2014) Early Eocene fossils suggest that the mammalian order Perissodactyla originated in India. *Nature Communications,* **5**: 5570

Strömberg, C.A.E. (2004) Using phytolith assemblages to reconstruct the origin and spread of grass-dominated habitats in the Great Plains of North America during the late Eocene to early Miocene. *Palaeogeography, Palaeoclimatology, Palaeoecology,* **207**: 239–275

Chapter 13

Allen, A.P., Gillooly, J.F., Savage, V.M. & Brown, J.H. (2006) Kinetic effects of temperature on rates of genetic divergence and speciation. *Proceedings of the National Academy of Sciences of the United States of America,* **103**: 9130–9135

Berman, T., Chava, S., Kaplan, B. & Wynne, D. (1991) Dissolved organic substrates as phosphorus and nitrogen sources for axenic batch cultures of freshwater green algae. *Phycologia,* **30**(4): 339–345

Blois, J.L., Zarnetske, P.L., Fitzpatrick, M.C. & Finnegan, S. (2013) Climate change and the past, present, and future of biotic interactions. *Science,* **341**: 499–504

Collinson, M.E., Manchester, S.R. & Wilde, V. (2012) Fossil fruits and seeds of the Middle Eocene Messel biota, Germany. *Abhandlungen der Senckenberg Gesellschaft für Naturforschung,* **570**: 1–251

Dunne, J.A., Labandeira, C.C. & Williams, R.J. (2014) Highly resolved early Eocene food webs show development of modern trophic structure after the end-Cretaceous extinction. *Proceedings of the Royal Society of London Series B,* **281**: 20133280

Emmons, L.H. & Gentry, A.H. (1983) Tropical forest structure and the distribution of gliding and prehensile-tailed vertebrates. *American Naturalist,* **121**, 513–524

Goth, K. (1990) Der Messeler Ölschiefer- ein Algenlaminit. *Courier Forschungsinstitut Senckenberg,* **131**: 1–141

Gould, S.J. (1989) Wonderful Life: The Burgess Shale and the Nature of History. W. W. Norton, New York

Greene, H.W. (1983) Dietary correlates of the origin and radiation of snakes. *American Zoologist,* **23**: 431–441

Grein, M., Utescher, T., Wilde, V. & Roth-Nebelsick, A. (2011) Reconstruction of the middle Eocene climate of Messel using palaeobotanical data. *Neues Jahrbuch für Geologie und Paläontologie, Abhandlungen,* **260**: 305–318

Gruber, G. & Micklich, N. (Ed.) (2007) Messel: Schätze der Urzeit. Hessisches Landesmuseum, Darmstadt

Hawkins, B.A., Field, R., Cornell, H.V., Currie, D.J., Guégan, J.-F., Kaufman, D.M., Kerr, J.T., Mittelbach, G.G., Oberdorff, T., O'brien, E.M., Porter, E.E. & Turner, J.R.G. (2003) Energy, water, and broad-scale geographic patterns of species richness. *Ecology,* **84**: 3105–3117

Heymann, E.W. & Hsia, S.S. (2015) Unlike fellows – a review of primate-non-primate associations. *Biological Reviews,* **90**: 142–156

Jablonski, D., Roy, K. & Valentine, J.W. (2006) Out of the tropics: evolutionary dynamics of the latitudinal diversity gradient. *Science,* **314**: 102–106

Jetz, W. & Fine, P.V.A. (2012) Global gradients in vertebrate diversity predicted by historical area-productivity dynamics and contemporary environment. *PLoS Biology,* **10**, e1001292

Kidwell, S.M. & Bosence, D.W.J. (1991) Taphonomy and time-averaging of marine shelly faunas. – In: Allison, P.A. & Briggs, D.E.G. (Eds.) Taphonomy: Releasing the Data Locked in the Fossil Record. 115–209; Plenum Press, New York

Lenz, O.K., Wilde, V. & Riegel, W. (2007a) Recolonization of a Middle Eocene volcanic site: quantitative palynology of the initial phase of the maar lake of Messel (Germany). *Review of Palaeobotany and Palynology,* **145**: 217–242

Lenz, O.K., Heinrichs, T. & Riegel, W. (2007b) Distribution and paleoecologic significance of the freshwater dinoflagellate cyst *Messelodinium thielepfeifferae* gen. et sp. nov (from the middle Eocene of Lake Messel, Germany. *Palynology,* **31**, 119–134

Lindeman, R.L. (1942) The trophic-dynamic aspect of ecology. *Ecology,* **23**: 399–417

Lutz, H. (1990) Systematische und palökologische Untersuchungen an Insekten aus dem Mittel-Eozän der Grube Messel bei Darmstadt. *Courier Forschungsinstitut Senckenberg,* **124**: 1–165

Mayr, G. (2017) The early Eocene birds of the Messel fossil site: a 48 million-year-old bird community adds a temporal perspective to the evolution of tropical avifaunas. *Biological Reviews,* **92**: 1174–1188

Micklich, N. (2012) Peculiarities of the Messel fish fauna and their palaeoecological implications: a case study. *Palaeobiodiversity and Palaeoenvironments,* **92**: 585–629

Morlo, M. (2004) Diet of *Messelornis* (Aves, Gruiformes), an Eocene bird from Germany. *Courier Forschungsinstitut Senckenberg,* **252**: 29–33

Pianka, E.R. (1973) The structure of lizard communities. *Annual Review of Ecology and Systematics,* **4**: 53–74

Rabenstein, R. & Schöller, M. (1996) Chrysomelid beetles on Vitaceae in Peninsular Malaysia with special regard to *Parastetha nigricornis* Baly 1879. *Senckenbergiana Biologica,* **76**: 39–45

Rabenstein, R., Habersetzer, J. & Mohrmann, E. (2007) Rekonstruktion des Messeler Halbaffen *Europolemur kelleri* im Rahmen senckenbergischer Multimedia-Projekte. *Hallesches Jahrbuch für Geowissenschaften, Beiheft,* **23**: 97–109

Richter, G. & Baszio, S. (2001a) Traces of a limnic food web in the Eocene Lake Messel – a preliminary report based on fish coprolite analyses. *Palaeogeography, Palaeoclimatology, Palaeoecology,* **166**: 345–368

Richter, G. & Baszio, S. (2001b) First proof of planctivory/insectivory in a fossil fish: *Thaumaturus intermedius* from the Eocene Lake Messel (FRG). *Palaeogeography, Palaeoclimatology, Palaeoecology,* **173**: 75–85

Richter, G. & Baszio, S. (2009) Geographic and stratigraphic distribution of spongillids (Porifera) and the leit value of spiculites in the Messel Pit Fossil Site. *Palaeobiodiversity and Palaeoenvironments,* **89**: 53–66

Richter, G. & Wedmann, S. (2005) Ecology of the Eocene Lake Messel revealed by analysis of small fish coprolites and sediments from a drilling core. *Palaeogeography, Palaeoclimatology, Palaeoecology,* **223**: 147–161

Richter, G., Baszio, S. & Wuttke, M. (Im Druck) Discontinuities in the microfossil record of middle Eocene Lake Messel: clues for ecological changes in lake's history? *Palaeobiodiversity and Palaeoenvironments*

Rietschel, S. (1988) Gastropod excrements, evidence of life in the Messel lake. *Courier Forschungsinstitut Senckenberg,* **107**: 163–168

Schweizer, M., Steele, A., Toporski, J.K.W. & Fogel, M.L. (2007) Stable isotopic evidence for fossil food webs in Eocene Lake Messel. *Paleobiology,* **33**: 590–609

Shurin, J.B., Gruner, D.S. & Hillebrand, H. (2006) All wet or dried up? Real differences between aquatic and terrestrial food webs. *Proceedings of the Royal Society of London Series B,* **273**: 1–9

Smith, K.T. (2009) A new lizard assemblage from the earliest Eocene (zone Wa0) of the Bighorn Basin, Wyoming, USA: Biogeography during the warmest interval of the Cenozoic. *Journal of Systematic Palaeontology,* **7**: 299–358

Smith, S.Y., Collinson, M.E., Rudall, P.J., Simpson, D.A., Marone, F. & Stampanoni, M. (2009) Virtual taphonomy using synchrotron tomographic microscopy reveals cryptic features and internal structure of modern and fossil plants. *Proceedings of the National Academy of Sciences of the United States of America,* **106**: 12013–12018

Tütken, T. (2014) Isotope compositions (C, O, Sr, Nd) of vertebrate fossils from the Middle Eocene oil shale of Messel, Germany: implications for their taphonomy and palaeoenvironment. *Palaeogeography, Palaeoclimatology, Palaeoecology,* **416**: 92–109

Wall, D., & Briand, F. (1979) Response of lake phytoplankton communities to in situ manipulations of light intensity and colour. *Journal of Plankton Research,* **1** (1): 103–112

Wilde, V. (1989) Untersuchungen zur Systematik der Blattreste aus dem Mitteleozän der Grube Messel bei Darmstadt (Hessen, Bundesrepublik Deutschland). *Courier Forschungsinstitut Senckenberg,* **115**: 1–213

Wilde, V. & Riegel, W. (2010) "Affenhaar" revisited – facies context of in situ preserved latex from the Middle Eocene of Central Germany. *International Journal of Coal Geology,* **83**: 182–194

Wilson, D.E. (1990) Mammals of La Selva, Costa Rica. In: A. H. Gentry (ed) *Four Neotropical Rainforests,* 273–286. Yale University Press, New Haven, Connecticut

Zachos, J., Pagani, M., Sloan, L., Thomas, E. & Billups, K. (2001) Trends, rhythms, and aberrations in global climate 65 Ma to present. *Science,* **292**: 686–693

Zachos, J., Dickens, G.R. & Zeebe, R.E. (2008) An early Cenozoic perspective on greenhouse warming and carbon-cycle dynamics. *Nature,* **451**: 279–283

List of authors

Christopher A. Brochu – A native of the East Coast, he received his BS at the University of Iowa (1989) and a PhD at the University of Texas (1997). He worked at the Field Museum in Chicago for three years before returning to the University of Iowa in 2001, where he currently works as a professor in the Department of Geo- and Environmental Sciences.

Georg N. Büchel – Senior Professor at the Institute of Geosciences of the Friedrich Schiller University in Jena. Prof. Büchel's work involves the formation of ground water-influenced volcanoes, including the origin of the Messel Maar. Additional areas of research involve bio-remediation and metallic raw materials exploration.

Edwin. A. Cadena – School of Geological Science and Engineering, Yachay Tech, San Miguel de Urcuquí, Ecuador. Dr. Cadena is a geologist and paleontologist who studies fossil turtles, linking their development with geographical, climatic and ecological events.

Andrej Čerňanský – Dept. of Ecology, Comenius University in Bratislava. Dr. Čerňanský studies the development of the main morphological traits in reptiles, particularly in lizards, as well as the consequences of these innovations for the temporal and geographical diversification within these groups.

Jens L. Franzen – Department of Messel Research and Mammalogy, Senckenberg Research Institute, Frankfurt am Main. Dr. Franzen's special research areas include the primates and odd-toed ungulates from the Messel Pit, their taxonomy, and the state of preservation in pregnant prehistoric horse mares.

Philip D. Gingerich – Professor Emeritus of Paleontology, University of Michigan, Ann Arbor. Prof. Gingerich is a vertebrate paleontologist who studies the evolution of primates and other Eocene mammals.

Gregg F. Gunnell (†) – Director of the Division of Fossil Primates, Duke Lemur Center. Dr. Gunnell studied the origin and diversification of modern mammalian groups, particularly chiropterans and primates. His field research took him to Asia, Africa, Indo-Pakistan, North America and the Caribbean.

Jörg Habersetzer – Department of Messel Research and Mammalogy, Senckenberg Research Institute, Frankfurt am Main. Dr. Habersetzer was until 2017 the head of the Section for Vertebrate Radiography; he studies both modern as well as fossil bats and other mammals from Messel, using 3D tomography and 2D radiographic techniques, which he partly developed himself.

List of authors

Walter G. Joyce – Department of Geosciences, University of Freiburg, Switzerland. Prof. Dr. Joyce is a vertebrate paleontologist with a special interest in the morphology, paleoecology and the phylogenetic relations of turtles from the time of their origin in the Triassic until today.

Gerald Mayr – Head of the Ornithology Section, Senckenberg Research Institute, Frankfurt am Main. Dr. Mayr is a biologist who studies the early evolution and major group systematics of birds.

Wighart v. Koenigswald – Professor Emeritus of Paleontology at the University of Bonn. His main area of interest involves the evolution of mammals. From 1977 until 1987 he worked at the Hessian State Museum in Darmstadt, and he described numerous mammals from Messel and studied their paleobiology.

Norbert Micklich – Natural History Department of the Hessian State Museum in Darmstadt. Dr. Micklich is the former head of the Messel Project at the State Museum. Besides the ichthyofauna at Messel, he also works on freshwater and ocean species from other Paleogene and Neogene fossil sites.

Cornelia Kurz – Municipal Museums, Natural History Museum in the Ottoneum, Kassel. Dr. Kurz is a custodian for geology and paleontology. Besides the historical preparation of the regional geological collections, her main interest is focused on the functional morphology and ecology of fossil mammals from the Cenozoic.

Jessica Miller-Camp – Department of Earth Sciences, University of California, Riverside. Dr. Miller-Camp is a paleontologist who heads a museum department and studies the evolution of alligators and dicynodonts in the fields of phylogeny, biogeography, ecomorphology and diversity dynamics.

Thomas Lehmann – Department of Messel Research and Mammalogy, Senckenberg Research Institute, Frankfurt am Main. Dr. Lehmann is an expert on fossil mammals and studies the systematics and paleoecology at crucial moments in the course of their evolution, such as during the Eocene (Messel) and the Miocene (Africa).

Michael Morlo – Department of Messel Research and Mammalogy, Senckenberg Research Institute, Frankfurt am Main. Dr. Morlo is a specialist for fossil predatory mammals. Besides evolution and taxonomy, his research also emphasizes functional morphology, distributional history and ecology.

Olaf K. Lenz – Currently works at the Directorate General of the Senckenberg Research Institute, Frankfurt am Main. Dr. Lenz is a paleontologist and sedimentologist. For several years, he has studied the effects of climate change in the Cenozoic on the development of the flora by means of analyzing pollen and spores.

Renate Rabenstein – Department of Messel Research and Mammalogy, Senckenberg Research Institute, Frankfurt am Main. Using radiographic techniques, Dr. Rabenstein studies the taphonomy and morphology of recent and fossil bats. In addition, she serves as the radiation protection commissioner who organizes the 2D and 3D radiology laboratories.

List of authors

Walter Riegel – Section for Paleobotany, Senckenberg Research Institute, Frankfurt am Main. Prof. Riegel studied geology and Paleontology and earned his doctorate in Philosophy at the Pennsylvania State University, USA. He tought in Göttingen, where he currently continues his work on the older Tertiary and the Devonian.

Kenneth D. Rose – Professor Emeritus of Functional Anatomy and Evolution at the Johns Hopkins University School of Medicine. Dr. Rose specializes in the anatomy and evolution of Eocene mammals, with an emphasis on the origin and early diversification of the modern orders of placental mammals.

Irina Ruf – Department of Messel Research and Mammalogy, Senckenberg Research Institute, Frankfurt am Main. PD Dr. Ruf is a biologist and the head of the Section for Recent Mammals. She studies the evolution, ontogenesis and functional morphology of recent and fossil mammals, primarily by means of computer tomography.

Agustín Scanferla – Consejo Nacional de Investigaciones Científicas y Técnicas, Argentina. Dr. Scanferla is a specialist on fossil snakes. He currently studies the evolution of macrostomatic snakes ("large-mouth snakes"), which encompass the snake fauna at Messel as well as most modern snake species around the world.

Stephan F.K. Schaal – Head of the Department of Messel Research and Mammalogy, Senckenberg Research Institute, Frankfurt am Main. Dr. Schaal is a geologist and paleontologist. His scope of work encompasses the study of Eocene fossil sites as well as the management of surface mining activities at the Messel Pit.

Krister T. Smith – Department of Messel Research and Mammalogy, Senckenberg Research Institute, Frankfurt am Main. Dr. Smith is an expert on fossil lizards and snakes. For years he has studied the mutual interrelations between fossil biodiversity and climate change, particularly during the Eocene.

Jakob Vinther – University of Bristol, United Kingdom. Dr. Vinther studies the preservation of exceptional fossils. His work with fossil pigments from Messel and other fossil sites makes it possible to reconstruct the coloration of extinct animals and to draw conclusions regarding their ecology and evolution.

Sonja Wedmann – Research Station Messel Pit. Department of Messel Research and Mammalogy, Senckenberg Research Institute. Dr. Wedmann is an entomologist who studies the systematics and paleobiogeography of fossil insects, primarily from the Eocene at the Messel Pit.

Volker Wilde – Paleobotany Section, Senckenberg Research Institute, Frankfurt am Main. Dr. habil. Wilde is a geologist and paleontologist. His main area of work involves the paleoecology of terrestrial and terrestrially influenced sediments, particularly from the Paleogene.

Michael Wuttke – Department of Messel Research and Mammalogy, Senckenberg Research Institute, Frankfurt am Main. Dr. Wuttke studies the genesis of Paleogene terrestrial fossil sites and their fossils, with an emphasis on taphonomy, frogs, and freshwater sponges.

Index

μCT 37, 38
acoustic specializations 249, 256, 259, 260
Acrostichum 49
Actinopterygii 105, 106
Adapiformes 241
adaptive radiation 132, 223
Aequornithes 169
aerial hawking 255, 260, 261
Afrotheria 215, 224, 235, 271
Ailuravinae 268, 269
Ailuravus 264–269, 313
Alangiaceae 54
algae 10, 19, 21, 30, 43, 47–49, 66, 72, 92, 305, 307
alligators 159, 162
Ameridelphia 217
Amiidae 105
Amniota 121
Amphilemuridae 235, 239
Amphiperatherium 218, 220, 221
Amphiperca 25, 105, 107–111, 307
Anacardiaceae 46, 60
Anguidae 134
Anguilla 105–107, 109
Anguilliformes 105
Anguinae 134
anthropoids 241
Apatemyidae 223, 231, 232
Apatemys 232, 233
Aphidina 77
Apidae 82, 83
Apodiformes 182, 185, 188
Araceae 46, 48, 52, 61
Araneae 67
Araneidae 67
arboreal 125, 127, 128, 131–133, 147, 170, 174, 176, 188, 191, 194, 199, 200, 209, 211, 223, 232, 237, 265, 267, 269, 271, 276, 277, 282, 311
Archaeonycteris 250, 251, 254–256, 261
Archosauria 121
Arctic Canada 283
Arecaceae 48, 53, 61
armadillos 215, 278
Arthropoda 66, 67, 69
aspect ratio 250, 261
Atractosteus 105, 107, 110
Auchenorrhyncha 71, 76
autotomy 129
Aves 121 ff.

baculum 242, 244, 246, 268, 277
basilar membrane 249, 255
bats 28, 29, 249
Belostomatidae 78
benthic 305, 307
Betulaceae 57, 59, 61
Bibionidae 100, 101
Bilateria 63
bipeds 133, 226
birds 25, 26, 27, 31, 32, 39, 59, 74, 84, 108, 121, 136, 145, 147, 169–191, 193–195, 197–200, 203–206, 208–213, 246, 307, 309, 311
bite mark 244
black pelites 9, 10, 11, 13
Blattodea 71, 75, 76
Blechnaceae 49
body mass 245, 260, 272, 276
Boidae 140
Boinae 140
Bombyliidae 101
Boreoeutheria 215
Botryococcus 17, 48
Brachycera 101
Buprestidae 71, 93, 94
Buxolestes 223, 224, 227, 228, 232, 233, 246, 307

callus tissue 246
caniniforms 159, 160, 165
Carabidae 90
carapace 69, 149, 150
Carettochelyidae 151, 154
carnassials 271, 276, 277
Carnivora 227, 271, 274, 277, 283
Carnivoraformes 271, 276, 283
Carnivoramorpha 271
carnivores 215, 242, 277, 307, 309
carnivorous 125, 145, 151, 154, 170, 174, 224, 246, 247, 271, 277, 283
Castor 228
Centrarchidae 105
Cephalotaxus 50
Cerambycidae 95, 96
Cercamoniinae 241
CF sounds 258
Chalcidoidea 81
Charadriiformes 181
Chascacocolius 191, 192
Chelotriton 118, 119

Chloranthaceae 51
Chrysomelidae 96, 97, 307
Cicadoidea 76
Cimolesta 224
clade 39–40
Cladocera 68, 69
claws 125, 159, 165, 166, 176, 189, 194, 228–231, 237, 239, 250, 265, 267, 268, 272, 277–279
climax community 19
climbing 17, 49, 52, 54, 61, 125, 203, 215, 217, 221, 223, 229–231, 237, 245, 265, 267–269, 272, 277, 282
Clupeocephala 105, 111
Coccina 77, 78
cochlea see inner ear
Coelastrum 49
Coleoptera 71, 90, 91, 95
Coliiformes 190, 193
Comptonia 57
consumers 305, 306, 307, 309
continental drift 7
convergent evolution 39, 123, 223, 232
coprolites 72, 99, 100, 115, 246, 305
Cretaceous 10
crocodiles 26, 35, 106, 119, 121, 138, 159, 164, 165, 211, 212, 233, 244, 246, 300, 307
Crocodylia 121, 159, 166
crocodyliforms 159, 167
crown group 40, 127, 131, 132, 149, 170, 210
Crustacea 69
Cryptodira 149, 153
Cryptolacerta 128, 131, 140
cryptozoic 131, 141, 142, 311
Cunninghamia 50
Cupedidae 90
Cupressaceae 50
Curculionoidea 71, 96, 98
Cyclanthaceae 46, 53, 61
Cyclurus 105, 106, 110, 236
Cydnidae 79, 80
Cyperaceae 46, 53, 61
Cypseloramphus 185, 187

Dactylopsila 232, 233
Darwinius 241, 244–247
Daubentonia 232, 233
Decapoda 69

deciduous 18, 74, 102, 274, 276, 277, 299, 309
Decodon 61, 309
Dermaptera 73
dermis 149
Deuterostomia 63, 64
Didelphidae 217, 336
Didelphimorphia 217, 331
dinosaurs 121, 328, 349
diplospondylia 106
Diptera 71, 100, 101
Doliostrobus 50, 61, 309
Doppler-shift effects 258
drilling 4, 9–13, 17, 19, 23, 46–49
Dryopteridaceae 49
dwarf boas 140
Dynamopterus 175–177, 205, 208, 211

Early Eocene Climatic Optimum 17
eccentricity 20
Ecdysozoa 63
echolocation 249, 254–261
ecosystem 41, 72, 127, 144, 154, 164, 167, 169, 180, 209, 210, 303, 304, 305, 307, 308, 309, 311
edentulous 277, 278, 279, 281
Elaeocarpaceae 56
Elateridae 71, 94
Ellesmere Island 7, 174, 233, 283
El Niño 21, 22
Elopomorpha 105
embryo 121, 141, 144, 215
embryophytes 43
Enantiornithes 170, 188, 189
endocarp 244, 245
endocasts 249
ENSO 22
Eocene 10
Eocoracias 171, 200, 202, 203
Eoglaucidium 191, 192
Eogliravus 264, 265, 267, 268, 269
Eohiodon 110
Eolacerta 127, 128, 130
Eomanis 271, 277, 278, 279, 281, 282, 283
Eopachypteryx 205
Eopelobates 113–115, 117
Ephemeroptera 72
epidermis 46, 149, 277
epiphyses 246, 281
epipubic bone 217, 218
Equidae 293, 294
equilibrium vestibular organ 254
Equoidea 293, 294, 299
Ericaceae 55, 61
Ericales 55
Erinaceomorpha 235, 239
Erycinae 140, 142, 145
Euarchontoglires 215, 223
Eulipotyphla 235

Euphorbiaceae 55, 57
Eurheloderma 136–139
Eurofluvioviridavis 171, 195, 196
Eurohippus 293–299, 307
Euromanis 271, 279, 281
Europolemur 26, 241–244, 246, 247, 307
Eurotamandua 37, 38, 271, 279–282
external ears 236, 245, 295

Fabaceae 57, 58
Fagaceae 57, 59, 61, 309
Fagales 57
Ferae 271, 272
flying foxes 256, 257
folivorous 245
food chain 25, 147, 307, 309
food webs 102, 190, 275, 305, 307, 309, 313
Formicidae 71, 84, 85, 86, 103
Formiciinae 84
Formicinae 87
Foro 206
fossorial 271
Fulgoromorpha 76
fur 25, 28, 226, 236, 238, 239, 245, 246, 252, 277
Fur Formation 199, 206

Galloanseres 169, 170, 174, 205, 208
Gastornis 173, 174, 206, 213, 309
gastralia 149
Geiseltal 66, 71, 93, 94, 119, 127, 132, 133, 135, 151, 159, 165, 167, 173, 174, 176, 191, 193, 194, 199, 200, 206, 213, 217, 232, 261, 281, 283, 288–290
Geiseltaliellus 25, 129, 132–134, 144, 147
Gekkota 123, 324
gender 246, 297
generalists 162, 226
Genetta 272
Geoemydidae 151
gestation 215
Gleicheniaceae 49
Gliridae 264, 265, 267–269
Glyptosaurinae 134, 135
Gracilitarsus 203, 204
Grande Coupure 239, 290
greenhouse 17, 18, 19, 20, 23, 127, 303, 313
Greenland 7, 221, 239
Green River Formation 108, 110, 174, 188, 189, 194, 195, 197, 199, 200, 206, 207, 208, 211, 212, 213, 250, 253
gut contents 38, 46, 125, 127, 224, 236, 237, 239, 244, 249, 251, 253, 254, 263, 265, 266, 268, 269, 275, 278, 287–289, 290, 305
hair 29, 31, 36, 215, 230, 231, 235, 237, 265, 267, 276, 295

Halcyornithidae 194, 195, 206
Hallensia 294–297
Haplorhini 241
Hartenbergeromys 264, 265, 267, 268
Hassianycteris 28, 32, 33, 249, 251, 253, 254, 261
Hassiavis 32, 184, 186
Heloderma 136–138
herbivores 307, 309
herbivorous 74, 102, 125, 151, 166, 230, 266, 279, 291, 309
Herpestes 272
Herpetotheriidae 218, 219, 221
Hessisches Landesmuseum 2, 242
Heterohyus 230, 231, 232, 233
Heteroptera 71, 77
Hexapoda 70
high duty cycle 258
Hipposideros 251, 258
holomictic 303
Holostei 105
holotype 38
horseshoe bats 256, 260
hunting styles 251, 261
Hyaenodonta 271, 335
Hydrocharitaceae 52, 61
Hydrophilidae 91, 92
Hymenoptera 71, 79, 82, 84, 103
hyperthermals 17
Hyrachyus 299, 301

Icacinaceae 55, 61
Icaronycteris 250, 253
Ichneumonidae 81, 82
Iguania 123
Iguanidae 123, 132
Initial Lake Phase 303, 304, 305, 309
inner ear 215, 226, 227, 249, 254–261, 274, 277
Insecta 66, 70
Insectivora 235
insectivorous 33, 125, 132, 188, 208, 215, 224, 235, 245, 255, 275, 291, 309
insects 25, 33, 35, 41, 59, 63, 67, 70–72, 74–79, 84, 88–90, 92, 96, 99–103, 108, 114, 115, 118, 125, 138, 142, 145, 147, 174, 175, 182–185, 188, 199, 200, 203, 211, 221, 224, 232, 237, 239, 249, 252–256, 261, 271, 278, 279, 282, 307, 309, 311
interactions 70, 102, 180, 307
Ischyromyidae 265, 268

Juglandaceae 17, 46, 59, 61, 83
Juncitarsus 181, 182, 211
juveniles 11, 69, 108, 110, 139, 144, 147, 152, 154, 159, 160, 167, 213, 217, 241, 246, 272, 274, 275, 277, 300
keratin 31, 32, 123, 198
Kopidodon 223, 224, 229, 230, 232

Lacertidae 125, 127, 128, 131, 132, 146, 311
Lake Messel 1, 8–15, 18–21, 303–305, 307, 308
land bridges 7, 8, 113, 152, 170, 173, 174, 221, 233, 268, 283
Lapillavis 205, 206
lapilli 11–13
Lauraceae 46, 48, 51, 52, 61, 266, 290
Laurasiatheria 215, 223, 224, 235, 271
Lecythidaceae 55
Leguminosae 57, 58, 61, 304
Lepidoptera 71, 99, 254, 320
Lepidosauria 121
Lepisosteidae 105
Leptictidae 223, 226, 232, 233
Leptictidium 223–227, 232
Lesmesodon 145, 271–275, 283
lianas 309, 313
limblessness 136
limiting nutrients 304
limnocyonines 283
Lipotyphla 235
Lithornithidae 171, 172, 173, 206
London Clay 183, 188, 194, 195, 197, 199, 200, 206, 208, 211, 213
Lophiodon 300
Lophiodontidae 299, 300
Lophotrochozoa 63
Loranthaceae 54
Lutetiobatrachus 117, 118
Lutra 228
Lythraceae 61

maar lake 1, 7–9, 11–13, 303, 304
Maastrichtidelphys 221
Macrocranion 235, 236, 237, 239, 307
macroscelideans 235
macrostomy 142
Magnoliaceae 51
Main Lake Phase 305
Malvales 46, 48
Mammalia 121, 215
mammary glands 215
Marsupialia 215, 217
marsupials 217–219, 221, 311
Masillacolius 191, 192, 204
Masillamys 263–268
Masillapodargus 184, 185
Masillaraptor 189, 191
Masillastega 181, 183, 213
Masillatrogon 199, 200
Masillosteus 106, 107, 110
Mastixiaceae 54
Mecoptera 101, 103
Megachilidae 82
Menispermaceae 54, 61
meromictic 10, 304
Messelastur 194, 195

Messelirrisor 198, 200
Messelogale 271, 276, 277, 283
Messelophis 140–142, 144
Messelornis 27, 177–180, 307
Metazoa 63
miacid 274, 283
Micropteropus 257, 258
Micro-X-ray and micro-computed tomography 37, 38
middle-ear 215
Milankovitch cycles 19, 20, 23
Mioplosus 111
mixolimnion 304, 305
Mollusca 65
monimolimnion 304
Monotremata 215
Mormoopidae 260
Moronidae 105
Museumsverein Messel 2, 349
mustached bats 260
Myricaceae 46, 48, 57
Myristicaceae 51
Myrmeciinae 87
Myrmecophagidae 281
myrmecophagous 278, 279, 281
Myrtaceae 46

nanoCT device 37
Necrosaurus 123, 139, 140
Neognathae 169
Neornithes 121, 169, 170, 188
Neuroptera 88, 101
niche 256, 265, 285, 290, 291, 311
niche differentiation 311
nitrogen 304
Notonectidae 78
Nyctaginaceae 54
Nymphaeales 43, 51
Nyssaceae 54

obliquity 20
oceanic circulation 23
Odonata 72, 73
oil shale 1–4, 9–11, 17–19
Old World leaf-nosed bats 251, 256, 258
Oligocene 10
omnivorous 125, 152, 153, 237, 239, 291, 299
omomyiforms 247
ontogenetic 147, 160, 274, 281
Onychonycteris 250, 253
Oodectes 283
Ophisauriscus 134, 135
Opiliones 68, 69
Ornatocephalus 124, 125, 127, 128, 309, 313
Orthoptera 74
Osmundaceae 49
osteoderms 125, 127, 134–136, 138, 139, 149, 159–162

Ostracoda 69
oxygen isotope 17

Palaeanodon 283
Palaeanodonta 223, 224, 271, 277–279, 281–283
Palaeobatrachus 114, 116, 117
Palaeochiropteryx 28, 32, 33, 250, 253–255
Palaeoglaux 189, 190, 213
Palaeognathae 169, 170, 173
Palaeoperca 105, 107–111
Palaeopsittacus 183, 184
Palaeopython 25, 133, 144–147, 275, 307
Palaeosinopa 224, 232, 233
Palaeotheriidae 293, 294
Palaeotis 172, 173, 206, 211
Paleocene 10
Paleocene-Eocene Thermal Maximum 17, 146, 152
Pandanaceae 53, 61, 309
pangolins 223, 271, 277–279, 281, 283
Pantolestes 224, 232
Pantolestidae 223, 224, 227, 232, 233
Paraortygoides 174, 175, 207
Paraprefica 184, 186, 187, 198, 206, 208, 309
Parargornis 187, 188, 189, 208
paratropical flora 18
Paroodectes 271, 274, 275, 276, 277, 283
Paroxyclaenidae 223, 224, 229, 232
Patriomanis 281, 283
PDO (Pacific Decadal Oscillation) 22
pelagic 305
Pentatomidae 79, 80
Peradectes 217–219, 221, 311
Peradectidae 218, 221
Percichthyidae 105
Perciformes 105
Perissodactyla 293
Permian 9, 121, 149
Perplexicervix 204, 205
Phasmatodea 63, 74, 75, 101
Pholidocercus 235, 238, 239, 309
Pholidota 223, 271, 277, 281, 283
Pholidotamorpha 271
phosphorous 304
pioneer vegetation 19, 309
Pipistrellus 251
placenta 215, 298
Placentalia 215, 217, 218, 223
Placosauriops 135, 138
Planorbidae 65, 66
plantigrade 276, 277
plastron 149, 150, 154
Plecoptera 72, 73
Plesiocathartes 199, 206, 207, 209
Pleurodira 149, 153
Pleurodonta 132
Podocnemididae 150, 153

poisonous 246, 247
poisonous gases 246, 247
pollen 8, 17–22, 41, 43, 45–47, 50–55, 57, 59, 61, 82, 83, 88, 96, 180, 197, 254, 256, 309
Polychrus 134, 135
Polyspora 266
Porifera 63, 64, 307
precession 20
prehensile tail 125, 127, 128, 131, 217, 221, 265, 277, 313
Primobucco 200
Primozygodactylus 195, 196, 197, 210, 307, 309
Priscacara 111
proboscis 76, 77, 224, 237, 327
producers 305, 307, 309
Propalaeotherium 294, 295, 297
Properca 107
Protocypselomorphus 184, 185
Protostomia 63
Proviverrinae 271, 283
Psephenidae 92, 93
Pseudasturides 194
pseudoautotomy 129, 133
Pseudorhyncocyonidae 223, 224, 232
Psittacopes 197
Pteridaceae 49
Pteropodidae 256
Ptolemaiidae 224
Pumiliornis 180, 197

Quasisyndactylus 203
Quercitherium 272
Quercygale 271, 283

rainforest 84, 303
Reduviidae 79
regurgitated 124, 125, 145, 147, 190, 275
reptiles 84, 121, 123, 149, 275, 311
resegmentation 129
Restionaceae 17, 53, 61, 309
Rhenanoperca 105, 107–110, 305
rhinarium 241
Rhinolophidae 256, 260
Rhynchaeites 181, 183
Rieppelophis 140, 141, 142
ring wall 12, 13, 303
Rutaceae 46, 55

Salmila 176, 178
saltatorial 237
Sandcoleidae 191, 206
Sapotaceae 17, 55
sauropsids 121

scales 25, 50, 70, 99, 105, 106, 110, 111, 125, 127, 128, 131, 133–139, 149, 150, 152, 198, 238, 252, 253, 277
Scaniacypselus 169, 186, 188, 206, 309
scanning electron microscope see SEM
scansorial 268, 277
Scarabaeoidea 92
Schizaeaceae 17, 49
Scincoidea 124
sciuromorph 264
scutes 149, 153–155
sea level 303
Selmes 193
SEM 23, 32, 38, 64, 99, 244, 252, 256
semi-aquatic 138, 169, 211, 223, 228
semicircular canals 226, 254
semifossorial 131, 140
seriemas 174, 175, 178
Serudaptus 194, 195
sexually dimorphic 133
sheath-tailed bats 249
Shinisaurus 138, 140
siderite 11, 36, 65, 66
sinopines 283
Siricidae 81, 103
smectite 10
Smilacaceae 52, 61
soft tissue preservation 30–32
solar activity 20, 23
Soricomorpha 235
species richness 40, 41, 167, 303, 309, 311
spines 125, 145, 238, 279
Spongillidae 64, 65
spores 17–22, 45–47, 49, 87
Staatliches Museum für Naturkunde Karlsruhe 242, 350
Staphylinidae 91
Stehlinia 254, 255, 257, 261
stem group 40
Sternorrhyncha 76
Stratiomyidae 101
Strepsiptera 89
Strepsirrhini 241
Strigogyps 176, 178, 211
Strisores 32, 169, 182–184, 205, 206, 209
sub-Milankovitch cycles 19, 20
symplesiomorphies 39
Symplocaceae 55
synapomorphies 39
synapsids 121

Tachypteron 33, 249, 251, 254, 257, 307
taphonomic 25, 26, 28, 33, 60, 65, 69, 71, 84, 99, 100, 119, 212, 247, 304

Taphozous 251, 258
taxonomy 38–40
Teleostei 106
Telluraves 169, 170, 188, 189, 205
temperature 10, 17, 18, 20–23, 28, 31, 86, 103, 123, 141, 145, 208, 209, 215, 303
Tenebrionidae 94, 95
Testudines 121, 149
Testudinidae 151
Tetraedron 10, 21, 23, 48, 304, 305
tetrapods 121, 129, 136, 145, 309, 311, 313
Thaumaturus 105–107, 111, 305
Theaceae 55, 61, 266
Theria 215
third eye 124, 128, 131–133
thrips 76
Thule land bridge 233, 283
Thysanoptera 76
time-averaging 311
Tingidae 78
Tipuloidea 100
Trichoptera 97, 99
Trionychidae 151, 154
trochlea 149
Turgai Strait 7, 233, 268, 283

Ulmaceae 46, 48, 60
UNESCO World Heritage Site 3
Uromanis 278

varves 18
Vastanavis 194, 208
venom 119, 136, 139
Vespoidea 84
vibrissae 235–237
Viridiplantae 43, 44
Vitaceae 17, 46, 55, 56, 61, 307
Viviparidae 65, 66
viviparity 141, 142
Vulpavus 283

warm-bloodedness 121, 215
wing forms 249, 250
wing membranes 249, 250, 261
wing proportions 250, 311

Xenarthra 215, 281
xenarthrous 281

YTONG 1

Zingiberaceae 53, 54
Zygaenidae 99
Zygodactylidae 195–197, 203, 206, 210

The group photo shows the Senckenberg team of the Department of Messel Research and Mammalogy in the summer of 2017 in the Messel Pit. Top row, from left: S.F.K. Schaal, E.E. Brahm, A. Vogel, S. Köster, K.T. Smith, J. Habersetzer; middle row, from left: R. Rabenstein, R. Posch, S. Wedmann, U. Kiel, M. Kuhn, B. Behr; bottom row, from left: M. Geißler, M. Müller, T. Lehmann, J. Eberhardt, I. Ruf, K. Krohmann.

Acknowledgments and Image Credits

There would be no images or scientific publications about Messel fossils were it not for the uncountable hours spent by hundreds of persons in field work, documentation and preparation. First and foremost are the non-professionals – citizen scientists – of the early years, the technical assistants and preparators, and also the scientists of the research institutions and students of different universities who participated in excavations at Messel. They generate the foundation on which the scientific study and publication of the fossils rests.

In particular, we would like to emphasize the following amateur paleontologists from the State of Hesse, who have stood by us for decades and have continually and willingly loaned or donated fossils to us: Christa Behnke, Irmela and Otto Feist†, Dr. h.c. Thomas Perner, Dr. Burkhard Pohl, Karsten Gabriel, and Manfred Kellert†. We warmly thank the following institutions for the usage rights to images: the Hessisches Landesmuseum Darmstadt, the Museumsverein Messel, Natural History Museum Mainz, the Royal Belgian Institute of Natural Sciences, the Wyoming Dinosaur Center Thermopolis, and the owners of private collections of Messel fossils.

A comprehensive picture of the history and stories of the Lagerstätte Messel derives from our own experiences and through numerous conversations with private persons who were involved in the events, among them the late lamented citizens of Messel Elfriede Köhler, Willy Mößle and Dr. Michael Höllwarth as well as Dr. Reinhard Heil. We would like especially to thank Prof. Dr. Wolfgang Martin, who, together with like-minded individuals, fought by all legal means and with great personal commitment for the preservation of the Messel Pit.

We single out our technical assistant (TA) Anika Vogel, as photographer and image editor, for the completion and processing of all images for this Messel book. The following "Senckenbergers" also contributed images: Sven Tränkner (photographer), Sabine Köster (TA), Uta Kiel (TA), Martin Müller (TA), and Juliane Eberhardt (TA). Finally, we must mention our preparator and longtime excavation leader Michael Ackermann (Fig. 5.2). We extend our profound gratitude to each and every one mentioned here.

The Editors

AMNH	American Museum of Natural History, New York, USA
FMNH	Field Museum of Natural History, Chicago, USA
GMH	Geiseltalmuseum, Halle, Germany
HVBG	Hessische Verwaltung für Bodenmanagement und Geoinformation, Germany
HLMD-Be	Hessisches Landesmuseum Darmstadt, Behnke collection, Germany
HLMD-Me	Hessisches Landesmuseum Darmstadt, Messel collection, Germany
HNBGU	Hemvati Nandan Bahuguna Garhwal University, India
IRSNB	Institut Royal des Sciences Naturelles de Belgique, Brussels, Belgium
MMN	Museum Mensch und Natur, München, Germany
MNHM	Naturhistorisches Museum, Mainz, Germany
PBP-MES	Private collection of Burkhard Pohl, Messel collection
PMO	Geological Museum, University of Oslo (previously: Palaeontological Museum Oslo), Norwegen
Priv. Coll.	Private collection
ROM	Royal Ontario Museum, Toronto, Canada
SGN	Senckenberg Gesellschaft für Naturforschung, Frankfurt, Germany
SGPIMH	Institut für Geologie, Hamburg (previously: Geologisch-Paläontologisches Institut und Museum Hamburg), Germany
SF	Senckenberg Gesellschaft für Naturforschung, Entomology collection, Frankfurt, Germany
SMF	Senckenberg Gesellschaft für Naturforschung, Mammalogy collection, Frankfurt, Germany
SMF-ME	Senckenberg Gesellschaft für Naturforschung, Messel collection, Frankfurt, Germany
SMF-MEA	Senckenberg Gesellschaft für Naturforschung, Messel cast collection, Frankfurt, Germany
SMF-MEI	Senckenberg Gesellschaft für Naturforschung, Messel insect collection, Messel, Germany
SMNK-PAL	Staatliches Museum für Naturkunde Karlsruhe, Paleontology collection, Germany
SM.B.Me	Senckenberg Gesellschaft für Naturforschung, Messel botany collection, Frankfurt, Germany
SNSD	Senckenberg Naturhistorische Sammlungen Dresden, Germany
SI	Steinmann-Institut für Geologie, Mineralogie und Paläontologie, Bonn, Germany
TNM	Tanzanian National Museum, Dar es Salaam, Tanzania
YPM	Yale Peabody Museum of Natural History, New Haven, USA

Image Credits

Key to the image credits: **image number:** copyright holder, collections number (if applicable), photographer or illustrator.

1.1:	SGN, – , Sven Tränkner
1.2:	Museumsverein Messel e.V., – , Stephan Schaal
1.3:	SGN, SMF-ME 10 a, Anika Vogel
1.4:	SGN, – , Sven Tränkner
1.5:	SGN, – , Stephan Schaal
1.6:	SGN, – , Sven Tränkner; Architekturbüro: Landau + Kindelbacher, München
2.1:	Georg Büchel, – , Jörn Engelhardt; based on the Digital Elevation Model DGM1 of the HVBG
2.2:	Georg Büchel, – , Jörn Engelhardt; based on the *Paleomap Project* of Scotese (2010)
2.3:	SGN, – , Eveline Junqueira / Anika Vogel; after a reconstruction by Ron Blakey (http://jan.ucc.nau.edu/~rcb7/50moll.jpg, accessed: 23.04.2008)
2.4:	Besucherzentrum Grube Messel, – , Georg Büchel; based on the Digital Elevation Model DGM5 of the HVBG
2.5:	SGN, – , Anika Vogel; data from Ogg et al. (2016)
2.6:	SGN, – , photos: Sven Tränkner / composition: Anika Vogel
2.7:	SGN, – , Stephan Schaal
2.8:	Georg Büchel, – , Jörn Engelhardt; based on the Digital Elevation Model DGM1 of the HVBG
2.9:	Georg Büchel, – , draft: authors / drawing: Jörn Engelhardt
2.10:	Georg Büchel, – , Georg Büchel 2014
3.1:	SGN, – , Olaf Lenz
3.2:	SGN, – , Volker Wilde / Olaf Lenz
3.3:	SGN, – , Volker Wilde / Olaf Lenz
3.4:	SGN, – , Volker Wilde
3.5:	left: SGN, – , Deutsche Montan Technologie; right: SGN,-, Volker Wilde / Olaf Lenz
Box 3.1:	SGN, – , Olaf Lenz
Box 3.2:	SGN, – , Olaf Lenz; after http//:octopus.gma.org/surfing/weather/elnino.html
4.1:	SGN, SMF-ME 2882, Anika Vogel
4.2:	SGN, left: SMF-ME 1128 b right: SMF-ME 11016 a, Anika Vogel
4.3:	SGN, SMF-ME 11534 a, Anika Vogel
4.4:	SGN, SMF-ME 807 a, Anika Vogel
4.5:	SGN, SMF-ME 1414 a, Erwin Haupt
4.6:	SGN, top: SMF-ME 1008 Middle: SMF-ME 1351 bottom left: SMF-ME 1107 a bottom right: SMF-ME 3365, Anika Vogel
4.7:	SGN, (CE-B), Renate Rabenstein
4.8:	SGN, top: SMF-ME 11412 a bottom: SMF-ME 11413 a, Marie-Louise Tritz / Renate Rabenstein
5.1:	SGN, – , Sonja Wedmann
5.2:	SGN, – , Sonja Wedmann
5.3:	Berthold Steinhilber, – , Berthold Steinhilber
5.4:	Berthold Steinhilber, – , Berthold Steinhilber
5.5:	SGN, – , graphic: Sabine Köster / drawing: Uta Kiel
5.6:	SGN, – , Jörg Habersetzer
5.7:	SGN, – , Jörg Habersetzer / Regina Posch
5.8:	SGN, – , Juliane Eberhardt
5.9:	SGN, – , graphic: Sabine Köster / drawing: Juliane Eberhardt
5.10:	SGN, – , graphic: Sabine Köster / drawing: Juliane Eberhardt
5.11:	SGN, – , graphic: Sabine Köster / drawing: Juliane Eberhardt
5.12:	SGN, – , Olaf Lenz
6.1:	SGN, SM.B.Me 10738, Martin Müller
6.2:	SGN, – , draft: Krister Smith / graphic: Sabine Köster
6.3:	SGN, SM.B.Me 7334, Martin Müller
6.4:	SGN, SM.B.Me 28520, Martin Müller
6.5:	SGN, SM.B.Me 19839, Volker Wilde
6.6:	SGN, SM.B.Me 19838, Volker Wilde
6.7:	SGN, SM.B.Me 2059, Martin Müller
6.8:	SGN, SM.B.Me unnumbered, Volker Wilde
6.9:	SGN, SM.B.Me 28700, Martin Müller
6.10:	SGN, SM.B.Me 13468, Martin Müller
6.11:	SGN, SM.B.Me 20694, Martin Müller
6.12:	SGN, SM.B.Me 17801, Martin Müller
6.13:	SGN, SM.B.Me 18801, Sven Tränkner
6.14:	SGN, SM.B.Me 1396, Martin Müller
6.15:	SGN, SM.B.Me 24734, Martin Müller
6.16:	SGN, SM.B.Me 19802, Martin Müller
6.17:	SGN, SM.B.Me 17630, Martin Müller
6.18:	SGN, SM.B.Me 21233, Martin Müller
6.19:	SGN, SM.B.Me 19093, Sven Tränkner
6.20:	SGN, SM.B.Me 7975, Martin Müller
6.21:	SGN, SM.B.Me 19705, Sven Tränkner
6.22:	SGN, SM.B.Me 4006, Martin Müller
6.23:	SGN, SM.B.Me 17492, Martin Müller
6.24:	SGN, SM.B.Me 2054 a, Sven Tränkner
6.25:	SGN, SM.B.Me 27989, Martin Müller
6.26:	SGN, SM.B.Me 22918, Martin Müller
6.27:	SGN, SM.B.Me 20422, Martin Müller
6.28:	SGN, SM.B.Me 3086, Martin Müller
6.29:	SGN, SM.B.Me 8912, Martin Müller
6.30:	SGN, SM.B.Me 28454, Martin Müller
6.31:	SGN, SM.B.Me 15974, Volker Wilde
6.32:	SGN, SM.B.Me 14755, Martin Müller
7.1:	top: SGN, SMF-MEI 12560, Sonja Wedmann; bottom: Sebastian Köpcke / Volker Weinhold
7.2:	SGN, – , draft: Krister Smith / graphic: Sabine Köster
7.3:	SGN, SMF-MEI 191, left: Uta Kiel right: Sabine Köster
7.4:	SGN, SMF-MEI 5143, left: Uta Kiel right: Sabine Köster
7.5:	SGN, SMF-MEI undescribed, Uta Kiel
7.6:	SGN, SMF-MEI undescribed, Uta Kiel
7.7:	SGN, – , draft: Krister Smith / graphic: Sabine Köster
7.8:	SGN, SMF-MEI 975, Sonja Wedmann
7.9:	SGN, SMF-MEI undescribed, Anika Vogel
7.10:	SGN, SMF-MEI undescribed, Übersicht: Uta Kiel / Detail: Sonja Wedmann
7.11:	SGN, SMF-MEI undescribed, left: Sonja Wedmann right: Uta Kiel

7.12:	SGN, SMF-MEI undescribed, Anika Vogel
7.13:	SGN, SMF-MEI undescribed, Anika Vogel
7.14:	SGN, SMF-MEI undescribed, Uta Kiel
7.15:	SGN, SMF-MEI 13209, Uta Kiel
7.16:	SGN, SMF-MEI undescribed, Uta Kiel
7.17:	SGN, SMF-MEI under study, Uta Kiel
7.18:	SGN, SMF-MEI undescribed, Sonja Wedmann
7.19:	SGN, SMF-MEI under study, Erwin Haupt
7.20:	SGN, SMF-MEI undescribed, Uta Kiel
7.21:	left: SGN, SMF-MEI 12560, Sonja Wedmann; right: HLMD, unnumbered, Georg Oleschinski
7.22:	SGN, SMF-MEI under study, Uta Kiel
7.23:	SGN, SMF-MEI undescribed, Uta Kiel
7.24:	SGN, SMF-MEI undescribed, Erwin Haupt
7.25:	SGN, SMF-MEI 1512, Uta Kiel
7.26:	SGN, SMF-MEI undescribed, Uta Kiel
7.27:	SGN, SMF-MEI undescribed, Sonja Wedmann
7.28:	SGN, SMF-MEI 10508, Uta Kiel
7.29:	left: SGN, SMF-MEI under study, Uta Kiel; right: SGN, coll. Heteroptera SF, Sven Tränkner
7.30:	SGN, SMF-MEI undescribed, Uta Kiel
7.31:	SGN, SMF-MEI undescribed, Uta Kiel
7.32:	left: SGN, SMF-MEI 11628, Sonja Wedmann; right: SGN, coll. Heteroptera SF, Sven Tränkner
7.33:	top: SGN, SMF-MEI 14895, Sonja Wedmann; bottom: SGN, coll. Hymenoptera SF, Sven Tränkner
7.34:	SGN, SMF-MEI undescribed, Sonja Wedmann
7.35:	top: SGN, SMF-MEI under study, Sonja Wedmann; bottom: SGN, coll. Hymenoptera SF, Sven Tränkner
7.36:	SGN, SMF-MEI 6388, Uta Kiel
7.37:	SGN, SMF-MEI 10890, top: Uta Kiel bottom: Sonja Wedmann
7.38:	HLMD, HLMD-Me 13627, Torsten Wappler
7.39:	SGN, SMF-MEI 11036, Sonja Wedmann
7.40:	SGN, SMF-MEI under study, Sonja Wedmann
7.41:	SGN, SMF-MEI 998, Uta Kiel
7.42:	SGN, SMF-MEI 1006, Uta Kiel
7.43:	SGN, SMF-MEI 10841, Sonja Wedmann
7.44:	SGN, SMF-MEI 10999, Uta Kiel
7.45:	SGN, SMF-MEI 11953, Uta Kiel
7.46:	left: SGN, SMF-MEI 6343, Uta Kiel; right: private, – , Sonja Wedmann
7.47:	top: SGN, SMF-MEI 8384, Uta Kiel; bottom: SGN, coll. Neuroptera SF, Sven Tränkner
7.48:	SGN, SMF-MEI 196, Uta Kiel
7.49:	SGN, SMF-MEI 2806, Uta Kiel
7.50:	SGN, SMF-MEI under study, Sonja Wedmann
7.51:	SGN, SMF-MEI undescribed, Sonja Wedmann
7.52:	left: SGN, SMF-MEI 5484, Sonja Wedmann; right: SGN, coll. Coleoptera SF, Sven Tränkner
7.53:	SGN, SMF-MEI undescribed, Uta Kiel
7.54:	SGN, SMF-MEI undescribed, Uta Kiel
7.55:	SGN, SMF-MEI 14050, Uta Kiel
7.56:	left: SGN, SMF-MEI undescribed, Sonja Wedmann; right: SGN, coll. Coleoptera SF, Sven Tränkner
7.57:	left: SGN, SMF-MEI 6799, Uta Kiel; right: SGN, coll. Coleoptera SF, Sven Tränkner
7.58:	SGN, SMF-MEI 2627, Uta Kiel
7.59:	left: SGN, SMF-MEI 1593, Uta Kiel; right: SGN, coll. Coleoptera SF, Sven Tränkner
7.60:	SGN, SMF-MEI 14796, Sonja Wedmann
7.61:	SGN, SMF-MEI undescribed, Uta Kiel
7.62:	left: SGN, SMF-MEI undescribed, Sonja Wedmann; right: SGN, coll. Coleoptera SF, Sven Tränkner
7.63:	left, SGN, SMF-MEI undescribed, Uta Kiel; right: SGN, coll. Coleoptera SF, Sven Tränkner
7.64:	SGN, SMF-MEI undescribed, Uta Kiel / Sonja Wedmann
7.65:	SGN, SMF-MEI undescribed, Sonja Wedmann
7.66:	SGN, SMF-MEI undescribed, Sonja Wedmann
7.67:	PLoS Biology, SMF-MEI 12269, top: Sonja Wedmann bottom: Maria McNamara
7.68:	SGN, SMF-MEI undescribed, Sonja Wedmann
7.69:	SGN, SMF-MEI 11199, Uta Kiel
7.70:	SGN, SMF-MEI undescribed, Sonja Wedmann
7.71:	SGN, SMF-MEI undescribed, Sonja Wedmann
7.72:	top: SGN, SMF-MEI 12852, Uta Kiel; bottom: Jiri Lochmann, – , Jiri Lochmann
7.73:	SGN, SMF-MEI undescribed, Uta Kiel
Box 7.1:	SGN, top: SMF-MEI undescribed bottom: SMF-MEI 12269, Sonja Wedmann
Box 7.2:	left, right top: SGN, SM.B.Me 10167, Uta Kiel; right bottom: David Hughes, – , David Hughes
Box 7.3:	SGN, top left: SM.B.Me 21184 top right: SM.B.Me 1534, bottom: SM.B.Me 3582, Torsten Wappler
8.1:	HLMD, HLMD-Me 15835, Wolfgang Fuhrmannek
8.2:	SGN, – , draft: Krister Smith / graphic: Sabine Köster
8.3:	HLMD, HLMD-Me 15575a, Wolfgang Fuhrmannek
8.4:	left: SGN, SMF-ME 719, Erwin Haupt; right: HLMD, HLMD-Me 12616, Wolfgang Fuhrmannek
8.5:	MNHM, PW 1981-4, Bastian Lischewsky
8.6:	HLMD, HLMD-Be 17, Wolfgang Fuhrmannek
8.7:	HLMD, top: HLMD-Me 8958 middle: HLMD-Me 17780 bottom: HLMD-Me 10485, Wolfgang Fuhrmannek
8.8:	SGN, SMF-ME 432, Jörg Habersetzer
8.9:	HLMD, left: HLMD-Me 13781 right: HLMD-Me 13783, Wolfgang Fuhrmannek
8.10:	HLMD, HLMD-Me 12505, Wolfgang Fuhrmannek
8.11:	HLMD, – , Norbert Micklich
9.1:	SGN, SMF-ME 1788, Anika Vogel
9.2:	SGN, – , draft: Krister Smith / graphic: Sabine Köster
9.3:	SGN, SMF-ME 11106 b, Martin Müller
9.4:	SGN, SMF-ME 1381 a, Anika Vogel
9.5:	IRSNB, unnumbered, Erwin Haupt
9.6:	SGN, SMF-ME 2590 a+b, Anika Vogel
9.7:	SGN, SMF-ME 3592 a+b, Anika Vogel
9.8:	SGN, SMF-ME 11279 a, Anika Vogel
9.9:	SGN, SMF-ME 476, Anika Vogel
9.10:	SGN, SMF-ME 2605 a, Anika Vogel
9.11:	SGN, SMF-ME under study, Anika Vogel
10.1:	SGN, – , draft: Krister Smith / graphic: Sabine Köster
10.1.1:	SGN, SMF-ME 10954 a, Anika Vogel
10.1.2:	SGN, – , draft: Krister Smith / graphic: Sabine Köster
10.1.3:	SGN, SMF-ME under study, Sabine Köster
10.1.4:	SGN, top: SMF-ME 3516 bottom: SMF-ME 11522 a-b, Anika Vogel / Sven Tränkner

10.1.5:	Wolfgang Weber, – , Wolfgang Weber	11.1:	SGN, SMF-ME 3576 a, Sven Tränkner
10.1.6:	SGN, PBP-MES-351, Anika Vogel	11.2:	SGN, – , draft: Krister Smith / graphic: Sabine Köster
10.1.7:	SGN, SMF-ME 11241, Anika Vogel	11.3:	SGPIMH, SGPIMH-MEV 1a, Sven Tränkner
10.1.8:	SGN, SMF-ME under study, Anika Vogel	11.4:	IRSNB, IRSNB-Av 82, Sven Tränkner
10.1.9:	SGN, SMF-ME 2604, Anika Vogel	11.5:	SGN, SMF-ME 1578, Anika Vogel
10.1.10:	SGN, PBP-MES-504, Anika Vogel	11.6:	HLMD, HLMD-Me 6116, Wolfgang Fuhrmannek
10.1.11:	HLMD, HLMD-Me 10207, Wolfgang Fuhrmannek	11.7:	SGN, – , Gerald Mayr
10.1.12:	SGN, – , Juliane Eberhardt	11.8:	SGN, left/middle: SMF-ME 11112 a right: SMF-ME 11112 b, left/top: Sven Tränkner middle/bottom: Jörg Habersetzer
10.1.13:	SGN, PBP-MES-353, Anika Vogel		
10.1.14:	SGN, PBP-MES-350, Anika Vogel		
10.1.15:	IRSNB, IRSNB-Be 594, Thierry Hubin	11.9:	SGN, left: SMF-ME 1577, Sven Tränkner; right: SMF-ME 3437 a, Anika Vogel
10.1.16:	SGN, SMF-ME under study, Sven Tränkner		
10.1.17:	Georg Stelzner, – , Georg Stelzner courtesy of Michael Zollweg	11.10:	IRSNB, IRSNB-Av 127 a, left: Sven Tränkner right: Erwin Haupt
10.1.18:	SGN, SMF-ME 11403 a, Anika Vogel	11.11:	SGN, SMF-ME 1818 a+b, Anika Vogel
10.1.19:	SGN, PBP-MES-355, Anika Vogel	11.12:	HLMD, HLMD-Be 161, Sven Tränkner
10.1.20:	top: SGN, SMF-ME 2379, Anika Vogel; bottom: HLMD, HLMD-Me 7915, Wolfgang Fuhrmannek	11.13:	SGN, top: SMF-ME 610 bottom right: SMF-ME 3551, Sven Tränkner; SGN, bottom left: SMF-ME 1275 inset: SMF-ME 11596, Anika Vogel
10.1.21:	SGN, SMF-ME 992, Anika Vogel		
10.1.22:	HLMD, HLMD-Me under study, Anika Vogel	11.14:	SGN, SMF-MEA 295, Anika Vogel
10.1.23:	SGN, SMF-ME 11398 a+b, Anika Vogel	11.15:	SI, 140 b, Sven Tränkner
10.1.24:	SGN, SMF-ME 2074, Sven Tränkner	11.16:	SMNK, SMNK-PAL 725, Sven Tränkner
Box 10.1.1:	SGN, – , draft: Krister Smith / graphic: Sabine Köster	11.17:	SMNK, SMNK-PAL 3834 a, Sven Tränkner
Box 10.1.2:	Sebastian Lotzkat, – , top: Sebastian Lotzkat / Andreas Nöllert, bottom: Andreas Nöllert	11.18:	SMNK, SMNK-PAL 1083, Sven Tränkner
		11.19:	SGN, SMF-ME 11043, Sven Tränkner
Box 10.1.3:	Agustin Scanferla, – , left: Augustin Scanferla right: Heidi Ann	11.20:	SGN, SMF-ME 3545, Sven Tränkner
		11.21:	SGN, SMF-ME 3727 a, Sven Tränkner
Box 10.1.4:	SGN, SMF-ME 11332 a, photo: Sven Tränkner / drawing: Krister Smith and Anika Vogel and Juliane Eberhardt	11.22:	HLMD, HLMD-Be 164, Wolfgang Fuhrmannek
		11.23:	SGN, SMF-ME 11081, Sven Tränkner
		11.24:	left: SMNK, SMNK-PAL 301, Sven Tränkner; right: SGN, SMF-ME 11345 a, Michael Ackermann
10.2.1:	SGN, SMF-ME 185, Anika Vogel		
10.2.2:	SNSD, D32832, Sabine Köster and Markus Auer; graphic: Krister Smith	11.25:	HLMD, HLMD-Be 193, left: Wolfgang Fuhrmannek right: Sven Tränkner
10.2.3:	SGN, – , draft: Krister Smith / graphic: Sabine Köster	11.26:	SGN, SMF-ME 1144 a, Anika Vogel
10.2.4:	SGN, left: SMF-ME 3777, Anika Vogel; right: SMF-ME 11558, Sven Tränkner	11.27:	SGN, SMF-ME 11042, Anika Vogel
		11.28:	left: SMNK, SMNK-PAL 553 a, Sven Tränkner; middle: SGN, SMF-ME 8 a, Sven Tränkner; right: SGN, PBP-MES-400, Anika Vogel
10.2.5:	SGN, left: SMF-ME 11532 a right: SMF-ME 11532 b, Anika Vogel		
10.2.6:	SGN, SMF-ME 715, Sven Tränkner	11.29:	SGN, SMF-ME 3790, Anika Vogel
10.2.7:	SGN, SMF-ME 611, Anika Vogel	11.30:	HLMD, HLMD-Me 10472, Sven Tränkner
10.2.8:	SGN, SMF-MEA 93, Anika Vogel	11.31:	left: SGN, PBP-MES-401 right: SGN, SMF-ME 2375, Anika Vogel
Box 10.2.1:	SGN, SMF-ME 2449, Anika Vogel		
		11.32:	SMNK, SMNK-PAL 2373 a, Sven Tränkner
10.3.1:	SGN, SMF-ME 10876, Anika Vogel	11.33:	SGN, PBP-MES-402, Anika Vogel
10.3.2:	SGN, – , draft: Krister Smith / graphic: Sabine Köster	11.34:	SGN, SMF-ME 11348, Anika Vogel
10.3.3:	SGN, top: SMF-ME 900, Sven Tränkner; bottom: SMF-ME 896, Anika Vogel	11.35:	SMNK, SMNK-PAL 3835 1 b, Sven Tränkner
		11.36:	SGN, left: SMF-ME 11091 a right: SMF-ME 1074, Sven Tränkner
10.3.4:	top: SMNK, SMNK-PAL 966 bottom: HLMD, HLMD V3975; Christopher Brochu	11.37:	SGN, SMF-ME 1279, Sven Tränkner inset: Anika Vogel
10.3.5:	SGN, top: SMF-ME 899 bottom: SMF-ME 2609, Anika Vogel	11.38:	HLMD, HLMD-Be 162, Wolfgang Fuhrmannek
		11.39:	IRSNB, IRSNB-Av 81, Sven Tränkner
10.3.6:	HLMD, HLMD-Me 9119, Wolfgang Fuhrmannek	11.40:	top: HLMD, HLMD-Be 178 bottom left: SGN, SMF-ME 10956 a bottom right: SMNK, SMNK-PAL 3802; Sven Tränkner
10.3.7:	top: SGN, SMF-ME 1801, Erwin Haupt; bottom: SGN, PBP-MES-352, Anika Vogel		
10.3.8:	left: AMNH, AMNH 2090 middle: GMH, GMH Leo I-3631a-1932 right: HLMD, HLMD-Me 5346 bottom: FMNH, FMNH PR 399; Christopher Brochu	11.41:	SGN, SMF-ME 11349 a, Anika Vogel
		11.42:	SMNK, SMNK-PAL 2663 a, Sven Tränkner
		11.43:	SGN, SMF-ME 3543 a, Sven Tränkner
10.3.9:	HLMD, HLMD-Me 7003, Erwin Haupt	11.44:	SGN, SMF-ME 3547, Sven Tränkner

11.45:	SGN, SMF-ME 11211 a, Anika Vogel; inset: SGN, SMF-ME 3548, Sven Tränkner
11.46:	SGN, SMF-ME 2426, Anika Vogel
11.47:	SGN, SMF-ME 3547, Sven Tränkner
11.48:	HLMD, HLMD-Me 14972, Sven Tränkner
11.49:	left: SGN, SMF-ME 1303 a, Anika Vogel; right: SGN, PBP-MES-403, Sven Tränkner
11.50:	left: SGN, SMF-ME 3639 right: SGN, PBP-MES-404; Sven Tränkner
11.51:	SGN, SMF-ME 11286 Inset: HNBGU, No. 1809, Sven Tränkner
11.52:	SGN, SMF-ME 1758 a, Anika Vogel
Box 11.1:	SMNK, SMNK-PAL 3835, Sven Tränkner
Box 11.2:	SGN, SMF-ME 11593 a, left: Sven Tränkner right: Michael Ackermann
Box 11.3:	SGN, SMF-ME 11309 a, Sven Tränkner
Box 11.4:	SGN, SMF-ME 11414 a, left: Sven Tränkner right: Volker Wilde
Box 11.5:	SGN, top left: SMF-ME 1645 a top right: SMF-ME 1231 bottom left: SMF-ME 707 a bottom right: SMF-ME 778, Sven Tränkner
Box 11.6:	SGN, left: SMF-ME 3811 top right: SMF-ME 3850 a bottom middle: SMF-ME 3093 bottom right: SMF-ME 3829, Anika Vogel
Box 11.7:	SGN, SMF-ME 815 a, Sven Tränkner
12.1:	SGN, – , draft: Krister Smith / graphic: Sabine Köster
12.1.1:	HLMD, HLMD-Me 8035, Anika Vogel
12.1.2:	SGN, – , draft: Krister Smith / graphic: Sabine Köster
12.1.3:	HLMD, HLMD-Me 17001, Anika Vogel
12.1.4:	HLMD, HLMD-Me 17001, Renate Rabenstein
12.1.5:	SMNK, SMNK-PAL 983, Sven Tränkner Inset: Anika Vogel
12.2.1:	SGN, SMF-ME 2401, Erwin Haupt
12.2.2:	SGN, – , draft: Krister Smith / graphic: Sabine Köster
12.2.3:	SGN, – , Juliane Eberhardt
12.2.4:	HLMD, HLMD-Me 8011, Wolfgang Fuhrmannek
12.2.5:	SGN, SMF-ME 11377 a, Anika Vogel
12.2.6:	HLMD, HLMD-Me 7431, Werner Kumpf
12.2.7:	left: SMNK, SMNK-PAL 464 a, Sven Tränkner; right: HLMD, HLMD-Me 8086, Wolfgang Fuhrmannek
12.2.8:	Museumsverein Messel e.V., unnumbered, Anika Vogel
12.2.9:	HLMD, HLMD-Me 8850, Georg Oleschinski
12.2.10:	SGN, SMF-ME 10939 a, Beate Simon
12.2.11:	HLMD, HLMD-Me 8850, Dorothea Kranz
12.2.12:	Dorothea Kranz, – , Dorothea Kranz
12.2.13:	left: FMNH, PM 61092, John Weinstein; right: HLMD, PBP-MES-502, Georg Oleschinski
Box 12.2.1:	SGN, SMF-ME 11377 a, Virginie Volpato
12.3.1:	SGN, SMF-ME 11412 a, Anika Vogel inset: Uta Kiel
12.3.2:	SGN, – , draft: Krister Smith / graphic: Sabine Köster
12.3.3:	SGN, Priv. coll. Gabriel unnumbered, Anika Vogel
12.3.4:	SGN, – , Juliane Eberhardt
12.3.5:	HLMD, left: HLMD-Be 190 right: HLMD-Be 191, Wolfgang Fuhrmannek
12.3.6:	SGN, SMF-ME 11277, Anika Vogel
12.3.7:	SGN, SMF-ME 758 b, Anika Vogel
12.3.8:	SGN, SMF-ME 758 a, Anika Vogel
Box 12.3.1:	SGN, Priv. coll. Gabriel unnumbered, Anika Vogel / graphic: Juliane Eberhardt
12.4.1:	PLoS ONE, PMO 214.214, Per Aas doi.org/10.1371/journal.pone.0005723
12.4.2:	SGN, – , draft: Krister Smith / graphic: Sabine Köster
12.4.3:	HLMD, HLMD-Me 7430, Beate Simon
12.4.4:	SGN, SMF-ME 1228 b, Anika Vogel
12.4.5:	SGN, SMF-ME 1228 a, left: Anika Vogel right: Jörg Habersetzer
12.4.6:	SMNK, SMNK-PAL 684, Volker Griener
12.4.7:	SGN, SMF-ME 1683, left: Jens Lorenz Franzen right: Christine Hemm
12.4.8:	SMNK, SMNK-PAL 1125, left: Anika Vogel right: Jörg Habersetzer
12.4.9:	SGN, – , Marie-Louise Tritz und Jens Lorenz Franzen
12.4.10:	PLoS ONE, left: PMO 214.214 middle and right: PBP-MES-500, middle: Jörg Habersetzer right: Anika Vogel
12.4.11:	SGN, PBP-MES-500, left and middle: Jens Lorenz Franzen right: Volker Wilde
12.4.12:	PLoS ONE, PMO 214.214, left: Per Aas right: Jörg Habersetzer
12.4.13:	PLoS ONE, PMO 214.214, Jörg Habersetzer doi.org/10.1371/journal.pone.0005723
12.4.14:	SGN, top: SMF-ME 2986 a, Uta Kiel; bottom: SMF-ME 2986 b, Anika Vogel
12.4.15:	SGN, left: SMF-ME 3379 a right: SMF-ME 3379 b, Anika Vogel
12.5.1:	IRSNB, IRScNB 4119 a, left: Jörg Habersetzer right: Sven Tränkner 12.5.2: SGN, – , draft: Jörg Habersetzer and Renate Rabenstein / graphic: Sabine Köster
12.5.3:	SGN, SMF-ME 1469 a, Anika Vogel
12.5.4:	SGN, SMF-ME 1789 b, Anika Vogel
12.5.5:	SGN, SMF-ME 2022, Anika Vogel
12.5.6:	SGN, top: SMF-ME 11432 a bottom: SMF-ME 11432 b, Anika Vogel
12.5.7:	ROM, ROM-Z 5534, Anika Vogel
12.5.8:	YPM, YPM VPPU 18150, Anika Vogel
12.5.9:	SGN, – , draft: Jörg Habersetzer / graphic: Anika Vogel and Regina Posch
12.5.10:	SGN, – , photos: Jörg Habersetzer / graphic: Regina Posch
12.5.11:	SGN, left column: SMF-ME 508 middle top: SMF-ME 496 middle middle: Priv. coll. Bastelberger middle bottom: SMF-ME 2003 right top: SMF-ME 1469 right middle: Priv. coll. Bastelberger right bottom: SMF-ME unbekannt, SEM images: Gotthard Richter / graphic: Renate Rabenstein and Regina Posch
12.5.12:	SGN, – , Jörg Habersetzer
12.5.13:	TNM, TNM-MP-207, top: Anika Vogel bottom: Jörg Habersetzer
12.5.14:	SGN, left: SMF 33148 middle: SMF 46333 right: SMF 60990, Jörg Habersetzer and Renate Rabenstein / graphic: Renate Rabenstein and Regina Posch

12.5.15:	SGN, – , draft: Jörg Habersetzer and Renate Rabenstein / graphic: Regina Posch	12.8.4:	SGN, SMF-ME 510, Anika Vogel
12.5.16:	top: TNM, TNM-MP-207 bottom: ROM, ROM-Z 5534, draft: Jörg Habersetzer / graphic: Anika Vogel and Regina Posch	12.8.5:	SGN, – , Juliane Eberhardt
		12.8.6:	SGN, SMF-ME 1527 a, Anika Vogel
		12.8.7:	SGN, SMF-ME 1001, Anika Vogel
		12.8.8:	HLMD, HLMD-Me 18500, Wolfgang Fuhrmannek
12.6.1:	SGN, SMF-ME 11295 a inset: SMF-ME 11295 b, Sven Tränkner	Box 12.8.1:	SGN, SMF-ME 510, Sabine Köster
12.6.2:	SGN, – , draft: Krister Smith / graphic: Sabine Köster	12.9.1:	SGN, PBP-MES-503, Sven Tränkner
12.6.3:	SGN, SMF-ME 11500 a, Sven Tränkner	12.9.2:	SGN, left: SMF 6730 right: SMF 95622, Anika Vogel
12.6.4:	SGN, – , Juliane Eberhardt	12.9.3:	SGN, – , draft: Krister Smith / graphic: Sabine Köster
12.6.5:	SGN, SMF-ME 11500 a, Anika Vogel	12.9.4:	SGN, SMF-ME 12, Sven Tränkner
12.6.6:	HLMD, HLMD-Me 11015, Wolfgang Fuhrmannek	12.9.5:	HLMD, HLMD-Be 136, Wolfgang Fuhrmannek
12.6.7:	SGN, PBP-MES-501, Sven Tränkner	12.9.6:	HLMD, HLMD-Me 12000, Wolfgang Fuhrmannek
Box 12.6.1:	SGN, top left: SMF 63525 top right: SMF 20087 bottom left: SMF 33567 bottom right: SMF 92678, Anika Vogel; middle: SGN, SMF-ME 11295, Sven Tränkner; graphic: Juliane Eberhardt	12.9.7:	SGN, Priv. coll. Perner unnumbered, Sven Tränkner
		12.9.8:	SMNK, SMNK-PAL 4010, Sven Tränkner
		12.9.9:	MMN, BSP-1985-I-62, Jens Lorenz Franzen
		12.9.10:	IRSNB, IRSNB-M-1465, Jens Lorenz Franzen
		12.9.11:	top: Vera Kassühlke, – , Vera Kassühlke; SGN, left: SMF-ME 2947 right: SMF-ME 11034 a, Sven Tränkner
12.7.1:	HLMD, HLMD-Me 17000, Anika Vogel		
12.7.2:	SGN, – , draft: Krister Smith / graphic: Sabine Köster	12.9.12:	SGN, SMF-ME 11034 a, Sven Tränkner inset: Jörg Habersetzer
12.7.3:	HLMD, HLMD-Be 156, Wolfgang Fuhrmannek		
12.7.4:	HLMD, HLMD-Me 14590 a, Anika Vogel	12.9.13:	SGN, PBP-MES-503, photo: Sven Tränkner / graphic: Jens Lorenz Franzen
12.7.5:	HLMD, HLMD-Me 13710, Wolfgang Fuhrmannek; inset: SGN, SMF-MEA 142, Anika Vogel		
		12.9.14:	SGN, SMF-ME 1931, Anika Vogel
12.7.6:	SGN, SMF-ME 1284, top: Anika Vogel bottom: Thomas Lehmann	12.9.15:	SGN, SMF-ME 3515, Sven Tränkner
		Box 12.9.1:	SGN, SMF-ME 12, Friedemann Schaarschmidt
12.7.7:	SGN, SMF-MEA 263, Anika Vogel		
12.7.8:	SGN, SMF-ME 1573 e, Anika Vogel	13.1:	HLMD, – , Oscar Sanisidro
12.7.9:	SGN, SMF-MEA 264, Anika Vogel	13.2:	Georg Büchel, – , Georg Büchel
12.7.10:	SGN, HLMD-Me 17000, Renate Rabenstein and Jörg Habersetzer and Regina Posch	13.3:	SGN, – , Krister Smith
		13.4:	SGN, – , Krister Smith
		13.5:	HLMD, – , Oscar Sanisidro
12.7.11:	SGN, HLMD-Me 17000, Renate Rabenstein and Jörg Habersetzer and Regina Posch	13.6:	HLMD, – , Oscar Sanisidro
		13.7:	SGN, – , draft: Jörg Habersetzer / graphic: Anika Vogel
12.7.12:	SGN, HLMD-Me 17000, Renate Rabenstein and Jörg Habersetzer and Regina Posch		
		Box 13.1:	SGN, – , drafts: Renate Rabenstein and Jörg Habersetzer / models: Claudia Weißbrod / graphic: Anika Vogel
Box 12.7.1:	SGN, SMF-ME 3843 a, Anika Vogel insets: Thomas Lehmann		
12.8.1:	SGN, SMF-ME 1001, Anika Vogel		
12.8.2:	SGN, left: SMF 94726 right: SMF 95623, Anika Vogel		
12.8.3:	SGN, – , draft: Krister Smith / graphic: Sabine Köster		

Group picture: SGN, – , Sven Tränkner